S0-ACP-211

Study and Solutions Guide

TRIGONOMETRY

FIFTH EDITION

Larson/Hostetler

Dianna L. Zook
Indiana University—
Purdue University at
Fort Wayne, Indiana

HOUGHTON MIFFLIN COMPANY   **Boston**   **New York**

Editor-in-Chief: Jack Shira
Managing Editor: Cathy Cantin
Development Manager: Maureen Ross
Associate Editor: Laura Wheel
Assistant Editor: Carolyn Johnson
Supervising Editor: Karen Carter
Project Editor: Patty Bergin
Art Supervisor: Gary Crespo
Marketing Manager: Michael Busnach
Senior Manufacturing Coordinator: Sally Culler
Composition and Art: Meridian Creative Group

Copyright © 2001 by Houghton Mifflin Company. All rights reserved.

No part of this work may be reproduced or transmitted in any form or by
any means, electronic or mechanical, including photocopying and recording,
or by any information storage or retrieval system without the prior
permission of Houghton Mifflin Company unless such copying is expressly
permitted by federal copyright law. Address inquiries to College Permissions,
Houghton Mifflin Company, 222 Berkeley Street, Boston, MA 02116-3764.

Printed in the United States of America

ISBN: 0-618-07279-9

23456789-CRS-04 03 02 01

CONTENTS

TO THE STUDENT

The *Study and Solutions Guide for Trigonometry* is a supplement to the textbook *Trigonometry*, Fifth Edition, by Ron Larson, Robert P. Hostetler, and Bruce H. Edwards.

As a mathematics instructor, I often have students come to us with questions about assigned homework. When I ask to see their work, the reply often is "I didn't know where to start." The purpose of the *Study Guide* is to provide brief summaries of the topics covered in the textbook and enough detailed solutions to problems so that you will be able to work the remaining exercises.

A special thanks to Meridian Creative Group for typing this guide. Also, I would like to thank my husband, Edward Schlindwein, for his support during the several months I worked on this project.

If you have any corrections or suggestions for improving this *Study Guide*, I would appreciate hearing from you.

Good luck with your study of trigonometry.

Dianna L. Zook
Indiana University
Purdue University
Fort Wayne, IN 46805
Zook@ipfw.edu

STUDY STRATEGIES

- Attend all classes and come prepared. Have your homework completed. Bring the text, paper, pen or pencil, and a calculator (scientific or graphing) to each class.

- Read the section in the text that is to be covered before class. Make notes about any questions that you have and, if not answered during the lecture, ask them at the appropriate time.

- Participate in class. As mentioned above, ask questions — and answer them.

- Take notes on all definitions, concepts, rules, formulas, and examples. After class, read your notes and fill in any gaps, or make notations of any questions that you have.

- DO THE HOMEWORK!!! You learn mathematics by doing it yourself. Allow at least two hours outside of each class for homework. Do not fall behind.

- Seek help when needed. Visit your instructor during office hours and come prepared with specific questions; check with your school's tutoring service; find a study partner in class; check additional books in the library for more examples – just do something before the problem becomes insurmountable.

- Do not cram for exams. Each chapter in the text contains a chapter review and a chapter test and this *Study Guide* contains a practice test at the end of each chapter. (The answers are at the end of Part I.) Work these problems a few days before the exam and review any areas of weakness.

PART I

CHAPTER P
Prerequisites

CHAPTER P
Prerequisites

Section P.1   Real Numbers

■ You should know the following sets.

(a) The set of real numbers includes the rational numbers and the irrational numbers.

(b) The set of rational numbers includes all real numbers that can be written as the ratio p/q of two integers, where $q \neq 0$.

(c) The set of irrational numbers includes all real numbers which are not rational.

(d) The set of integers: $\{\ldots, -3, -2, -1, 0, 1, 2, 3, \ldots\}$

(e) The set of whole numbers: $\{0, 1, 2, 3, 4, \ldots\}$

(f) The set of natural numbers: $\{1, 2, 3, 4, \ldots\}$

■ The real number line is used to represent the real numbers.

■ Know the inequality symbols.

(a) $a < b$ means a is less than b. (b) $a \leq b$ means a is less than or equal to b.

(c) $a > b$ means a is greater than b. (d) $a \geq b$ means a is greater than or equal to b.

■ You should know that
$$|a| = \begin{cases} a, \text{if } a \geq 0 \\ -a, \text{if } a < 0. \end{cases}$$

■ Know the properties of absolute value.

(a) $|a| \geq 0$ (b) $|-a| = |a|$ (c) $|ab| = |a|\,|b|$ (d) $\left|\dfrac{a}{b}\right| = \dfrac{|a|}{|b|}, b \neq 0$

■ The distance between a and b on the real line is $d(a, b) = |b - a| = |a - b|$.

■ You should be able to identify the terms in an algebraic expression.

■ You should know and be able to use the basic rules of algebra.

■ Commutative Property

(a) Addition: $a + b = b + a$ (b) Multiplication: $a \cdot b = b \cdot a$

■ Associative Property

(a) Addition: $(a + b) + c = a + (b + c)$ (b) Multiplication: $(ab)c = a(bc)$

■ Identity Property

(a) Addition: 0 is the identity; $a + 0 = 0 + a = a$.   (b) Multiplication: 1 is the identity; $a \cdot 1 = 1 \cdot a = a$.

■ Inverse Property

(a) Addition: $-a$ is the additive inverse of a; $a + (-a) = -a + a = 0$.

(b) Multiplication: $1/a$ is the multiplicative inverse of a, $a \neq 0$; $a(1/a) = (1/a)a = 1$.

■ Distributive Property

(a) $a(b + c) = ab + ac$ (b) $(a + b)c = ac + bc$

—CONTINUED—

2

- **Properties of Negation**

  (a) $(-1)a = -a$

  (b) $-(-a) = a$

  (c) $(-a)b = a(-b) = -ab$

  (d) $(-a)(-b) = ab$

  (e) $-(a + b) = (-a) + (-b) = -a - b$

- **Properties of Equality**

  (a) If $a = b$, then $a + c = b + c$.

  (b) If $a = b$, then $ac = bc$.

  (c) If $a + c = b + c$, then $a = b$.

  (d) If $ac = bc$ and $c \neq 0$, then $a = b$.

- **Properties of Zero**

  (a) $a \pm 0 = a$

  (b) $a \cdot 0 = 0$

  (c) $0 \div a = 0/a = 0, a \neq 0$

  (d) $a/0$ is undefined.

  (e) If $ab = 0$, then $a = 0$ or $b = 0$.

- **Properties of Fractions** ($b \neq 0, d \neq 0$)

  (a) Equivalent Fractions: $a/b = c/d$ if and only if $ad = bc$.

  (b) Rule of Signs: $-a/b = a/-b = -(a/b)$ and $-a/-b = a/b$

  (c) Equivalent Fractions: $a/b = ac/bc, c \neq 0$

  (d) Addition and Subtraction

   1. Like Denominators: $(a/b) \pm (c/b) = (a \pm c)/b$

   2. Unlike Denominators: $(a/b) \pm (c/d) = (ad \pm bc)/bd$

  (e) Multiplication: $(a/b) \cdot (c/d) = (ac)/(bd)$

  (f) Division: $(a/b) \div (c/d) = (a/b) \cdot (d/c) = (ad)/(bc)$ if $c \neq 0$.

Solutions to Odd-Numbered Exercises

1. $-9, -\frac{7}{2}, 5, \frac{2}{3}, \sqrt{2}, 0, 1, -4, 2, -11$

   (a) Natural numbers: $5, 1, 2$

   (b) Integers: $-9, 5, 0, 1, -4, 2, -11$

   (c) Rational numbers: $-9, -\frac{7}{2}, 5, \frac{2}{3}, 0, 1, -4, 2, -11$

   (d) Irrational numbers: $\sqrt{2}$

3. $2.01, 0.666 \ldots, -13, 0.010110111 \ldots, 1, -6$

   (a) Natural numbers: 1

   (b) Integers: $-13, 1, -6$

   (c) Rational numbers: $2.01, 0.666 \ldots, -13, 1, -6$

   (d) Irrational numbers: $0.010110111 \ldots$

5. $-\pi, -\frac{1}{3}, \frac{6}{3}, \frac{1}{2}\sqrt{2}, -7.5, -1, 8, -22$

   (a) Natural numbers: $\frac{6}{3}$ (since it equals 2), 8

   (b) Integers: $\frac{6}{3}, -1, 8, -22$

   (c) Rational numbers: $-\frac{1}{3}, \frac{6}{3}, -7.5, -1, 8, -22$

   (d) Irrational numbers: $-\pi, \frac{1}{2}\sqrt{2}$

7. $\frac{5}{8} = 0.625$

9. $\frac{41}{333} = 0.\overline{123}$

11. $4.1 = \frac{41}{10}$

13. $10.\overline{2} = \frac{92}{9}$

15. $-2.01\overline{2} = -\frac{1811}{900}$

17. $-1 < 2.5$

19. $-4 > -8$

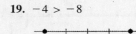

21. $\frac{3}{2} < 7$

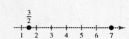

23. $\frac{5}{6} > \frac{2}{3}$

25. The inequality $x \leq 5$ denotes the set of all real numbers less than or equal to 5. The interval is unbounded.

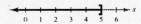

27. The inequality $x < 0$ denotes the set of all negative real numbers. The interval is unbounded.

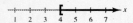

29. The inequality $x \geq 4$ denotes the set of all real numbers greater than or equal to 4. The interval is unbounded.

31. The inequality $-2 < x < 2$ denotes the set of all real numbers greater than -2 and less than 2. The interval is bounded.

33. The inequality $-1 \leq x < 0$ denotes the set of all negative real numbers greater than or equal to -1. The interval is bounded.

35. $\frac{127}{90} \approx 1.41111, \frac{584}{413} \approx 1.41404, \frac{7071}{5000} = 1.41420, \sqrt{2} \approx 1.41421, \frac{47}{33} \approx 1.42424$

37. $-2 < x \leq 4$ **39.** $y \geq 0$ **41.** $10 \leq t \leq 22$ **43.** $W > 65$

45. This interval consists of all real numbers greater than or equal to zero, but less than 8.

47. This interval consists of all real numbers greater than -6.

49. $|-10| = -(-10) = 10$

51. If $x \leq 3$, then $|3 - x| = 3 - x$
If $x > 3$, then $|3 - x| = -(3 - x) = -3 + x = x - 3$

53. $|-1| - |-2| = 1 - 2 = -1$

55. $\dfrac{-5}{|-5|} = \dfrac{-5}{-(-5)} = \dfrac{-5}{5} = -1$

57. If $x < -2$, then $x + 2$ is negative.
Thus $\dfrac{|x + 2|}{x + 2} = \dfrac{-(x + 2)}{x + 2} = -1$

59. $|-3| > -|-3|$ since $3 > -3$.

61. $-5 = -|5|$ since $-5 = -5$.

63. $-|-2| = -|2|$ since $-2 = -2$.

65. $d(-1, 3) = |3 - (-1)| = |3 + 1| = 4$

67. $d(126, 75) = |75 - 126| = 51$

69. $d\left(-\frac{5}{2}, 0\right) = \left|0 - \left(-\frac{5}{2}\right)\right| = \frac{5}{2}$

71. $d\left(\frac{16}{5}, \frac{112}{75}\right) = \left|\frac{112}{75} - \frac{16}{5}\right| = \frac{128}{75}$

73. (a) Since $A > 0, -A < 0$. The expression is negative.

(b) Since $B < A, B - A < 0$. The expression is negative.

75. $d(18, 7) = |7 - 18| = 11$ miles

77. $d(23°, 60°) = |60° - 23°| = 37°$

79. $d(x, 5) = |x - 5|$ and $d(x, 5) \leq 3$, thus $|x - 5| \leq 3$.   **81.** $d(y, 0) = |y - 0| = |y|$ and $d(y, 0) \geq 6$, thus $|y| \geq 6$.

83.

Budgeted Expense, b	*Actual Expense, a*	$|a - b|$	$0.05b$
$112,700	$113,356	$656	$5635

The actual expense difference is greater than $500 (but is less than 5% of the budget) so the actual expense does not pass the test.

85.

Budgeted Expense, b	*Actual Expense, a*	$|a - b|$	$0.05b$
$37,640	$37,335	$305	$1882

Since $305 < $500 and $305 < $1882, the actual expense passes the "budget variance test."

87. Receipts = $92.5 billion; $|\text{Receipts} - \text{Outlay}| = |92.5 - 92.2| = \$0.3$ billion surplus

89. Receipts = $1032.0 billion; $|\text{Receipts} - \text{Outlay}| = |1032.0 - 1253.2| = \$221.2$ billion deficit

91. $7x + 4$

Terms: $7x, 4$

Coefficient: 7

93. $\sqrt{3}x^2 - 8x - 11$

Terms: $\sqrt{3}x^2, -8x, -11$

Coefficients: $\sqrt{3}, -8$

95. $4x^3 + \dfrac{x}{2} - 5$

Terms: $4x^3, \dfrac{x}{2}, -5$

Coefficients: $4, \dfrac{1}{2}$

97. $4x - 6$

(a) $4(-1) - 6 = -4 - 6 = -10$

(b) $4(0) - 6 = 0 - 6 = -6$

99. $x^2 - 3x + 4$

(a) $(-2)^2 - 3(-2) + 4 = 4 + 6 + 4 = 14$

(b) $(2)^2 - 3(2) + 4 = 4 - 6 + 4 = 2$

101. $\dfrac{x + 1}{x - 1}$

(a) $\dfrac{1 + 1}{1 - 1} = \dfrac{2}{0}$

Division by zero is undefined

(b) $\dfrac{-1 + 1}{-1 - 1} = \dfrac{0}{-2} = 0$

103. $x + 9 = 9 + x$

Commutative Property of Addition

105. $\dfrac{1}{(h + 6)}(h + 6) = 1, h \neq -6$

Multiplicative Inverse Property

107. $2(x + 3) = 2x + 6$

Distributive Property

109. $1 \cdot (1 + x) = 1 + x$

Multiplicative Identity Property

111. $x(3y) = (x \cdot 3)y$ Associative Property of Multiplication

       $= (3x)y$ Commutative Property of Multiplication

113. $\dfrac{3}{16} + \dfrac{5}{16} = \dfrac{8}{16} = \dfrac{1}{2}$

115. $\dfrac{5}{8} - \dfrac{5}{12} + \dfrac{1}{6} = \dfrac{15}{24} - \dfrac{10}{24} + \dfrac{4}{24} = \dfrac{9}{24} = \dfrac{3}{8}$

117. $12 \div \dfrac{1}{4} = 12 \cdot \dfrac{4}{1} = 12 \cdot 4 = 48$

119. $\dfrac{2x}{3} - \dfrac{x}{4} = \dfrac{8x}{12} - \dfrac{3x}{12} = \dfrac{5x}{12}$

121. $-3 + \dfrac{3}{7} \approx -2.57$

123. $\dfrac{11.46 - 5.37}{3.91} \approx 1.56$

125. (a)

n	1	0.5	0.01	0.0001	0.000001
$5/n$	5	10	500	50,000	5,000,000

(b) The value of $\dfrac{5}{n}$ approaches infinity as n approaches 0.

127. (a) $|u + v| \neq |u| + |v|$ if u is positive and v is negative or vice versa.

(b) $|u + v| \leq |u| + |v|$

They are equal when u and v have the same sign. If they differ in sign, $|u + v|$ is less than $|u| + |v|$.

129. The only even prime number is 2, because its factors are itself and 1.

131. False. The denominators cannot be added when adding fractions.

133. Yes, if a is a negative number, then $-a$ is positive. Thus, $|a| = -a$ if a is negative.

Section P.2   Solving Equations

- ■ You should know how to solve linear equations.

  $ax + b = 0$

- ■ An identity is an equation whose solution consists of every real number in its domain.

- ■ To solve an equation you can:

  (a) Add or subtract the same quantity from both sides.

  (b) Multiply or divide both sides by the same nonzero quantity.

- ■ To solve an equation that can be simplified to a linear equation:

  (a) Remove all symbols of grouping and all fractions.

  (b) Combine like terms.

  (c) Solve by algebra.

  (d) Check the answer.

- ■ A "solution" that does not satisfy the original equation is called an extraneous solution.

- ■ You should be able to solve a quadratic equation by factoring, if possible.

- ■ You should be able to solve a quadratic equation of the form $u^2 = d$ by extracting square roots.

- ■ You should be able to solve a quadratic equation by completing the square.

- ■ You should know and be able to use the Quadratic Formula: For $ax^2 + bx + c = 0, a \neq 0$,

  $$x = \frac{-b \pm \sqrt{b^2 - 4ac}}{2a}.$$

- ■ You should be able to solve polynomials of higher degree by factoring.

- ■ For equations involving radicals or fractional powers, raise both sides to the same power.

- ■ For equations with fractions, multiply both sides by the least common denominator to clear the fractions.

- ■ For equations involving absolute value, remember that the expression inside the absolute value can be positive or negative.

Solutions to Odd-Numbered Exercises

1. $2(x - 1) = 2x - 2$ is an *identity* by the Distributive Property. It is true for all real values of x.

3. $-6(x - 3) + 5 = -2x + 10$ is *conditional*. There are real values of x for which the equation is not true.

5. $4(x + 1) - 2x = 4x + 4 - 2x = 2x + 4 = 2(x + 2)$

This is an *identity* by simplification. It is true for all real values of x.

7. $x^2 - 8x + 5 = (x - 4)^2 - 11$ is an *identity* since $(x - 4)^2 - 11 = x^2 - 8x + 16 - 11 = x^2 - 8x + 5$.

9. $3 + \dfrac{1}{x + 1} = \dfrac{4x}{x + 1}$ is *conditional*. There are real values of x for which the equation is not true.

11.
$$x + 11 = 15$$
$$x + 11 - 11 = 15 - 11$$
$$x = 4$$

13.
$$7 - 2x = 25$$
$$7 - 7 - 2x = 25 - 7$$
$$-2x = 18$$
$$\frac{-2x}{-2} = \frac{18}{-2}$$
$$x = -9$$

15.
$$8x - 5 = 3x + 20$$
$$8x - 3x - 5 = 3x - 3x + 20$$
$$5x - 5 = 20$$
$$5x - 5 + 5 = 20 + 5$$
$$5x = 25$$
$$\frac{5x}{5} = \frac{25}{5}$$
$$x = 5$$

17.
$$2(x + 5) - 7 = 3(x - 2)$$
$$2x + 10 - 7 = 3x - 6$$
$$2x + 3 = 3x - 6$$
$$-x = -9$$
$$x = 9$$

19.
$$x - 3(2x + 3) = 8 - 5x$$
$$x - 6x - 9 = 8 - 5x$$
$$-5x - 9 = 8 - 5x$$
$$-5x + 5x - 9 = 8 - 5x + 5x$$
$$-9 \neq 8$$

No solution

21.
$$\frac{5x}{4} + \frac{1}{2} = x - \frac{1}{2}$$
$$4\left(\frac{5x}{4}\right) + 4\left(\frac{1}{2}\right) = 4(x) - 4\left(\frac{1}{2}\right)$$
$$5x + 2 = 4x - 2$$
$$x = -4$$

23.
$$\tfrac{3}{2}(z + 5) - \tfrac{1}{4}(z + 24) = 0$$
$$4\left(\tfrac{3}{2}\right)(z + 5) - 4\left(\tfrac{1}{4}\right)(z + 24) = 4(0)$$
$$6(z + 5) - (z + 24) = 0$$
$$6z + 30 - z - 24 = 0$$
$$5z = -6$$
$$z = -\tfrac{6}{5}$$

25.
$$0.25x + 0.75(10 - x) = 3$$
$$4(0.25x) + 4(0.75)(10 - x) = 4(3)$$
$$x + 3(10 - x) = 12$$
$$x + 30 - 3x = 12$$
$$-2x = -18$$
$$x = 9$$

27.
$$x + 8 = 2(x - 2) - x$$
$$x + 8 = 2x - 4 - x$$
$$x + 8 = x - 4$$
$$8 \neq -4$$

No solution

29.
$$\frac{100 - 4u}{3} = \frac{5u + 6}{4} + 6$$
$$12\left(\frac{100 - 4u}{3}\right) = 12\left(\frac{5u + 6}{4}\right) + 12(6)$$
$$4(100 - 4u) = 3(5u + 6) + 72$$
$$400 - 16u = 15u + 18 + 72$$
$$-31u = -310$$
$$u = 10$$

31. $\dfrac{5x - 4}{5x + 4} = \dfrac{2}{3}$

$3(5x - 4) = 2(5x + 4)$

$15x - 12 = 10x + 8$

$5x = 20$

$x = 4$

33. $10 - \dfrac{13}{x} = 4 + \dfrac{5}{x}$

$\dfrac{10x - 13}{x} = \dfrac{4x + 5}{x}$

$10x - 13 = 4x + 5$

$6x = 18$

$x = 3$

35. $\dfrac{x}{x + 4} + \dfrac{4}{x + 4} + 2 = 0$

$\dfrac{x + 4}{x + 4} + 2 = 0$

$1 + 2 = 0$

$3 \neq 0$

Contradiction : no solution
The variable is divided out.

37. $\dfrac{1}{x} + \dfrac{2}{x - 5} = 0$ Multiply both sides by $x(x - 5)$

$1(x - 5) + 2x = 0$

$3x - 5 = 0$

$3x = 5$

$x = \dfrac{5}{3}$

39. $\dfrac{2}{(x - 4)(x - 2)} = \dfrac{1}{x - 4} + \dfrac{2}{x - 2}$ Multiply both sides by $(x - 4)(x - 2)$.

$2 = 1(x - 2) + 2(x - 4)$

$2 = x - 2 + 2x - 8$

$2 = 3x - 10$

$12 = 3x$

$4 = x$

A check reveals that $x = 4$ is an extraneous solution–it makes the denominator zero. There is no real solution.

41. $\dfrac{1}{x - 3} + \dfrac{1}{x + 3} = \dfrac{10}{x^2 - 9}$

$\dfrac{(x + 3) + (x - 3)}{x^2 - 9} = \dfrac{10}{x^2 - 9}$

$2x = 10$

$x = 5$

43. $\dfrac{3}{x^2 - 3x} + \dfrac{4}{x} = \dfrac{1}{x - 3}$ Multiply both sides by $x(x - 3)$.

$3 + 4(x - 3) = x$

$3 + 4x - 12 = x$

$3x = 9$

$x = 3$

A check reveals that $x = 3$ is an extraneous solution, so there is no solution.

45. $(x + 2)^2 + 5 = (x + 3)^2$

$x^2 + 4x + 4 + 5 = x^2 + 6x + 9$

$4x + 9 = 6x + 9$

$-2x = 0$

$x = 0$

47. $(x + 2)^2 - x^2 = 4(x + 1)$

$x^2 + 4x + 4 - x^2 = 4x + 4$

$4 = 4$

The equation is an identity; every real number is a solution.

49. $2x^2 = 3 - 8x$

General form: $2x^2 + 8x - 3 = 0$

51. $(x - 3)^2 = 3$

$x^2 - 6x + 9 = 3$

General form: $x^2 - 6x + 6 = 0$

53. $\frac{1}{5}(3x^2 - 10) = 18x$

$3x^2 - 10 = 90x$

General form: $3x^2 - 90x - 10 = 0$

55. $6x^2 + 3x = 0$

$3x(2x + 1) = 0$

$3x = 0 \quad \text{or} \quad 2x + 1 = 0$

$x = 0 \quad \text{or} \qquad x = -\frac{1}{2}$

57. $x^2 - 2x - 8 = 0$

$(x - 4)(x + 2) = 0$

$x - 4 = 0 \quad \text{or} \quad x + 2 = 0$

$x = 4 \quad \text{or} \qquad x = -2$

59. $x^2 + 10x + 25 = 0$

$(x + 5)(x + 5) = 0$

$x + 5 = 0$

$x = -5$

61. $3 + 5x - 2x^2 = 0$

$(3 - x)(1 + 2x) = 0$

$3 - x = 0 \quad \text{or} \quad 1 + 2x = 0$

$x = 3 \quad \text{or} \qquad x = -\frac{1}{2}$

63. $x^2 + 4x = 12$

$x^2 + 4x - 12 = 0$

$(x + 6)(x - 2) = 0$

$x + 6 = 0 \quad \text{or} \quad x - 2 = 0$

$x = -6 \quad \text{or} \qquad x = 2$

65. $\frac{3}{4}x^2 + 8x + 20 = 0$

$4\left(\frac{3}{4}x^2 + 8x + 20\right) = 4(0)$

$3x^2 + 32x + 80 = 0$

$(3x + 20)(x + 4) = 0$

$3x + 20 = 0 \qquad \text{or} \quad x + 4 = 0$

$x = -\frac{20}{3} \quad \text{or} \qquad x = -4$

67. $x^2 + 2ax + a^2 = 0$

$(x + a)^2 = 0$

$x + a = 0$

$x = -a$

69. $x^2 = 49$

$x = \pm\sqrt{49}$

$= \pm 7$

$= \pm 7.00$

71. $x^2 = 11$

$x = \pm\sqrt{11}$

$x \approx \pm 3.32$

73. $3x^2 = 81$

$x^2 = 27$

$x = \pm\sqrt{27} = \pm 3\sqrt{3}$

$x \approx \pm 5.20$

75. $(x - 12)^2 = 16$

$x - 12 = \pm\sqrt{16}$

$x = 12 \pm 4$

$x = 16 \quad \text{or} \quad x = 8$

$x = 16.00 \quad \text{or} \quad x = 8.00$

77. $(x + 2)^2 = 14$

$x + 2 = \pm\sqrt{14}$

$x = -2 \pm \sqrt{14}$

$x \approx 1.74$ or $x \approx -5.74$

79. $(2x - 1)^2 = 18$

$2x - 1 = \pm\sqrt{18}$

$2x = 1 \pm 3\sqrt{2}$

$x = \dfrac{1 \pm 3\sqrt{2}}{2}$

$x \approx 2.62$ or $x \approx -1.62$

81. $(x - 7)^2 = (x + 3)^2$

$x - 7 = \pm(x + 3)$

$x - 7 = x + 3$ or $x - 7 = -x - 3$

$-7 \neq 3$ $2x = 4$

No solution $x = 2 = 2.00$

83. $x^2 - 2x = 0$

$x^2 - 2x + 1 = 0 + 1$

$(x - 1)^2 = 1$

$x - 1 = \pm\sqrt{1}$

$x = 1 \pm 1$

$x = 0$ or $x = 2$

85. $x^2 + 4x - 32 = 0$

$x^2 + 4x = 32$

$x^2 + 4x + 2^2 = 32 + 2^2$

$(x + 2)^2 = 36$

$x + 2 = \pm\sqrt{36}$

$x = -2 \pm 6$

$x = 4$ or $x = -8$

87. $x^2 + 6x + 2 = 0$

$x^2 + 6x = -2$

$x^2 + 6x + 3^2 = -2 + 3^2$

$(x + 3)^2 = 7$

$x + 3 = \pm\sqrt{7}$

$x = -3 \pm \sqrt{7}$

89. $9x^2 - 18x = -3$

$x^2 - 2x = -\dfrac{1}{3}$

$x^2 - 2x + 1 = -\dfrac{1}{3} + 1$

$(x - 1)^2 = \dfrac{2}{3}$

$x - 1 = \pm\sqrt{\dfrac{2}{3}}$

$x = 1 \pm \sqrt{\dfrac{6}{9}}$

$x = 1 \pm \dfrac{\sqrt{6}}{3}$

91. $8 + 4x - x^2 = 0$

$-x^2 + 4x + 8 = 0$

$x^2 - 4x - 8 = 0$

$x^2 - 4x = 8$

$x^2 - 4x + 2^2 = 8 + 2^2$

$(x - 2)^2 = 12$

$x - 2 = \pm\sqrt{12}$

$x = 2 \pm 2\sqrt{3}$

93. $2x^2 + x - 1 = 0$

$$x = \frac{-b \pm \sqrt{b^2 - 4ac}}{2a}$$

$$= \frac{-1 \pm \sqrt{1^2 - 4(2)(-1)}}{2(2)}$$

$$= \frac{-1 \pm 3}{4} = \frac{1}{2}, -1$$

95. $16x^2 + 8x - 3 = 0$

$$x = \frac{-b \pm \sqrt{b^2 - 4ac}}{2a}$$

$$= \frac{-8 \pm \sqrt{8^2 - 4(16)(-3)}}{2(16)}$$

$$= \frac{-8 \pm 16}{32} = \frac{1}{4}, -\frac{3}{4}$$

97. $2 + 2x - x^2 = 0$

$$x = \frac{-b \pm \sqrt{b^2 - 4ac}}{2a}$$

$$= \frac{-2 \pm \sqrt{2^2 - 4(-1)(2)}}{2(-1)}$$

$$= \frac{-2 \pm 2\sqrt{3}}{-2} = 1 \pm \sqrt{3}$$

99. $x^2 + 14x + 44 = 0$

$$x = \frac{-b \pm \sqrt{b^2 - 4ac}}{2a}$$

$$= \frac{-14 \pm \sqrt{14^2 - 4(1)(44)}}{2(1)}$$

$$= \frac{-14 \pm 2\sqrt{5}}{2} = -7 \pm \sqrt{5}$$

101. $x^2 + 8x - 4 = 0$

$$x = \frac{-b \pm \sqrt{b^2 - 4ac}}{2a}$$

$$= \frac{-8 \pm \sqrt{8^2 - 4(1)(-4)}}{2(1)}$$

$$= \frac{-8 \pm 4\sqrt{5}}{2}$$

$$= -4 \pm 2\sqrt{5}$$

103. $\qquad 12x - 9x^2 = -3$

$$-9x^2 + 12x + 3 = 0$$

$$x = \frac{-b \pm \sqrt{b^2 - 4ac}}{2a}$$

$$= \frac{-12 \pm \sqrt{12^2 - 4(-9)(3)}}{2(-9)}$$

$$= \frac{-12 \pm 6\sqrt{7}}{-18} = \frac{2}{3} \pm \frac{\sqrt{7}}{3}$$

105. $9x^2 + 24x + 16 = 0$

$$x = \frac{-b \pm \sqrt{b^2 - 4ac}}{2a}$$

$$= \frac{-24 \pm \sqrt{24^2 - 4(9)(16)}}{2(9)}$$

$$= \frac{-24 \pm 0}{18}$$

$$= -\frac{4}{3}$$

107. $4x^2 + 4x = 7$

$$4x^2 + 4x - 7 = 0$$

$$x = \frac{-b \pm \sqrt{b^2 - 4ac}}{2a}$$

$$= \frac{-4 \pm \sqrt{4^2 - 4(4)(-7)}}{2(4)}$$

$$= \frac{-4 \pm 8\sqrt{2}}{8} = -\frac{1}{2} \pm \sqrt{2}$$

109. $\qquad 28x - 49x^2 = 4$

$$-49x^2 + 28x - 4 = 0$$

$$x = \frac{-b \pm \sqrt{b^2 - 4ac}}{2a}$$

$$= \frac{-28 \pm \sqrt{28^2 - 4(-49)(-4)}}{2(-49)}$$

$$= \frac{-28 \pm 0}{-98} = \frac{2}{7}$$

111. $\qquad 8t = 5 + 2t^2$

$$-2t^2 + 8t - 5 = 0$$

$$t = \frac{-b \pm \sqrt{b^2 - 4ac}}{2a}$$

$$= \frac{-8 \pm \sqrt{8^2 - 4(-2)(-5)}}{2(-2)}$$

$$= \frac{-8 \pm 2\sqrt{6}}{-4} = 2 \pm \frac{\sqrt{6}}{2}$$

113.
$$(y - 5)^2 = 2y$$
$$y^2 - 12y + 25 = 0$$
$$x = \frac{-b \pm \sqrt{b^2 - 4ac}}{2a}$$
$$= \frac{-(-12) \pm \sqrt{(-12)^2 - 4(1)(25)}}{2(1)}$$
$$= \frac{12 \pm 2\sqrt{11}}{2} = 6 \pm \sqrt{11}$$

115. $\frac{1}{2}x^2 + \frac{3}{8}x = 2$
$$4x^2 + 3x = 16$$
$$4x^2 + 3x - 16 = 0$$
$$x = \frac{-b \pm \sqrt{b^2 - 4ac}}{2a}$$
$$= \frac{-3 \pm \sqrt{3^2 - 4(4)(-16)}}{2(4)}$$
$$= \frac{-3 \pm \sqrt{265}}{8} = -\frac{3}{8} \pm \frac{\sqrt{265}}{8}$$

117. $5.1x^2 - 1.7x - 3.2 = 0$
$$x = \frac{1.7 \pm \sqrt{(-1.7)^2 - 4(5.1)(-3.2)}}{2(5.1)}$$
$$x \approx 0.976, -0.643$$

119. $-0.067x^2 - 0.852x + 1.277 = 0$
$$x = \frac{-(-0.852) \pm \sqrt{(-0.852)^2 - 4(-0.067)(1.277)}}{2(-0.067)}$$
$$x \approx -14.071, 1.355$$

121. $422x^2 - 506x - 347 = 0$
$$x = \frac{506 \pm \sqrt{(-506)^2 - 4(422)(-347)}}{2(422)}$$
$$x \approx 1.687, -0.488$$

123. $12.67x^2 + 31.55x + 8.09 = 0$
$$x = \frac{-31.55 \pm \sqrt{(31.55)^2 - 4(12.67)(8.09)}}{2(12.67)}$$
$$x \approx -2.200, -0.290$$

125. $x^2 - 2x - 1 = 0$
$$x^2 - 2x = 1$$
$$x^2 - 2x + 1^2 = 1 + 1^2$$
$$(x - 1)^2 = 2$$
$$x - 1 = \pm\sqrt{2}$$
$$x = 1 \pm \sqrt{2}$$

127. $(x + 3)^2 = 81$
$$x + 3 = \pm 9$$
$$x + 3 = 9 \quad \text{or} \quad x + 3 = -9$$
$$x = 6 \quad \text{or} \quad x = -12$$

129. $x^2 - x - \frac{11}{4} = 0$ Complete the Square
$$x^2 - x = \frac{11}{4}$$
$$x^2 - x + \left(\frac{1}{2}\right)^2 = \frac{11}{4} + \left(\frac{1}{2}\right)^2$$
$$\left(x - \frac{1}{2}\right)^2 = \frac{12}{4}$$
$$x - \frac{1}{2} = \pm\sqrt{\frac{12}{4}}$$
$$x = \frac{1}{2} \pm \sqrt{3}$$

131. $(x + 1)^2 = x^2$ Extract Square Roots
$$x^2 = (x + 1)^2$$
$$x = \pm(x + 1)$$
For $x = +(x + 1)$:
$$0 \neq 1 \quad \text{No solution}$$
For $x = -(x + 1)$:
$$2x = -1$$
$$x = -\frac{1}{2}$$

133. $3x + 4 = 2x^2 - 7$ Quadratic Formula

$$0 = 2x^2 - 3x - 11$$

$$x = \frac{-(-3) \pm \sqrt{(-3)^2 - 4(2)(-11)}}{2(2)}$$

$$= \frac{3 \pm \sqrt{97}}{4} = \frac{3}{4} \pm \frac{\sqrt{97}}{4}$$

135. $4x^4 - 18x^2 = 0$

$$2x^2(2x^2 - 9) = 0$$

$$2x^2 = 0 \implies x = 0$$

$$2x^2 - 9 = 0 \implies x = \pm\frac{3\sqrt{2}}{2}$$

137. $x^4 - 81 = 0$

$$(x^2 + 9)(x + 3)(x - 3) = 0$$

$$x^2 + 9 = 0 \implies x = \pm 3i$$

$$x + 3 = 0 \implies x = -3$$

$$x - 3 = 0 \implies x = 3$$

139. $x^3 + 216 = 0$

$$x^3 + 6^3 = 0$$

$$(x + 6)(x^2 - 6x + 36) = 0$$

$$x + 6 = 0 \implies x = -6$$

$$x^2 - 6x + 36 = 0 \implies x = 3 \pm 3\sqrt{3}i$$

(By completing the square)

141. $5x^3 + 30x^2 + 45x = 0$

$$5x(x^2 + 6x + 9) = 0$$

$$5x(x + 3)^2 = 0$$

$$5x = 0 \implies x = 0$$

$$x + 3 = 0 \implies x = -3$$

143. $x^3 - 3x^2 - x + 3 = 0$

$$x^2(x - 3) - (x - 3) = 0$$

$$(x - 3)(x^2 - 1) = 0$$

$$(x - 3)(x + 1)(x - 1) = 0$$

$$x - 3 = 0 \implies x = 3$$

$$x + 1 = 0 \implies x = -1$$

$$x - 1 = 0 \implies x = 1$$

145. $x^4 - x^3 + x - 1 = 0$

$$x^3(x - 1) + (x - 1) = 0$$

$$(x - 1)(x^3 + 1) = 0$$

$$(x - 1)(x + 1)(x^2 - x + 1) = 0$$

$$x - 1 = 0 \implies x = 1$$

$$x + 1 = 0 \implies x = -1$$

$$x^2 - x + 1 = 0 \implies x = \frac{1}{2} \pm \frac{\sqrt{3}}{2}i \quad \text{(By the Quadratic Formula)}$$

147. $x^4 - 4x^2 + 3 = 0$

$$(x^2 - 3)(x^2 - 1) = 0$$

$$\left(x + \sqrt{3}\right)\left(x - \sqrt{3}\right)(x + 1)(x - 1) = 0$$

$$x + \sqrt{3} = 0 \implies x = -\sqrt{3}$$

$$x - \sqrt{3} = 0 \implies x = \sqrt{3}$$

$$x + 1 = 0 \implies x = -1$$

$$x - 1 = 0 \implies x = 1$$

149. $4x^4 - 65x^2 + 16 = 0$

$$(4x^2 - 1)(x^2 - 16) = 0$$

$$(2x + 1)(2x - 1)(x + 4)(x - 4) = 0$$

$$2x + 1 = 0 \implies x = -\tfrac{1}{2}$$

$$2x - 1 = 0 \implies x = \tfrac{1}{2}$$

$$x + 4 = 0 \implies x = -4$$

$$x - 4 = 0 \implies x = 4$$

151.
$$x^6 + 7x^3 - 8 = 0$$
$$(x^3 + 8)(x^3 - 1) = 0$$
$$(x + 2)(x^2 - 2x + 4)(x - 1)(x^2 + x + 1) = 0$$
$$x + 2 = 0 \implies x = -2$$
$$x^2 - 2x + 4 = 0 \implies x = 1 \pm \sqrt{3}\,i \text{ (By the Quadratic Formula)}$$
$$x - 1 = 0 \implies x = 1$$
$$x^2 + x + 1 = 0 \implies x = -\frac{1}{2} \pm \frac{\sqrt{3}}{2}\,i \text{ (By the Quadratic Formula)}$$

153.
$$\sqrt{2x} - 10 = 0$$
$$\sqrt{2x} = 10$$
$$2x = 100$$
$$x = 50$$

155.
$$\sqrt{x - 10} - 4 = 0$$
$$\sqrt{x - 10} = 4$$
$$x - 10 = 16$$
$$x = 26$$

157.
$$\sqrt[3]{2x + 5} + 3 = 0$$
$$\sqrt[3]{2x + 5} = -3$$
$$2x + 5 = -27$$
$$2x = -32$$
$$x = -16$$

159.
$$-\sqrt{26 - 11x} + 4 = x$$
$$4 - x = \sqrt{26 - 11x}$$
$$16 - 8x + x^2 = 26 - 11x$$
$$x^2 + 3x - 10 = 0$$
$$(x + 5)(x - 2) = 0$$
$$x + 5 = 0 \implies x = -5$$
$$x - 2 = 0 \implies x = 2$$

161.
$$\sqrt{x + 1} = \sqrt{3x + 1}$$
$$x + 1 = 3x + 1$$
$$-2x = 0$$
$$x = 0$$

163.
$$(x - 5)^{3/2} = 8$$
$$(x - 5)^3 = 8^2$$
$$x - 5 = 8^{2/3}$$
$$x = 5 + 4$$
$$x = 9$$

165.
$$(x + 3)^{2/3} = 8$$
$$(x + 3)^2 = 8^3$$
$$x + 3 = \pm\sqrt{8^3}$$
$$x + 3 = \pm\sqrt{512}$$
$$x = -3 \pm 16\sqrt{2}$$

167.
$$(x^2 - 5)^{3/2} = 27$$
$$x^2 - 5 = 27^{2/3}$$
$$x^2 = 5 + 9$$
$$x^2 = 14$$
$$x = \pm\sqrt{14}$$

169.
$$3x(x - 1)^{1/2} + 2(x - 1)^{3/2} = 0$$
$$(x - 1)^{1/2}[3x + 2(x - 1)] = 0$$
$$(x - 1)^{1/2}(5x - 2) = 0$$
$$(x - 1)^{1/2} = 0 \implies x - 1 = 0 \implies x = 1$$
$$5x - 2 = 0 \implies x = \tfrac{2}{5} \text{ which is extraneous.}$$

171. $\dfrac{20 - x}{x} = x$

$20 - x = x^2$

$0 = x^2 + x - 20$

$0 = (x + 5)(x - 4)$

$x + 5 = 0 \implies x = -5$

$x - 4 = 0 \implies x = 4$

173. $\dfrac{1}{x} - \dfrac{1}{x + 1} = 3$

$x(x + 1)\dfrac{1}{x} - x(x + 1)\dfrac{1}{x + 1} = x(x + 1)(3)$

$x + 1 - x = 3x(x + 1)$

$1 = 3x^2 + 3x$

$0 = 3x^2 + 3x - 1; \; a = 3, \; b = 3, \; c = -1$

$x = \dfrac{-3 \pm \sqrt{(3)^2 - 4(3)(-1)}}{2(3)} = \dfrac{-3 \pm \sqrt{21}}{6}$

175. $x = \dfrac{3}{x} + \dfrac{1}{2}$

$(2x)(x) = (2x)\left(\dfrac{3}{x}\right) + (2x)\left(\dfrac{1}{2}\right)$

$2x^2 = 6 + x$

$2x^2 - x - 6 = 0$

$(2x + 3)(x - 2) = 0$

$2x + 3 = 0 \implies x = -\dfrac{3}{2}$

$x - 2 = 0 \implies x = 2$

177. $\dfrac{4}{x + 1} - \dfrac{3}{x + 2} = 1$

$4(x + 2) - 3(x + 1) = (x + 1)(x + 2), x \neq -2, -1$

$4x + 8 - 3x - 3 = x^2 + 3x + 2$

$x^2 + 2x - 3 = 0$

$(x - 1)(x + 3) = 0$

$x - 1 = 0 \implies x = 1$

$x + 3 = 0 \implies x = -3$

179. $|2x - 1| = 5$

$2x - 1 = 5 \implies x = 3$

$-(2x - 1) = 5 \implies x = -2$

181. $|x| = x^2 + x - 3$

$x = x^2 + x - 3$ OR $-x = x^2 + x - 3$

$x^2 - 3 = 0$ $x^2 + 2x - 3 = 0$

$x = \pm\sqrt{3}$ $(x - 1)(x + 3) = 0$

$x - 1 = 0 \implies x = 1$

$x + 3 = 0 \implies x = -3$

Only $x = \sqrt{3}$, and $x = -3$ are solutions to the original equation. $x = -\sqrt{3}$ and $x = 1$ are extraneous.

183. $|x + 1| = x^2 - 5$

$x + 1 = x^2 - 5$ OR $-(x + 1) = x^2 - 5$

$x^2 - x - 6 = 0$ $-x - 1 = x^2 - 5$

$(x - 3)(x + 2) = 0$ $x^2 + x - 4 = 0$

$x - 3 = 0 \implies x = 3$

$x + 2 = 0 \implies x = -2$ $x = \dfrac{-1 \pm \sqrt{17}}{2}$

Only $x = 3$ and $x = \dfrac{-1 - \sqrt{17}}{2}$ are solutions to the original equation. $x = -2$ and $x = \dfrac{-1 + \sqrt{17}}{2}$ are extraneous.

185. $16 = 0.432x - 10.44$

$26.44 = 0.432x$

$\dfrac{26.44}{0.432} = x$

$x \approx 61.2$ inches

187. $10,000 = 0.32m + 2500$

$7,500 = 0.32m$

$\dfrac{7,500}{0.32} = m$

$m = 23,437.5$ miles

189. (a)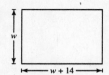

(b) $w(w + 14) = 1632$

$w^2 + 14w - 1632 = 0$

(c) $(w - 34)(w + 48) = 0$

$w = 34$ or $w = -48$ Extraneous

$l = 34 + 14 = 48$

width: 34 feet

length: 48 feet

191.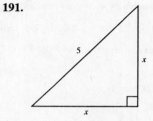

$x^2 + x^2 = 5^2$

$2x^2 = 25$

$x^2 = \dfrac{25}{2}$

$x = \sqrt{\dfrac{25}{2}} = \dfrac{5}{\sqrt{2}} = \dfrac{5\sqrt{2}}{2} \approx 3.54$ centimeters

193. Let r = speed of the eastbound plane and $r + 50$ = speed of the northbound plane.
After 3 hours the eastbound plane has traveled $3r$ miles and the northbound plane
has traveled $3(r + 50)$ miles.

$[3r^2] + [3(r + 50)]^2 = 2440^2$

$9r^2 + 9(r^2 + 100r + 2500) = 5,953,600$

$18r^2 + 900r - 5,931,100 = 0$

By the Quadratic Formula, $r \approx 550$ (discard the negative value of r as extraneous).

Speed of the eastbound plane: 550 miles per hour

Speed of the northbound plane: 600 miles per hour

195. $x(20 - 0.0002x) = 500{,}000$

$20x - 0.0002x^2 = 500{,}000$

$0 = 0.0002x^2 - 20x + 500{,}000$

By the Quadratic Formula, $x = 50{,}000$ units.

197. $240 = 75.82 - 2.11x + 43.51\sqrt{x}$

$2.11x - 43.51\sqrt{x} + 164.18 = 0$

Let $u = \sqrt{x}$, then we have

$2.11u^2 - 43.51u + 164.18 = 0$

By the Quadratic Formula, $u \approx 15.6484$ or $u \approx 4.9724$.

Since $5 \le x \le 40$, we discard the larger value of u.

$\sqrt{x} \approx 4.9724$

$x \approx 24.725$ pounds per square inch

199. False. The product must equal **zero** for the Zero Factor Property to be used.

201. The student should have subtracted $15x$ from both sides so that the equation is equal to zero. By factoring out an x, there are **two** solutions.

$x = 0 \quad \text{or} \quad x = 6$

203. Remove symbols of grouping, combine like terms, reduce fractions.

Add (or subtract) the same quantity to (from) both sides of the equation.

Multiply (or divide) both sides of the equation by the same nonzero quantity.

Interchange the two sides of the equation.

205. (a) $ax^2 + bx = 0$, $a \ne 0$, $b \ne 0$

$x(ax + b) = 0$

$x = 0 \quad \text{or} \quad ax + b = 0$

$ax = -b$

$x = -\dfrac{b}{a}$

(b) $ax^2 - ax = 0$, $a \ne 0$

$ax(x - 1) = 0$

$ax = 0 \quad \text{or} \quad x - 1 = 0$

$x = 0 \qquad\qquad x = 1$

207. Isolate the absolute value by subtracting x from both sides of the equation. The expression inside the absolute value signs can be positive or negative, so two separate equations must be solved.

Section P.3   The Cartesian Plane and Graphs of Equations

- You should be able to plot points.
- You should know that the distance between (x_1, y_1) and (x_2, y_2) in the plane is

  $$d = \sqrt{(x_2 - x_1)^2 + (y_2 - y_1)^2}.$$

- You should know that the midpoint of the line segment joining (x_1, y_1) and (x_2, y_2) is

  $$\left(\frac{x_1 + x_2}{2}, \frac{y_1 + y_2}{2}\right).$$

- You should be able to use the point-plotting method of graphing.
- You should be able to find x- and y-intercepts.

  (a) To find the x-intercepts, let $y = 0$ and solve for x.

  (b) To find the y-intercepts, let $x = 0$ and solve for y.

- You should be able to test for symmetry.

  (a) To test for x-axis symmetry, replace y with $-y$.

  (b) To test for y-axis symmetry, replace x with $-x$.

  (c) To test for origin symmetry, replace x with $-x$ and y with $-y$.

- You should know the standard equation of a circle with center (h, k) and radius r:

  $$(x - h)^2 + (y - k)^2 = r^2$$

Solutions to Odd-Numbered Exercises

1. Triangle

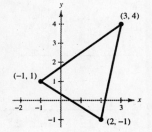

3. Square

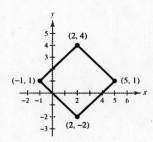

5. $A: (2, 6)$, $B: (-6, -2)$, $C: (4, -4)$, $D: (-3, 2)$

7. $(-3, 4)$

9. $(-5, -5)$

11. $x > 0$ and $y < 0$ in Quadrant IV.

13. $x = -4 \Rightarrow x$ is in Quadrant II or III.

$y > 0 \quad \Rightarrow y$ is in Quadrant I or II.

Both conditions are met in Quadrant II.

15. $y < -5 \Rightarrow y$ is in Quadrant III or IV.

17. $(x, -y)$ is in the second Quadrant means that (x, y) is in Quadrant III.

19. $xy > 0 \Rightarrow x$ and y are either both positive or are both negative. $\Rightarrow$ Quadrant I or III.

21.

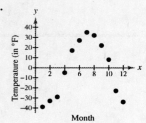

23. (a) The distance between $(0, 2)$ and $(4, 2)$ is 4.

The distance between $(4, 2)$ and $(4, 5)$ is 3.

The distance between $(0, 2)$ and $(4, 5)$ is
$$\sqrt{(4 - 0)^2 + (5 - 2)^2} = \sqrt{16 + 9}$$
$$= \sqrt{25} = 5.$$

(b) $4^2 + 3^2 = 16 + 9 = 25 = 5^2$

25. (a) The distance between $(-1, 1)$ and $(9, 1)$ is 10.

The distance between $(9, 1)$ and $(9, 4)$ is 3.

The distance between $(-1, 1)$ and $(9, 4)$ is
$$\sqrt{(9 - (-1))^2 + (4 - 1)^2} = \sqrt{100 + 9} = \sqrt{109}.$$

(b) $10^2 + 3^2 = 109 = \left(\sqrt{109}\right)^2$

27. (a)

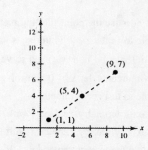

(b) $d = \sqrt{(9 - 1)^2 + (7 - 1)^2}$
$$= \sqrt{64 + 36} = 10$$

(c) $\left(\dfrac{9 + 1}{2}, \dfrac{7 + 1}{2}\right) = (5, 4)$

29. (a)

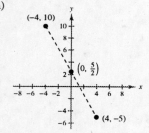

(b) $d = \sqrt{(4 + 4)^2 + (-5 - 10)^2}$
$$= \sqrt{64 + 225} = 17$$

(c) $\left(\dfrac{4 - 4}{2}, \dfrac{-5 + 10}{2}\right) = \left(0, \dfrac{5}{2}\right)$

31. (a)

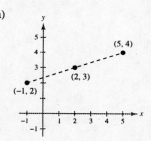

(b) $d = \sqrt{(5 + 1)^2 + (4 - 2)^2}$
$$= \sqrt{36 + 4} = 2\sqrt{10}$$

(c) $\left(\dfrac{-1 + 5}{2}, \dfrac{2 + 4}{2}\right) = (2, 3)$

33. (a)

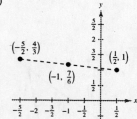

(b)
$$d = \sqrt{\left(\dfrac{1}{2} + \dfrac{5}{2}\right)^2 + \left(1 - \dfrac{4}{3}\right)^2}$$
$$d = \sqrt{9 + \dfrac{1}{9}} = \dfrac{\sqrt{82}}{3}$$

(c) $\left(\dfrac{-\frac{5}{2} + \frac{1}{2}}{2}, \dfrac{\frac{4}{3} + 1}{2}\right) = \left(-1, \dfrac{7}{6}\right)$

35. (a)

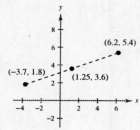

(b) $d = \sqrt{(6.2 + 3.7)^2 + (5.4 - 1.8)^2}$

$\quad = \sqrt{98.01 + 12.96}$

$\quad = \sqrt{110.97}$

(c) $\left(\dfrac{6.2 - 3.7}{2}, \dfrac{5.4 + 1.8}{2} \right) = (1.25, 3.6)$

37. (a)

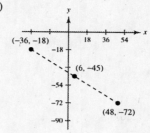

(b) $d = \sqrt{(48 + 36)^2 + (-72 + 18)^2}$

$\quad = \sqrt{7056 + 2916}$

$\quad = \sqrt{9972} = 6\sqrt{277}$

(c) $\left(\dfrac{-36 + 48}{2}, \dfrac{-18 - 72}{2} \right) = (6, -45)$

39. (a) The number of artists elected each year seems to be nearly steady except for the first few years. Between 6 and 8 artists will be elected in 2001.

(b) The Rock and Roll Hall of Fame began in 1986.

41. $d = \sqrt{(45 - 10)^2 + (40 - 15)^2} = \sqrt{35^2 + 25^2} = \sqrt{1850} = 5\sqrt{74} \approx 43$ yards

43. $\left(\dfrac{1996 + 1998}{2}, \dfrac{\$696.5 + \$1308.7}{2} \right) = (1997, \$1002.6)$

In 1997 sales were approximately \$1002.6 million.

45. $y = \sqrt{x + 4}$

(a) $(0, 2)$: $2 \overset{?}{=} \sqrt{0 + 4}$

$\qquad\qquad 2 = 2$

Yes, the point *is* on the graph.

(b) $(5, 3)$: $3 \overset{?}{=} \sqrt{5 + 4}$

$\qquad\qquad 3 = \sqrt{9}$

Yes, the point *is* on the graph.

47. $y = 4 - |x - 2|$

(a) $(1, 5)$: $5 \overset{?}{=} 4 - |1 - 2|$

$\qquad\qquad 5 \neq 4 - 1$

No, the point *is not* on the graph.

(b) $(6, 0)$: $0 \overset{?}{=} 4 - |6 - 2|$

$\qquad\qquad 0 = 4 - 4$

Yes, the point *is* on the graph.

49. $y = \frac{3}{4}x - 1$

x	-2	0	1	$\frac{4}{3}$	2
y	$-\frac{5}{2}$	-1	$-\frac{1}{4}$	0	$\frac{1}{2}$

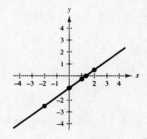

51. $y = 16 - 4x^2$

   x-intercepts: $0 = 16 - 4x^2$

$$4x^2 = 16$$

$$x^2 = 4$$

$$x = \pm 2$$

$$(-2, 0), (2, 0)$$

   y-intercept: $y = 16 - 4(0)^2 = 16$

$$(0, 16)$$

53. $y = 2x^3 - 5x^2$

   x-intercepts: $0 = 2x^3 - 5x^2$

$$0 = x^2(2x - 5)$$

$$x = 0 \text{ or } x = \tfrac{5}{2}$$

$$(0, 0), \left(\tfrac{5}{2}, 0\right)$$

   y-intercept: $y = 2(0)^3 - 5(0)^2 = 0$

$$(0, 0)$$

55. $y = 3 - \dfrac{1}{2}x$

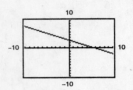

   Intercepts: $(6, 0), (0, 3)$

57. $y = \dfrac{2x}{x - 1}$

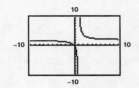

   Intercept: $(0, 0)$

59. $x^2 - y = 0$

   $(-x)^2 - y = 0 \implies x^2 - y = 0 \implies y$-axis symmetry

   $x^2 - (-y) = 0 \implies x^2 + y = 0 \implies$ No x-axis symmetry

   $(-x)^2 - (-y) = 0 \implies x^2 + y = 0 \implies$ No origin symmetry

61. $y = x^3$

   $y = (-x)^3 \implies y = -x^3 \implies$ No y-axis symmetry

   $-y = x^3 \implies y = -x^3 \implies$ No x-axis symmetry

   $-y = (-x)^3 \implies -y = -x^3 \implies y = x^3 \implies$ Origin symmetry

63. $y = \dfrac{x}{x^2 + 1}$

   $y = \dfrac{-x}{(-x)^2 + 1} \implies y = \dfrac{-x}{x^2 + 1} \implies$ No y-axis symmetry

   $-y = \dfrac{x}{x^2 + 1} \implies y = \dfrac{-x}{x^2 + 1} \implies$ No x-axis symmetry

   $-y = \dfrac{-x}{(-x)^2 + 1} \implies -y = \dfrac{-x}{x^2 + 1} \implies y = \dfrac{x}{x^2 + 1} \implies$ Origin symmetry

65. $xy^2 + 10 = 0$

   $(-x)y^2 + 10 = 0 \implies -xy^2 + 10 = 0 \implies$ No y-axis symmetry

   $x(-y)^2 + 10 = 0 \implies xy^2 + 10 = 0 \implies x$-axis symmetry

   $(-x)(-y)^2 + 10 = 0 \implies -xy^2 + 10 = 0 \implies$ No origin symmetry

67. *y*-axis symmetry

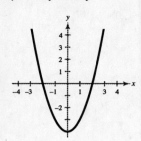

69. Origin symmetry

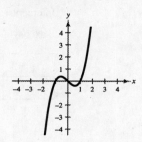

71. $y = 1 - x$ has intercepts $(1, 0)$ and $(0, 1)$.
Matches graph (c).

73. $y = x^3 - x + 1$ has a *y*-intercept of $(0, 1)$ and the points $(1, 1)$ and $(-2, -5)$ are on the graph. Matches graph (b).

75. $y = -3x + 1$

x-intercept: $\left(\frac{1}{3}, 0\right)$

y-intercept: $(0, 1)$

No axis or origin symmetry

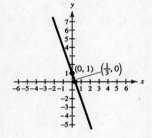

77. $y = x^2 - 2x$

Intercepts: $(0, 0)$, $(2, 0)$

No axis or origin symmetry

x	-1	0	1	2	3
y	3	0	-1	0	3

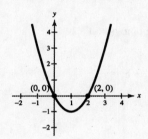

79. $y = x^3 + 3$

Intercepts: $(0, 3)$, $\left(\sqrt[3]{-3}, 0\right)$

No axis or origin symmetry

x	-2	-1	0	1	2
y	-5	2	3	4	11

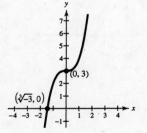

81. $y = \sqrt{x - 3}$

Domain: $[3, \infty)$

Intercept: $(3, 0)$

No axis or origin symmetry

x	3	4	7	12
y	0	1	2	3

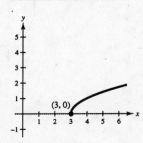

83. $y = |x - 6|$

Intercepts: $(0, 6)$, $(6, 0)$

No axis or origin symmetry

x	-2	0	2	4	6	8	10
y	8	6	4	2	0	2	4

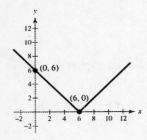

85. $x = y^2 - 1$

Intercepts: $(0, -1)$, $(0, 1)$, $(-1, 0)$

x-axis symmetry

x	-1	0	3
y	0	± 1	± 2

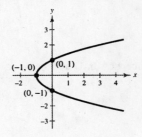

87. Center: $(0, 0)$; radius: 4

Standard form: $(x - 0)^2 + (y - 0)^2 = 4^2$

$$x^2 + y^2 = 16$$

89. Center: $(2, -1)$; radius: 4

Standard form: $(x - 2)^2 + (y - (-1))^2 = 4^2$

$$(x - 2)^2 + (y + 1)^2 = 16$$

91. Center: $(-1, 2)$; solution point: $(0, 0)$

$(x - (-1))^2 + (y - 2)^2 = r^2$

$(0 + 1)^2 + (0 - 2)^2 = r^2 \Longrightarrow 5 = r^2$

Standard form: $(x + 1)^2 + (y - 2)^2 = 5$

93. Endpoints of a diameter: $(0, 0)$, $(6, 8)$

Center: $\left(\dfrac{0 + 6}{2}, \dfrac{0 + 8}{2}\right) = (3, 4)$

$(x - 3)^2 + (y - 4)^2 = r^2$

$(0 - 3)^2 + (0 - 4)^2 = r^2 \Longrightarrow 25 = r^2$

Standard form: $(x - 3)^2 + (y - 4)^2 = 25$

95. $x^2 + y^2 = 25$

Center: $(0, 0)$

Radius: 5

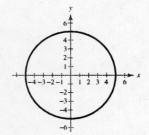

97. $(x - 1)^2 + (y + 3)^2 = 9$

Center: $(1, -3)$

Radius: 3

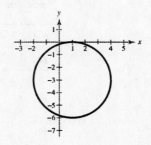

99. $\left(x - \frac{1}{2}\right)^2 + \left(y - \frac{1}{2}\right)^2 = \frac{9}{4}$

Center: $\left(\frac{1}{2}, \frac{1}{2}\right)$

Radius: $\frac{3}{2}$

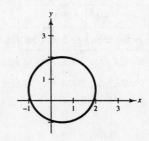

101. $y = 225{,}000 - 20{,}000t,\ 0 \le t \le 8$

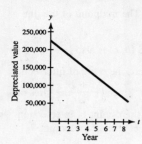

103. (a)

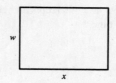

(b) $2x + 2w = 12 \Rightarrow w = 6 - x$
$A = x \cdot w = x(6 - x)$

(c)

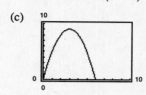

(d) The area is maximum when $x = 3$ and $w = 6 - 3 = 3$.

$x = 3$ meters

$w = 3$ meters

105. (a) and (b)

Year	1950	1960	1970	1980	1990	1994	1997	1998
Per Capita Debt	$1688	$1572	$1807	$3981	$12,848	$15,750	$20,063	$20,513
t	0	10	20	30	40	44	47	48
y	$1837.433	$1204.583	$1763.133	$4851.083	$11,806.433	$15,971.157	$19,692.78	$21,054.377

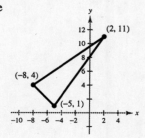

The curve seems to be a good fit for the data.

(c) for the year 2002, $t = 52$ and $y \approx \$27{,}142$

for the year 2004, $t = 54$ and $y \approx \$30{,}589$

107. True

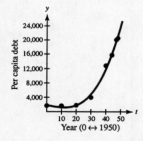

The distance between $(-8, 4)$ and $(2, 11)$ is:

$$d_1 = \sqrt{(2 - (-8))^2 + (11 - 4)^2} = \sqrt{149}$$

The distance between $(-5, 1)$ and $(2, 11)$ is:

$$d_2 = \sqrt{(2 - (-5))^2 + (11 - 1)^2} = \sqrt{149}$$

Since $d_1 = d_2$, the triangle is isosceles.

$\left(\text{Note: The length of the third side is } d_3 = \sqrt{18}.\right)$

109. True. Horizontal lines have no x-intercepts and all other lines cross the x-axis one time.

111. The midpoint of the line segment through $(0, 0)$ and $(a + b, c)$ is $\left(\dfrac{a + b}{2}, \dfrac{c}{2}\right)$. The midpoint of the line segment through $(a, 0)$ and (b, c) is also $\left(\dfrac{a + b}{2}, \dfrac{c}{2}\right)$. Since the diagonals intersect, the point of intersection must lie on both diagonals. Since $\left(\dfrac{a + b}{2}, \dfrac{c}{2}\right)$ lies on both diagonals, it must be their point of intersection. Thus, the diagonals intersect at their midpoints.

Section P.4   Linear Equations in Two Variables

You should know the following important facts about lines.

■   The graph of $y = mx + b$ is a straight line. It is called a linear equation in two variables.

■   The slope of the line through (x_1, y_1) and (x_2, y_2) is

$$m = \frac{y_2 - y_1}{x_2 - x_1} = \frac{\text{change in } y}{\text{change in } x} = \frac{\text{rise}}{\text{run}}.$$

■   (a) If $m > 0$, the line rises from left to right.

   (b) If $m = 0$, the line is horizontal.

   (c) If $m < 0$, the line falls from left to right.

   (d) If m is undefined, the line is vertical.

■   Equations of Lines

   (a) Slope-Intercept: $y = mx + b$

   (b) Point-Slope: $y - y_1 = m(x - x_1)$

   (c) Two-Point: $y - y_1 = \dfrac{y_2 - y_1}{x_2 - x_1}(x - x_1)$

   (d) General: $Ax + By + C = 0$

   (e) Vertical: $x = a$

   (f) Horizontal: $y = b$

■   Given two distinct nonvertical lines

   $L_1: y = m_1x + b_1$ and $L_2: y = m_2x + b_2$

   (a) L_1 is parallel to L_2 if and only if $m_1 = m_2$ and $b_1 \neq b_2$.

   (b) L_1 is perpendicular to L_2 if and only if $m_1 = -1/m_2$.

Solutions to Odd-Numbered Exercises

1. (a) $m = \frac{2}{3}$. Since the slope is positive,
   the line rises. Matches L_2.

   (b) m is undefined. The line is vertical. Matches L_3.

   (c) $m = -2$. The line falls. Matches L_1.

3.

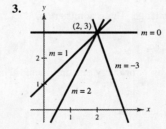

5. Two points on the line: $(0, 0)$ and $(5, 8)$

Slope $= \dfrac{\text{rise}}{\text{run}} = \dfrac{8}{5}$

7. Two points on the line: $(0, 3)$ and $(1, 3)$

Slope $= \dfrac{\text{rise}}{\text{run}} = \dfrac{0}{1} = 0$

9. Two points on the line: $(0, 8)$ and $(2, 0)$

Slope $= \dfrac{\text{rise}}{\text{run}} = \dfrac{-8}{2} = -4$

11.

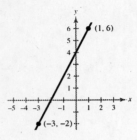

$m = \dfrac{6 - (-2)}{1 - (-3)} = \dfrac{8}{4} = 2$

13.

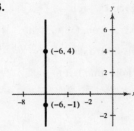

$m = \dfrac{4 - (-1)}{-6 - (-6)} = \dfrac{5}{0}$

m is undefined

15.

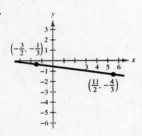

$m = \dfrac{-\dfrac{1}{3} - \left(-\dfrac{4}{3}\right)}{-\dfrac{3}{2} - \dfrac{11}{2}} = -\dfrac{1}{7}$

17.

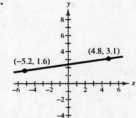

$m = \dfrac{1.6 - 3.1}{-5.2 - 4.8} = \dfrac{-1.5}{-10}$

$= 0.15$

19. Point: $(2, 1)$ Slope: $m = 0$

Since $m = 0$, y does not change. Three points are $(0, 1)$, $(3, 1)$, and $(-1, 1)$.

21. Point: $(5, -6)$ Slope: $m = 1$

Since $m = 1$, y increases by 1 for every one unit increase in x. Three points are $(6, -5)$, $(7, -4)$, and $(8, -3)$.

23. Point: $(-8, 1)$ Slope is undefined

Since m is undefined, x does not change. Three points are $(-8, 0)$, $(-8, 2)$, and $(-8, 3)$.

25. Point: $(-5, 4)$ Slope: $m = 2$

Since $m = 2 = \frac{2}{1}$, y increases by 2 for every one unit increase in x.
Three additional points are $(-4, 6)$, $(-3, 8)$, and $(-2, 10)$.

27. Point: $(7, -2)$ Slope: $m = \frac{1}{2}$

Since $m = \frac{1}{2}$, y increases by 1 unit for every two unit increase in x.
Three additional points are $(9, -1)$, $(11, 0)$, and $(13, 1)$.

29. Slope of L_1: $m = \dfrac{9 + 1}{5 - 0} = 2$

Slope of L_2: $m = \dfrac{1 - 3}{4 - 0} = -\dfrac{1}{2}$

L_1 and L_2 are perpendicular.

31. Slope of L_1: $m = \dfrac{0 - 6}{-6 - 3} = \dfrac{2}{3}$

Slope of L_2: $m = \dfrac{\frac{7}{3} + 1}{5 - 0} = \dfrac{2}{3}$

L_1 and L_2 are parallel.

33. (a) $m = 135$. The sales are increasing 135 units per year.

(b) $m = 0$. There is no change in sales.

(c) $m = -40$. The sales are decreasing 40 units per year.

35. (a) The greatest increase (largest positive slope) was from 1990 to 1991 and from 1996 to 1997.
The greatest decrease (largest negative slope) was from 1997 to 1998.

(b) $(1, 0.98)$ and $(11, 1.35)$

$$m = \frac{1.35 - 0.98}{11 - 1} = \frac{0.37}{10} = 0.037$$

(c) On average, the earnings per share increased by \$0.037 per year over this 10 year period.

37. (a) and (b)

x	300	600	900	1200	1500	1800	2100
y	-25	-50	-75	-100	-125	-150	-175

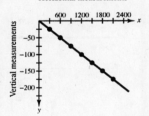

(c) $m = \dfrac{-50 - (-25)}{600 - 300} = \dfrac{-25}{300} = -\dfrac{1}{12}$

$y - (-50) = -\dfrac{1}{12}(x - 600)$

$y + 50 = -\dfrac{1}{12}x + 50$

$y = -\dfrac{1}{12}x$

(d) Since $m = -\frac{1}{12}$, for every 12 horizontal measurements
the vertical measurement decreases by 1.

(e) $\dfrac{1}{12} \approx 0.083 = 8.3\%$ grade

39. $\dfrac{\text{rise}}{\text{run}} = \dfrac{3}{4} = \dfrac{x}{\frac{1}{2}(32)}$

$\qquad \dfrac{3}{4} = \dfrac{x}{16}$

$\qquad 4x = 48$

$\qquad x = 12$

The maximum height in the attic is 12 feet.

41. $y = x - 10$

Slope: $m = 1$

y-intercept: $(0, -10)$

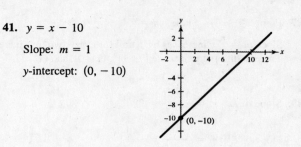

43. $3y + 5 = 0$

$\qquad 3y = -5$

$\qquad y = -\dfrac{5}{3}$

Slope: $m = 0$

y-intercept: $\left(0, -\dfrac{5}{3}\right)$

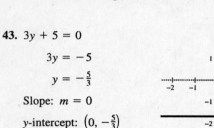

45. $2x + 3y = 9$

$\qquad 3y = -2x + 9$

$\qquad y = -\dfrac{2}{3}x + 3$

Slope: $m = -\dfrac{2}{3}$

y-intercept: $(0, 3)$

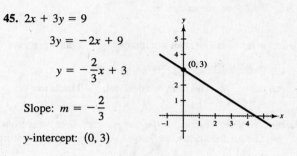

47. $m = -1, (0, 10)$

$\qquad y - 10 = -1(x - 0)$

$\qquad y - 10 = -x$

$\qquad y = -x + 10$

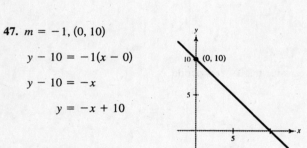

49. $m = 4, (0, 0)$

$\qquad y - 0 = 4(x - 0)$

$\qquad y = 4x$

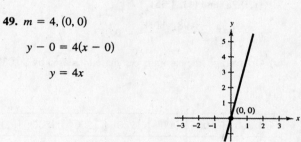

51. $m = \dfrac{3}{4}, (-2, -5)$

$\qquad y + 5 = \dfrac{3}{4}(x + 2)$

$\qquad 4y + 20 = 3x + 6$

$\qquad 4y = 3x - 14$

$\qquad y = \dfrac{3}{4}x - \dfrac{7}{2}$

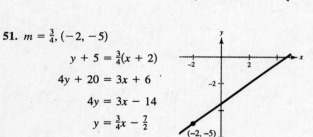

53. $m = 0, (-10, 4)$

$\qquad y - 4 = 0(x + 10)$

$\qquad y - 4 = 0$

$\qquad y = 4$

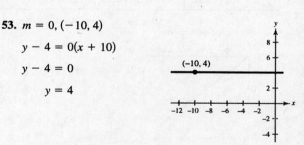

55. $m = -3, \left(-\dfrac{1}{2}, \dfrac{3}{2}\right)$

$\qquad y - \dfrac{3}{2} = -3\left(x + \dfrac{1}{2}\right)$

$\qquad y - \dfrac{3}{2} = -3x - \dfrac{3}{2}$

$\qquad y = -3x$

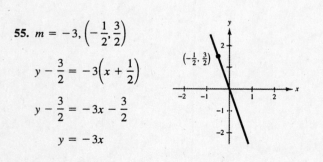

57. $m = -\dfrac{5}{2}, (2.3, -8.5)$

$\qquad y - (-8.5) = -\dfrac{5}{2}(x - 2.3)$

$\qquad y + 8.5 = -2.5x + 5.75$

$\qquad y = -2.5x - 2.75$

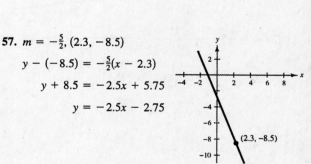

59. $(4, 3), (-4, -4)$

$$y - 3 = \frac{-4 - 3}{-4 - 4}(x - 4)$$

$$y - 3 = \frac{7}{8}(x - 4)$$

$$y - 3 = \frac{7}{8}x - \frac{7}{2}$$

$$y = \frac{7}{8}x - \frac{1}{2}$$

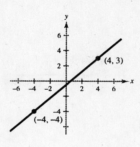

61. $(-1, 4), (6, 4)$

$$y - 4 = \frac{4 - 4}{6 - (-1)}(x + 1)$$

$$y - 4 = 0(x + 1)$$

$$y - 4 = 0$$

$$y = 4$$

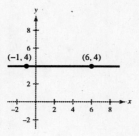

63. $(1, 1), \left(6, -\frac{2}{3}\right)$

$$y - 1 = \frac{-\frac{2}{3} - 1}{6 - 1}(x - 1)$$

$$y - 1 = -\frac{1}{3}(x - 1)$$

$$y - 1 = -\frac{1}{3}x + \frac{1}{3}$$

$$y = -\frac{1}{3}x + \frac{4}{3}$$

65. $\left(\frac{3}{4}, \frac{3}{2}\right), \left(-\frac{4}{3}, \frac{7}{4}\right)$

$$y - \frac{3}{2} = \frac{\frac{7}{4} - \frac{3}{2}}{-\frac{4}{3} - \frac{3}{4}}\left(x - \frac{3}{4}\right)$$

$$y - \frac{3}{2} = \frac{\frac{1}{4}}{-\frac{25}{12}}\left(x - \frac{3}{4}\right)$$

$$y - \frac{3}{2} = -\frac{3}{25}\left(x - \frac{3}{4}\right)$$

$$y - \frac{3}{2} = -\frac{3}{25}x + \frac{9}{100}$$

$$y = -\frac{3}{25}x + \frac{159}{100}$$

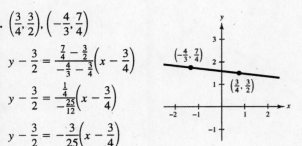

67. $(-8, 0.6), (2, -2.4)$

$$y - 0.6 = \frac{-2.4 - 0.6}{2 - (-8)}(x + 8)$$

$$y - 0.6 = -\frac{3}{10}(x + 8)$$

$$10y - 6 = -3(x + 8)$$

$$10y - 6 = -3x - 24$$

$$10y = -3x - 18$$

$$y = -\frac{3}{10}x - \frac{9}{5} \quad \text{or} \quad y = -0.3x - 1.8$$

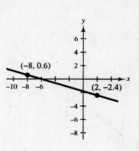

69. $(-3, 0), (0, 4)$

$$\frac{x}{-3} + \frac{y}{4} = 1$$

$$(-12)\frac{x}{-3} + (-12)\frac{y}{4} = (-12) \cdot 1$$

$$4x - 3y + 12 = 0$$

71. $\left(\frac{2}{3}, 0\right), (0, -2)$

$$\frac{x}{\frac{2}{3}} + \frac{y}{-2} = 1$$

$$\frac{3x}{2} - \frac{y}{2} = 1$$

$$3x - y - 2 = 0$$

73. $(d, 0), (0, d), (-3, 4)$

$$\frac{x}{d} + \frac{y}{d} = 1$$

$$x + y = d$$

$$-3 + 4 = d$$

$$1 = d$$

$$x + y = 1$$

$$x + y - 1 = 0$$

75. $x + y = 7$

$$y = -x + 7$$

Slope: $m = -1$

(a) $m = -1, (-3, 2)$

$$y - 2 = -1(x + 3)$$

$$y - 2 = -x - 3$$

$$y = -x - 1$$

(b) $m = 1, (-3, 2)$

$$y - 2 = 1(x + 3)$$

$$y = x + 5$$

77. $5x + 3y = 0$

$$3y = -5x$$

$$y = -\frac{5}{3}x$$

Slope: $m = -\frac{5}{3}$

(a) $m = -\frac{5}{3}, \left(\frac{7}{8}, \frac{3}{4}\right)$

$$y - \frac{3}{4} = -\frac{5}{3}\left(x - \frac{7}{8}\right)$$

$$24y - 18 = -40\left(x - \frac{7}{8}\right)$$

$$24y - 18 = -40x + 35$$

$$24y = -40x + 53$$

$$y = -\frac{5}{3}x + \frac{53}{24}$$

(b) $m = \frac{3}{5}, \left(\frac{7}{8}, \frac{3}{4}\right)$

$$y - \frac{3}{4} = \frac{3}{5}\left(x - \frac{7}{8}\right)$$

$$40y - 30 = 24\left(x - \frac{7}{8}\right)$$

$$40y - 30 = 24x - 21$$

$$40y = 24x + 9$$

$$y = \frac{3}{5}x + \frac{9}{40}$$

79. $x = 4$

m is undefined.

(a) $(2, 5)$, m is undefined.

$$x = 2$$

(b) $(2, 5)$, $m = 0$

$$y = 5$$

81. $6x + 2y = 9$

$$2y = -6x + 9$$

$$y = -3x + \frac{9}{2}$$

Slope: $m = -3$

(a) $(-3.9, -1.4)$, $m = -3$

$$y - (-1.4) = -3(x - (-3.9))$$

$$y + 1.4 = -3x - 11.7$$

$$y = -3x - 13.1$$

(b) $(-3.9, -1.4)$, $m = \frac{1}{3}$

$$y - (-1.4) = \frac{1}{3}(x - (-3.9))$$

$$y + 1.4 = \frac{1}{3}x + 1.3$$

$$y = \frac{1}{3}x - 0.1$$

83. (a) $y = \frac{2}{3}x$ (b) $y = -\frac{3}{2}x$ (c) $y = \frac{2}{3}x + 2$

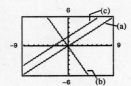

(a) is parallel to (c). (b) is perpendicular to (a) and (c).

85. (a) $y = x - 8$ (b) $y = x + 1$ (c) $y = -x + 3$

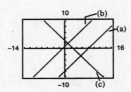

(a) is parallel to (b). (c) is perpendicular to (a) and (b).

87. $(1, 156)$, $m = 4.50$

$$V - 156 = 4.50(t - 1)$$

$$V - 156 = 4.50t - 4.50$$

$$V = 4.5t + 151.5$$

89. The y-intercept is 8.5 and the slope is 2, which represents the increase in hourly wage per unit produced. Matches graph (c).

91. The y-intercepts is 750 and the slope is -100, which represents the decrease in the value of the word processor each year. Matches graph (d).

93. Set the distance between $(6, 5)$ and (x, y) equal to the distance between $(1, -8)$ and (x, y).

$$\sqrt{(x - 6)^2 + (y - 5)^2} = \sqrt{(x - 1)^2 + (y - (-8))^2}$$
$$(x - 6)^2 + (y - 5)^2 = (x - 1)^2 + (y + 8)^2$$
$$x^2 - 12x + 36 + y^2 - 10y + 25 = x^2 - 2x + 1 + y^2 + 16y + 64$$
$$x^2 + y^2 - 12x - 10y + 61 = x^2 + y^2 - 2x + 16y + 65$$
$$-12x - 10y + 61 = -2x + 16y + 65$$
$$-10x - 26y - 4 = 0$$
$$-2(5x + 13y + 2) = 0$$
$$5x + 13y + 2 = 0$$
$$13y = -5x - 2$$
$$y = -\tfrac{5}{13}x - \tfrac{2}{13}$$

95. Set the distance between $\left(-\tfrac{1}{2}, -4\right)$ and (x, y) equal to the distance between $\left(\tfrac{7}{2}, \tfrac{5}{4}\right)$ and (x, y).

$$\sqrt{\left(x - \left(-\tfrac{1}{2}\right)\right)^2 + (y - (-4))^2} = \sqrt{\left(x - \tfrac{7}{2}\right)^2 + \left(y - \tfrac{5}{4}\right)^2}$$
$$\left(x + \tfrac{1}{2}\right)^2 + (y + 4)^2 = \left(x - \tfrac{7}{2}\right)^2 + \left(y - \tfrac{5}{4}\right)^2$$
$$x^2 + x + \tfrac{1}{4} + y^2 + 8y + 16 = x^2 - 7x + \tfrac{49}{4} + y^2 - \tfrac{5}{2}y + \tfrac{25}{16}$$
$$x^2 + y^2 + x + 8y + \tfrac{65}{4} = x^2 + y^2 - 7x - \tfrac{5}{2}y + \tfrac{221}{16}$$
$$x + 8y + \tfrac{65}{4} = -7x - \tfrac{5}{2}y + \tfrac{221}{16}$$
$$8x + \tfrac{21}{2}y + \tfrac{39}{16} = 0$$
$$128x + 168y + 39 = 0$$
$$168y = -128x - 39$$
$$y = -\tfrac{16}{21}x - \tfrac{13}{56}$$

97. $t = 0$ represents 1996

$(0, 3927)$ and $(1, 3981)$

$$m = \frac{3981 - 3927}{1 - 0} = 54$$

$N = 54t + 3927$

$t = 3$ represents 1999: $N = 54(3) + 3927 = 4089$ stores.

$t = 4$ represents 2000: $N = 54(4) + 3927 = 4143$ stores.

99. $F = \frac{9}{5}C + 32$

$F = 0°;$ $0 = \frac{9}{5}C + 32$ $C = -10°;$ $F = \frac{9}{5}(-10) + 32$

$\qquad -32 = \frac{9}{5}C$ $\qquad\qquad\qquad F = -18 + 32$

$\qquad -17.8 \approx C$ $\qquad\qquad\qquad F = 14$

$C = 10°;$ $F = \frac{9}{5}(10) + 32$ $F = 68°;$ $68 = \frac{9}{5}C + 32$

$\qquad\quad F = 18 + 32$ $\qquad\qquad\quad 36 = \frac{9}{5}C$

$\qquad\quad F = 50$ $\qquad\qquad\qquad 20 = C$

$F = 90°;$ $90 = \frac{9}{5}C + 32$ $C = 177°;$ $F = \frac{9}{5}(177) + 32$

$\qquad\quad 58 = \frac{9}{5}C$ $\qquad\qquad\qquad F = 318.6 + 32$

$\qquad\quad 32.2 \approx C$ $\qquad\qquad\qquad F = 350.6$

C	$-17.8°$	$-10°$	$10°$	$20°$	$32.2°$	$177°$
F	$0°$	$14°$	$50°$	$68°$	$90°$	$350.6°$

101. Let $t = 0$ represent 1998.

$(0, 2546)$ and $(2, 2702)$

$m = \dfrac{2702 - 2546}{2 - 0} = 78$

$N = 78t + 2546$

$t = 6$ represents 2004: $N = 78(6) + 2546 = 3014$ students.

103. $(0, 25{,}000)$ and $(10, 2000)$

$m = \dfrac{2000 - 25000}{10 - 0} = -2300$

$V = -2300t + 25{,}000, \quad 0 \le t \le 10$

105. $W = 0.75x + 11.50$

107. $(580, 50)$ and $(625, 47)$

(a) $m = \dfrac{47 - 50}{625 - 580} = \dfrac{-3}{45} = -\dfrac{1}{15}$

$x - 50 = -\dfrac{1}{15}(p - 580)$

$x - 50 = -\dfrac{1}{15}p + \dfrac{116}{3}$

$x = -\dfrac{1}{15}p + \dfrac{266}{3}$

(b) $x = -\dfrac{1}{15}(655) + \dfrac{266}{3} = 45$ units

(c) $x = -\dfrac{1}{15}(595) + \dfrac{266}{3} = 49$ units

109. $W = 0.07S + 2500$

111. Let x = amount invested in the $2\frac{1}{2}\%$ fund and z = amount invested in the 4% fund.

(a) $x + z = 12{,}000 \implies z = 12{,}000 - x$ in the 4% fund.

(b) $y = 0.025x + 0.04(12000 - x)$

$ = 0.025x + 480 - 0.04x$

$ = -0.015x + 480$

(c)

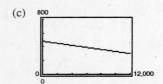

(d) As the amount invested at the lower interest rate increases, the annual interest decreases.

113.

x	18	10	19	16	13	15
y	87	55	96	79	76	82

(a) and (b)

(c) Answers will vary. One approximation is $y = 4x + 19$.

(d) Answers will vary, depending on the equation found in part (c).

$y = 4(17) + 19 = 87$

(e) If 4 points are added to each y-value, then each point would move up 4 units on the graph, and the y-intercept in the equation would increase by 4.

115. $(-8, 2)$ and $(-1, 4) : m_1 = \dfrac{4 - 2}{-1 - (-8)} = \dfrac{2}{7}$

$(0, -4)$ and $(-7, 7) : m_2 = \dfrac{7 - (-4)}{-7 - 0} = \dfrac{11}{-7}$

False, the lines are not parallel.

117. On a vertical line, all the points have the same x-value, so when you evaluate $m = \dfrac{y_2 - y_1}{x_2 - x_1}$, you would have a zero in the denominator, and division by zero is undefined.

119. Since $|-4| > \left|\frac{5}{2}\right|$, the steeper line is the one with a slope of -4.
The slope with the greatest magnitude corresponds to the steepest line.

121. Any pair of distinct points on a line can be used to calculate the slope of the line.
The rate of change remains constant on a straight line.

Section P.5   Functions

- Given a set or an equation, you should be able to determine if it represents a function.
- Given a function, you should be able to do the following.
  - (a) Find the domain and range.
  - (b) Evaluate it at specific values.
- You should be able to use function notation.

Solutions to Odd-Numbered Exercises

1. Yes, the relationship is a function. Each domain value is matched with only one range value.

3. No, the relationship is not a function. The domain values are each matched with three range values.

5. Yes, it does represent a function. Each input value is matched with only one output value.

7. No, it does not represent a function. The input values of 10 and 7 are each matched with two output values.

9. (a) Each element of A is matched with exactly one element of B, so it does represent a function.

    (b) The element 1 in A is matched with two elements, -2 and 1 of B, so it does not represent a function.

    (c) Each element of A is matched with exactly one element of B, so it does represent a function.

    (d) The element 2 in A is not matched with an element of B, so it does not represent a function.

11. Each is a function. For each year there corresponds one and only one circulation.

13. $x^2 + y^2 = 4 \implies y = \pm\sqrt{4 - x^2}$

    No, y *is not* a function of x.

15. $x^2 + y = 4 \implies y = 4 - x^2$

    Yes, y *is* a function of x.

17. $2x + 3y = 4 \implies y = \frac{1}{3}(4 - 2x)$

    Yes, y *is* a function of x.

19. $y^2 = x^2 - 1 \implies y = \pm\sqrt{x^2 - 1}$

    No, y *is not* a function of x.

21. $y = |4 - x|$

    Yes, y *is* a function of x.

23. $f(s) = \dfrac{1}{s + 1}$

    (a) $f(4) = \dfrac{1}{(4) + 1}$

    (b) $f(0) = \dfrac{1}{(0) + 1}$

    (c) $f(4x) = \dfrac{1}{(4x) + 1}$

    (d) $f(x + c) = \dfrac{1}{(x + c) + 1}$

25. $f(x) = 2x - 3$

    (a) $f(1) = 2(1) - 3 = -1$

    (b) $f(-3) = 2(-3) - 3 = -9$

    (c) $f(x - 1) = 2(x - 1) - 3 = 2x - 5$

27. $V(r) = \frac{4}{3}\pi r^3$

    (a) $V(3) = \frac{4}{3}\pi(3)^3 = \frac{4}{3}\pi(27) = 36\pi$

    (b) $V\left(\frac{3}{2}\right) = \frac{4}{3}\pi\left(\frac{3}{2}\right)^3 = \frac{4}{3}\pi\left(\frac{27}{8}\right) = \frac{9}{2}\pi$

    (c) $V(2r) = \frac{4}{3}\pi(2r)^3 = \frac{4}{3}\pi(8r^3) = \frac{32}{3}\pi r^3$

29. $f(y) = 3 - \sqrt{y}$

    (a) $f(4) = 3 - \sqrt{4} = 1$

    (b) $f(0.25) = 3 - \sqrt{0.25} = 2.5$

    (c) $f(4x^2) = 3 - \sqrt{4x^2} = 3 - 2|x|$

31. $q(x) = \dfrac{1}{x^2 - 9}$

    (a) $q(0) = \dfrac{1}{0^2 - 9} = -\dfrac{1}{9}$

    (b) $q(3) = \dfrac{1}{3^2 - 9}$ is undefined.

    (c) $q(y + 3) = \dfrac{1}{(y + 3)^2 - 9} = \dfrac{1}{y^2 + 6y}$

33. $f(x) = \dfrac{|x|}{x}$

    (a) $f(2) = \dfrac{|2|}{2} = 1$

    (b) $f(-2) = \dfrac{|-2|}{-2} = -1$

    (c) $f(x - 1) = \dfrac{|x - 1|}{x - 1}$

35. $f(x) = \begin{cases} 2x + 1, & x < 0 \\ 2x + 2, & x \geq 0 \end{cases}$

    (a) $f(-1) = 2(-1) + 1 = -1$

    (b) $f(0) = 2(0) + 2 = 2$

    (c) $f(2) = 2(2) + 2 = 6$

37. $f(x) = x^2 - 3$

x	-2	-1	0	1	2
$f(x)$	1	-2	-3	-2	1

39. $h(t) = \frac{1}{2}|t + 3|$

t	-5	-4	-3	-2	-1
$h(t)$	1	$\frac{1}{2}$	0	$\frac{1}{2}$	1

41. $f(x) = \begin{cases} -\frac{1}{2}x + 4, & x \leq 0 \\ (x - 2)^2, & x > 0 \end{cases}$

x	-2	-1	0	1	2
$f(x)$	5	$\frac{9}{2}$	4	1	0

43. $15 - 3x = 0$

    $3x = 15$

    $x = 5$

45. $\dfrac{3x - 4}{5} = 0$

    $3x - 4 = 0$

    $x = \dfrac{4}{3}$

47. $x^2 - 9 = 0$

    $x^2 = 9$

    $x = \pm 3$

49.
$$x^3 - x = 0$$
$$x(x^2 - 1) = 0$$
$$x(x + 1)(x - 1) = 0$$
$$x = 0, \; x = -1, \text{ or } x = 1$$

51.
$$f(x) = g(x)$$
$$x^2 = x + 2$$
$$x^2 - x - 2 = 0$$
$$(x + 1)(x - 2) = 0$$
$$x = -1 \text{ or } x = 2$$

53.
$$f(x) = g(x)$$
$$\sqrt{3x} + 1 = x + 1$$
$$\sqrt{3x} = x$$
$$3x = x^2$$
$$0 = x^2 - 3x$$
$$0 = x(x - 3)$$
$$x = 0 \text{ or } x = 3$$

55. $f(x) = 5x^2 + 2x - 1$

Since $f(x)$ is a polynomial, the domain is all real numbers x.

57. $h(t) = \dfrac{4}{t}$

Domain: All real numbers except $t = 0$

59. $g(y) = \sqrt{y - 10}$

Domain: $y - 10 \geq 0$
$$y \geq 10$$

61. $f(x) = \sqrt[4]{1 - x^2}$

Domain: $1 - x^2 \geq 0$
$$-x^2 \geq -1$$
$$x^2 \leq 1$$
$$x^2 - 1 \leq 0$$

Critical Numbers: $x = \pm 1$

Test Intervals: $(-\infty, -1), (-1, 1), (1, \infty)$

Test: Is $x^2 - 1 \leq 0$?

Solution: $[-1, 1]$ or $-1 \leq x \leq 1$

63. $g(x) = \dfrac{1}{x} - \dfrac{1}{x + 2}$

Domain: All real numbers except $x = 0, \; x = -2$

65. $f(s) = \dfrac{\sqrt{s - 1}}{s - 4}$

Domain: $s - 1 \geq 0 \Rightarrow s \geq 1$ and $s \neq 4$

The domain consists of all real numbers s, such that $s \geq 1$ and $s \neq 4$.

67. $f(x) = \dfrac{\sqrt[3]{x - 4}}{x}$

The domain is all real numbers except $x = 0$.

69. $f(x) = x^2$

$\{(-2, 4), (-1, 1), (0, 0), (1, 1), (2, 4)\}$

71. $f(x) = \sqrt{x + 2}$

$\left\{(-2, 0), (-1, 1), \left(0, \sqrt{2}\right), \left(1, \sqrt{3}\right), (2, 2)\right\}$

73. By plotting the points, we have a parabola, so $g(x) = cx^2$. Since $(-4, -32)$ is on the graph, we have $-32 = c(-4)^2 \implies c = -2$. Thus, $g(x) = -2x^2$.

75. Since the function is undefined at 0, we have $r(x) = c/x$. Since $(-4, -8)$ is on the graph, we have $-8 = c/-4 \implies c = 32$. Thus, $r(x) = 32/x$.

77.
$$f(x) = x^2 - x + 1$$
$$f(2 + h) = (2 + h)^2 - (2 + h) + 1$$
$$= 4 + 4h + h^2 - 2 - h + 1$$
$$= h^2 + 3h + 3$$
$$f(2) = (2)^2 - 2 + 1 = 3$$
$$f(2 + h) - f(2) = h^2 + 3h$$
$$\frac{f(2 + h) - f(2)}{h} = \frac{h^2 + 3h}{h} = h + 3, \ h \neq 0$$

79. $f(x) = x^3$
$$f(x + c) = (x + c)^3 = x^3 + 3x^2c + 3xc^2 + c^3$$
$$\frac{f(x + c) - f(x)}{c} = \frac{(x^3 + 3x^2c + 3xc^2 + c^3) - x^3}{c}$$
$$= \frac{c(3x^2 + 3xc + c^2)}{c}$$
$$= 3x^2 + 3xc + c^2, \ c \neq 0$$

81. $g(x) = 3x - 1$
$$\frac{g(x) - g(3)}{x - 3} = \frac{(3x - 1) - 8}{x - 3} = \frac{3x - 9}{x - 3} = \frac{3(x - 3)}{x - 3} = 3, \ x \neq 3$$

83. $f(x) = \sqrt{5x}$
$$\frac{f(x) - f(5)}{x - 5} = \frac{\sqrt{5x} - 5}{x - 5}$$

85. $A = s^2$ and $P = 4s \Longrightarrow \dfrac{P}{4} = s$
$$A = \left(\frac{P}{4}\right)^2 = \frac{P^2}{16}$$

87.

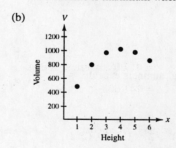

$$8^2 + \left(\frac{b}{2}\right)^2 = s^2$$
$$\frac{b^2}{4} = s^2 - 64$$
$$b^2 = 4(s^2 - 64)$$
$$b = 2\sqrt{s^2 - 64}$$

Thus, $A = \dfrac{1}{2}bh$
$$= \frac{1}{2}\left(2\sqrt{s^2 - 64}\right)(8)$$
$$= 8\sqrt{s^2 - 64} \text{ square inches.}$$

89. (a)

Height, x	Width	Volume, V
1	$24 - 2(1)$	$1[24 - 2(1)]^2 = 484$
2	$24 - 2(2)$	$2[24 - 2(2)]^2 = 800$
3	$24 - 2(3)$	$3[24 - 2(3)]^2 = 972$
4	$24 - 2(4)$	$4[24 - 2(4)]^2 = 1024$
5	$24 - 2(5)$	$5[24 - 2(5)]^2 = 980$
6	$24 - 2(6)$	$6[24 - 2(6)]^2 = 864$

The volume is maximum when $x = 4$.

(b)

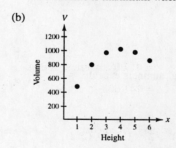

V is a function of x.

(c) $V = x(24 - 2x)^2$

Domain: $0 < x < 12$

91. $A = \frac{1}{2}bh = \frac{1}{2}xy$

Since $(0, y)$, $(2, 1)$, and $(x, 0)$ all lie on the same line, the slopes between any pair are equal.

$$\frac{1 - y}{2 - 0} = \frac{0 - 1}{x - 2}$$

$$\frac{1 - y}{2} = \frac{-1}{x - 2}$$

$$y = \frac{2}{x - 2} + 1$$

$$y = \frac{x}{x - 2}$$

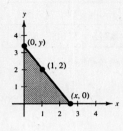

Therefore,

$$A = \frac{1}{2}x\left(\frac{x}{x - 2}\right) = \frac{x^2}{2(x - 2)}.$$

The domain of A includes x-values such that $x^2/[2(x - 2)] > 0$.
Using methods of Section 1.8 we find that the domain is $x > 2$.

93. $p(t) = \begin{cases} 17.27 + 1.036t, & -6 \le t \le 11 \\ -4.807 + 2.882t - 0.011t^2, & 12 \le t \le 17 \end{cases}$

where $t = 0$ represents 1980

1978: $t = -2$ and $p(-2) = 17.27 + 1.036(-2)$

$\qquad\qquad\qquad\qquad = 15.198$ thousand dollars

$\qquad\qquad\qquad\qquad = \$15,198$

1988: $t = 8$ and $p(8) = 17.27 + 1.036(8)$

$\qquad\qquad\qquad\qquad = 25.558$ thousand dollars

$\qquad\qquad\qquad\qquad = \$25,558$

1993: $t = 13$ and $p(13) = -4.807 + 2.882(13) - 0.011(13)^2$

$\qquad\qquad\qquad\qquad = 30.8$ thousand dollars

$\qquad\qquad\qquad\qquad = \$30,800$

1997: $t = 17$ and $p(17) = -4.807 + 2.882(17) - 0.011(17)^2$

$\qquad\qquad\qquad\qquad = 41.008$ thousand dollars

$\qquad\qquad\qquad\qquad = \$41,008$

95. (a) Cost = variable costs + fixed costs

$\qquad C = 12.30x + 98,000$

(b) Revenue = price per unit $\times$ number of units

$\qquad R = 17.98x$

(c) Profit = Revenue − Cost

$\qquad P = 17.98x - (12.30x + 98,000)$

$\qquad P = 5.68x - 98,000$

97. (a) $R = n(\text{rate}) = n[8.00 - 0.05(n - 80)], \; n \geq 80$

$R = 12.00n - 0.05n^2 = 12n - \dfrac{n^2}{20} = \dfrac{240n - n^2}{20}, \; n \geq 80$

(b)

n	90	100	110	120	130	140	150
$R(n)$	\$675	\$700	\$715	\$720	\$715	\$700	\$675

The revenue is maximum when 120 people take the trip.

99. (a)

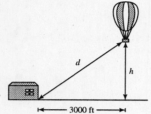

(b) $(3000)^2 + h^2 = d^2$

$h = \sqrt{d^2 - (3000)^2}$

Domain: $d \geq 3000$
(since both $d \geq 0$ and $d^2 - (3000)^2 \geq 0$)

101. $y = -\dfrac{1}{10}x^2 + 3x + 6$

$y(30) = -\dfrac{1}{10}(30)^2 + 3(30) + 6 = 6$ feet

If the child holds a glove at a height of 5 feet, then the ball *will* be over the child's head since it will be at a height of 6 feet.

103. True, the set represents a function.
Each x-value corresponds to one y-value.

105. The domain is the set of inputs of the function, and the range is the set of outputs.

Section P.6   Analyzing Graphs of Functions

- You should be able to determine the domain and range of a function from its graph.

- You should be able to use the vertical line test for functions.

- You should be able to find the zeros of a function.

- You should be able to determine when a function is constant, increasing, or decreasing.

- You should be able to approximate relative minimums and relative maximums from the graph of a function.

- You should know that f is

  (a) odd if $f(-x) = -f(x)$.

  (b) even if $f(-x) = f(x)$.

Solutions to Odd-Numbered Exercises

1. $f(x) = \dfrac{2}{3}x - 4$

Domain: All real numbers

Range: All real numbers

3. $f(x) = 1 - x^2$

Domain: All real numbers

Range: $(-\infty, 1]$

5. $f(x) = \sqrt{x^2 - 1}$

Domain: $(-\infty, -1] \cup [1, \infty)$

Range: $[0, \infty)$

7. $h(x) = \sqrt{16 - x^2}$

Domain: $[-4, 4]$

Range: $[0, 4]$

9. $y = \frac{1}{2}x^2$

A vertical line intersects the graph just once, so y is a function of x.

11. $x - y^2 = 1 \implies y = \pm\sqrt{x - 1}$

y is not a function of x.
Some vertical lines cross the graph twice.

13. $x^2 = 2xy - 1$

A vertical line intersects the graph just once, so y is a function of x.

15. $2x^2 - 7x - 30 = 0$

$(2x + 5)(x - 6) = 0$

$2x + 5 = 0 \quad$ or $\quad x - 6 = 0$

$x = -\frac{5}{2} \quad$ or $\qquad x = 6$

17. $\dfrac{x}{9x^2 - 4} = 0$

$x = 0$

19. $\frac{1}{2}x^3 - x = 0$

$x^3 - 2x = 2(0)$

$x(x^2 - 2) = 0$

$x = 0 \quad$ or $\quad x^2 - 2 = 0$

$x^2 = 2$

$x = \pm\sqrt{2}$

21. $\quad 4x^3 - 24x^2 - x + 6 = 0$

$4x^2(x - 6) - 1(x - 6) = 0$

$(x - 6)(4x^2 - 1) = 0$

$(x - 6)(2x + 1)(2x - 1) = 0$

$x - 6 = 0, \quad 2x + 1 = 0, \quad 2x - 1 = 0$

$x = 6, \qquad x = -\frac{1}{2}, \qquad x = \frac{1}{2}$

23. $3 + \dfrac{5}{x} = 0$

$3x + 5 = 0$

$x = -\dfrac{5}{3}$

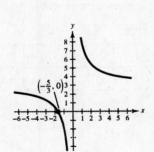

25. $\sqrt{2x + 11} = 0$

$2x + 11 = 0$

$x = -\dfrac{11}{2}$

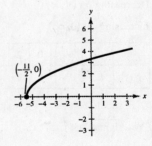

27. $\dfrac{3x - 1}{x - 6} = 0$

$3x - 1 = 0$

$x = \dfrac{1}{3}$

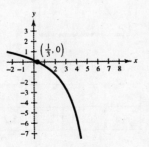

29. $f(x) = \frac{3}{2}x$

    (a) f is increasing on $(-\infty, \infty)$.

    (b) Since $f(-x) = -f(x)$, f is odd.

31. $f(x) = x^3 - 3x^2 + 2$

    (a) f is increasing on $(-\infty, 0)$ and $(2, \infty)$.

        f is decreasing on $(0, 2)$.

    (b) $f(-x) \neq -f(x)$

        $f(-x) \neq f(x)$

        f is neither odd nor even.

33. $f(x) = 3$

    (a)

    Constant on $(-\infty, \infty)$

    (b)

x	-2	-1	0	1	2
$f(x)$	3	3	3	3	3

35. $f(x) = 5 - 3x$

    (a)

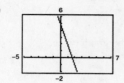

    Decreasing on $(-\infty, \infty)$

    (b)

x	-2	-1	0	1	2
$f(x)$	11	8	5	2	-1

37. $g(s) = \dfrac{s^2}{4}$

    (a)

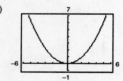

    Decreasing on $(-\infty, 0)$

    Increasing on $(0, \infty)$

    (b)

s	-4	-2	0	2	4
$g(s)$	4	1	0	1	4

39. $f(t) = -t^4$

    (a)

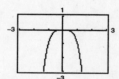

    Increasing on $(-\infty, 0)$

    Decreasing on $(0, \infty)$

    (b)

t	-2	-1	0	1	2
$f(t)$	-16	-1	0	-1	-16

41. $f(x) = \sqrt{1 - x}$

    (a)

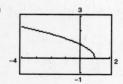

    Decreasing on $(-\infty, 1)$

    (b)

x	-3	-2	-1	0	1
$f(x)$	2	$\sqrt{3}$	$\sqrt{2}$	1	0

43. $f(x) = x^{3/2}$

    (a)

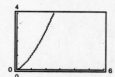

    Increasing on $(0, \infty)$

    (b)

x	0	1	2	3	4
$f(x)$	0	1	2.82	5.2	8

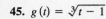

45. $g(t) = \sqrt[3]{t - 1}$

(a)

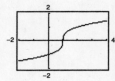

Increasing on $(-\infty, \infty)$

(b)

t	-2	-1	0	1	2
$g(t)$	-1.44	-1.26	-1	0	1

47. $f(x) = |x + 2|$

(a)

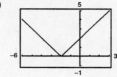

Decreasing on $(-\infty, -2)$

Increasing on $(-2, \infty)$

(b)

x	-6	-4	-2	0	2
$f(x)$	4	2	0	2	4

49. $f(x) = \begin{cases} x + 3, & x \le 0 \\ 3, & 0 < x < 2 \\ 2x - 1, & x > 2 \end{cases}$

(a)

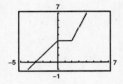

Increasing on $(-\infty, 0)$ and $(2, \infty)$

Constant on $(0, 2)$

(b)

x	-2	-1	0	1	2	3	4
$f(x)$	1	2	3	3	3	5	7

51. $f(x) = (x - 4)(x + 2)$

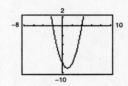

Relative Minimum at $(1, -9)$

53. $f(x) = x(x - 2)(x + 3)$

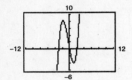

Relative Minimum at $(1.12, -4.06)$

Relative Maximum at $(-1.79, 8.21)$

55. $f(x) = 2x^3 - 5x^2 - 4x - 1$

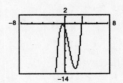

Relative Minimum at $(2, -13)$

Relative Maximum at $(-0.33, -0.30)$

57. $f(x) = 2x - 1$

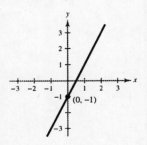

59. $f(x) = -x - \frac{3}{4}$

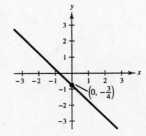

61. $f(x) = -\frac{1}{6}x - \frac{5}{2}$

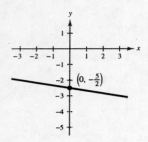

63. $f(x) = 2.5x - 1.8$

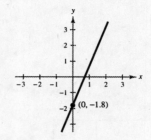

65. $f(1) = 4$, $f(0) = 6$

$(1, 4)$ and $(0, 6)$

$m = \dfrac{6 - 4}{0 - 1} = -2$

$y - 6 = -2(x - 0)$

$y = -2x + 6$

$f(x) = -2x + 6$

67. $f(5) = -4$, $f(-2) = 17$

$(5, -4)$ and $(-2, 17)$

$m = \dfrac{17 - (-4)}{-2 - 5} = \dfrac{21}{-7} = -3$

$y - (-4) = -3(x - 5)$

$y + 4 = -3x + 15$

$y = -3x + 11$

$f(x) = -3x + 11$

69. $f(-5) = -5$, $f(5) = -1$

$(-5, -5)$ and $(5, -1)$

$m = \dfrac{-1 - (-5)}{5 - (-5)} = \dfrac{4}{10} = \dfrac{2}{5}$

$y - (-5) = \dfrac{2}{5}(x - (-5))$

$y + 5 = \dfrac{2}{5}x + 2$

$y = \dfrac{2}{5}x - 3$

$f(x) = \dfrac{2}{5}x - 3$

71. $f\left(\dfrac{1}{2}\right) = -6, f(4) = -3$

$\left(\dfrac{1}{2}, -6\right)$ and $(4, -3)$

$m = \dfrac{-3 - (-6)}{4 - \dfrac{1}{2}} = \dfrac{3}{7/2} = \dfrac{6}{7}$

$y - (-3) = \dfrac{6}{7}(x - 4)$

$y + 3 = \dfrac{6}{7}x - \dfrac{24}{7}$

$y = \dfrac{6}{7}x - \dfrac{45}{7}$

$f(x) = \dfrac{6}{7}x - \dfrac{45}{7}$

73. Vertical shift 2 units downward.

$f(x) = [\![x]\!] - 2$

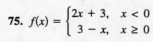

75. $f(x) = \begin{cases} 2x + 3, & x < 0 \\ 3 - x, & x \geq 0 \end{cases}$

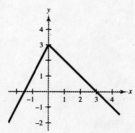

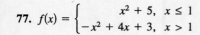

77. $f(x) = \begin{cases} x^2 + 5, & x \le 1 \\ -x^2 + 4x + 3, & x > 1 \end{cases}$

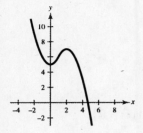

79. $f(x) = 4 - x$

$f(x) \ge 0$ on $(-\infty, 4]$.

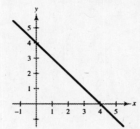

81. $f(x) = x^2 - 9$

$f(x) \ge 0$ on $(-\infty, -3]$ and $[3, \infty)$.

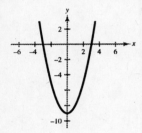

83. $f(x) = 1 - x^4$

$f(x) \ge 0$ on $[-1, 1]$.

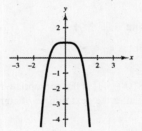

85. $f(x) = x^2 + 1$

$f(x) \ge 0$ on $(-\infty, \infty)$.

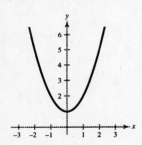

87. $f(x) = -5$, $f(x) < 0$ for all x.

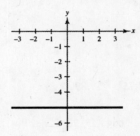

89. $f(x) = \begin{cases} 1 - 2x^2, & x \le -2 \\ -x + 8, & x > -2 \end{cases}$

$f(x) \ge 0$ on $(-2, 8]$

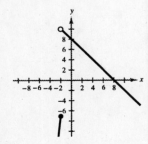

91. $s(x) = 2\left(\frac{1}{4}x - \left[\!\left[\frac{1}{4}x\right]\!\right]\right)$

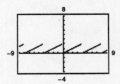

Domain: $(-\infty, \infty)$

Range: $[0, 2)$

Sawtooth pattern

93. $f(x) = x^6 - 2x^2 + 3$

$f(-x) = (-x)^6 - 2(-x)^2 + 3$

$= x^6 - 2x^2 + 3$

$= f(x)$

f is even.

95. $g(x) = x^3 - 5x$

$g(-x) = (-x)^3 - 5(-x)$

$= -x^3 + 5x$

$= -g(x)$

g is odd.

97. $f(t) = t^2 + 2t - 3$

$f(-t) = (-t)^2 + 2(-t) - 3$

$= t^2 - 2t - 3$

$\neq f(t),\ \neq -f(t)$

f is neither even nor odd.

99. $\left(-\frac{3}{2}, 4\right)$

(a) If f is even, another point is $\left(\frac{3}{2}, 4\right)$.

(b) If f is odd, another point is $\left(\frac{3}{2}, -4\right)$.

101. $(4, 9)$

(a) If f is even, another point is $(-4, 9)$

(b) If f is odd, another point is $(-4, -9)$

103. (a) $C_2(t) = 1.05 - 0.38[\![-(t - 1)]\!]$ is the appropriate model since the cost does not increase until after the next minute of conversation has started.

(b)

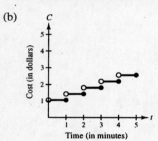

$C = 1.05 - 0.38[\![-17.75]\!] = \$7.89$

105. $L = -0.294x^2 + 97.744x - 664.875,\ 20 \le x \le 90$

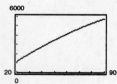

$L = 2000$ when $x \approx 29.9645 \approx 30$ watts

107. $h = \text{top} - \text{bottom}$

$= 3 - (4x - x^2)$

$= 3 - 4x + x^2$

109. $h = \text{top} - \text{bottom}$

$= 2 - \sqrt[3]{x}$

111. $L = \text{right} - \text{left}$

$= 2 - \sqrt[3]{2y}$

113. $L = \text{right} - \text{left}$

$= \frac{2}{y} - 0$

$= \frac{2}{y}$

115.

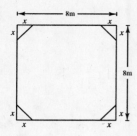

(a) $A = (8)(8) - 4\left(\frac{1}{2}\right)(x)(x)$

$\quad = 64 - 2x^2$

Domain: $0 \le x \le 4$

(b)

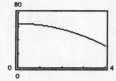

Range: $32 \le A \le 64$

(c) When $x = 4$, the resulting figure is a square.

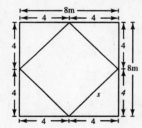

By the Pythagorean Theorem, $4^2 + 4^2 = s^2 \implies s = \sqrt{32} = 4\sqrt{2}$ meters.

117.

Interval	Intake Pipe	Drainpipe 1	Drainpipe 2
[0, 5]	Open	Closed	Closed
[5, 10]	Open	Open	Closed
[10, 20]	Closed	Closed	Closed
[20, 30]	Closed	Closed	Open
[30, 40]	Open	Open	Open
[40, 45]	Open	Closed	Open
[45, 50]	Open	Open	Open
[50, 60]	Open	Open	Closed

119. False. A piecewise-defined function is a function that is defined by two or more equations over a specified domain. That domain may or may not include x- and y-intercepts.

121. $f(x) = a_{2n}x^{2n} + a_{2n-2}x^{2n-2} + \cdots + a_2x^2 + a_0$

$f(-x) = a_{2n}(-x)^{2n} + a_{2n-2}(-x)^{2n-2} + \cdots + a_2(-x)^2 + a_0$

$\qquad = a_{2n}x^{2n} + a_{2n-2}x^{2n-2} + \cdots + a_2x^2 + a_0$

$\qquad = f(x)$

Therefore, $f(x)$ is even.

123. Yes, the graph in Exercise 11 does represent x as a function of y.
Each y-value corresponds to only one x-value.

125. (a) $y = x$ (b) $y = x^2$ (c) $y = x^3$

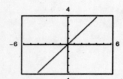 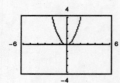 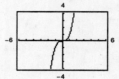

(d) $y = x^4$ (e) $y = x^5$ (f) $y = x^6$

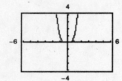 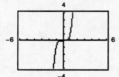 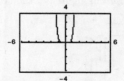

All the graphs pass through the origin. The graphs of the odd powers of x are symmetric with respect to
the origin and the graphs of the even powers are symmetric with respect to the y-axis. As the powers increase,
the graphs become flatter in the interval $-1 < x < 1$.

Section P.7   Shifting, Reflecting, and Stretching Graphs

■   You should know the basic types of transformations.

Let $y = f(x)$ and let c be a positive real number.

1. $h(x) = f(x) + c$ Vertical shift c units upward

2. $h(x) = f(x) - c$ Vertical shift c units downward

3. $h(x) = f(x - c)$ Horizontal shift c units to the right

4. $h(x) = f(x + c)$ Horizontal shift c units to the left

5. $h(x) = -f(x)$ Reflection in the x-axis

6. $h(x) = f(-x)$ Reflection in the y-axis

7. $h(x) = cf(x), c > 1$ Vertical stretch

8. $h(x) = cf(x), 0 < c < 1$ Vertical shrink

Solutions to Odd-Numbered Exercises

1. (a) $f(x) = x^3 + c$

$\quad c = -2 : f(x) = x^3 - 2$ Vertical shift 2 units downward

$\quad c = 0 : f(x) = x^3$ Basic cubic function

$\quad c = 2 : f(x) = x^3 + 2$ Vertical shift 2 units upward

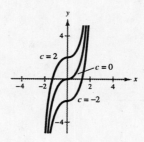

(b) $f(x) = (x - c)^3$

$\quad c = -2 : f(x) = (x + 2)^3$ Horizontal shift 2 units to the left

$\quad c = 0 : f(x) = x^3$ Basic cubic function

$\quad c = 2 : f(x) = (x - 2)^3$ Horizontal shift 2 units to the right

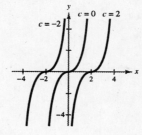

3. (a) $f(x) = \sqrt{x} + c$ Vertical shifts

$\quad c = -3 : f(x) = \sqrt{x} - 3$ 3 units downward

$\quad c = -1 : f(x) = \sqrt{x} - 1$ 1 unit downward

$\quad c = 1 : f(x) = \sqrt{x} + 1$ 1 unit upward

$\quad c = 3 : f(x) = \sqrt{x} + 3$ 3 units upward

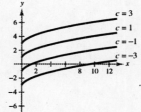

(b) $f(x) = \sqrt{x - c}$ Horizontal shifts

$\quad c = -3 : f(x) = \sqrt{x + 3}$ 3 units to the left

$\quad c = -1 : f(x) = \sqrt{x + 1}$ 1 unit to the left

$\quad c = 1 : f(x) = \sqrt{x - 1}$ 1 unit to the right

$\quad c = 3 : f(x) = \sqrt{x - 3}$ 3 units to the right

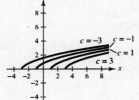

(c) $f(x) = \sqrt{x - 3} + c$ Horizontal shift 3 units to the right and a vertical shift

$\quad c = -3 : f(x) = \sqrt{x - 3} - 3$ 3 units downward

$\quad c = -1 : f(x) = \sqrt{x - 3} - 1$ 1 unit downward

$\quad c = 1 : f(x) = \sqrt{x - 3} + 1$ 1 unit upward

$\quad c = 3 : f(x) = \sqrt{x - 3} + 3$ 3 units upward

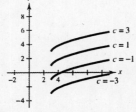

5. (a) $y = f(x) + 2$

Vertical shift 2 units upward.

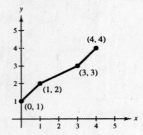

(b) $y = f(x - 2)$

Horizontal shift 2 units to the right.

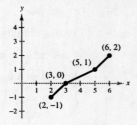

(c) $y = 2f(x)$

Vertical stretch by a factor of 2.

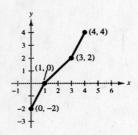

(d) $y = -f(x)$

Reflection in the x-axis.

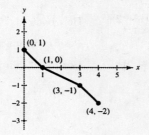

(e) $y = f(x + 3)$

Horizontal shift 3 units to the left.

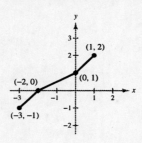

(f) $y = f(-x)$

Reflection in the y-axis.

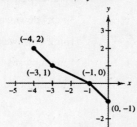

7. (a) $y = f(x) - 1$

Vertical shift 1 unit downward.

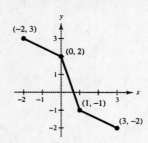

(b) $y = f(x - 1)$

Horizontal shift 1 unit to the right.

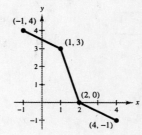

— **CONTINUED** —

7. — CONTINUED —

(c) $y = f(-x)$

Reflection about the y-axis.

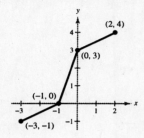

(d) $y = f(x + 1)$

Horizontal shift 1 unit to the left.

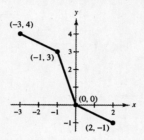

(e) $y = -f(x - 2)$

Reflection about the x-axis and a horizontal shift 2 units to the right.

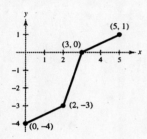

(f) $y = \frac{1}{2} f(x)$

Vertical shrink by a factor of $\frac{1}{2}$.

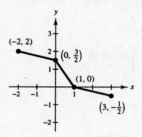

9. (a) Vertical shift 1 unit downward.

$f(x) = x^2 - 1$

(b) Reflection about the x-axis, horizontal shift 1 unit to the left, and a vertical shift 1 unit upward.

$f(x) = -(x + 1)^2 + 1$

(c) Reflection about the x-axis, horizontal shift 2 units to the right, and a vertical shift 6 units upward.

$f(x) = -(x - 2)^2 + 6$

(d) Horizontal shift 5 units to the right and a vertical shift 3 units downward.

$f(x) = (x - 5)^2 - 3$

11. (a) Vertical shift 5 units upward.

$f(x) = |x| + 5$

(b) Reflection in the x-axis and a horizontal shift 3 units to the left.

$f(x) = -|x + 3|$

(c) Horizontal shift 2 units to the right and a vertical shift 4 units downward.

$f(x) = |x - 2| - 4$

(d) Reflection in the x-axis, horizontal shift 6 units to the right, and a vertical shift 1 unit downward.

$f(x) = -|x - 6| - 1$

13. Common function: $f(x) = x^3$

Horizontal shift 2 units to the right: $y = (x - 2)^3$

15. Common function: $f(x) = x^2$

Reflection in the x-axis: $y = -x^2$

17. Common function: $f(x) = \sqrt{x}$

Reflection in the x-axis and a vertical shift 1 unit upward: $y = -\sqrt{x} + 1$

19. $f(x) = 12 - x^2$

Common function: $g(x) = x^2$

Reflection in the x-axis and a vertical shift
12 units upward.

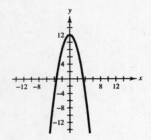

21. $f(x) = x^3 + 7$

Common function: $g(x) = x^3$

Vertical shift 7 units upward.

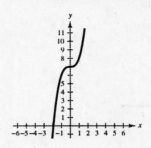

23. $f(x) = 2 - (x + 5)^2$

Common function: $g(x) = x^2$

Reflection in the x-axis, horizontal shift
5 units to the left, and a vertical shift
2 units upward.

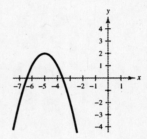

25. $f(x) = (x - 1)^3 + 2$

Common function: $g(x) = x^3$

Horizontal shift 1 unit to the right and
a vertical shift 2 units upward.

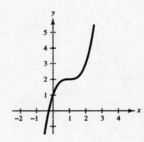

27. $f(x) = -|x| - 2$

Common function: $g(x) = |x|$

Reflection in the x-axis and a vertical shift
2 units downward.

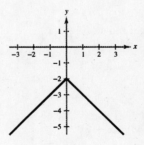

29. $f(x) = -|x + 4| + 8$

Common function: $g(x) = |x|$

Reflection in the x-axis, horizontal shift 4 units to the left, and a vertical shift 8 units upward.

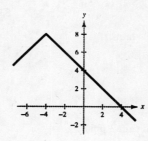

31. $f(x) = \sqrt{x - 9}$

Common function: $g(x) = \sqrt{x}$

Horizontal shift 9 units to the right.

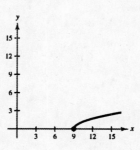

33. $f(x) = \sqrt{7 - x} - 2$ or $f(x) = \sqrt{-(x - 7)} - 2$

Reflection in the y-axis, horizontal shift 7 units to the right, and a vertical shift 2 units downward.

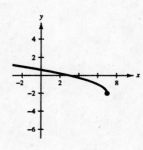

35. $f(x) = x^2$ moved 2 units to the right and 8 units down.

$g(x) = (x - 2)^2 - 8$

37. $f(x) = x^3$ moved 13 units to the right.

$g(x) = (x - 13)^3$

39. $f(x) = |x|$ moved 10 units up and reflected about the x-axis.

$g(x) = -(|x| + 10) = -|x| - 10$

41. $f(x) = \sqrt{x}$ moved 6 units to the left and reflected in both the x and y axes.

$g(x) = -\sqrt{-x + 6}$

43. $f(x) = x^2$

(a) Reflection in the x-axis and a vertical stretch by a factor of 3.

$g(x) = -3x^2$

(b) Vertical shift 3 units upward and a vertical stretch by a factor of 4.

$g(x) = 4x^2 + 3$

45. $f(x) = |x|$

(a) Reflection in the x-axis and a vertical shrink by a factor of $\frac{1}{2}$.

$g(x) = -\frac{1}{2}|x|$

(b) Vertical stretch by a factor of 3 and a vertical shift 3 units downward.

$g(x) = 3|x| - 3$

47. Common function: $f(x) = x^3$

Vertical stretch by a factor of $\frac{3}{2}$: $g(x) = \frac{3}{2}x^3$

49. Common function: $f(x) = x^2$

Reflection in the x-axis and a vertical shrink by a factor of $\frac{1}{2}$: $g(x) = -\frac{1}{2}x^2$

51. Common function: $f(x) = \sqrt{x}$

Reflection in the y-axis and a vertical shrink by a factor or $\frac{1}{2}$: $g(x) = \frac{1}{2}\sqrt{-x}$

53. Common function: $f(x) = x^3$

Reflection in the x-axis, horizontal shift 2 units to the right and a vertical shift 2 units upward: $g(x) = -(x - 2)^3 + 2$

55. Common function: $f(x) = \sqrt{x}$

Reflection in the x-axis and a vertical shift 3 units downward: $g(x) = -\sqrt{x} - 3$

57. (a) $g(x) = f(x) + 2$

Vertical shift 2 unit upward.

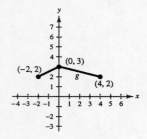

(b) $g(x) = f(x) - 1$

Vertical shift 1 unit downward.

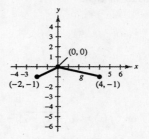

(c) $g(x) = f(-x)$

Reflection in the y-axis.

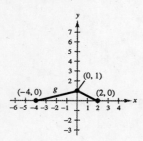

(d) $g(x) = -2f(x)$

Reflection in the x-axis and a vertical stretch by a factor of 2.

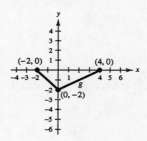

59. $F = f(t) = 20.46 + 0.04t^2$

(a) Common function: $f(x) = x^2$
Vertical shrink by a factor of 0.04 and a vertical shift of 20.46 units.

(b) This represents a horizontal shift 10 units to the left, so
$g(t) = f(t + 10) = 20.46 + 0.04(t + 10)^2$.

61. True, since $|x| = |-x|$, the graphs of $f(x) = |x| + 6$ and $f(x) = |-x| + 6$ are identical.

63. (a) The profits were only $\frac{3}{4}$ as large as expected: $g(t) = \frac{3}{4}f(t)$

(b) The profits were \$10,000 greater than predicted: $g(t) = f(t) + 10,000$

(c) There was a 2-year delay: $g(t) = f(t - 2)$

65. $y = f(x + 2) - 1$

Horizontal shift 2 units to the left and a vertical shift 1 unit downward.

$(0, 1) \rightarrow (0 - 2, 1 - 1) = (-2, 0)$

$(1, 2) \rightarrow (1 - 2, 2 - 1) = (-1, 1)$

$(2, 3) \rightarrow (2 - 2, 3 - 1) = (0, 2)$

Section P.8   Combinations of Functions

■   Given two functions, f and g, you should be able to form the following functions (if defined):

   1. Sum: $(f + g)(x) = f(x) + g(x)$

   2. Difference: $(f - g)(x) = f(x) - g(x)$

   3. Product: $(fg)(x) = f(x)g(x)$

   4. Quotient: $(f/g)(x) = f(x)/g(x), g(x) \neq 0$

   5. Composition of f with g: $(f \circ g)(x) = f(g(x))$

   6. Composition of g with f: $(g \circ f)(x) = g(f(x))$

Solutions to Odd-Numbered Exercises

1.

x	0	1	2	3
f	2	3	1	2
g	-1	0	$\frac{1}{2}$	0
$f + g$	1	3	$\frac{3}{2}$	2

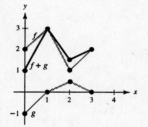

3.

x	-2	0	1	2	4
f	2	0	1	2	4
g	4	2	1	0	2
$f + g$	6	2	2	2	6

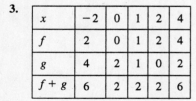

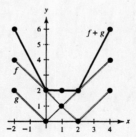

5. $f(x) = x + 2, g(x) = x - 2$

   (a) $(f + g)(x) = f(x) + g(x) = (x + 2) + (x - 2) = 2x$

   (b) $(f - g)(x) = f(x) - g(x) = (x + 2) - (x - 2) = 4$

   (c) $(fg)(x) = f(x) \cdot g(x) = (x + 2)(x - 2) = x^2 - 4$

   (d) $\left(\dfrac{f}{g}\right)(x) = \dfrac{f(x)}{g(x)} = \dfrac{x + 2}{x - 2}$

   Domain: all real numbers except $x = 2$

7. $f(x) = x^2, g(x) = 2 - x$

   $(f + g)(x) = f(x) + g(x) = x^2 + (2 - x) = x^2 - x + 2$

   $(f - g)(x) = f(x) - g(x) = x^2 - (2 - x) = x^2 + x - 2$

   $(fg)(x) = f(x) \cdot g(x) = x^2(2 - x) = 2x^2 - x^3$

   $\left(\dfrac{f}{g}\right)(x) = \dfrac{f(x)}{g(x)} = \dfrac{x^2}{2 - x}$

   Domain: all real numbers except $x = 2$

9. $f(x) = x^2 + 6, g(x) = \sqrt{1 - x}$

$(f + g)(x) = f(x) + g(x) = (x^2 + 6) + \sqrt{1 - x}$

$(f - g)(x) = f(x) - g(x) = (x^2 + 6) - \sqrt{1 - x}$

$(fg)(x) = f(x) \cdot g(x) = (x^2 + 6)\sqrt{1 - x}$

$\left(\dfrac{f}{g}\right)(x) = \dfrac{f(x)}{g(x)} = \dfrac{x^2 + 6}{\sqrt{1 - x}}$,   Domain: $x < 1$

11. $f(x) = \dfrac{1}{x}, g(x) = \dfrac{1}{x^2}$

$(f + g)(x) = f(x) + g(x) = \dfrac{1}{x} + \dfrac{1}{x^2} = \dfrac{x + 1}{x^2}$

$(f - g)(x) = f(x) - g(x) = \dfrac{1}{x} - \dfrac{1}{x^2} = \dfrac{x - 1}{x^2}$

$(fg)(x) = f(x) \cdot g(x) = \dfrac{1}{x}\left(\dfrac{1}{x^2}\right) = \dfrac{1}{x^3}$

$\left(\dfrac{f}{g}\right)(x) = \dfrac{f(x)}{g(x)} = \dfrac{1/x}{1/x^2} = \dfrac{x^2}{x} = x, \ x \neq 0$

For Exercises 13–24, $f(x) = x^2 + 1$ and $g(x) = x - 4$

13. $(f + g)(2) = f(2) + g(2) = (2^2 + 1) + (2 - 4) = 3$

15. $(f - g)(0) = f(0) - g(0) = (0^2 + 1) - (0 - 4) = 5$

17. $(f - g)(3t) = f(3t) - g(3t) = [(3t)^2 + 1] - (3t - 4)$
$$= 9t^2 - 3t + 5$$

19. $(fg)(6) = f(6)g(6) = (6^2 + 1)(6 - 4) = 74$

21. $\left(\dfrac{f}{g}\right)(5) = \dfrac{f(5)}{g(5)} = \dfrac{5^2 + 1}{5 - 4} = 26$

23. $\left(\dfrac{f}{g}\right)(-1) - g(3) = \dfrac{f(-1)}{g(-1)} - g(3)$

$$= \dfrac{(-1)^2 + 1}{-1 - 4} - (3 - 4)$$

$$= -\dfrac{2}{5} + 1 = \dfrac{3}{5}$$

25. $f(x) = \frac{1}{2}x, g(x) = x - 1, (f + g)(x) = \frac{3}{2}x - 1$

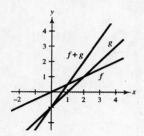

27. $f(x) = x^2, g(x) = -2x, (f + g)(x) = x^2 - 2x$

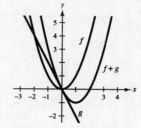

29. $f(x) = 3x, g(x) = -\frac{x^3}{10}, (f + g)(x) = 3x - \frac{x^3}{10}$

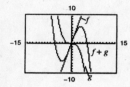

For $0 \le x \le 2$, $f(x)$ contributes most to the magnitude.

For $x > 6$, $g(x)$ contributes most to the magnitude.

31. $T(x) = R(x) + B(x) = \frac{3}{4}x + \frac{1}{15}x^2$

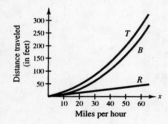

33.

Year	1990	1991	1992	1993	1994	1995	1996
y_1	144.4	151.6	159.5	163.6	164.8	166.7	171.2
y_2	238.6	259.4	282.5	303.3	315.6	326.9	337.3
y_3	21.8	24.0	25.1	27.3	29.6	31.7	32.4

$y_1 = -0.59x^2 + 7.66x + 144.90$

$y_2 = 16.58x + 245.06$

$y_3 = 1.85x + 21.88$

35. (a) T is a function of t since for each time t there corresponds one and only one temperature T.

(b) $T(4) = 60°$

$T(15) = 72°$

(c) $H(t) = T(t - 1)$; All the temperature changes would be one hour later.

(d) $H(t) = T(t) - 1$; The temperature would be decreased by one degree.

37. $f(x) = x^2, g(x) = x - 1$

(a) $(f \circ g)(x) = f(g(x)) = f(x - 1) = (x - 1)^2$

(b) $(g \circ f)(x) = g(f(x)) = g(x^2) = x^2 - 1$

(c) $(f \circ f)(x) = f(f(x)) = f(x^2) = (x^2)^2 = x^4$

39. $f(x) = 3x + 5, g(x) = 5 - x$

(a) $(f \circ g)(x) = f(g(x)) = f(5 - x) = 3(5 - x) + 5 = 20 - 3x$

(b) $(g \circ f)(x) = g(f(x)) = g(3x + 5) = 5 - (3x + 5) = -3x$

(c) $(f \circ f)(x) = f(f(x)) = f(3x + 5) = 3(3x + 5) + 5 = 9x + 20$

41. $f(x) = \sqrt{x + 4}$ Domain: $x \geq -4$

$g(x) = x^2$ Domain: all real numbers

(a) $(f \circ g)(x) = f(g(x)) = f(x^2) = \sqrt{x^2 + 4}$

Domain: all real numbers

(b) $(g \circ f)(x) = g(f(x)) = g(\sqrt{x + 4}) = (\sqrt{x + 4})^2 = x + 4$

Domain: $x \geq -4$

43. $f(x) = \frac{1}{3}x - 3$ Domain: all real numbers

$g(x) = 3x + 1$ Domain: all real numbers

(a) $(f \circ g)(x) = f(g(x)) = f(3x + 1) = \frac{1}{3}(3x + 1) - 3 = x - \frac{8}{3}$ Domain: all real numbers

(b) $(g \circ f)(x) = g(f(x)) = g(\frac{1}{3}x - 3) = 3(\frac{1}{3}x - 3) + 1 = x - 8$ Domain: all real numbers

45. $f(x) = x^4$ Domain: all real numbers

$g(x) = x^4$ Domain: all real numbers

(a) and (b) $(f \circ g)(x) = (g \circ f)(x) = (x^4)^4 = x^{16}$

Domain: all real numbers

47. $f(x) = |x|$ Domain: all real numbers

$g(x) = x + 6$ Domain: all real numbers

(a) $(f \circ g)(x) = f(g(x)) = f(x + 6) = |x + 6|$ Domain: all real numbers

(b) $(g \circ f)(x) = g(f(x)) = g(|x|) = |x| + 6$ Domain: all real numbers

49. $f(x) = \frac{1}{x}$ Domain: all real numbers except $x = 0$

$g(x) = x + 3$ Domain: all real numbers

(a) $(f \circ g)(x) = f(g(x)) = f(x + 3) = \frac{1}{x + 3}$ (b) $(g \circ f)(x) = g(f(x)) = g(\frac{1}{x}) = \frac{1}{x} + 3$

Domain: all real numbers except $x = -3$ Domain: all real numbers except $x = 0$

51. (a) $(f + g)(3) = f(3) + g(3) = 2 + 1 = 3$ **53.** (a) $(f \circ g)(2) = f(g(2)) = f(2) = 0$

(b) $\left(\dfrac{f}{g}\right)(2) = \dfrac{f(2)}{g(2)} = \dfrac{0}{2} = 0$ (b) $(g \circ f)(2) = g(f(2)) = g(0) = 4$

55. Let $f(x) = x^2$ and $g(x) = 2x + 1$, then $(f \circ g)(x) = h(x)$. This is not a unique solution.

For example, if $f(x) = (x + 1)^2$ and $g(x) = 2x$, then $(f \circ g)(x) = h(x)$ as well.

57. Let $f(x) = \sqrt[3]{x}$ and $g(x) = x^2 - 4$, then $(f \circ g)(x) = h(x)$.

This answer is not unique. Other possibilities may be:

$f(x) = \sqrt[3]{x - 4}$ and $g(x) = x^2$

or $f(x) = \sqrt[3]{-x}$ and $g(x) = 4 - x^2$

or $f(x) = \sqrt[9]{x}$ and $g(x) = (4 - x^2)^3$

59. Let $f(x) = 1/x$ and $g(x) = x + 2$, then $(f \circ g)(x) = h(x)$.

This is not a unique solution. Other possibilities may be:

$f(x) = \dfrac{1}{x + 2}$ and $g(x) = x$

or $f(x) = \dfrac{1}{x + 1}$ and $g(x) = x + 1$

or $f(x) = \dfrac{1}{x^2 + 2}$ and $g(x) = \sqrt{x}$

61. Let $f(x) = \dfrac{x + 3}{4 + x}$ and $g(x) = -x^2$, then $(f \circ g)(x) = h(x)$. This answer is not unique.

Other possibilities may be:

$f(x) = \dfrac{x + 1}{x + 2}$ and $g(x) = -x^2 + 2$

or $f(x) = x^2$ and $g(x) = \sqrt{\dfrac{-x^2 + 3}{4 - x^2}}$

or $f(x) = \sqrt{x}$ and $g(x) = \left(\dfrac{-x^2 + 3}{4 - x^2}\right)^2$

63. (a) $r(x) = \dfrac{x}{2}$

(b) $A(r) = \pi r^2$

(c) $(A \circ r)(x) = A(r(x)) = A\left(\dfrac{x}{2}\right) = \pi\left(\dfrac{x}{2}\right)^2$

$(A \circ r)(x)$ represents the area of the circular base of the tank on the square foundation with side length x.

65. $(C \circ x)(t) = C(x(t))$

$= 60(50t) + 750$

$= 3000t + 750$

$(C \circ x)(t)$ represents the cost after t production hours.

67. True. The range of g must be a subset of the domain of f for $(f \circ g)(x)$ to be defined. Since $(f \circ g)(x) = f(g(x))$ and since $g(x)$ represents the range of g, then f is being evaluated with values from g's range.

69. Let $f(x)$ and $g(x)$ be two odd functions and define $h(x) = f(x)g(x)$. Then

$h(-x) = f(-x)g(-x)$

$= [-f(x)][-g(x)]$ Since f and g are odd

$= f(x)g(x)$

$= h(x)$

Thus, $h(x)$ is even.

Let $f(x)$ and $g(x)$ be two even functions and define $h(x) = f(x)g(x)$. Then

$h(-x) = f(-x)g(-x)$

$= f(x)g(x)$ Since f and g are even

$= h(x)$

Thus, $h(x)$ is even.

Section P.9   Inverse Functions

■ Two functions f and g are inverses of each other if $f(g(x)) = x$ for every x in the domain of g and $g(f(x)) = x$ for every x in the domain of f.

■ A function f has an inverse function if and only if no **horizontal** line crosses the graph of f at more than one point.

■ Be able to find the inverse of a function, if it exists.

1. Use the Horizontal Line Test to see if f^{-1} exists.

2. Replace $f(x)$ with y.

3. Interchange x and y and solve for y.

4. Replace y with $f^{-1}(x)$.

Solutions to Odd-Numbered Exercises

1. The inverse is a line through $(-1, 0)$.

Matches graph (c).

3. The inverse is half a parabola starting at $(1, 0)$.

Matches graph (a).

5. $f^{-1}(x) = \dfrac{x}{6} = \dfrac{1}{6}x$

$f(f^{-1}(x)) = f\left(\dfrac{x}{6}\right) = 6\left(\dfrac{x}{6}\right) = x$

$f^{-1}(f(x)) = f^{-1}(6x) = \dfrac{6x}{6} = x$

7. $f^{-1}(x) = x - 9$

$f(f^{-1}(x)) = f(x - 9) = (x - 9) + 9 = x$

$f^{-1}(f(x)) = f^{-1}(x + 9) = (x + 9) - 9 = x$

9. $f^{-1}(x) = \dfrac{x - 1}{3}$

$f(f^{-1}(x)) = f\left(\dfrac{x - 1}{3}\right) = 3\left(\dfrac{x - 1}{3}\right) + 1 = x$

$f^{-1}(f(x)) = f^{-1}(3x + 1) = \dfrac{(3x + 1) - 1}{3} = x$

11. $f^{-1}(x) = x^3$

$f(f^{-1}(x)) = f(x^3) = \sqrt[3]{x^3} = x$

$f^{-1}(f(x)) = f^{-1}(\sqrt[3]{x}) = \left(\sqrt[3]{x}\right)^3 = x$

13. (a) $f(g(x)) = f\left(\dfrac{x}{2}\right) = 2\left(\dfrac{x}{2}\right) = x$

$g(f(x)) = g(2x) = \dfrac{2x}{2} = x$

(b)

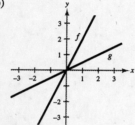

15. (a) $f(g(x)) = f\left(\dfrac{x - 1}{5}\right) = 5\left(\dfrac{x - 1}{5}\right) + 1 = x$

(a) $g(f(x)) = g(5x + 1) = \dfrac{(5x + 1) - 1}{5} = x$

(b)

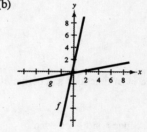

17. (a) $f(g(x)) = f(\sqrt[3]{x}) = (\sqrt[3]{x})^3 = x$

$g(f(x)) = g(x^3) = \sqrt[3]{x^3} = x$

(b)

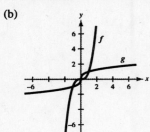

19. (a) $f(g(x)) = f(x^2 + 4), \; x \geq 0$

$\qquad = \sqrt{(x^2 + 4) - 4} = x$

$g(f(x)) = g(\sqrt{x - 4})$

$\qquad = (\sqrt{x - 4})^2 + 4 = x$

(b)

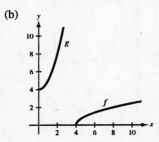

21. (a) $f(g(x)) = f(\sqrt{9 - x}), \; x \leq 9$

$\qquad = 9 - (\sqrt{9 - x})^2 = x$

$g(f(x)) = g(9 - x^2), \; x \geq 0$

$\qquad = \sqrt{9 - (9 - x^2)} = x$

(b)

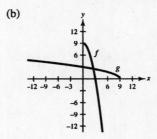

23. (a) $f(g(x)) = f\left(-\dfrac{5x + 1}{x - 1}\right)$

$$= \frac{\left(-\dfrac{5x + 1}{x - 1} - 1\right)}{\left(-\dfrac{5x + 1}{x - 1} + 5\right)} \cdot \frac{x - 1}{x - 1}$$

$$= \frac{-(5x + 1) - (x - 1)}{-(5x + 1) + 5(x - 1)}$$

$$= \frac{-6x}{-6}$$

$$= x$$

$g(f(x)) = g\left(\dfrac{x - 1}{x + 5}\right)$

$$= -\frac{\left[5\left(\dfrac{x - 1}{x + 5}\right) + 1\right]}{\left[\dfrac{x - 1}{x + 5} - 1\right]} \cdot \frac{x + 5}{x + 5}$$

$$= -\frac{5(x - 1) + (x + 5)}{(x - 1) - (x + 5)}$$

$$= -\frac{6x}{-6}$$

$$= x$$

(b)

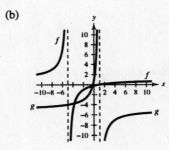

25. No, $\{(-2, -1), (1, 0), (2, 1), (1, 2), (-2, 3), (-6, 4)\}$ does not represent a function.
-2 and 1 are paired with two different values.

27.

x	-2	0	2	4	6	8
$f^{-1}(x)$	-2	-1	0	1	2	3

29. Since no horizontal line crosses the graph of f at more than one point, f **has** an inverse.

31. Since some horizontal lines cross the graph of f twice, f does **not** have an inverse.

33. $g(x) = \dfrac{4 - x}{6}$

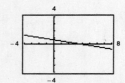

g passes the horizontal line test, so g **has** an inverse.

35. $h(x) = |x + 4| - |x - 4|$

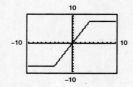

h does not pass the horizontal line test, so h does **not** have an inverse.

37. $f(x) = -2x\sqrt{16 - x^2}$

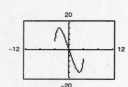

f does not pass the horizontal line test, so f does **not** have an inverse.

39. $f(x) = 2x - 3$

$y = 2x - 3$

$x = 2y - 3$

$y = \dfrac{x + 3}{2}$

$f^{-1}(x) = \dfrac{x + 3}{2}$

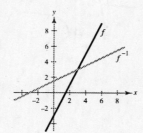

41. $f(x) = x^5 - 2$

$y = x^5 - 2$

$x = y^5 - 2$

$y = \sqrt[5]{x + 2}$

$f^{-1}(x) = \sqrt[5]{x + 2}$

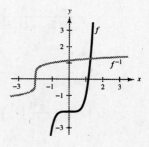

43. $f(x) = \sqrt{x}$

$y = \sqrt{x}$

$x = \sqrt{y}$

$y = x^2$

$f^{-1}(x) = x^2, \; x \geq 0$

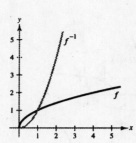

45. $f(x) = \sqrt{4 - x^2}, \; 0 \leq x \leq 2$

$y = \sqrt{4 - x^2}$

$x = \sqrt{4 - y^2}$

$f^{-1}(x) = \sqrt{4 - x^2}, \; 0 \leq x \leq 2$

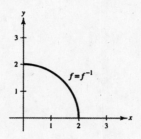

47. $f(x) = \dfrac{4}{x}$

$y = \dfrac{4}{x}$

$x = \dfrac{4}{y}$

$xy = 4$

$y = \dfrac{4}{x}$

$f^{-1}(x) = \dfrac{4}{x}$

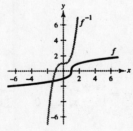

49. $f(x) = \dfrac{x + 1}{x - 2}$

$y = \dfrac{x + 1}{x - 2}$

$x = \dfrac{y + 1}{y - 2}$

$x(y - 2) = y + 1$

$xy - 2x = y + 1$

$xy - y = 2x + 1$

$y(x - 1) = 2x + 1$

$y = \dfrac{2x + 1}{x - 1}$

$f^{-1}(x) = \dfrac{2x + 1}{x - 1}$

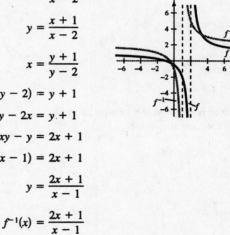

51. $f(x) = \sqrt[3]{x - 1}$

$y = \sqrt[3]{x - 1}$

$x = \sqrt[3]{y - 1}$

$x^3 = y - 1$

$y = x^3 + 1$

$f^{-1}(x) = x^3 + 1$

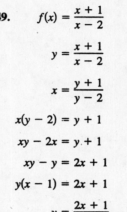

53.
$$f(x) = \frac{6x + 4}{4x + 5}$$

$$y = \frac{6x + 4}{4x + 5}$$

$$x = \frac{6y + 4}{4y + 5}$$

$$x(4y + 5) = 6y + 4$$

$$4xy + 5x = 6y + 4$$

$$4xy - 6y = -5x + 4$$

$$y(4x - 6) = -5x + 4$$

$$y = \frac{-5x + 4}{4x - 6}$$

$$f^{-1}(x) = \frac{-5x + 4}{4x - 6} = \frac{5x - 4}{6 - 4x}$$

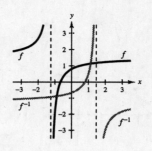

55.
$$f(x) = x^4$$

$$y = x^4$$

$$x = y^4$$

$$y = \pm\sqrt[4]{x}$$

This does not represent y as a function of x.
f does not have an inverse.

57.
$$g(x) = \frac{x}{8}$$

$$y = \frac{x}{8}$$

$$x = \frac{y}{8}$$

$$y = 8x$$

This is a function of x, so g has an inverse.
$g^{-1}(x) = 8x$

59.
$$p(x) = -4$$

$$y = -4$$

Since $y = -4$ for all x, the graph is a horizontal line and fails the horizontal line test. p does not have an inverse.

61.
$$f(x) = (x + 3)^2, \ x \geq -3 \ \Rightarrow \ y \geq 0$$

$$y = (x + 3)^2, \ x \geq -3, \ y \geq 0$$

$$x = (y + 3)^2, \ y \geq -3, \ x \geq 0$$

$$\sqrt{x} = y + 3, \ y \geq -3, \ x \geq 0$$

$$y = \sqrt{x} - 3, \ x \geq 0, \ y \geq -3$$

This is a function of x, so f has an inverse.
$$f^{-1}(x) = \sqrt{x} - 3, \ x \geq 0$$

63. $f(x) = \begin{cases} x + 3, & x < 0 \\ 6 - x, & x \geq 0 \end{cases}$

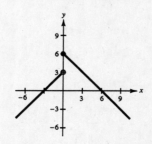

The graph fails the horizontal line test, so $f(x)$ does not have an inverse.

65. $h(x) = \dfrac{1}{x}$

$y = \dfrac{1}{x}$

$xy = 1$

$y = \dfrac{1}{x}$

This is a function of x, so h has an inverse.

$h^{-1}(x) = \dfrac{1}{x}$

67. $f(x) = \sqrt{2x + 3} \;\Rightarrow\; x \geq -\dfrac{3}{2},\; y \geq 0$

$y = \sqrt{2x + 3},\; x \geq -\dfrac{3}{2},\; y \geq 0$

$x = \sqrt{2y + 3},\; y \geq -\dfrac{3}{2},\; x \geq 0$

$x^2 = 2y + 3,\; x \geq 0,\; y \geq -\dfrac{3}{2}$

$y = \dfrac{x^2 - 3}{2},\; x \geq 0,\; y \geq -\dfrac{3}{2}$

This is a function of x, so f has an inverse.

$f^{-1}(x) = \dfrac{x^2 - 3}{2},\; x \geq 0$

In Exercises 69, 71, and 73, $f(x) = \frac{1}{8}x - 3$, $f^{-1}(x) = 8(x + 3)$, $g(x) = x^3$, $g^{-1}(x) = \sqrt[3]{x}$.

69. $(f^{-1} \circ g^{-1})(1) = f^{-1}(g^{-1}(1)) = f^{-1}(\sqrt[3]{1}) = 8(\sqrt[3]{1} + 3) = 32$

71. $(f^{-1} \circ f^{-1})(6) = f^{-1}(f^{-1}(6)) = f^{-1}(8[6 + 3]) = 8[8(6 + 3) + 3] = 600$

73. $(f \circ g)(x) = f(g(x)) = f(x^3) = \frac{1}{8}x^3 - 3$

$y = \frac{1}{8}x^3 - 3$

$x = \frac{1}{8}y^3 - 3$

$x + 3 = \frac{1}{8}y^3$

$8(x + 3) = y^3$

$\sqrt[3]{8(x + 3)} = y$

$(f \circ g)^{-1}(x) = 2\sqrt[3]{x + 3}$

In Exercises 75 and 77, $f(x) = x + 4$, $f^{-1}(x) = x - 4$, $g(x) = 2x - 5$, $g^{-1}(x) = \dfrac{x + 5}{2}$.

75. $(g^{-1} \circ f^{-1})(x) = g^{-1}(f^{-1}(x)) = g^{-1}(x - 4) = \dfrac{(x - 4) + 5}{2} = \dfrac{x + 1}{2}$

77. $(f \circ g)(x) = f(g(x)) = f(2x - 5) = (2x - 5) + 4 = 2x - 1$

$(f \circ g)^{-1}(x) = \dfrac{x + 1}{2}$

Note: Comparing Exercises 75 and 77, we see that $(f \circ g)^{-1}(x) = (g^{-1} \circ f^{-1})(x)$.

79. (a) $y = 8 + 0.75x$

$x = 8 + 0.75y$

$x - 8 = 0.75y$

$\dfrac{x - 8}{0.75} = y$

$f^{-1}(x) = \dfrac{x - 8}{0.75}$

(b) x = hourly wage

y = number of units produced

(c) $y = \dfrac{22.25 - 8}{0.75} = 19$ units

81. (a) $y = 0.03x^2 + 245.50, \ 0 < x < 100$

$x = 0.03y^2 + 245.50$

$x - 245.50 = 0.03y^2$

$\dfrac{x - 245.50}{0.03} = y^2$

$\sqrt{\dfrac{x - 245.50}{0.03}} = y, \ 245.50 < x < 545.50$

$f^{-1}(x) = \sqrt{\dfrac{x - 245.50}{0.03}}$

x = temperature in degrees Fahrenheit

y = percent load for a diesel engine

(b)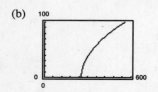

(c) $0.03x^2 + 245.50 < 500$

$0.03x^2 < 254.50$

$x^2 < 92.11$

$x < 92.11$

Thus, $0 < x < 92.11$.

83. No, since both 1994 and 1998 would be paired with the same y-value, the inverse would not exist. It would not pass the Horizontal Line Test.

85. (a) Yes, f^{-1} exists. It would represent the year for a given per capita consumption of regular soft drinks.

(b) $f^{-1}(39.8) = 5$ which represents 1995.

87. True. If $f(x) = x - 6$ and $f^{-1}(x) = x + 6$, then the y-intercept of f is $(0, -6)$ and the x-intercept of f^{-1} is $(-6, 0)$.

89. False. Some examples:

$$f(x) = f^{-1}(x) = x$$

$$f(x) = f^{-1}(x) = \frac{1}{x}$$

$$f(x) = f^{-1}(x) = \sqrt{4 - x^2}, \quad 0 \le x \le 2$$

91.

x	$f(x)$
-2	-5
-1	-2
1	2
3	3

x	$f^{-1}(x)$
-5	-2
-2	-1
2	1
3	3

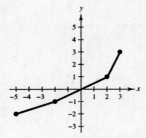

93.

x	$f(x)$
-4	3
-2	4
0	0
3	-1

The graph does not pass the Horizontal Line Test, so $f^{-1}(x)$ does not exist.

Review Exercises for Chapter P

Solutions to Odd-Numbered Exercises

1. $\left\{11, -14, -\frac{8}{9}, \frac{5}{2}, \sqrt{6}, 0.4\right\}$

   (a) Natural numbers: 11

   (b) Integers: $11, -14$

   (c) Rational numbers:
     $11, -14, -\frac{8}{9}, \frac{5}{2}, 0.4$

   (d) Irrational numbers: $\sqrt{6}$

3. (a) $\frac{5}{6} = 0.8\overline{3}$

   (b) $\frac{7}{8} = 0.875$

$$\frac{5}{6} < \frac{7}{8}$$

5. $x \le 7$

The set consists of all real numbers less than or equal to 7.

7. $d(-92, 63) = |63 - (-92)| = 155$

9. $d(x, 25) = |x - 25|$ and $d(x, 25) \le 10$, thus, $|x - 25| \le 10$.

11. $-x^2 + x - 1$

   (a) $-(1)^2 + 1 - 1 = -1$

   (b) $-(-1)^2 + (-1) - 1 = -3$

13. $2x + (3x - 10) = (2x + 3x) - 10$

Illustrates the Associative Property of Addition

15. $\dfrac{2}{y + 4} \cdot \dfrac{y + 4}{2} = 1, \quad y \ne -4$

Illustrates the Multiplicative Inverse Property

17. $|-3| + 4(-2) - 6 = 3 - 8 - 6 = -11$

19. $\dfrac{5}{18} \div \dfrac{10}{3} = \dfrac{\cancel{5}}{\cancel{18}\,_6} \cdot \dfrac{\cancel{3}}{\cancel{10}\,_2} = \dfrac{1}{12}$

21. $6[4 - 2(6 + 8)] = 6[4 - 2(14)] = 6[4 - 28] = 6(-24) = -144$

23. $6 - (x - 2)^2 = 2 + 4x - x^2$

     $6 - (x^2 - 4x + 4) = 2 + 4x - x^2$

       $2 + 4x - x^2 = 2 + 4x - x^2$

             $0 = 0$ Identity

All real numbers are solutions.

25. $3x - 2(x + 5) = 10$

     $3x - 2x - 10 = 10$

           $x = 20$

27. $4(x + 3) - 3 = 2(4 - 3x) - 4$

    $4x + 12 - 3 = 8 - 6x - 4$

        $4x + 9 = -6x + 4$

          $10x = -5$

            $x = -\frac{1}{2}$

29. Let $x =$ the number of liters of pure antifreeze.

  30% of $(10 - x) + 100\%$ of $x = 50\%$ of 10

     $0.30(10 - x) + 1.00x = 0.50(10)$

        $3 - 0.30x + 1.00x = 5$

               $0.70x = 2$

                  $x = \dfrac{2}{0.70} = \dfrac{20}{7} = 2\dfrac{6}{7}$ liters

31. $2x^2 - x - 28 = 0$

    $(2x + 7)(x - 4) = 0$

   $2x + 7 = 0$ or $x - 4 = 0$

        $2x = -7$ $x = 4$

           $x = -\dfrac{7}{2}$

33. $(x - 8)^2 = 15$

$\quad x - 8 = \pm\sqrt{15}$

$\quad\quad x = 8 \pm \sqrt{15}$

35. $-2x^2 - 5x + 27 = 0$

$\quad\quad 2x^2 + 5x - 27 = 0$

$\quad\quad x^2 + \dfrac{5}{2}x = \dfrac{27}{2}$

$\quad\quad x^2 + \dfrac{5}{2}x + \dfrac{25}{16} = \dfrac{27}{2} + \dfrac{25}{16}$

$\quad\quad \left(x + \dfrac{5}{4}\right)^2 = \dfrac{241}{16}$

$\quad\quad x + \dfrac{5}{4} = \pm\sqrt{\dfrac{241}{16}}$

$\quad\quad x = -\dfrac{5}{4} \pm \dfrac{\sqrt{241}}{4}$

37. $\quad 4x^3 - 6x^2 = 0$

$\quad 2x^2(2x - 3) = 0$

$\quad 2x^2 = 0$ or $2x - 3 = 0$

$\quad x = 0$ or $\quad\quad x = \dfrac{3}{2}$

39. $\sqrt{3x - 2} = 4 - x$

$\quad 3x - 2 = 16 - 8x + x^2$

$\quad\quad 0 = x^2 - 11x + 18$

$\quad\quad 0 = (x - 9)(x - 2)$

$\quad x - 9 = 0$ or $x - 2 = 0$

$\quad\quad x = 9 \quad\quad\quad x = 2$

The solution $x = 9$ is extraneous, only $x = 2$ is a valid solution.

41. $(x + 2)^{3/4} = 27$

$\quad x + 2 = 27^{4/3}$

$\quad x + 2 = 81$

$\quad\quad x = 79$

43. $|x^2 - 6| = x$

$\quad x^2 - 6 = x \quad$ or $\quad x^2 - 6 = -x$

$\quad x^2 - x - 6 = 0 \quad\quad\quad x^2 + x - 6 = 0$

$(x - 3)(x + 2) = 0 \quad\quad (x + 3)(x - 2) = 0$

$x = 3$ or $x = -2 \quad\quad\quad x = -3$ or $x = 2$

$\quad\quad$ Extraneous $\quad\quad\quad\quad$ Extraneous

The only solutions to the original equation are $x = 2$ or $x = 3$.

45.

$d_1 = \sqrt{(13 - 5)^2 + (11 - 22)^2} = \sqrt{64 + 121} = \sqrt{185}$

$d_2 = \sqrt{(2 - 13)^2 + (3 - 11)^2} = \sqrt{121 + 64} = \sqrt{185}$

$d_3 = \sqrt{(2 - 5)^2 + (3 - 22)^2} = \sqrt{9 + 361} = \sqrt{370}$

$d_1^2 + d_2^2 = 185 + 185 = 370 = d_3^2$

Thus, the triangle is a right triangle.

47. $x > 0$ and $y = -2$ in Quadrant IV.

49. $(-x, y)$ is in the third quadrant means that (x, y) is in Quadrant IV.

51. (a)

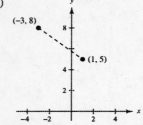

(b) $d = \sqrt{(-3 - 1)^2 + (8 - 5)^2} = \sqrt{16 + 9} = 5$

53.

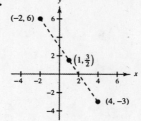

Midpoint: $\left(\dfrac{-2 + 4}{2}, \dfrac{6 + (-3)}{2}\right) = \left(1, \dfrac{3}{2}\right)$

55. $y - 2x - 3 = 0$

$y = 2x + 3$

Line with x-intercept $\left(-\dfrac{3}{2}, 0\right)$ and y-intercept $(0, 3)$

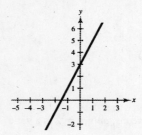

57. $y + 2x^2 = 0$

$y = -2x^2$ is a parabola.

x	0	± 1	± 2
y	0	-2	-8

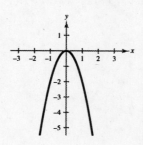

59. $y = 2x - 9$

x-intercept: $0 = 2x - 9 \implies x = \dfrac{9}{2}$

$\left(\dfrac{9}{2}, 0\right)$ is the x-intercept

y-intercept: $y = 2(0) - 9 = -9$

$(0, -9)$ is the y-intercept

61. $y = x\sqrt{9 - x^2}$

x-intercepts: $0 = x\sqrt{9 - x^2} \implies 0 = x^2(9 - x^2)$

$\qquad\qquad x = 0 \quad \text{or} \quad x = \pm 3$

The x-intercepts are $(0, 0)$, $(\pm 3, 0)$

y-intercept: $y = 0\sqrt{9 - 0^2} = 0$

The y-intercept is $(0, 0)$

63. $y = 5 - x^2$

Intercepts: $(\pm\sqrt{5}, 0)$, $(0, 5)$

y-axis symmetry

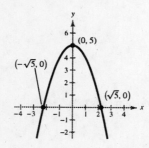

65. $y = \sqrt{x + 5}$

Domain: $[-5, \infty)$

Intercepts: $(-5, 0)$, $\left(0, \sqrt{5}\right)$

No axis or origin symmetry

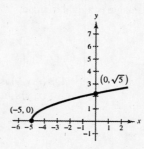

67. $(x + 2)^2 + y^2 = 16$

$(x - (-2))^2 + (y - 0)^2 = 4^2$

Center: $(-2, 0)$

Radius: 4

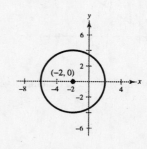

69. $\left(x - \frac{1}{2}\right)^2 + (y + 1)^2 = 36$

$\left(x - \frac{1}{2}\right)^2 + (y - (-1))^2 = 6^2$

Center: $\left(\frac{1}{2}, -1\right)$

Radius: 6

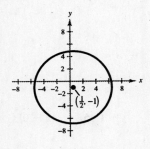

71. Endpoints of a diameter: $(0, 0)$ and $(4, -6)$

Center: $\left(\dfrac{0 + 4}{2}, \dfrac{0 + (-6)}{2}\right) = (2, -3)$

Radius: $r = \sqrt{(2 - 0)^2 + (-3 - 0)^2} = \sqrt{4 + 9} = \sqrt{13}$

Standard form: $(x - 2)^2 + (y - (-3))^2 = \left(\sqrt{13}\right)^2$

$(x - 2)^2 + (y + 3)^2 = 13$

73. $y = -2x - 7$

y-intercept: $(0, -7)$

Slope: $m = -2 = -\frac{2}{1}$

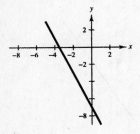

75. $y = 6$

Horizontal line

y-intercept: $(0, 6)$

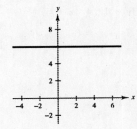

77. $y = 3x + 13$

y-intercept: $(0, 13)$

Slope: $m = 3 = \frac{3}{1}$

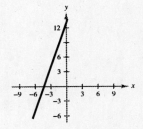

79. $y = -\frac{5}{2}x - 1$

y-intercept: $(0, -1)$

Slope: $m = -\frac{5}{2}$

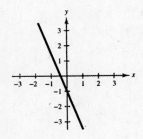

81. Point: $(2, -1)$

Slope: $m = \dfrac{1}{4} = \dfrac{\text{rise}}{\text{run}}$

$(2 + 4, -1 + 1) = (6, 0)$

$(6 + 4, 0 + 1) = (10, 1)$

$(2 - 4, -1 - 1) = (-2, -2)$

83. $(3, -4)$ and $(-7, 1)$

$m = \dfrac{1 - (-4)}{-7 - 3} = \dfrac{5}{-10} = -\dfrac{1}{2}$

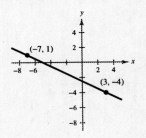

85. $(-4.5, 6)$, $(2.1, 3)$

$$m = \frac{3 - 6}{2.1 - (-4.5)} = \frac{-3}{6.6} = -\frac{30}{66} = -\frac{5}{11}$$

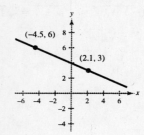

87. $(-2, 5)$, $(0, t)$, $(1, 1)$ are collinear.

$$\frac{t - 5}{0 - (-2)} = \frac{1 - 5}{1 - (-2)}$$

$$\frac{t - 5}{2} = \frac{-4}{3}$$

$$3(t - 5) = -8$$

$$3t - 15 = -8$$

$$3t = 7$$

$$t = \frac{7}{3}$$

89. $(0, 0)$, $(0, 10)$

$$m = \frac{10 - 0}{0 - 0} = \frac{10}{0} \quad \text{undefined}$$

The line is vertical.

$x = 0$

91. $(-1, 4)$, $(2, 0)$

$$m = \frac{0 - 4}{2 - (-1)} = -\frac{4}{3}$$

$$y - 4 = -\frac{4}{3}(x - (-1)) \quad \text{or}$$

$$y = -\frac{4}{3}x + \frac{8}{3}$$

$$-3y + 12 = 4x + 4$$

$$4x + 3y - 8 = 0$$

93.
$$y - (-5) = \frac{3}{2}(x - 0)$$

$$y + 5 = \frac{3}{2}x$$

$$y = \frac{3}{2}x - 5 \quad \text{or}$$

$$3x - 2y - 10 = 0$$

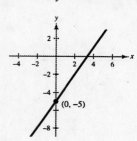

95.
$$y - (-3) = -\frac{1}{2}(x - 10)$$

$$y + 3 = -\frac{1}{2}x + 5$$

$$y = -\frac{1}{2}x + 2 \quad \text{or}$$

$$x + 2y - 4 = 0$$

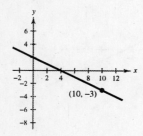

97. $5x - 4y = 8 \implies y = \frac{5}{4}x - 2$ and $m = \frac{5}{4}$

(a) Parallel slope: $m = \frac{5}{4}$

$$y - (-2) = \frac{5}{4}(x - 3)$$

$$4y + 8 = 5x - 15$$

$$5x - 4y - 23 = 0$$

(b) Perpendicular slope: $m = -\frac{4}{5}$

$$y - (-2) = -\frac{4}{5}(x - 3)$$

$$5y + 10 = -4x + 12$$

$$4x + 5y - 2 = 0$$

99. $(0, 12{,}500)$ $m = 850$

$$V - 12{,}500 = 850(t - 0)$$

$$V - 12{,}500 = 850t$$

$$V = 850t + 12{,}500, \quad 0 \le t \le 5$$

101. $(2, 160{,}000)$, $(3, 185{,}000)$

$$m = \frac{185{,}000 - 160{,}000}{3 - 2} = 25{,}000$$

$$S - 160{,}000 = 25{,}000(t - 2)$$

$$S = 25{,}000t + 110{,}000$$

For the fourth quarter let $t = 4$. Then we have

$$S = 25{,}000(4) + 110{,}000 = \$210{,}000.$$

103. $A = \{10, 20, 30, 40\}$ and

$B = \{0, 2, 4, 6\}$

(a) 20 is matched with two elements in the range so it is **not** a function.

(b) function

(c) function

(d) 30 is not matched with any element of B so it is **not** a function.

105. $16x - y^4 = 0$

$y^4 = 16x$

$y = \pm 2\sqrt[4]{x}$

y is **not** a function of x. Some x-values correspond to two y-values.

107. $y = \sqrt{1 - x}$

Each x-value, $x \leq 1$, corresponds to only one y-value so y **is** a function of x.

109. $f(x) = x^2 + 1$

(a) $f(2) = (2)^2 + 1 = 5$

(b) $f(-4) = (-4)^2 + 1 = 17$

(c) $f(t^2) = (t^2)^2 + 1 = t^4 + 1$

(d) $-f(x) = -(x^2 + 1) = -x^2 - 1$

111. $f(x) = \sqrt{25 - x^2}$

Domain: $\qquad 25 - x^2 \geq 0$

$\qquad\qquad (5 + x)(5 - x) \geq 0$

Critical Numbers: $x = \pm 5$

Test intervals: $(-\infty, -5), \ (-5, 5), \ (5, \infty)$

Solution set: $-5 \leq x \leq 5$

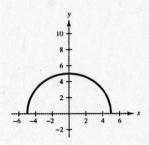

113. $g(s) = \dfrac{5}{3s - 9} = \dfrac{5}{3(s - 3)}$

Domain: All real numbers except $s = 3$.

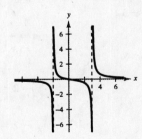

115. $h(x) = \dfrac{x}{x^2 - x - 6} = \dfrac{x}{(x + 2)(x - 3)}$

Domain: All real numbers except $x = -2, 3$.

117. $v(t) = -32t + 48$

    (a) $v(1) = 16$ ft/sec

    (b) $0 = -32t + 48$

        $t = \frac{48}{32} = 1.5$ sec

    (c) $v(2) = -16$ ft/sec

119.

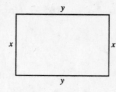

    (a) $2x + 2y = 24$

        $y = 12 - x$

      $A = xy = x(12 - x)$

    (b) Since x and y cannot be negative, we have $0 < x < 12$. The domain is $0 < x < 12$.

121. $y = (x - 3)^2$ passes the Vertical Line Test so y is a function of x.

123. $x - 4 = y^2$ does not pass the Vertical Line Test so y is not a function of x.

125. $3x^2 - 16x + 21 = 0$

    $(3x - 7)(x - 3) = 0$

    $3x - 7 = 0$ or $x - 3 = 0$

        $x = \frac{7}{3}$ or $x = 3$

127. $\dfrac{8x + 3}{11 - x} = 0$

    $8x + 3 = 0$

           $x = -\dfrac{3}{8}$

129. $g(x) = |x + 2| - |x - 2|$

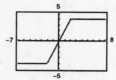

Increasing on $(-2, 2)$

Constant on $(-\infty, -2]$ and $[2, \infty)$

131. $h(x) = 4x^3 - x^4$

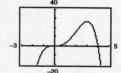

Increasing on $(-\infty, 3)$

Decreasing on $(3, \infty)$

133. $f(2) = -6, f(-1) = 3$

Points: $(2, -6), (-1, 3)$

$m = \dfrac{3 - (-6)}{-1 - 2} = \dfrac{9}{-3} = -3$

$y - (-6) = -3(x - 2)$

    $y + 6 = -3x + 6$

       $y = -3x$

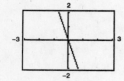

135. $f\left(-\dfrac{4}{5}\right) = 2, f\left(\dfrac{11}{5}\right) = 7$

Points: $\left(-\dfrac{4}{5}, 2\right), \left(\dfrac{11}{5}, 7\right)$

$m = \dfrac{7 - 2}{\dfrac{11}{5} - \left(-\dfrac{4}{5}\right)} = \dfrac{5}{3}$

$y - 2 = \dfrac{5}{3}\left(x - \left(-\dfrac{4}{5}\right)\right)$

$y - 2 = \dfrac{5}{3}x + \dfrac{4}{3}$

    $y = \dfrac{5}{3}x + \dfrac{10}{3}$

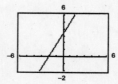

137. $f(x) = \begin{cases} 5x - 3, & x \geq -1 \\ -4x + 5, & x < -1 \end{cases}$

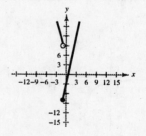

139. $f(x) = x^5 + 4x - 7$

$f(-x) = (-x)^5 + 4(-x) - 7$

$\qquad = -x^5 - 4x - 7$

$\qquad \neq f(x)$

$\qquad \neq -f(x)$

Neither even nor odd

141. $f(x) = 2x\sqrt{x^2 + 3}$

$f(-x) = 2(-x)\sqrt{(-x)^2 + 3}$

$\qquad = -2x\sqrt{x^2 + 3}$

$\qquad = -f(x)$

f is **odd**

143. Basic function: $f(x) = x^3$

Vertical shift 4 units upward and a horizontal shift 4 units to the left.

145. Basic function: $f(x) = |x|$

Horizontal shift 6 units to the right.

147. $f(x) = x^2$

$\quad h(x) = x^2 - 9$

Vertical shift 9 units downward.

149. $f(x) = \sqrt{x}$

$\quad h(x) = \sqrt{x - 7}$

Horizontal shift 7 units to the right.

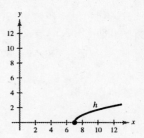

151. $f(x) = x^2$

$\quad h(x) = -(x + 3)^2 + 1$

Reflection in the x-axis, a horizontal shift 3 units to the left, and a vertical shift 1 unit upward.

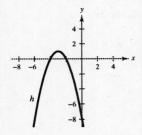

153. $f(x) = \sqrt{x}$

$\quad h(x) = -\sqrt{x + 1} + 9$

Reflection in the x-axis, a horizontal shift 1 unit to the left, and a vertical shift 9 units upward.

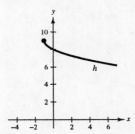

155. $f(x) = x^2$

$\quad h(x) = -2(x + 1)^2 - 3$

Reflection in the x-axis, a vertical stretch by a factor of 2, a horizontal shift 1 unit to the left and a vertical shift 3 units downward.

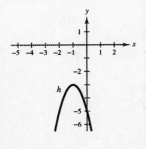

157. $f(x) = \sqrt{x}$

$\quad h(x) = -2\sqrt{x - 4}$

Reflection in the x-axis, a vertical stretch by a factor of 2, and a horizontal shift 4 units to the right.

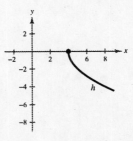

For Exercises 159-165, let $f(x) = 3 - 2x$, $g(x) = \sqrt{x}$, and $h(x) = 3x^2 + 2$.

159. $(f - g)(4) = f(4) - g(4)$

$\qquad = [3 - 2(4)] - \sqrt{4}$

$\qquad = -5 - 2$

$\qquad = -7$

161. $(fh)(1) = f(1)h(1)$

$\qquad = [3 - 2(1)][3(1)^2 + 2]$

$\qquad = (1)(5)$

$\qquad = 5$

163. $(h \circ g)(7) = h(g(7))$

$\qquad = h(\sqrt{7})$

$\qquad = 3(\sqrt{7})^2 + 2$

$\qquad = 23$

165. $(g \circ f)(-2) = g(f(-2))$

$\qquad = g(7)$

$\qquad = \sqrt{7}$

167. $y_1 \approx 0.80t^2 + 3.33t + 24.23$

$\quad y_2 \approx -0.43t^2 + 18.14t + 62.89$

169. $\qquad f(x) = 6x$

$\qquad f^{-1}(x) = \dfrac{x}{6}$

$\qquad f(f^{-1}(x)) = f\left(\dfrac{x}{6}\right) = 6\left(\dfrac{x}{6}\right) = x$

$\qquad f^{-1}(f(x)) = f^{-1}(6x) = \dfrac{6x}{6} = x$

171. $\qquad f(x) = x - 7$

$\qquad f^{-1}(x) = x + 7$

$\qquad f(f^{-1}(x)) = f(x + 7) = (x + 7) - 7 = x$

$\qquad f^{-1}(f(x)) = f^{-1}(x - 7) = (x - 7) + 7 = x$

173. $f(x) = 3x^3 - 5$

$f(x)$ passes the Horizontal Line Test, so $f(x)$ **has** an inverse.

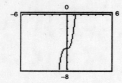

175. $f(x) = -\sqrt{4 - x}$

$f(x)$ passes the Horizontal Line Test, so $f(x)$ **has** an inverse.

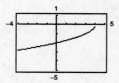

177. $f(x) = -|x + 2| + |7 - x|$

$f(x)$ does not pass the Horizontal Line Test, so $f(x)$ does **not** have an inverse.

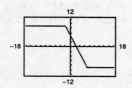

179. (a)
$$f(x) = \tfrac{1}{2}x - 3$$
$$y = \tfrac{1}{2}x - 3$$
$$x = \tfrac{1}{2}y - 3$$
$$x + 3 = \tfrac{1}{2}y$$
$$2(x + 3) = y$$
$$f^{-1}(x) = 2x + 6$$

(b)
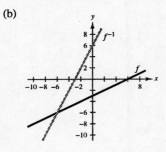

(c)
$$f^{-1}(f(x)) = f^{-1}\left(\tfrac{1}{2}x - 3\right)$$
$$= 2\left(\tfrac{1}{2}x - 3\right) + 6$$
$$= x - 6 + 6$$
$$= x$$
$$f(f^{-1}(x)) = f(2x + 6)$$
$$= \tfrac{1}{2}(2x + 6) - 3$$
$$= x + 3 - 3$$
$$= x$$

181. (a)
$$f(x) = \sqrt{x + 1}$$
$$y = \sqrt{x + 1}$$
$$x = \sqrt{y + 1}$$
$$x^2 = y + 1$$
$$x^2 - 1 = y$$
$$f^{-1}(x) = x^2 - 1, \ x \geq 0$$

Note: The inverse must have a restricted domain.

(b)
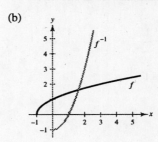

(c)
$$f^{-1}(f(x)) = f^{-1}\left(\sqrt{x + 1}\right)$$
$$= \left(\sqrt{x + 1}\right)^2 - 1$$
$$= x + 1 - 1$$
$$= x$$
$$f(f^{-1}(x)) = f(x^2 - 1)$$
$$= \sqrt{(x^2 - 1) + 1}$$
$$= \sqrt{x^2} = x \ \text{for} \ x \geq 0.$$

183. $f(x) = 2(x - 4)^2$ is increasing on $[4, \infty)$.

Let
$$f(x) = 2(x - 4)^2, \ x \geq 4 \ \text{and} \ y \geq 0$$
$$y = 2(x - 4)^2$$
$$x = 2(y - 4)^2, \ x \geq 0, \ y \geq 4$$
$$\frac{x}{2} = (y - 4)^2$$
$$\sqrt{\frac{x}{2}} = y - 4$$
$$\sqrt{\frac{x}{2}} + 4 = y$$
$$f^{-1}(x) = \sqrt{\frac{x}{2}} + 4, \ x \geq 0$$

185. False. $\dfrac{x^3 - 1}{x - 1} = \dfrac{(x - 1)(x^2 + x + 1)}{x - 1}$
$$= x^2 + x + 1, x \neq 1$$

187. False. The graph is reflected in the x-axis first, then shifted 9 units to the left, and 13 units down.

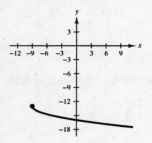

189. There is no solution. The absolute value of an expression is always nonnegative.

191. The Vertical Line Test is used to determine if a graph of y is a function of x. The Horizontal Line Test is used to determine if a function has an inverse function.

CHAPTER 1
Trigonometry

CHAPTER 1
Trigonometry

Section 1.1   Radian and Degree Measure

You should know the following basic facts about angles, their measurement, and their applications.

- Types of Angles:

  (a) Acute: Measure between 0° and 90°.

  (b) Right: Measure 90°.

  (c) Obtuse: Measure between 90° and 180°.

  (d) Straight: Measure 180°.

- α and β are complementary if $\alpha + \beta = 90°$. They are supplementary if $\alpha + \beta = 180°$.

- Two angles in standard position that have the same terminal side are called coterminal angles.

- To convert degrees to radians, use $1° = \pi/180$ radians.

- To convert radians to degrees, use 1 radian $= (180/\pi)°$.

- $1' =$ one minute $= 1/60$ of $1°$.

- $1'' =$ one second $= 1/60$ of $1' = 1/3600$ of $1°$.

- The length of a circular arc is $s = r\theta$ where θ is measured in radians.

- Linear speed $= \dfrac{\text{arc length}}{\text{time}} = \dfrac{s}{t}$

- Angular speed $= \theta/t = s/rt$

Solutions to Odd-Numbered Exercises

1.

The angle shown is approximately 2 radians.

3.

The angle shown is approximately -3 radians.

5.

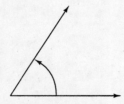

The angle shown is approximately 1 radian.

7. (a) Since $0 < \dfrac{\pi}{5} < \dfrac{\pi}{2}$; $\dfrac{\pi}{5}$ lies in Quadrant I.

  (b) Since $\pi < \dfrac{7\pi}{5} < \dfrac{3\pi}{2}$; $\dfrac{7\pi}{5}$ lies in Quadrant III.

9. (a) Since $-\dfrac{\pi}{2} < -\dfrac{\pi}{12} < 0$; $-\dfrac{\pi}{12}$ lies in Quadrant IV.

  (b) Since $-\dfrac{3\pi}{2} < -\dfrac{11\pi}{9} < -\pi$; $-\dfrac{11\pi}{9}$ lies in Quadrant II.

11. (a) Since $\pi < 3.5 < \dfrac{3\pi}{2}$; 3.5 lies in Quadrant III.

(b) Since $\dfrac{\pi}{2} < 2.25 < \pi$; 2.25 lies in Quadrant II.

13. (a)

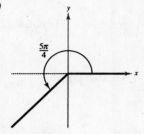

(b)

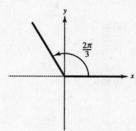

15. (a)

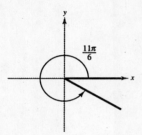

(b)

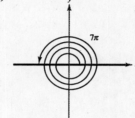

17. (a) Coterminal angles for $\dfrac{\pi}{6}$

$$\frac{\pi}{6} + 2\pi = \frac{13\pi}{6}$$

$$\frac{\pi}{6} - 2\pi = -\frac{11\pi}{6}$$

(b) Coterminal angles for $\dfrac{5\pi}{6}$

$$\frac{5\pi}{6} + 2\pi = \frac{17\pi}{6}$$

$$\frac{5\pi}{6} - 2\pi = -\frac{7\pi}{6}$$

19. (a) Coterminal angles for $\dfrac{2\pi}{3}$

$$\frac{2\pi}{3} + 2\pi = \frac{8\pi}{3}$$

$$\frac{2\pi}{3} - 2\pi = -\frac{4\pi}{3}$$

(b) Coterminal angles for $\dfrac{\pi}{12}$

$$\frac{\pi}{12} + 2\pi = \frac{25\pi}{12}$$

$$\frac{\pi}{12} - 2\pi = -\frac{23\pi}{12}$$

21. (a) Complement: $\dfrac{\pi}{2} - \dfrac{\pi}{3} = \dfrac{\pi}{6}$

Supplement: $\pi - \dfrac{\pi}{3} = \dfrac{2\pi}{3}$

(b) Complement: Not possible; $\dfrac{3\pi}{4}$ is greater than $\dfrac{\pi}{2}$.

Supplement: $\pi - \dfrac{3\pi}{4} = \dfrac{\pi}{4}$

23. (a) Complement: $\dfrac{\pi}{2} - 1 \approx 0.57$

Supplement: $\pi - 1 \approx 2.14$

(b) Complement: Not possible. 2 is greater than $\dfrac{\pi}{2}$.

Supplement: $\pi - 2 \approx 1.14$

25. (a) $30° = 30\left(\dfrac{\pi}{180}\right) = \dfrac{\pi}{6}$

(b) $150° = 150\left(\dfrac{\pi}{180}\right) = \dfrac{5\pi}{6}$

27. (a) $-20° = -20\left(\dfrac{\pi}{180}\right) = -\dfrac{\pi}{9}$

(b) $-240° = -240\left(\dfrac{\pi}{180}\right) = -\dfrac{4\pi}{3}$

29. $115° = 115\left(\dfrac{\pi}{180}\right) \approx 2.007$ radians

31. $-216.35° = -216.35\left(\dfrac{\pi}{180}\right) \approx -3.776$ radians

33. $532° = 532\left(\dfrac{\pi}{180}\right) \approx 9.285$ radians

35. $-0.83° = -0.83\left(\dfrac{\pi}{180}\right) \approx -0.014$ radian

37. (a) $\dfrac{3\pi}{2} = \dfrac{3\pi}{2}\left(\dfrac{180}{\pi}\right)° = 270°$

(b) $\dfrac{7\pi}{6} = \dfrac{7\pi}{6}\left(\dfrac{180}{\pi}\right)° = 210°$

39. (a) $\dfrac{7\pi}{3} = \dfrac{7\pi}{3}\left(\dfrac{180}{\pi}\right)° = 420°$

(b) $-\dfrac{11\pi}{30} = -\dfrac{11\pi}{30}\left(\dfrac{180}{\pi}\right)° = -66°$

41. $\dfrac{\pi}{7} = \dfrac{\pi}{7}\left(\dfrac{180}{\pi}\right)° \approx 25.714°$

43. $\dfrac{15\pi}{8} = \dfrac{15\pi}{8}\left(\dfrac{180}{\pi}\right)° = 337.5°$

45. $-4.2\pi = -4.2\pi\left(\dfrac{180}{\pi}\right)° = -756°$

47. $-2 = -2\left(\dfrac{180}{\pi}\right)° \approx -114.592°$

49.

The angle shown is approximately $210°$.

51.

The angle shown is approximately $-60°$.

53.

The angle shown is approximately $165°$.

55. (a) Since $90° < 130° < 180°$; $130°$ lies in Quadrant II.

(b) Since $270° < 285° < 360°$; $285°$ lies in Quadrant IV.

57. (a) Since $-180° < -132°50' < -90°$; $-132°\ 50'$ lies in Quadrant III.

(b) Since $-360° < -336° < -270°$; $-336°$ lies in Quadrant I.

59. (a)

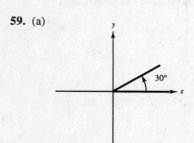

(b)

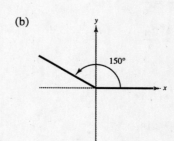

61. (a)

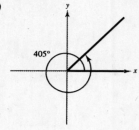

(b)

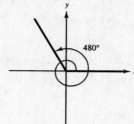

63. (a) Coterminal angles for 45°

$45° + 360° = 405°$

$45° - 360° = -315°$

(b) Coterminal angles for $-36°$

$-36° + 360° = 324°$

$-36° - 360° = -396°$

65. (a) Coterminal angles for 240°

$240° + 360° = 600°$

$240° - 360° = -120°$

(b) Coterminal angles for $-180°$

$-180° + 360° = 180°$

$-180° - 360° = -540°$

67. (a) Complement: $90° - 18° = 72°$

Supplement: $180° - 18° = 162°$

(b) Complement: Not possible; 115° is greater than 90°.

Supplement: $180° - 115° = 65°$

69. (a) Complement: $90° - 79° = 11°$

Supplement: $180° - 79° = 101°$

(b) Complement: Not possible. 150° is greater than 90°.

Supplement: $180° - 150° = 30°$

71. (a) $54° \, 45' = 54° + \left(\frac{45}{60}\right)° = 54.75°$

(b) $-128° \, 30' = -128° - \left(\frac{30}{60}\right)° = -128.5°$

73. (a) $85° \, 18' \, 30'' = \left(85 + \frac{18}{60} + \frac{30}{3600}\right)° \approx 85.308°$

(b) $330° \, 25'' = \left(330 + \frac{25}{3600}\right)° \approx 330.007°$

75. (a) $240.6° = 240° + 0.6(60)' = 240° \, 36'$

(b) $-145.8° = -[145° + 0.8(60')] = -145° \, 48'$

77. (a) $2.5° = 2° \, 30'$

(b) $-3.58° = -3° \, 30' \, 48''$

79. $s = r\theta$

$6 = 5\theta$

$\theta = \frac{6}{5}$ radians

81. $s = r\theta$

$32 = 7\theta$

$\theta = \frac{32}{7} = 4\frac{4}{7}$ radians

83. $s = r\theta$

$6 = 27\theta$

$\theta = \frac{6}{27} = \frac{2}{9}$ radian

85. $s = r\theta$

$25 = 14.5\theta$

$\theta = \frac{25}{14.5} = \frac{50}{29}$ radians

87. $s = r\theta, \ \theta$ in radians

$s = 15(180)\left(\frac{\pi}{180}\right) = 15\pi$ inches

≈ 47.12 inches

89. $s = r\theta, \ \theta$ in radians

$s = 3(1) = 3$ meters

91. $\theta = 41° \ 15' \ 42'' - 32° \ 47' \ 9'' = 8° \ 28' \ 33'' \approx 8.47583° \approx 0.14793$ radian

$s = r\theta = 4000(0.14793) \approx 591.72$ miles

93. $\theta = 42° \ 7' \ 15'' - 25° \ 46' \ 37'' = 16° \ 20' \ 38'' \approx 0.285255$ radian

$s = r\theta = 4000(0.285255) \approx 1141.02$ miles

95. $\theta = \dfrac{s}{r} = \dfrac{400}{6378} \approx 0.063$ radian $\approx 3.59°$

97. $\theta = \dfrac{s}{r} = \dfrac{2.5}{6} = \dfrac{25}{60} = \dfrac{5}{12}$ radian

99. (a) 65 miles per hour $= \dfrac{65(5280)}{60} = 5720$ feet per minute

The circumference of the tire is $C = 2.5\pi$ feet.

The number of revolutions per minute is $r = \dfrac{5720}{2.5\pi} \approx 728.3$ rev/min.

(b) The angular speed is $\dfrac{\theta}{t}$.

$\theta = \dfrac{5720}{2.5\pi}(2\pi) = 4576$ radians

Angular speed $= \dfrac{4576 \text{ radians}}{1 \text{ minute}} = 4576$ rad/min

101. Circumference: $C = 2\pi(1.68) = 3.36\pi$ inches

360 rev/min $= 6$ rev/sec

Linear speed: $(3.36\pi)(6) = 20.16\pi$ inches/sec

103. False. A measurement of 4π radians corresponds to two complete revolutions from the initial to the terminal side of an angle.

105. (a) An angle is in standard position if its vertex is at the origin and its initial side is on the positive x-axis.

(b) A negative angle is generated by a clockwise rotation of the terminal side.

(c) Two angles in standard position with the same terminal sides are coterminal.

(d) An obtuse angle measures between 90° and 180°.

107. 1 Radian $= \left(\dfrac{180}{\pi}\right)° \approx 57.3°$, so one radian is much larger than one degree.

109. The area of a circle is $A = \pi r^2 \implies \pi = \dfrac{A}{r^2}$.

The circumference of a circle is $C = 2\pi r$.

$C = 2\left(\dfrac{A}{r^2}\right)r$

$C = \dfrac{2A}{r}$

$\dfrac{Cr}{2} = A$

For a sector, $C = s = r\theta$

Thus, $A = \dfrac{(r\theta)r}{2} = \dfrac{1}{2}\theta r^2$ for a sector.

111. $f(x) = x^5 - 4$

Vertical shift 4 units downward

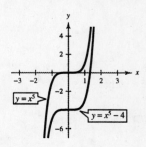

113. $f(x) = -(x + 3)^5$

Reflection in the *x*-axis and a horizontal shift 3 units to the left.

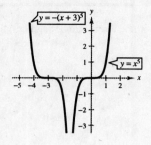

115. $\dfrac{2}{\sqrt{3}} = \dfrac{2}{\sqrt{3}} \cdot \dfrac{\sqrt{3}}{\sqrt{3}} = \dfrac{2\sqrt{3}}{3}$

117. $\dfrac{5\sqrt{5}}{2\sqrt{10}} = \dfrac{5}{2}\sqrt{\dfrac{5}{10}} = \dfrac{5}{2}\sqrt{\dfrac{1}{2}} = \dfrac{5}{2\sqrt{2}} \cdot \dfrac{\sqrt{2}}{\sqrt{2}} = \dfrac{5\sqrt{2}}{4}$

119. $\sqrt{18^2 + 12^2} = \sqrt{324 + 144} = \sqrt{468}$
$= \sqrt{36 \cdot 13} = 6\sqrt{13}$

121. $\sqrt{17^2 - 9^2} = \sqrt{289 - 81} = \sqrt{208}$
$= \sqrt{16 \cdot 13} = 4\sqrt{13}$

Section 1.2   Trigonometric Functions: The Unit Circle

■ You should know the definition of the trigonometric functions in terms of the unit circle. Let t be a real number and (x, y) the point on the unit circle corresponding to t.

$$\sin t = y \qquad\qquad \csc t = \frac{1}{y}, \quad y \neq 0$$

$$\cos t = x \qquad\qquad \sec t = \frac{1}{x}, \quad x \neq 0$$

$$\tan t = \frac{y}{x}, \quad x \neq 0 \qquad\qquad \cot t = \frac{x}{y}, \quad y \neq 0$$

■ The cosine and secant functions are even.

$$\cos(-t) = \cos t \qquad\qquad \sec(-t) = \sec t$$

■ The other four trigonometric functions are odd.

$$\sin(-t) = -\sin t \qquad\qquad \csc(-t) = -\csc t$$

$$\tan(-t) = -\tan t \qquad\qquad \cot(-t) = -\cot t$$

■ Be able to evaluate the trigonometric functions with a calculator.

Solutions to Odd-Numbered Exercises

1. $x = -\dfrac{8}{17}, \quad y = \dfrac{15}{17}$

$\sin t = y = \dfrac{15}{17} \qquad\qquad \csc t = \dfrac{1}{y} = \dfrac{17}{15}$

$\cos t = x = -\dfrac{8}{17} \qquad\qquad \sec t = \dfrac{1}{x} = -\dfrac{17}{8}$

$\tan t = \dfrac{y}{x} = -\dfrac{15}{8} \qquad\qquad \cot t = \dfrac{x}{y} = -\dfrac{8}{15}$

3. $x = \dfrac{12}{13}, \quad y = -\dfrac{5}{13}$

$\sin t = y = -\dfrac{5}{13} \qquad\qquad \csc t = \dfrac{1}{y} = -\dfrac{13}{5}$

$\cos t = x = \dfrac{12}{13} \qquad\qquad \sec t = \dfrac{1}{x} = \dfrac{13}{12}$

$\tan t = \dfrac{y}{x} = -\dfrac{5}{12} \qquad\qquad \cot t = \dfrac{x}{y} = -\dfrac{12}{5}$

5. $t = \dfrac{\pi}{4}$ corresponds to $\left(\dfrac{\sqrt{2}}{2}, \dfrac{\sqrt{2}}{2}\right)$.

7. $t = \dfrac{7\pi}{6}$ corresponds to $\left(-\dfrac{\sqrt{3}}{2}, -\dfrac{1}{2}\right)$.

9. $t = \dfrac{4\pi}{3}$ corresponds to $\left(-\dfrac{1}{2}, -\dfrac{\sqrt{3}}{2}\right)$.

11. $t = \dfrac{3\pi}{2}$ corresponds to $(0, -1)$.

13. $t = \dfrac{\pi}{4}$ corresponds to $\left(\dfrac{\sqrt{2}}{2}, \dfrac{\sqrt{2}}{2}\right)$.

$\sin t = y = \dfrac{\sqrt{2}}{2}$

$\cos t = x = \dfrac{\sqrt{2}}{2}$

$\tan t = \dfrac{y}{x} = 1$

15. $t = -\dfrac{\pi}{6}$ corresponds to $\left(\dfrac{\sqrt{3}}{2}, -\dfrac{1}{2}\right)$.

$\sin t = y = -\dfrac{1}{2}$

$\cos t = x = \dfrac{\sqrt{3}}{2}$

$\tan t = \dfrac{y}{x} = -\dfrac{1}{\sqrt{3}} = -\dfrac{\sqrt{3}}{3}$

17. $t = -\dfrac{7\pi}{4}$ corresponds to $\left(\dfrac{\sqrt{2}}{2}, \dfrac{\sqrt{2}}{2}\right)$.

$\sin t = y = \dfrac{\sqrt{2}}{2}$

$\cos t = x = \dfrac{\sqrt{2}}{2}$

$\tan t = \dfrac{y}{x} = 1$

19. $t = \dfrac{11\pi}{6}$ corresponds to $\left(\dfrac{\sqrt{3}}{2}, -\dfrac{1}{2}\right)$.

$\sin t = y = -\dfrac{1}{2}$

$\cos t = x = \dfrac{\sqrt{3}}{2}$

$\tan t = \dfrac{y}{x} = -\dfrac{1}{\sqrt{3}} = -\dfrac{\sqrt{3}}{3}$

21. $t = -\dfrac{3\pi}{2}$ corresponds to $(0, 1)$.

$\sin t = y = 1$

$\cos t = x = 0$

$\tan t = \dfrac{y}{x}$ is undefined.

23. $t = \dfrac{3\pi}{4}$ corresponds to $\left(-\dfrac{\sqrt{2}}{2}, \dfrac{\sqrt{2}}{2}\right)$.

$\sin t = y = \dfrac{\sqrt{2}}{2}$ $\csc t = \dfrac{1}{y} = \sqrt{2}$

$\cos t = x = -\dfrac{\sqrt{2}}{2}$ $\sec t = \dfrac{1}{x} = -\sqrt{2}$

$\tan t = \dfrac{y}{x} = -1$ $\cot t = \dfrac{x}{y} = -1$

25. $t = \dfrac{\pi}{2}$ corresponds to $(0, 1)$.

$\sin t = y = 1$ $\csc t = \dfrac{1}{y} = 1$

$\cos t = x = 0$ $\sec t = \dfrac{1}{x}$ is undefined.

$\tan t = \dfrac{y}{x}$ is undefined. $\cot t = \dfrac{x}{y} = 0$

27. $t = -\dfrac{\pi}{3}$ corresponds to $\left(\dfrac{1}{2}, -\dfrac{\sqrt{3}}{2}\right)$.

$\sin t = y = -\dfrac{\sqrt{3}}{2}$ $\csc t = \dfrac{1}{y} = -\dfrac{2\sqrt{3}}{3}$

$\cos t = x = \dfrac{1}{2}$ $\sec t = \dfrac{1}{x} = 2$

$\tan t = \dfrac{y}{x} = -\sqrt{3}$ $\cot t = \dfrac{x}{y} = -\dfrac{\sqrt{3}}{3}$

29. $\sin 5\pi = \sin \pi = 0$

31. $\cos \dfrac{8\pi}{3} = \cos \dfrac{2\pi}{3} = -\dfrac{1}{2}$

33. $\cos(-3\pi) = \cos \pi = -1$

35. $\sin\left(-\dfrac{9\pi}{4}\right) = \sin\left(\dfrac{7\pi}{4}\right) = -\dfrac{\sqrt{2}}{2}$

37. $\sin t = \dfrac{1}{3}$

  (a) $\sin(-t) = -\sin t = -\dfrac{1}{3}$

  (b) $\csc(-t) = -\csc t = -3$

39. $\cos(-t) = -\dfrac{1}{5}$

  (a) $\cos t = \cos(-t) = -\dfrac{1}{5}$

  (b) $\sec(-t) = \dfrac{1}{\cos(-t)} = -5$

41. $\sin y = \dfrac{4}{5}$

(a) $\sin(\pi - t) = \sin t = \dfrac{4}{5}$

(b) $\sin(t + \pi) = -\sin t = -\dfrac{4}{5}$

43. $\sin \dfrac{\pi}{4} \approx 0.7071$

45. $\csc 1.3 = \dfrac{1}{\sin 1.3} \approx 1.0378$

47. $\cos(-1.7) \approx -0.1288$

49. $\csc 0.8 = \dfrac{1}{\sin 0.8} \approx 1.3940$

51. $\sec 22.8 = \dfrac{1}{\cos 22.8} \approx -1.4486$

53. (a) $\sin 5 \approx -1$

(b) $\cos 2 \approx -0.4$

55. (a) $\sin t = 0.25$

$t \approx 0.25$ or 2.89

(b) $\cos t = -0.25$

$t \approx 1.82$ or 4.46

57. $\cos 1.5 \approx 0.0707$

$2 \cos 0.75 \approx 1.4634$

$\cos 2t \neq 2 \cos t$

59. $y(t) = \dfrac{1}{4} \cos 6t$

(a) $y(0) = \dfrac{1}{4} \cos 0 = 0.2500$ feet

(b) $y\left(\dfrac{1}{4}\right) = \dfrac{1}{4} \cos \dfrac{3}{2} \approx 0.0177$ feet

(c) $y\left(\dfrac{1}{2}\right) = \dfrac{1}{4} \cos 3 \approx -0.2475$ feet

61. False. $\sin(-t) = -\sin t$ means the function is odd, not that the sine of a negative angle is a negative number.

For example: $\sin\left(-\dfrac{3\pi}{2}\right) = -\sin\left(\dfrac{3\pi}{2}\right) = -(-1) = 1.$

Even though the angle is negative, the sine value is positive.

63. (a) The points have y-axis symmetry.

(b) $\sin t_1 = \sin(\pi - t_1)$ since they have the same y-value.

(c) $\cos(\pi - t_1) = -\cos t_1$ since the x-values have the opposite signs.

65. $\cos \theta = x = \cos(-\theta)$

$\sec \theta = \dfrac{1}{x} = \sec(-\theta)$

$\sin \theta = y$

$\sin(-\theta) = -y = -\sin \theta$

$\csc \theta = \dfrac{1}{y}$

$\csc(-\theta) = -\dfrac{1}{y} = -\csc \theta$

$\tan \theta = \dfrac{y}{x}$

$\tan(-\theta) = \dfrac{-y}{x} = -\tan \theta$

$\cot \theta = \dfrac{x}{y}$

$\cot(-\theta) = \dfrac{x}{-y} = -\cot \theta$

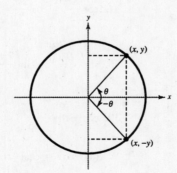

67. $f(t) = \sin t$ and $g(t) = \tan t$

Both f and g are odd functions.

$h(t) = f(t)g(t) = \sin t \tan t$

$h(-t) = \sin(-t)\tan(-t)$

$\qquad = (-\sin t)(-\tan t)$

$\qquad = \sin t \tan t = h(t)$

The function $h(t) = f(t)g(t)$ is even.

69. $f(x) = \frac{1}{4}x^3 + 1$

$\qquad y = \frac{1}{4}x^3 + 1$

$\qquad x = \frac{1}{4}y^3 + 1$

$\qquad x - 1 = \frac{1}{4}y^3$

$\qquad 4(x - 1) = y^3$

$\qquad y = \sqrt[3]{4(x - 1)}$

$\qquad f^{-1}(x) = \sqrt[3]{4(x - 1)}$

71. $f(x) = \dfrac{2x}{x + 1}, x > -1$

$\qquad y = \dfrac{2x}{x + 1}, x > -1$

$\qquad x = \dfrac{2y}{y + 1}$

$\qquad xy + x = 2y$

$\qquad x = 2y - xy$

$\qquad x = y(2 - x)$

$\qquad \dfrac{x}{2 - x} = y, x < 2$

$\qquad f^{-1}(x) = \dfrac{x}{2 - x}, x < 2$

Section 1.3   Right Triangle Trigonometry

- You should know the right triangle definition of trigonometric functions.

  (a) $\sin \theta = \dfrac{\text{opp}}{\text{hyp}}$ (b) $\cos \theta = \dfrac{\text{adj}}{\text{hyp}}$ (c) $\tan \theta = \dfrac{\text{opp}}{\text{adj}}$

  (d) $\csc \theta = \dfrac{\text{hyp}}{\text{opp}}$ (e) $\sec \theta = \dfrac{\text{hyp}}{\text{adj}}$ (f) $\cot \theta = \dfrac{\text{adj}}{\text{opp}}$

- You should know the following identities.

  (a) $\sin \theta = \dfrac{1}{\csc \theta}$ (b) $\csc \theta = \dfrac{1}{\sin \theta}$ (c) $\cos \theta = \dfrac{1}{\sec \theta}$

  (d) $\sec \theta = \dfrac{1}{\cos \theta}$ (e) $\tan \theta = \dfrac{1}{\cot \theta}$ (f) $\cot \theta = \dfrac{1}{\tan \theta}$

  (g) $\tan \theta = \dfrac{\sin \theta}{\cos \theta}$ (h) $\cot \theta = \dfrac{\cos \theta}{\sin \theta}$ (i) $\sin^2 \theta + \cos^2 \theta = 1$

  (j) $1 + \tan^2 \theta = \sec^2 \theta$ (k) $1 + \cot^2 \theta = \csc^2 \theta$

- You should know that two acute angles α and β are complementary if $\alpha + \beta = 90°$, and that cofunctions of complementary angles are equal.

- You should know the trigonometric function values of $30°$, $45°$, and $60°$, or be able to construct triangles from which you can determine them.

Solutions to Odd-Numbered Exercises

1. $\text{hyp} = \sqrt{6^2 + 8^2} = \sqrt{36 + 64} = \sqrt{100} = 10$

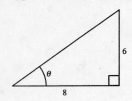

$$\sin \theta = \frac{\text{opp}}{\text{hyp}} = \frac{6}{10} = \frac{3}{5} \qquad \csc \theta = \frac{\text{hyp}}{\text{opp}} = \frac{10}{6} = \frac{5}{3}$$

$$\cos \theta = \frac{\text{adj}}{\text{hyp}} = \frac{8}{10} = \frac{4}{5} \qquad \sec \theta = \frac{\text{hyp}}{\text{adj}} = \frac{10}{8} = \frac{5}{4}$$

$$\tan \theta = \frac{\text{opp}}{\text{adj}} = \frac{6}{8} = \frac{3}{4} \qquad \cot \theta = \frac{\text{adj}}{\text{opp}} = \frac{8}{6} = \frac{4}{3}$$

3. $\text{adj} = \sqrt{41^2 - 9^2} = \sqrt{1681 - 81} = \sqrt{1600} = 40$

$$\sin \theta = \frac{\text{opp}}{\text{hyp}} = \frac{9}{41} \qquad \csc \theta = \frac{\text{hyp}}{\text{opp}} = \frac{41}{9}$$

$$\cos \theta = \frac{\text{adj}}{\text{hyp}} = \frac{40}{41} \qquad \sec \theta = \frac{\text{hyp}}{\text{adj}} = \frac{41}{40}$$

$$\tan \theta = \frac{\text{opp}}{\text{adj}} = \frac{9}{40} \qquad \cot \theta = \frac{\text{adj}}{\text{opp}} = \frac{40}{9}$$

5. $\text{adj} = \sqrt{3^2 - 1^2} = \sqrt{8} = 2\sqrt{2}$

$$\sin \theta = \frac{\text{opp}}{\text{hyp}} = \frac{1}{3} \qquad \csc \theta = \frac{\text{hyp}}{\text{opp}} = 3$$

$$\cos \theta = \frac{\text{adj}}{\text{hyp}} = \frac{2\sqrt{2}}{3} \qquad \sec \theta = \frac{\text{hyp}}{\text{adj}} = \frac{3}{2\sqrt{2}} = \frac{3\sqrt{2}}{4}$$

$$\tan \theta = \frac{\text{opp}}{\text{adj}} = \frac{1}{2\sqrt{2}} = \frac{\sqrt{2}}{4} \qquad \cot \theta = \frac{\text{adj}}{\text{opp}} = 2\sqrt{2}$$

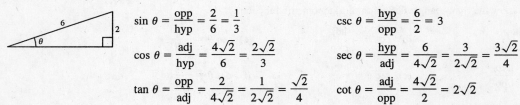

$\text{adj} = \sqrt{6^2 - 2^2} = \sqrt{32} = 4\sqrt{2}$

$$\sin \theta = \frac{\text{opp}}{\text{hyp}} = \frac{2}{6} = \frac{1}{3} \qquad \csc \theta = \frac{\text{hyp}}{\text{opp}} = \frac{6}{2} = 3$$

$$\cos \theta = \frac{\text{adj}}{\text{hyp}} = \frac{4\sqrt{2}}{6} = \frac{2\sqrt{2}}{3} \qquad \sec \theta = \frac{\text{hyp}}{\text{adj}} = \frac{6}{4\sqrt{2}} = \frac{3}{2\sqrt{2}} = \frac{3\sqrt{2}}{4}$$

$$\tan \theta = \frac{\text{opp}}{\text{adj}} = \frac{2}{4\sqrt{2}} = \frac{1}{2\sqrt{2}} = \frac{\sqrt{2}}{4} \qquad \cot \theta = \frac{\text{adj}}{\text{opp}} = \frac{4\sqrt{2}}{2} = 2\sqrt{2}$$

The function values are the same since the triangles are similar and the corresponding sides are proportional.

7. $\text{opp} = \sqrt{5^2 - 4^2} = 3$

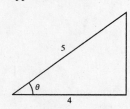

$$\sin \theta = \frac{\text{opp}}{\text{hyp}} = \frac{3}{5} \qquad \csc \theta = \frac{\text{hyp}}{\text{opp}} = \frac{5}{3}$$

$$\cos \theta = \frac{\text{adj}}{\text{hyp}} = \frac{4}{5} \qquad \sec \theta = \frac{\text{hyp}}{\text{adj}} = \frac{5}{4}$$

$$\tan \theta = \frac{\text{opp}}{\text{adj}} = \frac{3}{4} \qquad \cot \theta = \frac{\text{adj}}{\text{opp}} = \frac{4}{3}$$

— CONTINUED —

7. — CONTINUED —

$\text{opp} = \sqrt{1.25^2 - 1^2} = 0.75$

$$\sin \theta = \frac{\text{opp}}{\text{hyp}} = \frac{0.75}{1.25} = \frac{3}{5} \qquad \csc \theta = \frac{\text{hyp}}{\text{opp}} = \frac{1.25}{0.75} = \frac{5}{3}$$

$$\cos \theta = \frac{\text{adj}}{\text{hyp}} = \frac{1}{1.25} = \frac{4}{5} \qquad \sec \theta = \frac{\text{hyp}}{\text{adj}} = \frac{1.25}{1} = \frac{5}{4}$$

$$\tan \theta = \frac{\text{opp}}{\text{adj}} = \frac{0.75}{1} = \frac{3}{4} \qquad \cot \theta = \frac{\text{adj}}{\text{opp}} = \frac{1}{0.75} = \frac{4}{3}$$

The function values are the same since the triangles are similar and the corresponding sides are proportional.

9. Given: $\sin \theta = \dfrac{3}{4} = \dfrac{\text{opp}}{\text{hyp}}$

$3^2 + (\text{adj})^2 = 4^2$

$\quad\quad \text{adj} = \sqrt{7}$

$\cos \theta = \dfrac{\sqrt{7}}{4}$

$\tan \theta = \dfrac{3\sqrt{7}}{7}$

$\cot \theta = \dfrac{\sqrt{7}}{3}$

$\sec \theta = \dfrac{4\sqrt{7}}{7}$

$\csc \theta = \dfrac{4}{3}$

11. Given: $\sec \theta = 2 = \dfrac{2}{1} = \dfrac{\text{hyp}}{\text{adj}}$

$(\text{opp})^2 + 1^2 = 2^2$

$\quad\quad \text{opp} = \sqrt{3}$

$\sin \theta = \dfrac{\sqrt{3}}{2}$

$\cos \theta = \dfrac{1}{2}$

$\tan \theta = \sqrt{3}$

$\cot \theta = \dfrac{\sqrt{3}}{3}$

$\csc \theta = \dfrac{2\sqrt{3}}{3}$

13. Given: $\tan \theta = 3 = \dfrac{3}{1} = \dfrac{\text{opp}}{\text{adj}}$

$3^2 + 1^2 = (\text{hyp})^2$

$\quad\quad \text{hyp} = \sqrt{10}$

$\sin \theta = \dfrac{3\sqrt{10}}{10}$

$\cos \theta = \dfrac{\sqrt{10}}{10}$

$\cot \theta = \dfrac{1}{3}$

$\sec \theta = \sqrt{10}$

$\csc \theta = \dfrac{\sqrt{10}}{3}$

15. Given: $\cot \theta = \dfrac{3}{2} = \dfrac{\text{adj}}{\text{opp}}$

$2^2 + 3^2 = (\text{hyp})^2$

$\quad\quad \text{hyp} = \sqrt{13}$

$\sin \theta = \dfrac{2}{\sqrt{13}} = \dfrac{2\sqrt{13}}{13}$

$\cos \theta = \dfrac{3}{\sqrt{13}} = \dfrac{3\sqrt{13}}{13}$

$\tan \theta = \dfrac{2}{3}$

$\csc \theta = \dfrac{\sqrt{13}}{2}$

$\sec \theta = \dfrac{\sqrt{13}}{3}$

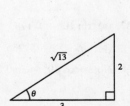

17. $\sin 60° = \dfrac{\sqrt{3}}{2}, \ \cos 60° = \dfrac{1}{2}$

(a) $\tan 60° = \dfrac{\sin 60°}{\cos 60°} = \sqrt{3}$

(b) $\sin 30° = \cos 60° = \dfrac{1}{2}$

(c) $\cos 30° = \sin 60° = \dfrac{\sqrt{3}}{2}$

(d) $\cot 60° = \dfrac{\cos 60°}{\sin 60°} = \dfrac{1}{\sqrt{3}} = \dfrac{\sqrt{3}}{3}$

19. $\csc \theta = \dfrac{\sqrt{13}}{2}$, $\sec \theta = \dfrac{\sqrt{13}}{3}$

 (a) $\sin \theta = \dfrac{1}{\csc \theta} = \dfrac{2}{\sqrt{13}} = \dfrac{2\sqrt{13}}{13}$

 (b) $\cos \theta = \dfrac{1}{\sec \theta} = \dfrac{3}{\sqrt{13}} = \dfrac{3\sqrt{13}}{13}$

 (c) $\tan \theta = \dfrac{\sin \theta}{\cos \theta} = \dfrac{\dfrac{2\sqrt{13}}{13}}{\dfrac{3\sqrt{13}}{13}} = \dfrac{2}{3}$

 (d) $\sec(90° - \theta) = \csc \theta = \dfrac{\sqrt{13}}{2}$

21. $\cos \alpha = \dfrac{1}{3}$

 (a) $\sec \alpha = \dfrac{1}{\cos \alpha} = 3$

 (b) $\sin^2 \alpha + \cos^2 \alpha = 1$

 $\sin^2 \alpha + \left(\dfrac{1}{3}\right)^2 = 1$

 $\sin^2 \alpha = \dfrac{8}{9}$

 $\sin \alpha = \dfrac{2\sqrt{2}}{3}$

 (c) $\cot \alpha = \dfrac{\cos \alpha}{\sin \alpha} = \dfrac{\dfrac{1}{3}}{\dfrac{2\sqrt{2}}{3}} = \dfrac{1}{2\sqrt{2}} = \dfrac{\sqrt{2}}{4}$

 (d) $\sin(90° - \alpha) = \cos \alpha = \dfrac{1}{3}$

23. (a) $\cos 60° = \dfrac{1}{2}$

 (b) $\csc 30° = 2$

 (c) $\tan 60° = \sqrt{3}$

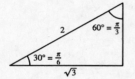

25. (a) $\sin 45° = \dfrac{1}{\sqrt{2}} = \dfrac{\sqrt{2}}{2}$

 (b) $\cos 30° = \dfrac{\sqrt{3}}{2}$

 (c) $\tan 30° = \dfrac{1}{\sqrt{3}} = \dfrac{\sqrt{3}}{3}$

27. (a) $\sin 10° \approx 0.1736$

 (b) $\cos 80° \approx 0.1736$

 Note: $\cos 80° = \sin(90° - 80°) = \sin 10°$

29. (a) $\sin 16.35° \approx 0.2815$

 (b) $\csc 16.35° = \dfrac{1}{\sin 16.35°} \approx 3.5523$

31. (a) $\sec 42°12' = \sec 42.2° = \dfrac{1}{\cos 42.2°} \approx 1.3499$

 (b) $\csc 48°7' = \dfrac{1}{\sin\left(48 + \frac{7}{60}\right)°} \approx 1.3432$

33. (a) $\cot 11°15' = \dfrac{1}{\tan 11.25°} \approx 5.0273$

 (b) $\tan 11°15' = \tan 11.25° \approx 0.1989$

35. (a) $\csc 32°40'3'' = \dfrac{1}{\sin 32.6675°} \approx 1.8527$

 (b) $\tan 44°28'16'' \approx \tan 44.4711° \approx 0.9817$

37. (a) $\sin \theta = \dfrac{1}{2} \implies \theta = 30° = \dfrac{\pi}{6}$

 (b) $\csc \theta = 2 \implies \theta = 30° = \dfrac{\pi}{6}$

39. (a) $\sec \theta = 2 \implies \theta = 60° = \dfrac{\pi}{3}$

 (b) $\cot \theta = 1 \implies \theta = 45° = \dfrac{\pi}{4}$

41. (a) $\csc \theta = \dfrac{2\sqrt{3}}{3} \implies \theta = 60° = \dfrac{\pi}{3}$

(b) $\sin \theta = \dfrac{\sqrt{2}}{2} \implies \theta = 45° = \dfrac{\pi}{4}$

43. (a) $\sin \theta = 0.0145 \implies \theta \approx 0.83° \approx 0.015$ radian

(b) $\sin \theta = 0.4565 \implies \theta \approx 27° \approx 0.474$ radian

45. (a) $\tan \theta = 0.0125 \implies \theta \approx 0.72° \approx 0.012$ radian

(b) $\tan \theta = 2.3545 \implies \theta \approx 67° \approx 1.169$ radians

47.

$\tan 30° = \dfrac{30}{x}$

$\dfrac{1}{\sqrt{3}} = \dfrac{30}{x}$

$x = 30\sqrt{3}$

49.

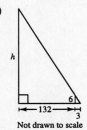

$\tan 60° = \dfrac{32}{x}$

$\sqrt{3} = \dfrac{32}{x}$

$\sqrt{3}\, x = 32$

$x = \dfrac{32}{\sqrt{3}} = \dfrac{32\sqrt{3}}{3}$

51. $\tan \theta \cot \theta = \tan \theta \left(\dfrac{1}{\tan \theta} \right) = 1$

53. $\tan \alpha \cos \alpha = \left(\dfrac{\sin \alpha}{\cos \alpha} \right) \cos \alpha = \sin \alpha$

55. $(1 + \cos \theta)(1 - \cos \theta) = 1 - \cos^2 \theta$

$= (\sin^2 \theta + \cos^2 \theta) - \cos^2 \theta$

$= \sin^2 \theta$

57. $(\sec \theta + \tan \theta)(\sec \theta - \tan \theta) = \sec^2 \theta - \tan^2 \theta$

$= (1 + \tan^2 \theta) - \tan^2 \theta$

$= 1$

59. $\dfrac{\sin \theta}{\cos \theta} + \dfrac{\cos \theta}{\sin \theta} = \dfrac{\sin^2 \theta + \cos^2 \theta}{\sin \theta \cos \theta}$

$= \dfrac{1}{\sin \theta \cos \theta}$

$= \dfrac{1}{\sin \theta} \cdot \dfrac{1}{\cos \theta}$

$= \csc \theta \sec \theta$

61. (a)

(b) $\dfrac{6}{3} = \dfrac{h}{135}$

(c) $2(135) = h$

$h = 270$ feet

63. (a)

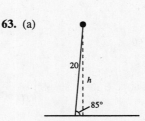

(b) $\sin 85° = \dfrac{h}{20}$

(c) $h = 20 \sin 85° \approx 19.9$ meters

65. Let $x = $ distance from the boat to the shoreline

$$\tan 4° = \frac{40}{x}$$

$$x = \frac{40}{\tan 4°} \approx 572 \text{ feet}$$

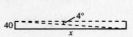

67.

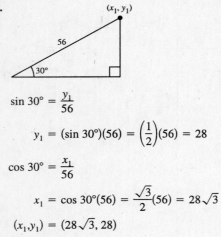

$$\sin 30° = \frac{y_1}{56}$$

$$y_1 = (\sin 30°)(56) = \left(\frac{1}{2}\right)(56) = 28$$

$$\cos 30° = \frac{x_1}{56}$$

$$x_1 = \cos 30°(56) = \frac{\sqrt{3}}{2}(56) = 28\sqrt{3}$$

$$(x_1, y_1) = (28\sqrt{3}, 28)$$

$$\sin 60° = \frac{y_2}{56}$$

$$y_2 = \sin 60°(56) = \left(\frac{\sqrt{3}}{2}\right)(56) = 28\sqrt{3}$$

$$\cos 60° = \frac{x_2}{56}$$

$$x_2 = (\cos 60°)(56) = \left(\frac{1}{2}\right)(56) = 28$$

$$(x_2, y_2) = (28, 28\sqrt{3})$$

69. $x \approx 9.397, \ y \approx 3.420$

$$\sin 20° = \frac{y}{10} \approx 0.34$$

$$\cos 20° = \frac{x}{10} \approx 0.94$$

$$\tan 20° = \frac{y}{x} \approx 0.36$$

$$\cot 20° = \frac{x}{y} \approx 2.75$$

$$\sec 20° = \frac{10}{x} \approx 1.06$$

$$\csc 20° = \frac{10}{y} \approx 2.92$$

71. True, $\csc x = \dfrac{1}{\sin x} \ \Rightarrow \ \sin 60° \csc 60° = \sin 60°\left(\dfrac{1}{\sin 60°}\right) = 1$

73. False, $\dfrac{\sqrt{2}}{2} + \dfrac{\sqrt{2}}{2} = \sqrt{2} \neq 1$

75. False, $\dfrac{\sin 60°}{\sin 30°} = \dfrac{\cos 30°}{\sin 30°} = \cot 30° \approx 1.7321; \ \sin 2° \approx 0.0349$

77. This is true because the corresponding sides of similar triangles are proportional.

79. (a)

θ	0.1	0.2	0.3	0.4	0.5
$\sin \theta$	0.0998	0.1987	0.2955	0.3894	0.4794

(b) As $\theta \rightarrow 0$, $\sin \theta \rightarrow 0$

81. $\dfrac{x^2 - 6x}{x^2 + 4x - 12} \cdot \dfrac{x^2 + 12x + 36}{x^2 - 36} = \dfrac{x\cancel{(x - 6)}}{\cancel{(x + 6)}(x - 2)} \cdot \dfrac{\cancel{(x + 6)}(x + 6)}{\cancel{(x + 6)}\cancel{(x - 6)}}$

$$= \dfrac{x}{x - 2}, x \neq \pm 6$$

83. $\dfrac{3}{x + 2} - \dfrac{2}{x - 2} + \dfrac{x}{x^2 + 4x + 4} = \dfrac{3(x + 2)(x - 2) - 2(x + 2)^2 + x(x - 2)}{(x - 2)(x + 2)^2}$

$$= \dfrac{3(x^2 - 4) - 2(x^2 + 4x + 4) + x^2 - 2x}{(x - 2)(x + 2)^2}$$

$$= \dfrac{2x^2 - 10x - 20}{(x - 2)(x + 2)^2} = \dfrac{2(x^2 - 5x - 10)}{(x - 2)(x + 2)^2}$$

85. $\dfrac{4}{x - 4} = \dfrac{12x}{24 - x}$

$4(24 - x) = 12x(x - 4)$

$96 - 4x = 12x^2 - 48x$

$0 = 12x^2 - 44x - 96$

$0 = 4(3x^2 - 11x - 24)$

$x = \dfrac{-(-11) \pm \sqrt{(-11)^2 - 4(3)(-24)}}{2(3)}$

$\quad = \dfrac{11 \pm \sqrt{409}}{6}$

87. $\dfrac{2}{x + 3} + \dfrac{4}{x - 2} = \dfrac{12}{x^2 + x - 6}$

$2(x - 2) + 4(x + 3) = 12$

$2x - 4 + 4x + 12 = 12$

$6x + 8 = 12$

$6x = 4$

$x = \dfrac{2}{3}$

Section 1.4   Trigonometric Functions of Any Angle

■ Know the Definitions of Trigonometric Functions of Any Angle.

If θ is in standard position, (x, y) a point on the terminal side and $r = \sqrt{x^2 + y^2} \neq 0$, then

$$\sin \theta = \frac{y}{r} \qquad\qquad \csc \theta = \frac{r}{y}, \ y \neq 0$$

$$\cos \theta = \frac{x}{r} \qquad\qquad \sec \theta = \frac{r}{x}, \ x \neq 0$$

$$\tan \theta = \frac{y}{x}, \ x \neq 0 \qquad\qquad \cot \theta = \frac{x}{y}, \ y \neq 0$$

■ You should know the signs of the trigonometric functions in each quadrant.

■ You should know the trigonometric function values of the quadrant angles 0, $\dfrac{\pi}{2}$, π, and $\dfrac{3\pi}{2}$.

■ You should be able to find reference angles.

■ You should be able to evaluate trigonometric functions of any angle. (Use reference angles.)

■ You should know that the period of sine and cosine is 2π.

Solutions to Odd-Numbered Exercises

1. (a) $(x, y) = (4, 3)$

$r = \sqrt{16 + 9} = 5$

$\sin \theta = \dfrac{y}{r} = \dfrac{3}{5} \qquad \csc \theta = \dfrac{r}{y} = \dfrac{5}{3}$

$\cos \theta = \dfrac{x}{r} = \dfrac{4}{5} \qquad \sec \theta = \dfrac{r}{x} = \dfrac{5}{4}$

$\tan \theta = \dfrac{y}{x} = \dfrac{3}{4} \qquad \cot \theta = \dfrac{x}{y} = \dfrac{4}{3}$

(b) $(x, y) = (8, -15)$

$r = \sqrt{64 + 225} = 17$

$\sin \theta = \dfrac{y}{r} = -\dfrac{15}{17} \qquad \csc \theta = \dfrac{r}{y} = -\dfrac{17}{15}$

$\cos \theta = \dfrac{x}{r} = \dfrac{8}{17} \qquad \sec \theta = \dfrac{r}{x} = \dfrac{17}{8}$

$\tan \theta = \dfrac{y}{x} = -\dfrac{15}{8} \qquad \cot \theta = \dfrac{x}{y} = -\dfrac{8}{15}$

3. (a) $(x, y) = \left(-\sqrt{3}, -1\right)$

$r = \sqrt{3 + 1} = 2$

$\sin \theta = \dfrac{y}{r} = -\dfrac{1}{2} \qquad \csc \theta = \dfrac{r}{y} = -2$

$\cos \theta = \dfrac{x}{r} = -\dfrac{\sqrt{3}}{2} \qquad \sec \theta = \dfrac{r}{x} = -\dfrac{2\sqrt{3}}{3}$

$\tan \theta = \dfrac{y}{x} = \dfrac{\sqrt{3}}{3} \qquad \cot \theta = \dfrac{x}{y} = \sqrt{3}$

(b) $(x, y) = (-4, 1)$

$r = \sqrt{16 + 1} = \sqrt{17}$

$\sin \theta = \dfrac{y}{r} = \dfrac{\sqrt{17}}{17} \qquad \csc \theta = \dfrac{r}{y} = \sqrt{17}$

$\cos \theta = \dfrac{x}{r} = -\dfrac{4\sqrt{17}}{17} \qquad \sec \theta = \dfrac{r}{x} = -\dfrac{\sqrt{17}}{4}$

$\tan \theta = \dfrac{y}{x} = -\dfrac{1}{4} \qquad \cot \theta = \dfrac{x}{y} = -4$

5. $(x, y) = (7, 24)$

$r = \sqrt{49 + 576} = 25$

$\sin \theta = \dfrac{y}{r} = \dfrac{24}{25}$ $\csc \theta = \dfrac{r}{y} = \dfrac{25}{24}$

$\cos \theta = \dfrac{x}{r} = \dfrac{7}{25}$ $\sec \theta = \dfrac{r}{x} = \dfrac{25}{7}$

$\tan \theta = \dfrac{y}{x} = \dfrac{24}{7}$ $\cot \theta = \dfrac{x}{y} = \dfrac{7}{24}$

7. $(x, y) = (-4, 10)$

$r = \sqrt{16 + 100} = 2\sqrt{29}$

$\sin \theta = \dfrac{y}{r} = \dfrac{5\sqrt{29}}{29}$ $\csc \theta = \dfrac{r}{y} = \dfrac{\sqrt{29}}{5}$

$\cos \theta = \dfrac{x}{r} = -\dfrac{2\sqrt{29}}{29}$ $\sec \theta = \dfrac{r}{x} = -\dfrac{\sqrt{29}}{2}$

$\tan \theta = \dfrac{y}{x} = -\dfrac{5}{2}$ $\cot \theta = \dfrac{x}{y} = -\dfrac{2}{5}$

9. $(x, y) = (-3.5, 6.8)$

$r = \sqrt{12.25 + 46.24} \approx 7.65$

$\sin \theta = \dfrac{y}{r} = \dfrac{6.8}{7.65} \approx 0.9$ $\csc \theta = \dfrac{r}{y} = \dfrac{7.65}{6.8} \approx 1.1$

$\cos \theta = \dfrac{x}{r} = -\dfrac{3.5}{7.65} \approx -0.5$ $\sec \theta = \dfrac{r}{x} = -\dfrac{7.65}{3.5} \approx -2.2$

$\tan \theta = \dfrac{y}{x} = -\dfrac{6.8}{3.5} \approx -1.9$ $\cot \theta = \dfrac{x}{y} = -\dfrac{3.5}{6.8} \approx -0.5$

11. $\sin \theta < 0 \Longrightarrow \theta$ lies in Quadrant III or in Quadrant IV.

$\cos \theta < 0 \Longrightarrow \theta$ lies in Quadrant II or in Quadrant III.

$\sin \theta < 0$ *and* $\cos \theta < 0 \Longrightarrow \theta$ lies in Quadrant III.

13. $\sin \theta > 0 \Longrightarrow \theta$ lies in Quadrant I or in Quadrant II.

$\tan \theta < 0 \Longrightarrow \theta$ lies in Quadrant II or in Quadrant IV.

$\sin \theta > 0$ *and* $\tan \theta < 0 \Longrightarrow \theta$ lies in Quadrant II.

15. $\sin \theta = \dfrac{y}{r} = \dfrac{3}{5} \Longrightarrow x^2 = 25 - 9 = 16$

θ in Quadrant II $\Longrightarrow x = -4$

$\sin \theta = \dfrac{y}{r} = \dfrac{3}{5}$ $\csc \theta = \dfrac{r}{y} = \dfrac{5}{3}$

$\cos \theta = \dfrac{x}{r} = -\dfrac{4}{5}$ $\sec \theta = \dfrac{r}{x} = -\dfrac{5}{4}$

$\tan \theta = \dfrac{y}{x} = -\dfrac{3}{4}$ $\cot \theta = \dfrac{x}{y} = -\dfrac{4}{3}$

17. $\tan \theta = \dfrac{y}{x} = \dfrac{-15}{8}$

$\sin \theta < 0$ and $\tan \theta < 0 \Longrightarrow \theta$ is in Quadrant IV $\Longrightarrow$ $y < 0$ and $x > 0$.

$x = 8, y = -15, r = 17$

$\sin \theta = \dfrac{y}{r} = -\dfrac{15}{17}$ $\csc \theta = \dfrac{r}{y} = -\dfrac{17}{15}$

$\cos \theta = \dfrac{x}{r} = \dfrac{8}{17}$ $\sec \theta = \dfrac{r}{x} = \dfrac{17}{8}$

$\tan \theta = \dfrac{y}{x} = -\dfrac{15}{8}$ $\cot \theta = \dfrac{x}{y} = -\dfrac{8}{15}$

19. $\cot \theta = \dfrac{x}{y} = -\dfrac{3}{1} = \dfrac{3}{-1}$

$\cos \theta > 0 \Longrightarrow \theta$ is in Quadrant IV $\Longrightarrow$ x is positive; $x = 3, y = -1, r = \sqrt{10}$

$\sin \theta = \dfrac{y}{r} = -\dfrac{\sqrt{10}}{10}$ $\csc \theta = \dfrac{r}{y} = -\sqrt{10}$

$\cos \theta = \dfrac{x}{r} = \dfrac{3\sqrt{10}}{10}$ $\sec \theta = \dfrac{r}{x} = \dfrac{\sqrt{10}}{3}$

$\tan \theta = \dfrac{y}{x} = -\dfrac{1}{3}$ $\cot \theta = \dfrac{x}{y} = -3$

21. $\sec \theta = \dfrac{r}{x} = \dfrac{2}{-1} \Longrightarrow y^2 = 4 - 1 = 3$

$\sin \theta > 0 \Longrightarrow \theta$ is in Quadrant II $\Longrightarrow$ $y = \sqrt{3}$

$\sin \theta = \dfrac{y}{r} = \dfrac{\sqrt{3}}{2}$ $\csc \theta = \dfrac{r}{y} = \dfrac{2\sqrt{3}}{3}$

$\cos \theta = \dfrac{x}{r} = -\dfrac{1}{2}$ $\sec \theta = \dfrac{r}{x} = -2$

$\tan \theta = \dfrac{y}{x} = -\sqrt{3}$ $\cot \theta = \dfrac{x}{y} = -\dfrac{\sqrt{3}}{3}$

23. $\cot \theta$ is undefined, $\dfrac{\pi}{2} \le \theta \le \dfrac{3\pi}{2} \Longrightarrow y = 0 \Longrightarrow \theta = \pi$

$\sin \pi = 0$ $\csc \pi$ is undefined

$\cos \pi = -1$ $\sec \pi = -1$

$\tan \pi = 0$ $\cot \pi$ is undefined

25. To find a point on the terminal side of θ, use any point on the line $y = -x$ that lies in Quadrant II. $(-1, 1)$ is one such point.

$x = -1, y = 1, r = \sqrt{2}$

$\sin \theta = \dfrac{1}{\sqrt{2}} = \dfrac{\sqrt{2}}{2}$ $\csc \theta = \sqrt{2}$

$\cos \theta = -\dfrac{1}{\sqrt{2}} = -\dfrac{\sqrt{2}}{2}$ $\sec \theta = -\sqrt{2}$

$\tan \theta = -1$ $\cot \theta = -1$

27. To find a point on the terminal side of θ, use any point on the line $y = 2x$ that lies in Quadrant III. $(-1, -2)$ is one such point.

$x = -1, y = -2, r = \sqrt{5}$

$\sin \theta = -\dfrac{2}{\sqrt{5}} = -\dfrac{2\sqrt{5}}{5}$ $\csc \theta = \dfrac{\sqrt{5}}{-2} = -\dfrac{\sqrt{5}}{2}$

$\cos \theta = -\dfrac{1}{\sqrt{5}} = -\dfrac{\sqrt{5}}{5}$ $\sec \theta = \dfrac{\sqrt{5}}{-1} = -\sqrt{5}$

$\tan \theta = \dfrac{-2}{-1} = 2$ $\cot \theta = \dfrac{-1}{-2} = \dfrac{1}{2}$

29. $(x, y) = (-1, 0), r = 1$

$\cos \pi = \dfrac{x}{r} = \dfrac{-1}{1} = -1$

31. $(x, y) = (-1, 0), r = 1$

$\sec \pi = \dfrac{r}{x} = \dfrac{1}{-1} = -1$

33. $(x, y) = (0, 1), r = 1$

$\tan \dfrac{\pi}{2} = \dfrac{y}{x} = \dfrac{1}{0}$ undefined.

35. $(x, y) = (0, 1)$

$\cot \dfrac{\pi}{2} = \dfrac{x}{y} = \dfrac{0}{1} = 0$

37. $\theta = 203°$

$\theta' = 203° - 180° = 23°$

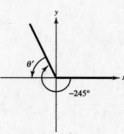

39. $\theta = -245°$

$360° - 245° = 115°$ (coterminal angle)

$\theta' = 180° - 115° = 65°$

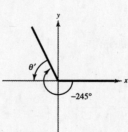

41. $\theta = \dfrac{2\pi}{3}$

$\theta' = \pi - \dfrac{2\pi}{3} = \dfrac{\pi}{3}$

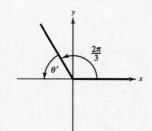

43. $\theta = 3.5$

$\theta' = 3.5 - \pi$

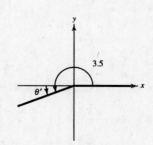

45. $\theta' = 45°$, Quadrant III

$\sin 225° = -\sin 45° = -\dfrac{\sqrt{2}}{2}$

$\cos 225° = -\cos 45° = -\dfrac{\sqrt{2}}{2}$

$\tan 225° = \tan 45° = 1$

47. $\theta' = 30°$, Quadrant I

$\sin 750° = \sin 30° = \dfrac{1}{2}$

$\cos 750° = \cos 30° = \dfrac{\sqrt{3}}{2}$

$\tan 750° = \tan 30° = \dfrac{\sqrt{3}}{3}$

49. $\theta' = 30°$, Quadrant III

$\sin(-150°) = -\sin 30° = -\dfrac{1}{2}$

$\cos(-150°) = -\cos 30° = -\dfrac{\sqrt{3}}{2}$

$\tan(-150°) = \tan 30° = \dfrac{\sqrt{3}}{3}$

51. $\theta' = \dfrac{\pi}{3}$, Quadrant III

$\sin \dfrac{4\pi}{3} = -\sin \dfrac{\pi}{3} = -\dfrac{\sqrt{3}}{2}$

$\cos \dfrac{4\pi}{3} = -\cos \dfrac{\pi}{3} = -\dfrac{1}{2}$

$\tan \dfrac{4\pi}{3} = \tan \dfrac{\pi}{3} = \sqrt{3}$

53. $\theta' = \dfrac{\pi}{6}$, Quadrant IV

$\sin\left(-\dfrac{\pi}{6}\right) = -\sin \dfrac{\pi}{6} = -\dfrac{1}{2}$

$\cos\left(-\dfrac{\pi}{6}\right) = \cos \dfrac{\pi}{6} = \dfrac{\sqrt{3}}{2}$

$\tan\left(-\dfrac{\pi}{6}\right) = -\tan \dfrac{\pi}{6} = -\dfrac{\sqrt{3}}{3}$

55. $\theta' = \dfrac{\pi}{4}$, Quadrant II

$\sin \dfrac{11\pi}{4} = \sin \dfrac{\pi}{4} = \dfrac{\sqrt{2}}{2}$

$\cos \dfrac{11\pi}{4} = -\cos \dfrac{\pi}{4} = -\dfrac{\sqrt{2}}{2}$

$\tan \dfrac{11\pi}{4} = -\tan \dfrac{\pi}{4} = -1$

57. $\theta' = \dfrac{\pi}{2}$

$\sin\left(-\dfrac{3\pi}{2}\right) = \sin \dfrac{\pi}{2} = 1$

$\cos\left(-\dfrac{3\pi}{2}\right) = \cos \dfrac{\pi}{2} = 0$

$\tan\left(-\dfrac{3\pi}{2}\right) = \tan \dfrac{\pi}{2}$ which is undefined

59. $\sin 10° \approx 0.1736$

61. $\cos(-110°) \approx -0.3420$

63. $\tan 4.5 \approx 4.6373$

65. $\tan \dfrac{\pi}{9} \approx 0.3640$

67. $\sin(-0.65) \approx -0.6052$

69. (a) $\sin \theta = \dfrac{1}{2} \implies$ reference angle is 30° or $\dfrac{\pi}{6}$ and θ is in Quadrant I or Quadrant II.

Values in degrees: 30°, 150°

Values in radians: $\dfrac{\pi}{6}, \dfrac{5\pi}{6}$

(b) $\sin \theta = -\dfrac{1}{2} \implies$ reference angle is 30° or $\dfrac{\pi}{6}$ and θ is in Quadrant III or Quadrant IV.

Values in degrees: 210°, 330°

Values in radians: $\dfrac{7\pi}{6}, \dfrac{11\pi}{6}$

71. (a) $\csc \theta = \dfrac{2\sqrt{3}}{3} \implies$ reference angle is 60° or $\dfrac{\pi}{3}$ and θ is in Quadrant I or Quadrant II.

Values in degrees: 60°, 120°

Values in radians: $\dfrac{\pi}{3}, \dfrac{2\pi}{3}$

(b) $\cot \theta = -1 \implies$ reference angle is 45° or $\dfrac{\pi}{4}$ and θ is in Quadrant II or Quadrant IV.

Values in degrees: 135°, 315°

Values in radians: $\dfrac{3\pi}{4}, \dfrac{7\pi}{4}$

73. (a) $\tan \theta = 1 \implies$ reference angle is 45° or $\dfrac{\pi}{4}$ and θ is in Quadrant I or Quadrant III.

Values in degrees: 45°, 225°

Values in radians: $\dfrac{\pi}{4}, \dfrac{5\pi}{4}$

(b) $\cot \theta = -\sqrt{3} \implies$ reference angle is 30° or $\dfrac{\pi}{6}$ and θ is in Quadrant II or Quadrant IV.

Values in degrees: 150°, 330°

Values in radians: $\dfrac{5\pi}{6}, \dfrac{11\pi}{6}$

75. $\sin \theta = 0.8191$

Quadrant I: $\theta = \sin^{-1} 0.8191 \approx 54.99°$

Quadrant II: $\theta = 180° - \sin^{-1} 0.8191 \approx 125.01°$

77. $\cos \theta = -0.4367 \implies \theta' \approx 64.11°$

Quadrant II: $\theta \approx 180° - 64.11° = 115.89°$

Quadrant III: $\theta \approx 180° + 64.11° = 244.11°$

79. $\cos \theta = 0.9848 \implies \theta' \approx 0.175$

Quadrant I: $\theta = \cos^{-1}(0.9848) \approx 0.175$

Quadrant IV: $\theta = 2\pi - \theta' \approx 6.109$

81. $\tan \theta = 1.192 \implies \theta' \approx 0.873$

Quadrant I: $\theta = \tan^{-1} 1.192 \approx 0.873$

Quadrant III: $\theta = \pi + \theta' \approx 4.014$

83. $\sec \theta = -2.6667 \Rightarrow \theta' = \cos^{-1}\left(\dfrac{1}{2.6667}\right) \approx 1.1864$

Quadrant II: $\theta = \pi - 1.1864 \approx 1.955$

Quadrant III: $\theta = \pi + 1.1864 \approx 4.328$

85.
$$\sin \theta = -\frac{3}{5}$$
$$\sin^2 \theta + \cos^2 \theta = 1$$
$$\cos^2 \theta = 1 - \sin^2 \theta$$
$$\cos^2 \theta = 1 - \left(-\frac{3}{5}\right)^2$$
$$\cos^2 \theta = 1 - \frac{9}{25}$$
$$\cos^2 \theta = \frac{16}{25}$$

$\cos \theta > 0$ in Quadrant IV.

$$\cos \theta = \frac{4}{5}$$

87. $\tan \theta = \dfrac{3}{2}$

$\sec^2 \theta = 1 + \tan^2 \theta$

$\sec^2 \theta = 1 + \left(\dfrac{3}{2}\right)^2$

$\sec^2 \theta = 1 + \dfrac{9}{4}$

$\sec^2 \theta = \dfrac{13}{4}$

$\sec \theta < 0$ in Quadrant III.

$\sec\theta = -\dfrac{\sqrt{13}}{2}$

89. $\cos \theta = \dfrac{5}{8}$

$\cos \theta = \dfrac{1}{\sec \theta} \Rightarrow \sec \theta = \dfrac{1}{\cos \theta}$

$\sec \theta = \dfrac{1}{\frac{5}{8}} = \dfrac{8}{5}$

91. (a) $t = 1$

$T = 45 - 23 \cos\left[\dfrac{2\pi}{365}(1 - 32)\right] \approx 25.2°\text{ F}$

(c) $t = 291$

$T = 45 - 23 \cos\left[\dfrac{2\pi}{365}(291 - 32)\right] \approx 50.8°\text{ F}$

(b) $t = 185$

$T = 45 - 23 \cos\left[\dfrac{2\pi}{365}(185 - 32)\right] \approx 65.1°\text{ F}$

93. $y(t) = 2 \cos 6t$

(a) $y(0) = 2 \cos 0 = 2$ centimeters

(b) $y\left(\dfrac{1}{4}\right) = 2 \cos\left(\dfrac{3}{2}\right) \approx 0.14$ centimeter

(c) $y\left(\dfrac{1}{2}\right) = 2 \cos 3 \approx -1.98$ centimeters

95. $d = \dfrac{v^2}{32} \sin \theta$

(a) $\theta = 30°$

$d = \dfrac{(100)^2}{32} \sin 30° = (312.5)(0.5) = 156.25$ feet

(b) $\theta = 50°$

$d = \dfrac{(100)^2}{32} \sin 50° = (312.5)(0.7660) \approx 239.39$ feet

(c) $\theta = 60°$

$d = \dfrac{(100)^2}{32} \sin 60° = (312.5)(0.8660) \approx 270.63$ feet

97. False. In each of the four quadrants, the sign of the secant function and the cosine function will be the same since they are reciprocals of each other.

99. As θ increases from $0°$ to $90°$, x decreases from 12 cm to 0 cm and y increases from 0 cm to 12 cm. Therefore, $\sin \theta = \dfrac{y}{12}$ increases from 0 to 1 and $\cos \theta = \dfrac{x}{12}$ decreases from 1 to 0. Thus, $\tan \theta = \dfrac{y}{x}$ increases without bound, and when $\theta = 90°$ the tangent is undefined.

101. First, determine a positive coterminal angle. Then determine the trigonometric function of the reference angle and prefix the appropriate sign.

103. $\dfrac{3}{t} + \dfrac{5}{t} = 1$

$\dfrac{8}{t} = 1$

$8 = t$

105. $\dfrac{12}{x} - 3 = \dfrac{4}{x} + 9$

$\dfrac{12}{x} - \dfrac{4}{x} = 9 + 3$

$\dfrac{8}{x} = 12$

$\dfrac{8}{12} = x$

$x = \dfrac{2}{3}$

107.   Slope $= \dfrac{9 - 0}{1 - 10} = \dfrac{9}{-9} = -1$

$m = -1$

$y - 0 = (-1)(x - 10)$

$y = -x + 10$

$x + y - 10 = 0$

109.   Slope $= \dfrac{-\frac{1}{3} - 3}{\frac{11}{2} - \left(-\frac{1}{2}\right)} = \dfrac{-\frac{10}{3}}{\frac{12}{2}} = -\dfrac{10}{3} \cdot \dfrac{1}{6} = -\dfrac{5}{9}$

$m = -\dfrac{5}{9}$

$y - 3 = -\dfrac{5}{9}\left(x - \left(-\dfrac{1}{2}\right)\right)$

$y - 3 = -\dfrac{5}{9}x - \dfrac{5}{18}$

$18y - 54 = -10x - 5$

$10x + 18y - 49 = 0$

Section 1.5   Graphs of Sine and Cosine Functions

- You should be able to graph $y = a \sin(bx - c)$ and $y = a \cos(bx - c)$. (Assume $b > 0$)
- Amplitude: $|a|$
- Period: $\dfrac{2\pi}{|b|}$
- Shift: Solve $bx - c = 0$ and $bx - c = 2\pi$.
- Key Increments: $\dfrac{1}{4}$ (period)

Solutions to Odd-Numbered Exercises

1. $y = 3 \sin 2x$

Period: $\dfrac{2\pi}{2} = \pi$

Amplitude: $|3| = 3$

3. $y = \dfrac{5}{2} \cos \dfrac{x}{2}$

Period: $\dfrac{2\pi}{\frac{1}{2}} = 4\pi$

Amplitude: $\left|\dfrac{5}{2}\right| = \dfrac{5}{2}$

5. $y = \dfrac{1}{2} \sin \dfrac{\pi x}{3}$

Period: $\dfrac{2\pi}{\frac{\pi}{3}} = 6$

Amplitude: $\left|\dfrac{1}{2}\right| = \dfrac{1}{2}$

7. $y = -2 \sin x$

Period: $\dfrac{2\pi}{1} = 2\pi$

Amplitude: $|-2| = 2$

9. $y = 3 \sin 10x$

Period: $\dfrac{2\pi}{10} = \dfrac{\pi}{5}$

Amplitude: $|3| = 3$

11. $y = \dfrac{1}{2} \cos \dfrac{2\pi}{3}$

Period: $\dfrac{2\pi}{\frac{2}{3}} = 3\pi$

Amplitude: $\left|\dfrac{1}{2}\right| = \dfrac{1}{2}$

13. $y = \dfrac{1}{4} \sin 2\pi x$

Period: $\dfrac{2\pi}{2\pi} = 1$

Amplitude: $\left|\dfrac{1}{4}\right| = \dfrac{1}{4}$

15. $f(x) = \sin x$

$g(x) = \sin(x - \pi)$

The graph of g is a horizontal shift to the right π units of the graph of f (a phase shift).

17. $f(x) = \cos 2x$

$g(x) = -\cos 2x$

The graph of g is a reflection in the x-axis of the graph of f.

19. $f(x) = \cos x$

$g(x) = \cos 2x$

The period of f is twice that of g.

21. $f(x) = \sin 2x$

$f(x) = 3 + \sin 2x$

The graph of g is a vertical shift 3 units upward of the graph of f.

23. The graph of g has twice the amplitude as the graph of f. The period is the same.

25. The graph of g is a horizontal shift π units to the right of the graph of f.

27. $f(x) = -2 \sin x$

Period: 2π

Amplitude: 2

$g(x) = 4 \sin x$

Period: 2π

Amplitude: 4

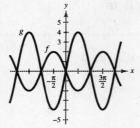

29. $f(x) = \cos x$

Period: 2π

Amplitude: 1

$g(x) = 1 + \cos x$

is a vertical shift of the graph of $f(x)$ one unit upward.

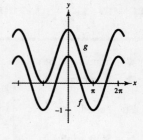

31. $f(x) = -\dfrac{1}{2} \sin \dfrac{x}{2}$

Period: 4π

Amplitude: $\dfrac{1}{2}$

$g(x) = 3 - \dfrac{1}{2} \sin \dfrac{x}{2}$ is the

graph of $f(x)$ shifted vertically three units upward.

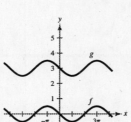

33. $f(x) = 2 \cos x$

Period: 2π

Amplitude: 2

$g(x) = 2 \cos(x + \pi)$ is the graph of $f(x)$ shifted π units to the left.

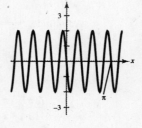

35. $y = -2 \sin 6x$; $a = -2$, $b = 6$, $c = 0$

Period: $\dfrac{2\pi}{6} = \dfrac{\pi}{3}$

Amplitude: $|-2| = 2$

Key points: $(0, 0), \left(\dfrac{\pi}{12}, -2\right), \left(\dfrac{\pi}{6}, 0\right), \left(\dfrac{\pi}{4}, 2\right), \left(\dfrac{\pi}{3}, 0\right)$

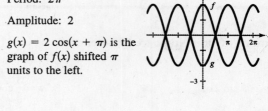

37. $y = \cos 2\pi x$

Period: $\dfrac{2\pi}{2\pi} = 1$

Amplitude: 1

Key points: $(0, 1), \left(\dfrac{1}{4}, 0\right), \left(\dfrac{1}{2}, -1\right), \left(\dfrac{3}{4}, 0\right)$

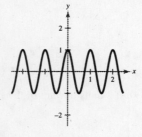

39. $y = -\sin \dfrac{2\pi x}{3}$; $a = -1$, $b = \dfrac{2\pi}{3}$, $c = 0$

Period: $\dfrac{2\pi}{\dfrac{2\pi}{3}} = 3$

Amplitude: 1

Key points: $(0, 0), \left(\dfrac{3}{4}, -1\right), \left(\dfrac{3}{2}, 0\right), \left(\dfrac{9}{4}, 1\right), (3, 0)$

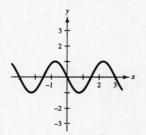

41. $y = \sin\left(x - \frac{\pi}{4}\right);\ a = 1,\ b = 1,\ c = \frac{\pi}{4}$

Period: 2π

Amplitude: 1

Shift: Set $x - \frac{\pi}{4} = 0$ and $x - \frac{\pi}{4} = 2\pi$

$$x = \frac{\pi}{4} \qquad x = \frac{9\pi}{4}$$

Key points: $\left(\frac{\pi}{4}, 0\right), \left(\frac{3\pi}{4}, 1\right), \left(\frac{5\pi}{4}, 0\right), \left(\frac{7\pi}{4}, -1\right), \left(\frac{9\pi}{4}, 0\right)$

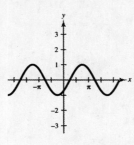

43. $y = 3\cos(x + \pi)$

Period: 2π

Amplitude: 3

Shift: Set $x + \pi = 0$ and $x + \pi = 2\pi$

$$x = -\pi \qquad x = \pi$$

Key points: $(-\pi, 3), \left(-\frac{\pi}{2}, 0\right), (0, -3), \left(\frac{\pi}{2}, 0\right), (\pi, 3)$

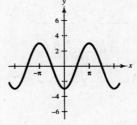

45. $y = 2 - \sin\frac{2\pi x}{3}$

Vertical shift 2 units upward of the graph in Exercise 39.

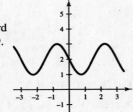

47. $y = 2 + \frac{1}{10}\cos 60\pi x$

Period: $\frac{2\pi}{60\pi} = \frac{1}{30}$

Amplitude: $\frac{1}{10}$

Vertical shift 2 units upward

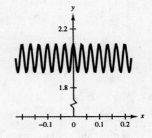

Key points:

$(0, 2.1), \left(\frac{1}{120}, 2\right), \left(\frac{1}{60}, 1.9\right), \left(\frac{1}{40}, 2\right), \left(\frac{1}{30}, 2.1\right)$

49. $y = 3\cos(x + \pi) - 3$

Vertical shift 3 units downward of the graph in Exercise 43.

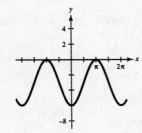

51. $y = \frac{2}{3} \cos\left(\frac{x}{2} - \frac{\pi}{4}\right)$; $a = \frac{2}{3}$, $b = \frac{1}{2}$, $c = \frac{\pi}{4}$

Period: 4π

Amplitude: $\frac{2}{3}$

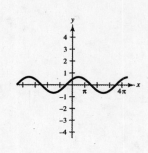

Shift: $\frac{x}{2} - \frac{\pi}{4} = 0$ and $\frac{x}{2} - \frac{\pi}{4} = 2\pi$

$\qquad x = \frac{\pi}{2} \qquad\qquad x = \frac{9\pi}{2}$

Key points: $\left(\frac{\pi}{2}, \frac{2}{3}\right)$, $\left(\frac{3\pi}{2}, 0\right)$, $\left(\frac{5\pi}{2}, \frac{-2}{3}\right)$, $\left(\frac{7\pi}{2}, 0\right)$, $\left(\frac{9\pi}{2}, \frac{2}{3}\right)$

53. $y = -2 \sin(4x + \pi)$

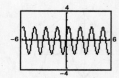

55. $y = \cos\left(2\pi x - \frac{\pi}{2}\right) + 1$

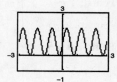

57. $y = 5 \sin(\pi - 2x) + 10$

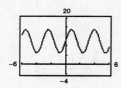

59. $y = -0.1 \sin\left(\frac{\pi x}{10} + \pi\right)$

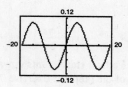

61. $f(x) = a \cos x + d$

Amplitude: $\frac{1}{2}[3 - (-1)] = 2 \implies a = 2$

Vertical shift 1 unit upward of $g(x) = 2 \cos x \implies d = 1$.
Thus, $f(x) = 2 \cos x + 1$.

63. $f(x) = a \cos x + d$

Amplitude: $\frac{1}{2}[8 - 0] = 4$

Since $f(x)$ is the graph of $g(x) = 4 \cos x$ reflected in the x-axis and shifted vertically 4 units upward, we have $a = -4$ and $d = 4$. Thus, $f(x) = -4 \cos x + 4$.

65. $y = a \sin(bx - c)$

Amplitude: $|a| = |3|$ Since the graph is reflected in the x-axis, we have $a = -3$.

Period: $\frac{2\pi}{b} = \pi \implies b = 2$

Phase shift: $c = 0$

Thus, $y = -3 \sin 2x$.

67. $y = a \sin(bx - c)$

Amplitude: $a = 2$

Period: $2\pi \implies b = 1$

Phase shift: $bx - c = 0$ when $x = -\frac{\pi}{4}$

$\qquad (1)\left(\frac{-\pi}{4}\right) - c = 0 \implies c = -\frac{\pi}{4}$

Thus, $y = 2 \sin\left(x + \frac{\pi}{4}\right)$.

69. $y_1 = \sin x$

$y_2 = -\dfrac{1}{2}$

In the interval $[-2\pi, 2\pi]$, $\sin x = -\dfrac{1}{2}$ when

$x = -\dfrac{5\pi}{6}, -\dfrac{\pi}{6}, \dfrac{7\pi}{6}, \dfrac{11\pi}{6}$.

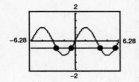

71. $y_1 = \cos x$

$y_2 = \dfrac{\sqrt{2}}{2}$

In the interval $[-2\pi, 2\pi]$, $\cos x = \dfrac{\sqrt{2}}{2}$

when $x = \pm\dfrac{\pi}{4}, \pm\dfrac{7\pi}{4}$.

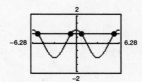

73. $y = 0.85 \sin \dfrac{\pi t}{3}$

(a) Time for one cycle $= \dfrac{2\pi}{\dfrac{\pi}{3}} = 6$ sec

(b) Cycles per min $= \dfrac{60}{6} = 10$ cycles per min

(c) Amplitude: 0.85

Period: 6

Key points: $(0, 0)$, $\left(\dfrac{3}{2}, 0.85\right)$, $(3, 0)$, $\left(\dfrac{9}{2}, -0.85\right)$, $(6, 0)$

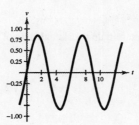

75. $y = 0.001 \sin 880\pi t$

(a) Period: $\dfrac{2\pi}{880\pi} = \dfrac{1}{440}$ seconds

(b) $f = \dfrac{1}{p} = 440$ cycles per second

77. (a) $C(t) = 56.35 + 27.35 \sin\left(\dfrac{\pi t}{6} + 4.19\right)$

(b)

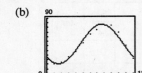

The model is a good fit for most months.

(c)

The model is a good fit.

(d) Use the constant term of each model to estimate the average annual temperature.

Honolulu: 84.40°

Chicago: 56.35°

(e) Each model has a period of 12. This corresponds to the 12 months in a year.

(f) Chicago has a greater variability in temperatures during the year. The amplitude of each model indicates this variability.

79. $S = 74.50 + 43.75 \sin \dfrac{\pi t}{6}$

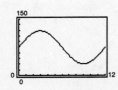

81. (a) and (c)

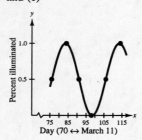

Day (70 ↔ March 11)

(d) For May 8, 2005, use $x = 128$

$y(128) \approx 0$

(b) Vertical shift: $\dfrac{1}{2} \Rightarrow d = \dfrac{1}{2}$

Amplitude: $\dfrac{1}{2} \Rightarrow a = \dfrac{1}{2}$

Period: $30 \Rightarrow \dfrac{2\pi}{b} = 30 \Rightarrow b = \dfrac{\pi}{15}$

Horizontal shift: $76 \Rightarrow \dfrac{\pi}{15}(76) + c = c \Rightarrow c = -\dfrac{76\pi}{15}$

$y = \dfrac{1}{2} + \dfrac{1}{2} \sin\left(\dfrac{\pi}{15}x - \dfrac{76\pi}{15}\right)$

$ = \dfrac{1}{2} + \dfrac{1}{2} \sin \dfrac{\pi}{15}(x - 76)$

The model is a good fit.

83. False. $y = \dfrac{1}{2} \cos 2x$ has an amplitude that is **half** that of $y = \cos x$. For $y = a \cos bx$, the amplitude is $|a|$.

85. $y = 2 + \sin x$

$y = 3.5 + \sin x$

$y = -2 + \sin x$

Each value of d produces a vertical shift of $y = \sin x$ upward (or downward) by d units.

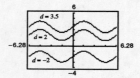

87. $y = \sin(x - 1)$

$y = \sin(x - 3)$

$y = \sin(x - (-2)) = \sin(x + 2)$

Each value of c produces a horizontal shift of $y = \sin x$ to the left (or right) by c units.

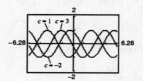

89. $f(x) = \sin x, \; g(x) = -\cos\left(x + \dfrac{\pi}{2}\right)$

x	0	$\dfrac{\pi}{2}$	π	$\dfrac{3\pi}{2}$	2π
$\sin x$	0	1	0	-1	0
$-\cos\left(x + \dfrac{\pi}{2}\right)$	0	1	0	-1	0

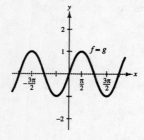

Conjecture: $\sin x = -\cos\left(x + \dfrac{\pi}{2}\right)$

91. $f(x) = \cos x, \; g(x) = -\cos(x - \pi)$

x	0	$\dfrac{\pi}{2}$	π	$\dfrac{3\pi}{2}$	2π
$\cos x$	1	0	-1	0	1
$-\cos(x - \pi)$	1	0	-1	0	1

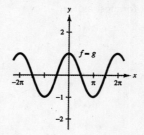

Conjecture: $\cos x = -\cos(x - \pi)$

93. (a) $\sin \dfrac{1}{2} \approx \dfrac{1}{2} - \dfrac{\left(\dfrac{1}{2}\right)^3}{3!} + \dfrac{\left(\dfrac{1}{2}\right)^5}{5!} \approx 0.4794$

$\sin \dfrac{1}{2} \approx 0.4794$ (by calculator)

(b) $\sin 1 \approx 1 - \dfrac{1}{3!} + \dfrac{1}{5!} \approx 0.8417$

$\sin 1 \approx 0.8415$ (by calculator)

(c) $\sin \dfrac{\pi}{6} \approx 1 - \dfrac{\left(\dfrac{\pi}{6}\right)^3}{3!} + \dfrac{\left(\dfrac{\pi}{6}\right)^5}{5!} \approx 0.5000$

$\sin \dfrac{\pi}{6} = 0.5$ (by calculator)

(d) $\cos(-0.5) \approx 1 - \dfrac{(-0.5)^2}{2!} + \dfrac{(-0.5)^4}{4!} \approx 0.8776$

$\cos(-0.5) \approx 0.8776$ (by calculator)

(e) $\cos 1 \approx 1 - \dfrac{1}{2!} + \dfrac{1}{4!} \approx 0.5417$

$\cos 1 \approx 0.5403$ (by calculator)

(f) $\cos \dfrac{\pi}{4} \approx 1 - \dfrac{\left(\dfrac{\pi}{4}\right)^2}{2!} + \dfrac{\left(\dfrac{\pi}{4}\right)^4}{4!} = 0.7074$

$\cos \dfrac{\pi}{4} \approx 0.7071$ (by calculator)

The error in the approximation is not the same in each case. The error appears to increase as x moves farther away from 0.

95. $f(x)$ is even $\Longrightarrow f(-x) = f(x)$

$g(x)$ is odd $\Longrightarrow g(-x) = -g(x)$

(a) $h(x) = [f(x)]^2$

$h(-x) = [f(-x)]^2$

$= [f(x)]^2$

$= h(x) \Longrightarrow h(x)$ is even

(b) $h(x) = [g(x)]^2$

$h(-x) = [g(-x)]^2$

$= [-g(x)]^2$

$= [g(x)]^2$

$= h(x) \Longrightarrow h(x)$ is even

97. $\dfrac{2}{x+5} - \dfrac{2}{x-5}$

$\dfrac{2}{x+5} - \dfrac{2}{x-5} = \dfrac{2(x-5) - 2(x+5)}{(x+5)(x-5)} = \dfrac{2x - 10 - 2x - 10}{(x+5)(x-5)} = -\dfrac{20}{(x+5)(x-5)}$

99. $\dfrac{x}{x-5} + \dfrac{1}{2}$

$\dfrac{x}{x-5} + \dfrac{1}{2} = \dfrac{2x + x - 5}{2(x-5)} = \dfrac{3x - 5}{2(x-5)}$

101. $f(x) = \dfrac{\sqrt{x-3}}{x-8}$

Domain: $x \geq 3, x \neq 8$

103. $f(x) = \sqrt[3]{4 - x^2}$

Domain: All real numbers

Section 1.6   Graphs of Other Trigonometric Functions

- ■ You should be able to graph

  $y = a \tan (bx - c)$ $y = a \cot (bx - c)$

  $y = a \sec (bx - c)$ $y = a \csc (bx - c)$

- ■ When graphing $y = a \sec (bx - c)$ or $y = a \csc (bx - c)$ you should first graph $y = a \cos (bx - c)$ or $y = a \sin (bx - c)$ because

  (a) The x-intercepts of sine and cosine are the vertical asymptotes of cosecant and secant.

  (b) The maximums of sine and cosine are the local minimums of cosecant and secant.

  (c) The minimums of sine and cosine are the local maximums of cosecant and secant.

- ■ You should be able to graph using a damping factor.

Solutions to Odd-Numbered Exercises

1. $y = \sec 2x$

Period: $\dfrac{2\pi}{2} = \pi$

Matches graph (e).

3. $y = \dfrac{1}{2} \cot \pi x$

Period: $\dfrac{\pi}{\pi} = 1$

Matches graph (a).

5. $y = -\csc x$

Period: 2π

Matches graph (d).

7. $y = \dfrac{1}{3} \tan x$

Period: π

Two consecutive asymptotes:

$x = -\dfrac{\pi}{2}$ and $x = \dfrac{\pi}{2}$

x	$-\dfrac{\pi}{4}$	0	$\dfrac{\pi}{4}$
y	$-\dfrac{1}{3}$	0	$\dfrac{1}{3}$

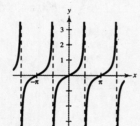

9. $y = \tan 3x$

Period: $\dfrac{\pi}{3}$

Two consecutive asymptotes:

$3x = -\dfrac{\pi}{2} \implies x = -\dfrac{\pi}{6}$

$3x - \dfrac{\pi}{2} \implies x = \dfrac{\pi}{6}$

x	$-\dfrac{\pi}{12}$	0	$\dfrac{\pi}{12}$
y	-1	0	1

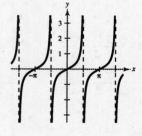

11. $y = -\dfrac{1}{2} \sec x$

Graph $y = -\dfrac{1}{2} \cos x$ first.

Period: 2π

One cycle: 0 to 2π

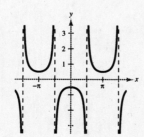

13. $y = \csc \pi x$

Graph $y = \sin \pi x$ first.

Period: $\dfrac{2\pi}{\pi} = 2$

One cycle: 0 to 2

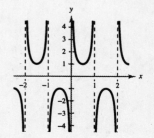

15. $y = \sec \pi x - 1$

Graph $y = \cos \pi x$ first

Period: $\dfrac{2\pi}{\pi} = 2$

One cycle: 0 to 2

Vertical shift 1 unit downward.

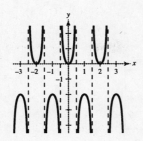

17. $y = \csc \dfrac{x}{2}$

Graph $y = \sin \dfrac{x}{2}$ first.

Period: $\dfrac{2\pi}{\frac{1}{2}} = 4\pi$

One cycle: 0 to 4π

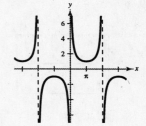

19. $y = \cot \dfrac{x}{2}$

Period: $\dfrac{\pi}{\frac{1}{2}} = 2\pi$

x	$\dfrac{\pi}{2}$	π	$\dfrac{3\pi}{2}$
y	1	0	-1

Two consecutive asymptotes: $\dfrac{x}{2} = 0 \Rightarrow x = 0$

$\dfrac{x}{2} = \pi \Rightarrow x = 2\pi$

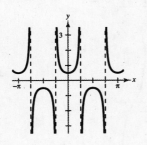

21. $y = \dfrac{1}{2} \sec 2x$

Graph $y = \dfrac{1}{2} \cos 2x$ first.

Period: $\dfrac{2\pi}{2} = \pi$

One cycle: 0 to π

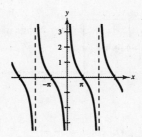

23. $y = \tan \dfrac{\pi x}{4}$

Period: $\dfrac{\pi}{\frac{\pi}{4}} = 4$

Two consecutive asymptotes:

$\dfrac{\pi x}{4} = -\dfrac{\pi}{2} \Rightarrow x = -2$

$\dfrac{\pi x}{4} = \dfrac{\pi}{2} \Rightarrow x = 2$

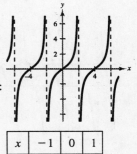

x	-1	0	1
y	-1	0	1

25. $y = \csc (\pi - x)$

Graph $y = \sin(\pi - x)$ first.

Period: 2π

Shift: Set $\pi - x = 0$ and $\pi - x = 2\pi$

$\qquad\qquad x = \pi \qquad\qquad x = -\pi$

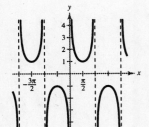

27. $y = \dfrac{1}{4} \csc\left(x + \dfrac{\pi}{4}\right)$

Graph $y = \dfrac{1}{4} \sin\left(x + \dfrac{\pi}{4}\right)$ first.

Period: 2π

Shift: Set $x + \dfrac{\pi}{4} = 0 \qquad$ and $\qquad x + \dfrac{\pi}{4} = 2\pi$

$\qquad\qquad x = -\dfrac{\pi}{4} \qquad$ to $\qquad\qquad x = \dfrac{7\pi}{4}$

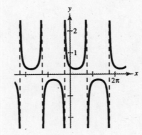

29. $y = \tan \dfrac{x}{3}$

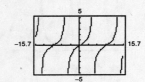

31. $y = -2 \sec 4x$

$$= \dfrac{-2}{\cos 4x}$$

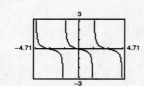

33. $y = \tan\left(x - \dfrac{\pi}{4}\right)$

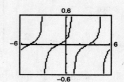

35. $y = \dfrac{1}{4} \cot\left(x - \dfrac{\pi}{2}\right)$

$$= \dfrac{1}{4 \tan\left(x - \dfrac{\pi}{2}\right)}$$

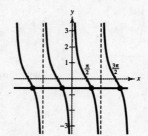

37. $y = 0.1 \tan\left(\dfrac{\pi x}{4} + \dfrac{\pi}{4}\right)$

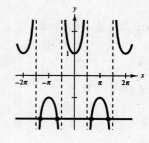

39. $\tan x = 1$

$$x = -\dfrac{7\pi}{4}, \ -\dfrac{3\pi}{4}, \ \dfrac{\pi}{4}, \ \dfrac{5\pi}{4}$$

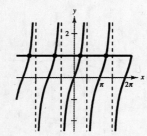

41. $\cot x = -\dfrac{\sqrt{3}}{3}$

$$x = -\dfrac{4\pi}{3}, \ -\dfrac{\pi}{3}, \ \dfrac{2\pi}{3}, \ \dfrac{5\pi}{3}$$

43. $\sec x = -2$

$$x = \pm\dfrac{2\pi}{3}, \ \pm\dfrac{4\pi}{3}$$

45. $\csc x = \sqrt{2}$

$$x = -\dfrac{7\pi}{4}, \ -\dfrac{5\pi}{4}, \ \dfrac{\pi}{4}, \ \dfrac{3\pi}{4}$$

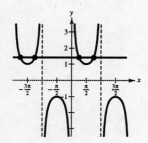

47. The graph of $f(x) = \sec x$ has y-axis symmetry. Thus, the function is even.

49. $f(x) = 2 \sin x$

$g(x) = \dfrac{1}{2} \csc x$

(a)

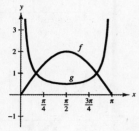

(b) $f > g$ on the interval, $\dfrac{\pi}{6} < x < \dfrac{5\pi}{6}$

(c) As $x \to \pi$, $f(x) = 2 \sin x \to 0$ and

$g(x) = \dfrac{1}{2} \csc x \to \pm\infty$ since $g(x)$ is the reciprocal of $f(x)$.

51. $y_1 = \sin x \csc x$ and $y_2 = 1$

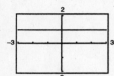

$\sin x \csc x = \sin x \left(\dfrac{1}{\sin x} \right) = 1,\ \sin x \neq 0$

The expressions are equivalent except when $\sin x = 0$ and y_1 is undefined.

53. $y_1 = \dfrac{\cos x}{\sin x}$ and $y_2 = \cot x = \dfrac{1}{\tan x}$

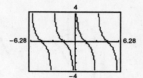

$\cot x = \dfrac{\cos x}{\sin x}$

The expressions are equivalent.

55. $f(x) = |x \cos x|$

As $x \to 0$, $f(x) \to 0$.

Matches graph (d).

57. $g(x) = |x| \sin x$

As $x \to 0$, $g(x) \to 0$.

Matches graph (b).

59. $f(x) = \sin x + \cos\left(x + \dfrac{\pi}{2} \right)$, $g(x) = 0$

$f(x) = g(x)$ The graph is the line $y = 0$.

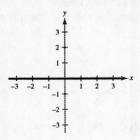

61. $f(x) = \sin^2 x$, $g(x) = \dfrac{1}{2}(1 - \cos 2x)$

$f(x) = g(x)$

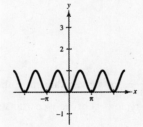

63. $f(x) = x \cos \pi x$

Damping factor: x

As $x \to \infty$, $f(x)$ oscillates and approaches $-\infty$ and ∞.

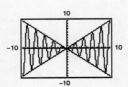

65. $g(x) = x^3 \sin x$

Damping factor: x^3

As $x \to \infty$, $g(x)$ oscillates and approaches $-\infty$ and ∞.

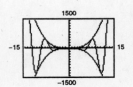

67. $y = \dfrac{6}{x} + \cos x,\ x > 0$

As $x \to 0,\ y \to \infty$.

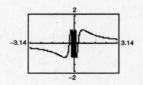

69. $g(x) = \dfrac{\sin x}{x}$

As $x \to 0,\ g(x) \to 1$.

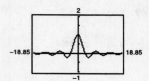

71. $f(x) = \sin \dfrac{1}{x}$

As $x \to 0, f(x)$ oscillates between -1 and 1.

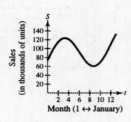

73. $\tan x = \dfrac{7}{d}$

$d = \dfrac{7}{\tan x} = 7 \cot x$

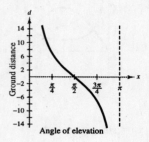

75. $S = 74 + 3x + 40 \sin \dfrac{\pi t}{6}$

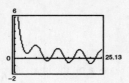

77. $H(t) = 54.33 - 20.38 \cos \dfrac{\pi t}{6} - 15.69 \sin \dfrac{\pi t}{6}$

$L(t) = 39.36 - 15.70 \cos \dfrac{\pi t}{6} - 14.16 \sin \dfrac{\pi t}{6}$

(a) Period of $\cos \dfrac{\pi t}{6}$: $\dfrac{2\pi}{\dfrac{\pi}{6}} = 12$

Period of $\sin \dfrac{\pi t}{6}$: $\dfrac{2\pi}{\dfrac{\pi}{6}} = 12$

Period of $H(t)$: 12

Period of $L(t)$: 12

(b) From the graph, it appears that the greatest difference between high and low temperatures occurs in summer. The smallest difference occurs in winter.

(c) The highest high and low temperatures appear to occur around the middle of July, roughly one month after the time when the sun is northernmost in the sky.

79. True. Since $y = \csc x = \dfrac{1}{\sin x}$, for a given value of x, the y-coordinate of $\csc x$ is the reciprocal of the y-coordinate of $\sin x$.

81. As $x \to \dfrac{\pi}{2}$ from the left, $f(x) = \tan x \to \infty$.

As $x \to \dfrac{\pi}{2}$ from the right, $f(x) = \tan x \to -\infty$.

83. $f(x) = x - \cos x$

(a)

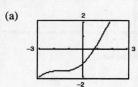

The zero between 0 and 1 appears to occur at
$x \approx 0.739$.

(b) $x_n = \cos(x_{n-1})$

$x_0 = 1$

$x_1 = \cos 1 \approx 0.5403$

$x_2 = \cos 0.5403 \approx 0.8576$

$x_3 = \cos 0.8576 \approx 0.6543$

$x_4 = \cos 0.6543 \approx 0.7935$

$x_5 = \cos 0.7935 \approx 0.7014$

$x_6 = \cos 0.7014 \approx 0.7640$

$x_7 = \cos 0.7640 \approx 0.7221$

$x_8 = \cos 0.7221 \approx 0.7504$

$x_9 = \cos 0.7504 \approx 0.7314$

$\vdots$

This sequence appears to be approaching the zero of
f: $x \approx 0.739$.

85. $y_1 = \sec x$

$y_2 = 1 + \dfrac{x^2}{2!} + \dfrac{5x^4}{4!}$

The approximation appears to
coincide on the interval
$-1.1 \le x \le 1.1$.

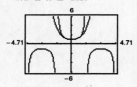

87. $x^2 = 64$

$x = \pm\sqrt{64} = \pm 8$

89. $4x^2 - 12x + 9 = 0$

$(2x - 3)^2 = 0$

$2x - 3 = 0$

$x = \dfrac{3}{2}$

91. $x^2 - 6x + 4 = 0$　　　　Complete the Square

$x^2 - 6x = -4$

$x^2 - 6x + 9 = -4 + 9$

$(x - 3)^2 = 5$

$x - 3 = \pm\sqrt{5}$

$x = 3 \pm \sqrt{5}$

93. $50 + 5x = 3x^2$

$0 = 3x^2 - 5x - 50$

$0 = (3x + 10)(x - 5)$

$3x + 10 = 0 \implies x = -\dfrac{10}{3}$

$x - 5 = 0 \implies x = 5$

Section 1.7 Inverse Trigonometric Functions

■ You should know the definitions, domains, and ranges of $y = \arcsin x$, $y = \arccos x$, and $y = \arctan x$.

Function	Domain	Range
$y = \arcsin x \implies x = \sin y$	$-1 \le x \le 1$	$-\dfrac{\pi}{2} \le y \le \dfrac{\pi}{2}$
$y = \arccos x \implies x = \cos y$	$-1 \le x \le 1$	$0 \le y \le \pi$
$y = \arctan x \implies x = \tan y$	$-\infty < x < \infty$	$-\dfrac{\pi}{2} < x < \dfrac{\pi}{2}$

■ You should know the inverse properties of the inverse trigonometric functions.

$$\sin(\arcsin x) = x \quad \text{and} \quad \arcsin(\sin y) = y,\ -\frac{\pi}{2} \le y \le \frac{\pi}{2}$$

$$\cos(\arccos x) = x \quad \text{and} \quad \arccos(\cos y) = y,\ 0 \le y \le \pi$$

$$\tan(\arctan x) = x \quad \text{and} \quad \arctan(\tan y) = y,\ -\frac{\pi}{2} < y < \frac{\pi}{2}$$

■ You should be able to use the triangle technique to convert trigonometric functions of inverse trigonometric functions into algebraic expressions.

Solutions to Odd-Numbered Exercises

1. $y = \arcsin \dfrac{1}{2} \implies \sin y = \dfrac{1}{2}$ for

$-\dfrac{\pi}{2} \le y \le \dfrac{\pi}{2} \implies y = \dfrac{\pi}{6}$

3. $y = \arccos \dfrac{1}{2} \implies \cos y = \dfrac{1}{2}$ for

$0 \le y \le \pi \implies y = \dfrac{\pi}{3}$

5. $y = \arctan \dfrac{\sqrt{3}}{3} \implies \tan y = \dfrac{\sqrt{3}}{3}$ for

$-\dfrac{\pi}{2} < y < \dfrac{\pi}{2} \implies y = \dfrac{\pi}{6}$

7. $y = \arccos\left(-\dfrac{\sqrt{3}}{2}\right) \implies \cos y = -\dfrac{\sqrt{3}}{2}$ for

$0 \le y \le \pi \implies y = \dfrac{5\pi}{6}$

9. $y = \arctan\left(-\sqrt{3}\right) \implies \tan y = -\sqrt{3}$ for

$-\dfrac{\pi}{2} < y < \dfrac{\pi}{2} \implies y = -\dfrac{\pi}{3}$

11. $y = \arccos\left(-\dfrac{1}{2}\right) \implies \cos y = -\dfrac{1}{2}$ for

$0 \le y \le \pi \implies y = \dfrac{2\pi}{3}$

13. $y = \arcsin \dfrac{\sqrt{3}}{2} \implies \sin y = \dfrac{\sqrt{3}}{2}$ for

$-\dfrac{\pi}{2} \le y \le \dfrac{\pi}{2} \implies y = \dfrac{\pi}{3}$

15. $y = \arctan 0 \implies \tan y = 0$ for $-\dfrac{\pi}{2} < y < \dfrac{\pi}{2} \implies y = 0$

17. $\arccos 0.28 = \cos^{-1} 0.28 \approx 1.29$

19. $\arcsin(-0.75) = \sin^{-1}(-0.75) \approx -0.85$

21. $\arctan(-3) = \tan^{-1}(-3) \approx -1.25$

23. $\arcsin 0.31 = \sin^{-1} 0.31 \approx 0.32$

25. $\arccos(-0.41) = \cos^{-1}(-0.41) \approx 1.99$

27. $\arctan 0.92 = \tan^{-1} 0.92 \approx 0.74$

29. $\arcsin\left(\dfrac{3}{4}\right) = \sin^{-1}(0.75) \approx 0.85$

31. $\arctan\left(\dfrac{7}{2}\right) = \tan^{-1}(3.5) \approx 1.29$

33. This is the graph of $y = \arctan x$. The coordinates are $\left(-\sqrt{3}, -\dfrac{\pi}{3}\right)$, $\left(-\dfrac{\sqrt{3}}{3}, -\dfrac{\pi}{6}\right)$, and $\left(1, \dfrac{\pi}{4}\right)$.

35. $f(x) = \tan x$ and $g(x) = \arctan x$

Graph $y_1 = \tan x$

Graph $y_2 = \tan^{-1} x$

Graph $y_3 = x$

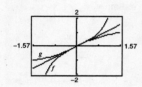

37. $\tan\theta = \dfrac{x}{4}$

$\theta = \arctan \dfrac{x}{4}$

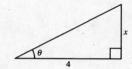

39. $\sin\theta = \dfrac{x+2}{5}$

$\theta = \arcsin\left(\dfrac{x+2}{5}\right)$

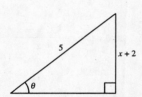

41. $\cos\theta = \dfrac{x+3}{2x}$

$\theta = \arccos\left(\dfrac{x+3}{2x}\right)$

43. $\sin(\arcsin 0.3) = 0.3$

45. $\cos[\arccos(-0.1)] = -0.1$

47. $\arcsin(\sin 3\pi) = \arcsin(0) = 0$

Note: 3π is not in the range of the arcsine function.

49. Let $y = \arctan\dfrac{3}{4}$. Then,

$\tan y = \dfrac{3}{4}$, $0 < y < \dfrac{\pi}{2}$

and $\sin y = \dfrac{3}{5}$.

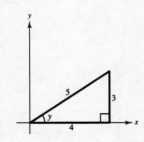

51. Let $y = \arctan 2$. Then,

$\tan y = 2 = \dfrac{2}{1}$, $0 < y < \dfrac{\pi}{2}$

and $\cos y = \dfrac{1}{\sqrt{5}} = \dfrac{\sqrt{5}}{5}$.

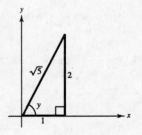

53. Let $y = \arcsin\dfrac{5}{13}$. Then,

$\sin y = \dfrac{5}{13}$, $0 < y < \dfrac{\pi}{2}$

and $\cos y = \dfrac{12}{13}$.

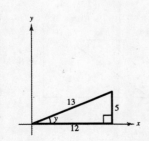

55. Let $y = \arctan\left(-\dfrac{3}{5}\right)$. Then,

$\tan y = -\dfrac{3}{5}$, $-\dfrac{\pi}{2} < y < 0$

and $\sec y = \dfrac{\sqrt{34}}{5}$.

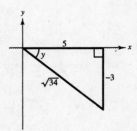

57. Let $y = \arccos\left(-\dfrac{2}{3}\right)$. Then,

$\cos y = -\dfrac{2}{3}, \dfrac{\pi}{2} < y < \pi$

and $\sin y = \dfrac{\sqrt{5}}{3}$.

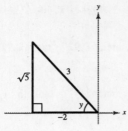

59. Let $y = \arctan x$. Then,

$\tan y = x = \dfrac{x}{1}$

and $\cot y = \dfrac{1}{x}$.

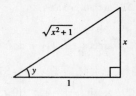

61. Let $y = \arcsin(2x)$. Then,

$\sin y = 2x = \dfrac{2x}{1}$

and $\cos y = \sqrt{1 - 4x^2}$.

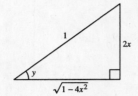

63. Let $y = \arccos x$. Then,

$\cos y = x = \dfrac{x}{1}$

and $\sin y = \sqrt{1 - x^2}$.

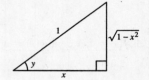

65. Let $y = \arccos\left(\dfrac{x}{3}\right)$. Then,

$\cos y = \dfrac{x}{3}$

and $\tan y = \dfrac{\sqrt{9 - x^2}}{x}$.

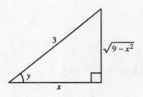

67. Let $y = \arctan \dfrac{x}{\sqrt{2}}$. Then,

$\tan y = \dfrac{x}{\sqrt{2}}$

and $\csc y = \dfrac{\sqrt{x^2 + 2}}{x}$.

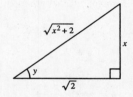

69. $f(x) = \sin(\arctan 2x)$, $g(x) = \dfrac{2x}{\sqrt{1 - 4x^2}}$

Let $y = \arctan 2x$. Then,

$\tan y = 2x = \dfrac{2x}{1}$

and $\sin y = \dfrac{2x}{\sqrt{1 + 4x^2}}$.

$g(x) = \dfrac{2x}{\sqrt{1 + 4x^2}} = f(x)$

The graph has horizontal asymptotes at $y = \pm 1$.

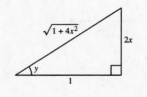

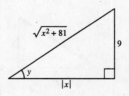

71. Let $y = \arctan \dfrac{9}{x}$. Then,

$\tan y = \dfrac{9}{x}$ and $\sin y = \dfrac{9}{\sqrt{x^2 + 81}}, x > 0; \dfrac{-9}{\sqrt{x^2 + 81}}, x < 0$

Thus, $\arcsin y = \dfrac{9}{\sqrt{x^2 + 81}}, x > 0$; $\arcsin y = \dfrac{-9}{\sqrt{x^2 + 81}}, x < 0$

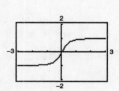

73. Let $y = \arccos \dfrac{3}{\sqrt{x^2 - 2x + 10}}$. Then,

$\cos y = \dfrac{3}{\sqrt{x^2 - 2x + 10}} = \dfrac{3}{\sqrt{(x - 1)^2 + 9}}$

and $\sin y = \dfrac{|x - 1|}{\sqrt{(x - 1)^2 + 9}}$.

Thus, $\arcsin y = \dfrac{|x - 1|}{\sqrt{(x - 1)^2 + 9}} = \arcsin \dfrac{|x - 1|}{\sqrt{x^2 - 2x + 10}}$.

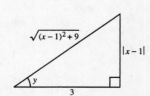

75. $y = 2 \arccos x$

Domain: $-1 \leq x \leq 1$

Range: $0 \leq y \leq 2\pi$

Vertical stretch of $f(x) = \arccos x$

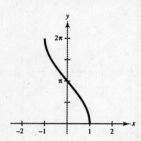

77. The graph of $f(x) = \arcsin(x - 1)$ is a horizontal translation of the graph of $y = \arcsin x$ by one unit.

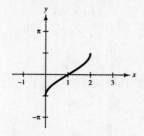

79. $f(x) = \arctan 2x$

Domain: all real numbers

Range: $-\dfrac{\pi}{2} < y < \dfrac{\pi}{2}$

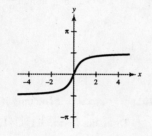

81. $h(v) = \tan(\arccos v) = \dfrac{\sqrt{1 - v^2}}{v}$

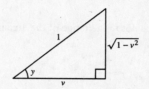

Domain: $-1 \leq v \leq 1, v \neq 0$

Range: all real numbers

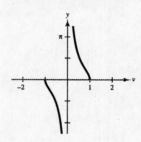

83. $f(x) = 2 \arccos (2x)$

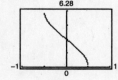

85. $f(x) = \arctan (2x - 3)$

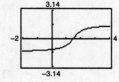

87. $f(x) = \pi - \arcsin\left(\dfrac{2}{3}\right) \approx 2.412$

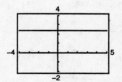

89. $f(t) = 3 \cos 2t + 3 \sin 2t = \sqrt{3^2 + 3^2} \sin\left(3t + \arctan \dfrac{3}{3}\right)$

$$= 3\sqrt{2} \sin(3t + \arctan 1)$$

$$= 3\sqrt{2} \sin\left(3t + \dfrac{\pi}{4}\right)$$

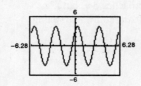

The graphs are the same.

91. (a) $\sin\theta = \dfrac{5}{s}$

$\theta = \arcsin\dfrac{5}{s}$

(b) $s = 40$: $\theta = \arcsin\dfrac{5}{40} \approx 0.13$

$s = 20$: $\theta = \arcsin\dfrac{5}{20} \approx 0.25$

93. $\beta = \arctan\dfrac{3x}{x^2 + 4}$

(a)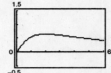

(b) β is maximum when $x = 2$.

(c) The graph has a horizontal asymptote at $\beta = 0$. As x increases, β decreases.

95. (a) $\tan\theta = \dfrac{x}{20}$

$\theta = \arctan\dfrac{x}{20}$

(b) $x = 5$: $\theta = \arctan\dfrac{5}{20} \approx 0.24$

$x = 12$: $\theta = \arctan\dfrac{12}{20} \approx 0.54$

97. False; $\arctan 1 = \dfrac{\pi}{4}$. $\dfrac{5\pi}{4}$ is not in the range of the arctangent function.

99. $y = \text{arccot}\, x$ if and only if $\cot y = x$.

Domain: $-\infty < x < \infty$

Range: $0 < x < \pi$

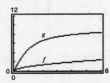

101. $y = \text{arccsc}\, x$ if and only if $\csc y = x$.

Domain: $(-\infty, -1] \cup [1, \infty)$

Range: $\left[-\dfrac{\pi}{2}, 0\right) \cup \left(0, \dfrac{\pi}{2}\right]$

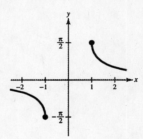

103. $f(x) = \sqrt{x}$

$g(x) = 6\arctan x$

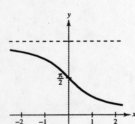

As x increases to infinity, g approaches 3π, but f has no maximum. Using the solve feature of the graphing utility, you find $a \approx 87.54$.

105. Let $y = \arcsin(-x)$. Then,

$\sin y = -x$

$-\sin y = x$

$\sin(-y) = x$

$-y = \arcsin x$

$y = -\arcsin x.$

Therefore, $\arcsin(-x) = -\arcsin x.$

107. $y = \pi - \arccos x$

$\cos y = \cos(\pi - \arccos x)$

$\cos y = \cos \pi \cos(\arccos x) + \sin \pi \sin(\arccos x)$

$\cos y = -x$

$y = \arccos(-x)$

Therefore, $\arccos(-x) = \pi - \arccos x$.

109.   Let $\alpha = \arcsin x$ and $\beta = \arccos x$, then

$\sin \alpha = x$ and $\cos \beta = x$. Thus, $\sin \alpha = \cos \beta$ which

implies that α and β are complementary angles and

we have

$$\alpha + \beta = \frac{\pi}{2}$$

$$\arcsin x + \arccos x = \frac{\pi}{2}.$$

111.   adj = 1, hyp = 2, opp = $\sqrt{4-1} = \sqrt{3}$

$\sin \theta = \dfrac{\text{opp}}{\text{hyp}} = \dfrac{\sqrt{3}}{2}$

$\cos \theta = \dfrac{\text{adj}}{\text{hyp}} = \dfrac{1}{2}$

$\tan \theta = \dfrac{\text{opp}}{\text{adj}} = \dfrac{\sqrt{3}}{1} = \sqrt{3}$

$\csc \theta = \dfrac{\text{hyp}}{\text{opp}} = \dfrac{2}{\sqrt{3}} = \dfrac{2\sqrt{3}}{3}$

$\sec \theta = \dfrac{\text{hyp}}{\text{adj}} = \dfrac{2}{1} = 2$

$\cot \theta = \dfrac{\text{adj}}{\text{opp}} = \dfrac{1}{\sqrt{3}} = \dfrac{\sqrt{3}}{3}$

113.   opp = 8, adj = 8, hyp = $\sqrt{64+64} = \sqrt{128} = 8\sqrt{2}$

$\sin \theta = \dfrac{\text{opp}}{\text{hyp}} = \dfrac{8}{8\sqrt{2}} = \dfrac{1}{\sqrt{2}} = \dfrac{\sqrt{2}}{2}$

$\cos \theta = \dfrac{\text{adj}}{\text{hyp}} = \dfrac{8}{8\sqrt{2}} = \dfrac{\sqrt{2}}{2}$

$\tan \theta = \dfrac{\text{opp}}{\text{adj}} = \dfrac{8}{8} = 1$

$\csc \theta = \dfrac{\text{hyp}}{\text{opp}} = \dfrac{8\sqrt{2}}{8} = \sqrt{2}$

$\sec \theta = \dfrac{\text{hyp}}{\text{adj}} = \dfrac{8\sqrt{2}}{8} = \sqrt{2}$

$\cot \theta = \dfrac{\text{adj}}{\text{opp}} = \dfrac{8}{8} = 1$

115. $\sin \theta = \dfrac{3}{4} = \dfrac{\text{opp}}{\text{hyp}}$

$(\text{adj})^2 + (3)^2 = (4)^2$

$(\text{adj})^2 + 9 = 16$

$(\text{adj})^2 = 7$

$\text{adj} = \sqrt{7}$

117. $\cos \theta = \dfrac{5}{6} = \dfrac{\text{adj}}{\text{hyp}}$

$(\text{opp})^2 + (5)^2 = (6)^2$

$(\text{opp})^2 + 25 = 36$

$(\text{opp})^2 = 11$

$\text{opp} = \sqrt{11}$

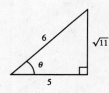

119.   Let x = the number of people presently in the group.

Each person's share is now $\dfrac{250,000}{x}$.   If two more join

the group, each person's share would then be $\dfrac{250,000}{x+2}$.

$\begin{matrix}\text{Share per person with} \\ \text{two more people}\end{matrix} = \begin{matrix}\text{Original share} \\ \text{per person}\end{matrix} - 6250$

$\dfrac{250,000}{x+2} = \dfrac{250,000}{x} - 6250$

$250,000x = 250,000(x+2) - 6250x(x+2)$

$250,000x = 250,000x + 500,000 - 6250x^2 - 12,500x$

$6250x^2 + 12,500x - 500,000 = 0$

$6250(x^2 + 2x - 80) = 0$

$6250(x + 10)(x - 8) = 0$

$x = -10$ or $x = 8$

$x = -10$ is not possible.

There were 8 people in the original group.

Section 1.8   Applications and Models

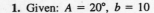

■ You should be able to solve right triangles.

■ You should be able to solve right triangle applications.

■ You should be able to solve applications of simple harmonic motion.

---

1. Given: $A = 20°$, $b = 10$

$$\tan A = \frac{a}{b} \implies a = b \tan A = 10 \tan 20° \approx 3.64$$

$$\cos A = \frac{b}{c} \implies c = \frac{b}{\cos A} = \frac{10}{\cos 20°} \approx 10.64$$

$$B = 90° - 20° = 70°$$

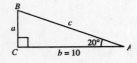

3. Given: $B = 71°$, $b = 24$

$$\tan B = \frac{b}{a} \implies a = \frac{b}{\tan B} = \frac{24}{\tan 71°} \approx 8.26$$

$$\sin B = \frac{b}{c} \implies c = \frac{b}{\sin B} = \frac{24}{\sin 71°} \approx 25.38$$

$$A = 90° - 71° = 19°$$

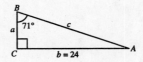

5. Given: $a = 6$, $b = 10$

$$c^2 = a^2 + b^2 \implies c = \sqrt{36 + 100}$$
$$= 2\sqrt{34} \approx 11.66$$

$$\tan A = \frac{a}{b} = \frac{6}{10} \implies A = \arctan \frac{3}{5} \approx 30.96°$$

$$B = 90° - 30.96° = 59.04°$$

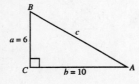

7. $b = 16$, $c = 52$

$$a = \sqrt{52^2 - 16^2}$$
$$= \sqrt{2448} = 12\sqrt{17} \approx 49.48$$

$$\cos A = \frac{16}{52}$$

$$A = \arccos \frac{16}{52} \approx 72.08°$$

$$B = 90° - 72.08° \approx 17.92°$$

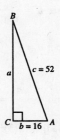

9. $A = 12°15'$, $c = 430.5$

$$B = 90° - 12°15' = 77°45'$$

$$\sin 12°15' = \frac{a}{430.5}$$

$$a = 430.5 \sin 12°15' \approx 91.34$$

$$\cos 12°15' = \frac{b}{430.5}$$

$$b = 430.5 \cos 12°15' \approx 420.70$$

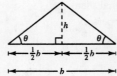

11. $\tan \theta = \dfrac{h}{\frac{1}{2}b} \implies h = \dfrac{1}{2}b \tan \theta$

$$h = \frac{1}{2}(4) \tan 52° \approx 2.56 \text{ inches}$$

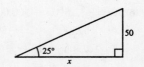

13. $\tan \theta = \dfrac{h}{\frac{1}{2}b} \implies h = \dfrac{1}{2}b \tan \theta$

$$h = \frac{1}{2}(46) \tan 41° \approx 19.99 \text{ inches}$$

15. $\tan 25° = \dfrac{50}{x}$

$$x = \frac{50}{\tan 25°} \approx 107.2 \text{ feet}$$

17. $\sin 80° = \dfrac{h}{20}$

$20 \sin 80° = h$

$h \approx 19.7$ feet

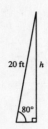

19. (a)

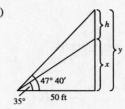

(b) Let the height of the church $= x$ and the height of the church and steeple $= y$. Then,

$\tan 35° = \dfrac{x}{50}$ and $\tan 47°40' = \dfrac{y}{50}$

$x = 50 \tan 35°$ and $y = 50 \tan 47°40'$

$h = y - x = 50 \left(\tan 47°40' - \tan 35°\right).$

(c) $h \approx 19.9$ feet

21. $\sin 34° = \dfrac{x}{4000}$

$x = 4000 \sin 34°$

≈ 2236.8 feet

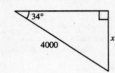

23. $\tan \theta = \dfrac{75}{50}$

$\theta = \arctan \dfrac{3}{2} \approx 56.3°$

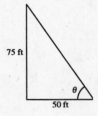

25. $10{,}900$ feet $= \dfrac{10{,}900}{5280}$ miles ≈ 2.0644 miles

$\sin \theta = \dfrac{4000}{4002.0644}$

$\theta = \arcsin \left(\dfrac{4000}{4002.0644}\right)$

$\theta \approx 88.16° \alpha$

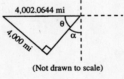

27. Since the airplane speed is

$\left(275 \dfrac{\text{ft}}{\text{sec}}\right)\left(60 \dfrac{\text{sec}}{\text{min}}\right) = 16{,}500 \dfrac{\text{ft}}{\text{min}},$

after one minute its distance travelled in $16{,}500$ feet.

$\sin 18° = \dfrac{a}{16{,}500}$

$a = 16{,}500 \sin 18°$

≈ 5099 ft

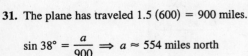

29. $\sin 10.5° = \dfrac{x}{4}$

$x = 4 \sin 10.5°$

≈ 0.73 mile

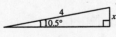

31. The plane has traveled $1.5 (600) = 900$ miles.

$\sin 38° = \dfrac{a}{900} \implies a \approx 554$ miles north

$\cos 38° = \dfrac{b}{900} \implies b \approx 709$ miles east

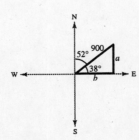

33. $\theta = 32°$, $\phi = 68°$

(a) $\alpha = 90° - 32° = 58°$

(b) Bearing from A to C: N 58° E

$$\beta = \theta = 32°$$

$$\gamma = 90° - \phi = 22°$$

$$C = \beta + \gamma = 54°$$

$$\tan C = \frac{d}{50} \implies \tan 54° = \frac{d}{50} \implies d \approx 68.82 \text{ meters}$$

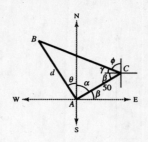

35. $\tan \theta = \frac{45}{30} \implies \theta \approx 56.3°$

Bearing: N 56.3° W

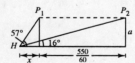

37. $\tan 6.5° = \frac{350}{d} \implies d \approx 3071.91 \text{ ft}$

$$\tan 4° = \frac{350}{D} \implies D \approx 5005.23 \text{ ft}$$

Distance between ships: $D - d \approx 1933.32 \text{ ft}$

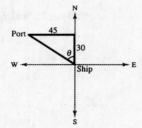

39. $\tan 57° = \frac{a}{x} \implies x = a \cot 57°$

$$\tan 16° = \frac{a}{x + \frac{55}{6}}$$

$$\tan 16° = \frac{a}{a \cot 57° + \frac{55}{6}}$$

$$\cot 16° = \frac{a \cot 57° + \frac{55}{6}}{a}$$

$$a \cot 16° - a \cot 57° = \frac{55}{6} \implies a \approx 3.23 \text{ miles}$$

$$\approx 17,054 \text{ ft}$$

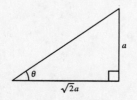

41. L_1: $3x - 2y = 5 \implies y = \frac{3}{2}x - \frac{5}{2} \implies m_1 = \frac{3}{2}$

L_2: $x + y = 1 \implies y = -x + 1 \implies m_2 = -1$

$$\tan \alpha = \left| \frac{-1 - \frac{3}{2}}{1 + (-1)\left(\frac{3}{2}\right)} \right| = \left| \frac{-\frac{5}{2}}{-\frac{1}{2}} \right| = 5$$

$$\alpha = \arctan 5 \approx 78.7°$$

43. The diagonal of the base has a length of $\sqrt{a^2 + a^2} = \sqrt{2}a$.

Now, we have $\tan \theta = \frac{a}{\sqrt{2}a} = \frac{1}{\sqrt{2}}$

$$\theta = \arctan \frac{1}{\sqrt{2}}$$

$$\theta \approx 35.3°.$$

45. $\sin 36° = \frac{d}{25} \implies d \approx 14.69$

Length of side: $2d \approx 29.4$ inches

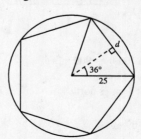

47. $\cos 30° = \dfrac{b}{r}$

$b = r \cos 30°$

$b = \dfrac{\sqrt{3}\,r}{2}$

$y = 2b = 2\left(\dfrac{\sqrt{3}\,r}{2}\right) = \sqrt{3}\,r$

49. $\tan 35° = \dfrac{b}{10}$

$b = 10 \tan 35° \approx 7$

$\cos 35° = \dfrac{10}{a}$

$a = \dfrac{10}{\cos 35°} \approx 12.2$

51. $d = 4 \cos 8\pi t$

(a) Maximum displacement = amplitude = 4

(b) Frequency $= \dfrac{\omega}{2\pi} = \dfrac{8\pi}{2\pi}$

$= 4$ cycles per unit of time

(c) $8\pi t = \dfrac{\pi}{2} \implies t = \dfrac{1}{16}$

53. $d = \dfrac{1}{16} \sin 120\pi t$

(a) Maximum displacement = amplitude $= \dfrac{1}{16}$

(b) Frequency $= \dfrac{\omega}{2\pi} = \dfrac{120\pi}{2\pi}$

$= 60$ cycles per unit of time

(c) $120\pi t = \pi \implies t = \dfrac{1}{120}$

55. $d = 0$ when $t = 0$, $a = 4$, Period $= 2$

Use $d = a \sin \omega t$ since $d = 0$ when $t = 0$.

$\dfrac{2\pi}{\omega} = 2 \implies \omega = \pi$

Thus, $d = 4 \sin \pi t$.

57. $d = 3$ when $t = 0$, $a = 3$, Period $= 1.5$

Use $d = a \cos \omega t$ since $d = 3$ when $t = 0$.

$\dfrac{2\pi}{\omega} = 1.5 \implies \omega = \dfrac{4\pi}{3}$

Thus, $d = 3 \cos\left(\dfrac{4\pi}{3}t\right) = 3 \cos\left(\dfrac{4\pi t}{3}\right)$.

59. $d = a \sin \omega t$

Period $= \dfrac{2\pi}{\omega} = \dfrac{1}{\text{frequency}}$

$\dfrac{2\pi}{\omega} = \dfrac{1}{264}$

$\omega = 2\pi(264) = 528\pi$

61. $y = \dfrac{1}{4} \cos 16t, \ t > 0$

(a)

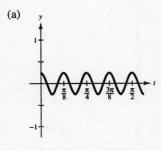

(b) Period: $\dfrac{2\pi}{16} = \dfrac{\pi}{8}$

(c) $\dfrac{1}{4} \cos 16t = 0$ when $16t = \dfrac{\pi}{2} \implies t = \dfrac{\pi}{32}$

63. False. One period is the time for one complete cycle of the motion.

65. (a) & (b)

Base 1	Base 2	Altitude	Area
8	$8 + 16 \cos 10°$	$8 \sin 10°$	22.1
8	$8 + 16 \cos 20°$	$8 \sin 20°$	42.5
8	$8 + 16 \cos 30°$	$8 \sin 30°$	59.7
8	$8 + 16 \cos 40°$	$8 \sin 40°$	72.7
8	$8 + 16 \cos 50°$	$8 \sin 50°$	80.5
8	$8 + 16 \cos 60°$	$8 \sin 60°$	83.1
8	$8 + 16 \cos 70°$	$8 \sin 70°$	80.7

The maximum occurs when $\theta = 60°$ and is approximately 83.1 square feet.

(c) $A(\theta) = \left[8 + (8 + 16 \cos \theta) \right] \left[\dfrac{8 \sin \theta}{2} \right]$

$\qquad = (16 + 16 \cos \theta)(4 \sin \theta)$

$\qquad = 64 (1 + \cos \theta)(\sin \theta)$

(d)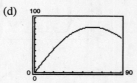

The maximum of 83.1 square feet occurs when

$\theta = \dfrac{\pi}{3} = 60°.$

67. (a)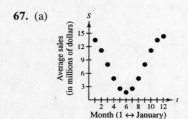

(b) $a = \dfrac{1}{2}(14.3 - 1.7) = 6.3$

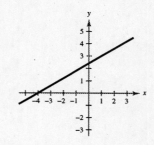

$\dfrac{2\pi}{b} = 12 \implies b = \dfrac{\pi}{6}$

Shift: $d = 14.3 - 6.3 = 8$

$S = d + a \cos bt$

$S = 8 + 6.3 \cos \left(\dfrac{\pi t}{6} \right)$

Note: Another model is $S = 8 + 6.3 \sin \left(\dfrac{\pi t}{6} + \dfrac{\pi}{2} \right)$

The model is a good fit.

(c) Period: $\dfrac{2\pi}{\pi/6} = 12$

This corresponds to the 12 months in a year. Since the sales of outerwear is seasonal this is reasonable.

(d) The amplitude represents the maximum displacement from average sales of 8 million dollars. Sales are greatest in December (cold weather + Christmas) and least in June.

69. $5y - 3x = 12$

$\quad 5y = 3x + 12$

$\quad y = \dfrac{3}{5}x + \dfrac{12}{5}$

The graph is a line with $m = \dfrac{3}{5}$ and y-intercept $\left(0, \dfrac{12}{5} \right)$

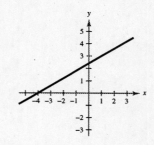

71. $(x + 3)^2 = 5y - 8$

$(x + 3)^2 = 5\left(y - \dfrac{8}{5} \right)$

x	0	-6
y	$\dfrac{17}{5}$	$\dfrac{17}{5}$

Parabola with vertex $\left(-3, \dfrac{8}{5} \right)$

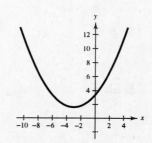

73. $2x^2 + y^2 - 4 = 0$

$2x^2 + y^2 = 4$

$\dfrac{x^2}{2} + \dfrac{y^2}{4} = 1$

Ellipse with center $(0, 0)$

Vertical major axis

$a = 2, b = \sqrt{2}$

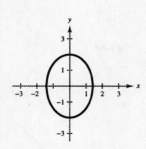

75. $\dfrac{y^2}{4} - \dfrac{(x+2)^2}{25} - 1 = 0$

$\dfrac{y^2}{4} - \dfrac{(x+2)^2}{25} = 1$

Hyperbola with center $(-2, 0)$

Vertical transverse axis

$a = 2, \ b = 5$

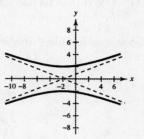

77. $(x - 2)^2 + y^2 = 25$

Circle with center $(2, 0)$ and radius $r = 5$

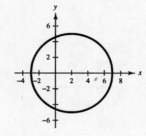

Review Exercises for Chapter 1

Solutions to Odd-Numbered Exercises

1. $\theta \approx 1$ radian

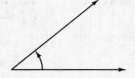

3. $\theta \approx 5$ radians

5. $\theta = \dfrac{11\pi}{4}$

Coterminal angles: $\dfrac{11\pi}{4} - 2\pi = \dfrac{3\pi}{4}$

$\dfrac{3\pi}{4} - 2\pi = -\dfrac{5\pi}{4}$

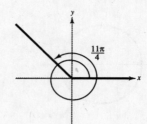

7. $\theta = -\dfrac{4\pi}{3}$

Coterminal angles: $-\dfrac{4\pi}{3} + 2\pi = \dfrac{2\pi}{3}$

$-\dfrac{4\pi}{3} - 2\pi = -\dfrac{10\pi}{3}$

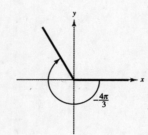

9. $\theta = 70°$

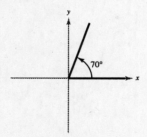

Coterminal angles: $70° + 360° = 430°$

$70° - 360° = -290°$

11. $\theta = -110°$

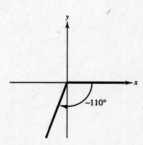

Coterminal angles: $-110° + 360° = 250°$

$-110° - 360° = -470°$

13. $\dfrac{5\pi \text{ rad}}{7} = \dfrac{5\pi \text{ rad}}{7} \cdot \dfrac{180°}{\pi \text{ rad}} \approx 128.57°$

15. $-3.5 \text{ rad} = -3.5 \text{ rad} \cdot \dfrac{180°}{\pi \text{ rad}} \approx -200.54°$

17. $480° = 480° \cdot \dfrac{\pi \text{ rad}}{180°} = \dfrac{8\pi}{3} \text{ radians} \approx 8.3776 \text{ radians}$

19. $-33°45' = -33.75° = -33.75° \cdot \dfrac{\pi \text{ rad}}{180°} = -\dfrac{3\pi}{16}$ radian ≈ -0.5890 radian

21. (a) Angular speed $= \dfrac{\left(33\frac{1}{3}\right)(2\pi) \text{ radians}}{1 \text{ minute}}$

$= 66\frac{2}{3}\pi$ radians per minute

(b) Linear speed $= \dfrac{6\left(66\frac{2}{3}\pi\right) \text{ inches}}{1 \text{ minute}}$

$= 400\pi$ inches per minute

23. $t = \dfrac{2\pi}{3}$ corresponds to the point $\left(-\dfrac{1}{2}, \dfrac{\sqrt{3}}{2}\right)$.

25. $t = \dfrac{5\pi}{6}$ corresponds to the point $\left(-\dfrac{\sqrt{3}}{2}, \dfrac{1}{2}\right)$.

27. $t = \dfrac{7\pi}{6}$ corresponds to the point $\left(-\dfrac{\sqrt{3}}{2}, -\dfrac{1}{2}\right)$.

$\sin\dfrac{7\pi}{6} = y = -\dfrac{1}{2}$ \qquad $\csc\dfrac{7\pi}{6} = \dfrac{1}{y} = -2$

$\cos\dfrac{7\pi}{6} = x = -\dfrac{\sqrt{3}}{2}$ \qquad $\sec\dfrac{7\pi}{6} = \dfrac{1}{x} = -\dfrac{2\sqrt{3}}{3}$

$\tan\dfrac{7\pi}{6} = \dfrac{y}{x} = \dfrac{1}{\sqrt{3}} = \dfrac{\sqrt{3}}{3}$ \qquad $\cot\dfrac{7\pi}{6} = \dfrac{x}{y} = \sqrt{3}$

29. $t = -\dfrac{2\pi}{3}$ corresponds to the point $\left(-\dfrac{1}{2}, -\dfrac{\sqrt{3}}{2}\right)$.

$\sin\left(-\dfrac{2\pi}{3}\right) = y = -\dfrac{\sqrt{3}}{2}$ \qquad $\csc\left(-\dfrac{2\pi}{3}\right) = \dfrac{1}{y} = -\dfrac{2\sqrt{3}}{3}$

$\cos\left(-\dfrac{2\pi}{3}\right) = x = -\dfrac{1}{2}$ \qquad $\sec\left(-\dfrac{2\pi}{3}\right) = \dfrac{1}{x} = -2$

$\tan\left(-\dfrac{2\pi}{3}\right) = \dfrac{y}{x} = \sqrt{3}$ \qquad $\cot\left(-\dfrac{2\pi}{3}\right) = \dfrac{x}{y} = \dfrac{\sqrt{3}}{3}$

31. $\sin\dfrac{11\pi}{4} = \sin\dfrac{3\pi}{4} = \dfrac{\sqrt{2}}{2}$

33. $\sin\left(-\dfrac{17\pi}{6}\right) = \sin\left(-\dfrac{5\pi}{6}\right) = -\dfrac{1}{2}$ \qquad **35.** $\tan 33 \approx -75.31$ \qquad **37.** $\sec\dfrac{12\pi}{5} = \dfrac{1}{\cos\left(\dfrac{12\pi}{5}\right)} \approx 3.24$

39. opp $= 4$, adj $= 5$, hyp $= \sqrt{4^2 + 5^2} = \sqrt{41}$

$\sin\theta = \dfrac{\text{opp}}{\text{hyp}} = \dfrac{4}{\sqrt{41}} = \dfrac{4\sqrt{41}}{41}$ \qquad $\csc\theta = \dfrac{\text{hyp}}{\text{opp}} = \dfrac{\sqrt{41}}{4}$

$\cos\theta = \dfrac{\text{adj}}{\text{hyp}} = \dfrac{5}{\sqrt{41}} = \dfrac{5\sqrt{41}}{41}$ \qquad $\sec\theta = \dfrac{\text{hyp}}{\text{adj}} = \dfrac{\sqrt{41}}{5}$

$\tan\theta = \dfrac{\text{opp}}{\text{adj}} = \dfrac{4}{5}$ \qquad $\cot\theta = \dfrac{\text{adj}}{\text{opp}} = \dfrac{5}{4}$

41. opp = 4, hyp = 8, adj = $\sqrt{8^2 - 4^2} = \sqrt{48} = 4\sqrt{3}$

$\sin \theta = \dfrac{\text{opp}}{\text{hyp}} = \dfrac{4}{8} = \dfrac{1}{2}$ $\qquad \csc \theta = \dfrac{\text{hyp}}{\text{opp}} = \dfrac{8}{4} = 2$

$\cos \theta = \dfrac{\text{adj}}{\text{hyp}} = \dfrac{4\sqrt{3}}{8} = \dfrac{\sqrt{3}}{2}$ $\qquad \sec \theta = \dfrac{\text{hyp}}{\text{adj}} = \dfrac{8}{4\sqrt{3}} = \dfrac{2\sqrt{3}}{3}$

$\tan \theta = \dfrac{\text{opp}}{\text{adj}} = \dfrac{4}{4\sqrt{3}} = \dfrac{\sqrt{3}}{3}$ $\qquad \cot \theta = \dfrac{\text{adj}}{\text{opp}} = \dfrac{4\sqrt{3}}{4} = \sqrt{3}$

43. $\sin \theta = \dfrac{1}{3}$

(a) $\csc \theta = \dfrac{1}{\sin \theta} = 3$

(b) $\sin^2 \theta + \cos^2 \theta = 1$

$\left(\dfrac{1}{3}\right)^2 + \cos^2\theta = 1$

$\cos^2 \theta = 1 - \dfrac{1}{9}$

$\cos^2 \theta = \dfrac{8}{9}$

$\cos \theta = \sqrt{\dfrac{8}{9}}$

$\cos \theta = \dfrac{2\sqrt{2}}{3}$

(c) $\sec \theta = \dfrac{1}{\cos \theta} = \dfrac{3}{2\sqrt{2}} = \dfrac{3\sqrt{2}}{4}$

(d) $\tan \theta = \dfrac{\sin \theta}{\cos \theta} = \dfrac{\frac{1}{3}}{\frac{2\sqrt{2}}{3}} = \dfrac{1}{2\sqrt{2}} = \dfrac{\sqrt{2}}{4}$

45. $\csc \theta = 4$

(a) $\sin \theta = \dfrac{1}{\csc \theta} = \dfrac{1}{4}$

(b) $\sin^2 \theta + \cos^2 \theta = 1$

$\left(\dfrac{1}{4}\right)^2 + \cos^2 \theta = 1$

$\cos^2 \theta = 1 - \dfrac{1}{16}$

$\cos^2 \theta = \dfrac{15}{16}$

$\cos \theta = \sqrt{\dfrac{15}{16}}$

$\cos \theta = \dfrac{\sqrt{15}}{4}$

(c) $\sec \theta = \dfrac{1}{\cos \theta} = \dfrac{4}{\sqrt{15}} = \dfrac{4\sqrt{15}}{15}$

(d) $\tan \theta = \dfrac{\sin \theta}{\cos \theta} = \dfrac{\frac{1}{4}}{\frac{\sqrt{15}}{4}} = \dfrac{1}{\sqrt{15}} = \dfrac{\sqrt{15}}{15}$

47. $\tan 33° \approx 0.65$

49. $\sin 34.2° \approx 0.56$

51. $\cot 15°14' \approx \cot 15.2333° = \dfrac{1}{\tan 15.2333°} \approx 3.67$

53. $\sin 1°10' = \dfrac{x}{3.5}$

$x = 3.5 \sin 1°10' \approx 0.07$ Kilometer

(Not drawn to scale)

55. $x = 12, y = 16, r = \sqrt{144 + 256} = \sqrt{400} = 20$

$\sin \theta = \dfrac{y}{r} = \dfrac{4}{5}$ $\qquad \csc \theta = \dfrac{r}{y} = \dfrac{5}{4}$

$\cos \theta = \dfrac{x}{r} = \dfrac{3}{5}$ $\qquad \sec \theta = \dfrac{r}{x} = \dfrac{5}{3}$

$\tan \theta = \dfrac{y}{x} = \dfrac{4}{3}$ $\qquad \cot \theta = \dfrac{x}{y} = \dfrac{3}{4}$

57. $x = \dfrac{2}{3}, y = \dfrac{5}{2}$

$$r = \sqrt{\left(\dfrac{2}{3}\right)^2 + \left(\dfrac{5}{2}\right)^2} = \dfrac{\sqrt{241}}{6}$$

$\sin \theta = \dfrac{y}{r} = \dfrac{\frac{5}{2}}{\frac{\sqrt{241}}{6}} = \dfrac{15}{\sqrt{241}} = \dfrac{15\sqrt{241}}{241}$ $\csc \theta = \dfrac{r}{y} = \dfrac{\frac{\sqrt{241}}{6}}{\frac{5}{2}} = \dfrac{2\sqrt{241}}{30} = \dfrac{\sqrt{241}}{15}$

$\cos \theta = \dfrac{x}{r} = \dfrac{\frac{2}{3}}{\frac{\sqrt{241}}{6}} = \dfrac{4}{\sqrt{241}} = \dfrac{4\sqrt{241}}{241}$ $\sec \theta = \dfrac{r}{x} = \dfrac{\frac{\sqrt{241}}{6}}{\frac{2}{3}} = \dfrac{\sqrt{241}}{4}$

$\tan \theta = \dfrac{y}{x} = \dfrac{\frac{5}{2}}{\frac{2}{3}} = \dfrac{15}{4}$ $\cot \theta = \dfrac{x}{y} = \dfrac{\frac{2}{3}}{\frac{5}{2}} = \dfrac{4}{15}$

59. $x = -0.5, y = 4.5$

$$r = \sqrt{(-0.5)^2 + 4.5^2} \approx 4.528$$

$\sin \theta = \dfrac{y}{r} \approx 1$ $\csc \theta = \dfrac{r}{y} \approx 1$

$\cos \theta = \dfrac{x}{r} \approx -0.1$ $\sec \theta = \dfrac{r}{x} \approx -9$

$\tan \theta = \dfrac{y}{x} \approx -9$ $\cot \theta = \dfrac{x}{y} \approx -0.1$

61. $(x, 4x), x > 0$

$x = x, y = 4x$

$$r = \sqrt{x^2 + (4x)^2} = \sqrt{17}\,x$$

$\sin \theta = \dfrac{y}{r} = \dfrac{4x}{\sqrt{17}x} = \dfrac{4\sqrt{17}}{17}$ $\csc \theta = \dfrac{r}{y} = \dfrac{\sqrt{17}x}{4x} = \dfrac{\sqrt{17}}{4}$

$\cos \theta = \dfrac{x}{r} = \dfrac{x}{\sqrt{17}x} = \dfrac{\sqrt{17}}{17}$ $\sec \theta = \dfrac{r}{x} = \dfrac{\sqrt{17}x}{x} = \sqrt{17}$

$\tan \theta = \dfrac{y}{x} = \dfrac{4x}{x} = 4$ $\cot \theta = \dfrac{x}{y} = \dfrac{x}{4x} = \dfrac{1}{4}$

63. $\sec \theta = \dfrac{6}{5}, \tan \theta < 0 \implies \theta$ is in Quadrant IV.

$$r = 6, x = 5, y = -\sqrt{36 - 25} = -\sqrt{11}$$

$\sin \theta = \dfrac{y}{r} = -\dfrac{\sqrt{11}}{6}$ $\csc \theta = \dfrac{r}{y} = -\dfrac{6\sqrt{11}}{11}$

$\cos \theta = \dfrac{x}{r} = \dfrac{5}{6}$ $\sec \theta = \dfrac{r}{x} = \dfrac{6}{5}$

$\tan \theta = \dfrac{y}{x} = -\dfrac{\sqrt{11}}{5}$ $\cot \theta = \dfrac{x}{y} = -\dfrac{5\sqrt{11}}{11}$

65. $\sin \theta = \dfrac{3}{8}$, $\cos \theta < 0 \implies \theta$ is in Quadrant II.

$y = 3, r = 8, x = -\sqrt{55}$

$$\sin \theta = \frac{y}{r} = \frac{3}{8} \qquad\qquad \csc \theta = \frac{8}{3}$$

$$\cos \theta = \frac{x}{r} = -\frac{\sqrt{55}}{8} \qquad\qquad \sec \theta = -\frac{8}{\sqrt{55}} = -\frac{8\sqrt{55}}{55}$$

$$\tan \theta = \frac{y}{x} = -\frac{3}{\sqrt{55}} = -\frac{3\sqrt{55}}{55} \qquad \cot \theta = -\frac{\sqrt{55}}{3}$$

67. $\cos \theta = \dfrac{x}{r} = \dfrac{-2}{5} \implies y^2 = 21$

$\sin \theta > 0 \implies \theta$ is in Quadrant II $\implies y = \sqrt{21}$

$$\sin \theta = \frac{y}{r} = \frac{\sqrt{21}}{5}$$

$$\tan \theta = \frac{y}{x} = -\frac{\sqrt{21}}{2}$$

$$\csc \theta = \frac{r}{y} = \frac{5}{\sqrt{21}} = \frac{5\sqrt{21}}{21}$$

$$\sec \theta = \frac{r}{x} = \frac{5}{-2} = -\frac{5}{2}$$

$$\cot \theta = \frac{x}{y} = \frac{-2}{\sqrt{21}} = -\frac{2\sqrt{21}}{21}$$

69. $\tan \dfrac{\pi}{3} = \sqrt{3}$

71. $\cos\left(-\dfrac{7\pi}{3}\right) = \cos \dfrac{\pi}{3} = \dfrac{1}{2}$

73. $\cos 495° = -\cos 45° = -\dfrac{\sqrt{2}}{2}$

75. $\sin 4 \approx -0.76$

77. $\sin(-3.2) \approx 0.06$

79. $\sin 3\pi = 0$

81. $\sec \dfrac{12\pi}{5} = \dfrac{1}{\cos\left(\dfrac{12\pi}{5}\right)} \approx 3.24$

83. $y = \sin x$

Amplitude: 1

Period: 2π

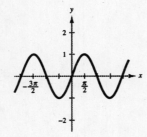

85. $f(x) = 5 \sin \dfrac{2x}{5}$

Amplitude: 5

Period: $\dfrac{2\pi}{\frac{2}{5}} = 5\pi$

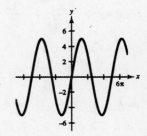

87. $y = 2 + \sin x$

Shift the graph of $y = \sin x$ two units upward.

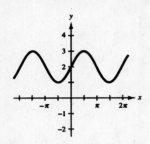

89. $g(t) = \dfrac{5}{2} \sin(t - \pi)$

Amplitude: $\dfrac{5}{2}$

Period: 2π

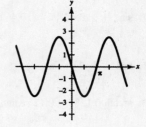

91. $y = a \sin bx$

(a) $a = 2, \dfrac{2\pi}{b} = \dfrac{1}{264} \Longrightarrow b = 528\pi$

$y = 2 \sin(528\pi x)$

(b) $f = \dfrac{1}{\frac{1}{264}} = 264$ cycles per second.

93. $f(x) = \tan x$

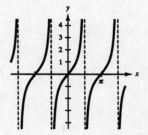

95. $f(x) = \cot x$

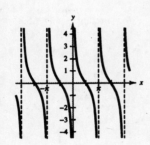

97. $f(x) = \sec x$

Graph $y = \cos x$ first.

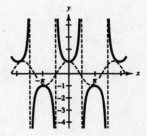

99. $f(x) = \csc x$

Graph $y = \sin x$ first.

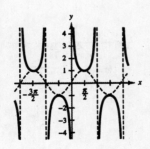

101. $f(x) = x^4 \cos x$

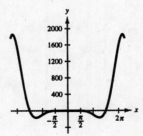

103. $\arcsin\left(-\dfrac{1}{2}\right) = -\arcsin \dfrac{1}{2} = -\dfrac{\pi}{6}$

105. $\arcsin 0.4 \approx 0.41$ radian

107. $\sin^{-1}(-0.44) \approx -0.46$ radian

109. $\arccos \dfrac{\sqrt{3}}{2} = \dfrac{\pi}{6}$

111. $\cos^{-1}(-1) = \pi$

113. $\arccos 0.324 \approx 1.24$ radians

115. $\arctan 0.123 \approx 0.12$ radian

117. $\arctan 5.783 \approx 1.40$ radians

119. $\tan^{-1}(-1.5) \approx -0.98$ radian

121. $\sin(\arcsin 0.72) = 0.72$

123. $\arctan(\tan \pi) = \arctan 0 = 0$

125. $\cos\left(\arctan \dfrac{3}{4}\right) = \dfrac{4}{5}$. Use a right triangle.

Let $\theta = \arctan \dfrac{3}{4}$

then $\tan \theta = \dfrac{3}{4}$

and $\cos \theta = \dfrac{4}{5}$

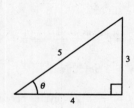

127. $\sec\left(\arctan \dfrac{12}{5}\right) = \dfrac{13}{5}$.   Use a right triangle.

Let $\theta = \arctan \dfrac{12}{5}$

then $\tan \theta = \dfrac{12}{5}$

and $\sec \theta = \dfrac{13}{5}$

129. $\tan \theta = \dfrac{70}{30}$

$\theta = \arctan\left(\dfrac{70}{30}\right) \approx 66.8°$

131. $\sin 48° = \dfrac{d_1}{650} \Longrightarrow d_1 \approx 483$

$\cos 25° = \dfrac{d_2}{810} \Longrightarrow d_2 \approx 734$ $\left.\rule{0pt}{22pt}\right\}$ $d_1 + d_2 \approx 1217$

$\cos 48° = \dfrac{d_3}{650} \Longrightarrow d_3 \approx 435$

$\sin 25° = \dfrac{d_4}{810} \Longrightarrow d_4 \approx 342$ $\left.\rule{0pt}{22pt}\right\}$ $d_3 - d_4 \approx 93$

$\tan \theta \approx \dfrac{93}{1217} \Longrightarrow \theta \approx 4.4°$

$\sec 4.4° \approx \dfrac{D}{1217} \Longrightarrow D \approx 1217 \sec 4.4° \approx 1221$

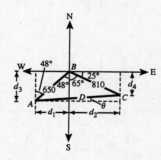

The distance is 1221 miles and the bearing is N 85.6° E.

133. False. The sine or cosine functions are often useful for modeling simple harmonic motion.

135. False. For each θ there corresponds exactly one value of y.

137. $y = 3 \sin x$

Amplitude: 3

Period: 2π

Matches graph (d)

139. $y = 2 \sin \pi x$

Amplitude: 2

Period: 2

Matches graph (b)

141. $f(\theta) = \sec \theta$ is undefined at the zeros of $g(\theta) = \cos \theta$ since $\sec \theta = \dfrac{1}{\cos \theta}$.

143. The ranges for the other four trigonometric functions are not bounded. For $y = \tan x$ and $y = \cot x$, the range is $(-\infty, \infty)$. For $y = \sec x$ and $y = \csc x$, the range is $(-\infty, -1] \cup [1, \infty)$

145. (a) $\arcsin x \approx x + \dfrac{x^3}{6} + \dfrac{3x^5}{40} + \dfrac{5x^7}{112}$

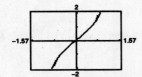

The approximation appears to be accurate over most of the domain of $\arcsin(x)$, $-1 \le x \le 1$.

(b) $\arctan x \approx x + \dfrac{x^3}{3} + \dfrac{x^5}{5} + \dfrac{x^7}{7}$

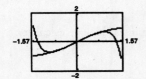

The approximation appears to be accurate over the interval $-1 < x < 1$.

(c) $\arctan x \approx x + \dfrac{x^3}{3} + \dfrac{x^5}{5} + \dfrac{x^7}{7} + \dfrac{x^9}{9}$

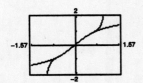

The accuracy improved.

Chapter 1   Practice Test

1. Express 350° in radian measure.

2. Express $(5\pi)/9$ in degree measure.

3. Convert $135°14'12''$ to decimal form.

4. Convert $-22.569°$ to D°M'S'' form.

5. If $\cos \theta = \frac{2}{3}$, use the trigonometric identities to find $\tan \theta$.

6. Find θ given $\sin \theta = 0.9063$.

7. Solve for x in the figure below.

8. Find the magnitude of the reference angle for $\theta = (6\pi)/5$.

9. Evaluate csc 3.92.

10. Find $\sec \theta$ given that θ lies in Quadrant III and $\tan \theta = 6$.

11. Graph $y = 3 \sin \dfrac{x}{2}$.

12. Graph $y = -2 \cos(x - \pi)$.

13. Graph $y = \tan 2x$.

14. Graph $y = -\csc\left(x + \dfrac{\pi}{4}\right)$.

15. Graph $y = 2x + \sin x$, using a graphing calculator.

16. Graph $y = 3x \cos x$, using a graphing calculator.

17. Evaluate arcsin 1.

18. Evaluate arctan (-3).

19. Evaluate $\sin\left(\arccos \dfrac{4}{\sqrt{35}}\right)$.

20. Write an algebraic expression for $\cos\left(\arcsin \dfrac{x}{4}\right)$.

For Exercises 21–23, solve the right triangle.

21. $A = 40°$, $c = 12$

22. $B = 6.84°$, $a = 21.3$

23. $a = 5$, $b = 9$

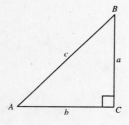

24. A 20-foot ladder leans against the side of a barn. Find the height of the top of the ladder if the angle of elevation of the ladder is 67°.

25. An observer in a lighthouse 250 feet above sea level spots a ship off the shore. If the angle of depression to the ship is 5°, how far out is the ship?

CHAPTER 2
Analytic Trigonometry

CHAPTER 2
Analytic Trigonometry

Section 2.1   Using Fundamental Identities

■ You should know the fundamental trigonometric identities.

(a) Reciprocal Identities

$$\sin u = \frac{1}{\csc u} \qquad\qquad \csc u = \frac{1}{\sin u}$$

$$\cos u = \frac{1}{\sec u} \qquad\qquad \sec u = \frac{1}{\cos u}$$

$$\tan u = \frac{1}{\cot u} = \frac{\sin u}{\cos u} \qquad\qquad \cot u = \frac{1}{\tan u} = \frac{\cos u}{\sin u}$$

(b) Pythagorean Identities

$$\sin^2 u + \cos^2 u = 1$$
$$1 + \tan^2 u = \sec^2 u$$
$$1 + \cot^2 u = \csc^2 u$$

(c) Cofunction Identities

$$\sin\left(\frac{\pi}{2} - u\right) = \cos u \qquad\qquad \cos\left(\frac{\pi}{2} - u\right) = \sin u$$

$$\tan\left(\frac{\pi}{2} - u\right) = \cot u \qquad\qquad \cot\left(\frac{\pi}{2} - u\right) = \tan u$$

$$\sec\left(\frac{\pi}{2} - u\right) = \csc u \qquad\qquad \csc\left(\frac{\pi}{2} - u\right) = \sec u$$

(d) Even/Odd Identities

$$\sin(-x) = -\sin x \qquad\qquad \csc(-x) = -\csc x$$
$$\cos(-x) = \cos x \qquad\qquad \sec(-x) = \sec x$$
$$\tan(-x) = -\tan x \qquad\qquad \cot(-x) = -\cot x$$

■ You should be able to use these fundamental identities to find function values.

■ You should be able to convert trigonometric expressions to equivalent forms by using the fundamental identities.

Solutions to Odd-Numbered Exercises

1. $\sin x = \dfrac{\sqrt{3}}{2}$, $\cos x = -\dfrac{1}{2}$ $\Longrightarrow$ x is in Quadrant II.

$$\tan x = \frac{\sin x}{\cos x} = \frac{\sqrt{3}/2}{-1/2} = -\sqrt{3}$$

$$\cot x = \frac{1}{\tan x} = -\frac{1}{\sqrt{3}} = -\frac{\sqrt{3}}{3}$$

$$\sec x = \frac{1}{\cos x} = \frac{1}{-1/2} = -2$$

$$\csc x = \frac{1}{\sin x} = \frac{1}{\sqrt{3}/2} = \frac{2}{\sqrt{3}} = \frac{2\sqrt{3}}{3}$$

3. $\sec \theta = \sqrt{2}$, $\sin \theta = -\dfrac{\sqrt{2}}{2}$ $\Longrightarrow$ θ is in Quadrant IV.

$$\cos \theta = \frac{1}{\sec \theta} = \frac{1}{\sqrt{2}} = \frac{\sqrt{2}}{2}$$

$$\tan \theta = \frac{\sin \theta}{\cos \theta} = \frac{-\sqrt{2}/2}{\sqrt{2}/2} = -1$$

$$\cot \theta = \frac{1}{\tan \theta} = -1$$

$$\csc \theta = \frac{1}{\sin \theta} = -\sqrt{2}$$

5. $\tan x = \dfrac{5}{12}$, $\sec x = -\dfrac{13}{12}$ $\Longrightarrow$ x is in

Quadrant III.

$$\cos x = \frac{1}{\sec x} = -\frac{12}{13}$$

$$\sin x = -\sqrt{1 - \cos^2 x} = -\sqrt{1 - \frac{144}{169}} = -\frac{5}{13}$$

$$\cot x = \frac{1}{\tan x} = \frac{12}{5}$$

$$\csc x = \frac{1}{\sin x} = -\frac{13}{5}$$

7. $\sec \phi = \dfrac{3}{2}$, $\csc \phi = -\dfrac{3\sqrt{5}}{5}$ $\Longrightarrow$ ϕ is in Quadrant IV.

$$\sin \phi = \frac{1}{\csc \phi} = \frac{1}{-3\sqrt{5}/5} = -\frac{\sqrt{5}}{3}$$

$$\cos \phi = \frac{1}{\sec \phi} = \frac{1}{3/2} = \frac{2}{3}$$

$$\tan \phi = \frac{\sin \phi}{\cos \phi} = \frac{-\sqrt{5}/3}{2/3} = -\frac{\sqrt{5}}{2}$$

$$\cot \phi = \frac{1}{\tan \phi} = \frac{1}{-\sqrt{5}/2} = -\frac{2}{\sqrt{5}} = -\frac{2\sqrt{5}}{5}$$

9. $\sin(-x) = -\dfrac{1}{3}$ $\Longrightarrow$ $\sin x = \dfrac{1}{3}$, $\tan x = -\dfrac{\sqrt{2}}{4}$ $\Longrightarrow$ x is

in Quadrant II.

$$\cos x = -\sqrt{1 - \sin^2 x} = -\sqrt{1 - \frac{1}{9}} = -\frac{2\sqrt{2}}{3}$$

$$\cot x = \frac{1}{\tan x} = \frac{1}{-\sqrt{2}/4} = -2\sqrt{2}$$

$$\sec x = \frac{1}{\cos x} = \frac{1}{-2\sqrt{2}/3} = -\frac{3\sqrt{2}}{4}$$

$$\csc x = \frac{1}{\sin x} = \frac{1}{1/3} = 3$$

11. $\tan \theta = 2$, $\sin \theta < 0$ $\Longrightarrow$ θ is in Quadrant III.

$$\sec \theta = -\sqrt{\tan^2 \theta + 1} = -\sqrt{4 + 1} = -\sqrt{5}$$

$$\cos \theta = \frac{1}{\sec \theta} = -\frac{1}{\sqrt{5}} = -\frac{\sqrt{5}}{5}$$

$$\sin \theta = -\sqrt{1 - \cos^2 \theta}$$

$$= -\sqrt{1 - \frac{1}{5}} = -\frac{2}{\sqrt{5}} = -\frac{2\sqrt{5}}{5}$$

$$\csc \theta = \frac{1}{\sin \theta} = -\frac{\sqrt{5}}{2}$$

$$\cot \theta = \frac{1}{\tan \theta} = \frac{1}{2}$$

13. $\sin \theta = -1$, $\cot \theta = 0$ $\Longrightarrow$ $\theta = \dfrac{3\pi}{2}$

$$\cos \theta = \sqrt{1 - \sin^2 \theta} = 0$$

$\sec \theta$ is undefined.

$\tan \theta$ is undefined.

$\csc \theta = -1$

15. $\sec x \cos x = \sec x \cdot \dfrac{1}{\sec x} = 1$

The expression is matched with (d).

17. $\cot^2 x - \csc^2 x = \cot^2 x - (1 + \cot^2 x) = -1$

The expression is matched with (b).

19. $\dfrac{\sin(-x)}{\cos(-x)} = \dfrac{-\sin x}{\cos x} = -\tan x$

The expression is matched with (e).

21. $\sin x \sec x = \sin x \cdot \dfrac{1}{\cos x} = \tan x$

The expression is matched with (b).

23. $\sec^4 x - \tan^4 x = (\sec^2 x + \tan^2 x)(\sec^2 x - \tan^2 x)$

$\qquad = (\sec^2 x + \tan^2 x)(1) = \sec^2 x + \tan^2 x$

The expression is matched with (f).

25. $\dfrac{\sec^2 x - 1}{\sin^2 x} = \dfrac{\tan^2 x}{\sin^2 x} = \dfrac{\sin^2 x}{\cos^2 x} \cdot \dfrac{1}{\sin^2 x} = \sec^2 x$

The expression is matched with (e).

27. $\cot \theta \sec \theta = \dfrac{\cos \theta}{\sin \theta} \cdot \dfrac{1}{\cos \theta} = \dfrac{1}{\sin \theta} = \csc \theta$

29. $\sin \phi \, (\csc \phi - \sin \phi) = (\sin \phi)\dfrac{1}{\sin \phi} - \sin^2 \phi$

$\qquad = 1 - \sin^2 \phi = \cos^2 \phi$

31. $\dfrac{\cot x}{\csc x} = \dfrac{\cos x / \sin x}{1/\sin x}$

$\qquad = \dfrac{\cos x}{\sin x} \cdot \dfrac{\sin x}{1} = \cos x$

33. $\dfrac{1 - \sin^2 x}{\csc^2 x - 1} = \dfrac{\cos^2 x}{\cot^2 x} = \cos^2 x \tan^2 x = (\cos^2 x)\dfrac{\sin^2 x}{\cos^2 x}$

$\qquad = \sin^2 x$

35. $\sec \alpha \, \dfrac{\sin \alpha}{\tan \alpha} = \dfrac{1}{\cos \alpha}\,(\sin \alpha) \cot \alpha$

$\qquad = \dfrac{1}{\cos \alpha}(\sin \alpha)\left(\dfrac{\cos \alpha}{\sin \alpha}\right) = 1$

37. $\cos\left(\dfrac{\pi}{2} - x\right)\sec x = (\sin x)(\sec x)$

$\qquad = (\sin x)\left(\dfrac{1}{\cos x}\right)$

$\qquad = \dfrac{\sin x}{\cos x}$

$\qquad = \tan x$

39. $\dfrac{\cos^2 y}{1 - \sin y} = \dfrac{1 - \sin^2 y}{1 - \sin y}$

$\qquad = \dfrac{(1 + \sin y)(1 - \sin y)}{1 - \sin y}$

$\qquad = 1 + \sin y$

41. $\sin \beta \tan \beta + \cos \beta = (\sin \beta)\dfrac{\sin \beta}{\cos \beta} + \cos \beta$

$$= \frac{\sin^2 \beta}{\cos \beta} + \frac{\cos^2 \beta}{\cos \beta}$$

$$= \frac{\sin^2 \beta + \cos^2 \beta}{\cos \beta}$$

$$= \frac{1}{\cos \beta}$$

$$= \sec \beta$$

43. $\cot u \sin u + \tan u \cos u = \dfrac{\cos u}{\sin u}(\sin u) + \dfrac{\sin u}{\cos u}(\cos u)$

$$= \cos u + \sin u$$

45. $\tan^2 x - \tan^2 x \sin^2 x = \tan^2 x(1 - \sin^2 x)$

$$= \tan^2 x \cos^2 x$$

$$= \frac{\sin^2 x}{\cos^2 x} \cdot \cos^2 x$$

$$= \sin^2 x$$

47. $\sin^2 x \sec^2 x - \sin^2 x = \sin^2 x(\sec^2 x - 1)$

$$= \sin^2 x \tan^2 x$$

49. $\dfrac{\sec^2 x - 1}{\sec x - 1} = \dfrac{(\sec x + 1)(\sec x - 1)}{\sec x - 1} = \sec x + 1$

51. $\tan^4 x + 2\tan^2 x + 1 = (\tan^2 x + 1)^2$

$$= (\sec^2 x)^2$$

$$= \sec^4 x$$

53. $\sin^4 x - \cos^4 x = (\sin^2 x + \cos^2 x)(\sin^2 x - \cos^2 x)$

$$= (1)(\sin^2 x - \cos^2 x)$$

$$= \sin^2 x - \cos^2 x$$

55. $\csc^3 x - \csc^2 x - \csc x + 1 = \csc^2 x(\csc x - 1) - 1(\csc x - 1)$

$$= (\csc^2 x - 1)(\csc x - 1)$$

$$= \cot^2 x(\csc x - 1)$$

57. $(\sin x + \cos x)^2 = \sin^2 x + 2\sin x \cos x + \cos^2 x$

$$= (\sin^2 x + \cos^2 x) + 2\sin x \cos x$$

$$= 1 + 2\sin x \cos x$$

59. $(2\csc x + 2)(2\csc x - 2) = 4\csc^2 x - 4 = 4(\csc^2 x - 1) = 4\cot^2 x$

61. $\dfrac{1}{1 + \cos x} + \dfrac{1}{1 - \cos x} = \dfrac{1 - \cos x + 1 + \cos x}{(1 + \cos x)(1 - \cos x)}$

$$= \frac{2}{1 - \cos^2 x}$$

$$= \frac{2}{\sin^2 x}$$

$$= 2\csc^2 x$$

63. $\dfrac{\cos x}{1 + \sin x} + \dfrac{1 + \sin x}{\cos x} = \dfrac{\cos^2 x + (1 + \sin x)^2}{\cos x(1 + \sin x)} = \dfrac{\cos^2 x + 1 + 2\sin x + \sin^2 x}{\cos x(1 + \sin x)}$

$$= \dfrac{2 + 2\sin x}{\cos x(1 + \sin x)}$$

$$= \dfrac{2(1 + \sin x)}{\cos x(1 + \sin x)}$$

$$= \dfrac{2}{\cos x}$$

$$= 2 \sec x$$

65. $\dfrac{\sin^2 y}{1 - \cos y} = \dfrac{1 - \cos^2 y}{1 - \cos y}$

$$= \dfrac{(1 + \cos y)(1 - \cos y)}{1 - \cos y}$$

$$= 1 + \cos y$$

67. $\dfrac{3}{\sec x - \tan x} \cdot \dfrac{\sec x + \tan x}{\sec x + \tan x} = \dfrac{3(\sec x + \tan x)}{\sec^2 x - \tan^2 x}$

$$= \dfrac{3(\sec x + \tan x)}{1}$$

$$= 3(\sec x + \tan x)$$

69. $y_1 = \cos\left(\dfrac{\pi}{2} - x\right),\ y_2 = \sin x$

x	0.2	0.4	0.6	0.8	1.0	1.2	1.4
y_1	0.1987	0.3894	0.5646	0.7174	0.8415	0.9320	0.9855
y_2	0.1987	0.3894	0.5646	0.7174	0.8415	0.9320	0.9855

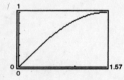

Conclusion: $y_1 = y_2$

71. $y_1 = \dfrac{\cos x}{1 - \sin x},\ y_2 = \dfrac{1 + \sin x}{\cos x}$

x	0.2	0.4	0.6	0.8	1.0	1.2	1.4
y_1	1.2230	1.5085	1.8958	2.4650	3.4082	5.3319	11.6814
y_2	1.2230	1.5085	1.8958	2.4650	3.4082	5.3319	11.6814

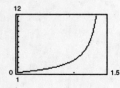

Conclusion: $y_1 = y_2$

73. $y_1 = \cos x \cot x + \sin x = \csc x$

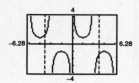

$$\cos x \cot x + \sin x = \cos x \left(\frac{\cos x}{\sin x}\right) + \sin x$$

$$= \frac{\cos^2 x}{\sin x} + \frac{\sin^2 x}{\sin x}$$

$$= \frac{\cos^2 x + \sin^2 x}{\sin x} = \frac{1}{\sin x} = \csc x$$

75. $y_1 = \dfrac{1}{\sin x}\left(\dfrac{1}{\cos x} - \cos x\right) = \tan x$

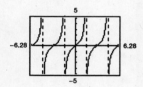

$$\frac{1}{\sin x}\left(\frac{1}{\cos x} - \cos x\right) = \frac{1}{\sin x \cos x} - \frac{\cos x}{\sin x}$$

$$= \frac{1 - \cos^2 x}{\sin x \cos x} = \frac{\sin^2 x}{\sin x \cos x} = \frac{\sin x}{\cos x} = \tan x$$

77. Let $x = 3 \cos \theta$, then

$$\sqrt{9 - x^2} = \sqrt{9 - (3 \cos \theta)^2} = \sqrt{9 - 9\cos^2 \theta} = \sqrt{9(1 - \cos^2 \theta)}$$

$$= \sqrt{9 \sin^2 \theta} = 3 \sin \theta$$

79. Let $x = 3 \sec \theta$, then

$$\sqrt{x^2 - 9} = \sqrt{(3 \sec \theta)^2 - 9}$$

$$= \sqrt{9 \sec^2 \theta - 9}$$

$$= \sqrt{9 (\sec^2 \theta - 1)}$$

$$= \sqrt{9 \tan^2 \theta}$$

$$= 3 \tan \theta$$

81. Let $x = 5 \tan \theta$, then

$$\sqrt{x^2 + 25} = \sqrt{(5 \tan \theta)^2 + 25}$$

$$= \sqrt{25 \tan^2 \theta + 25}$$

$$= \sqrt{25(\tan^2 \theta + 1)}$$

$$= \sqrt{25 \sec^2 \theta}$$

$$= 5 \sec \theta$$

83. Let $x = 3 \sin \theta$, then $\sqrt{9 - x^2} = 3$ becomes

$$\sqrt{9 - (3 \sin \theta)^2} = 3$$

$$\sqrt{9 - 9 \sin^2 \theta} = 3$$

$$\sqrt{9(1 - \sin^2 \theta)} = 3$$

$$\sqrt{9 \cos^2 \theta} = 3$$

$$3 \cos \theta = 3$$

$$\cos \theta = 1$$

$$\sin \theta = \sqrt{1 - \cos^2 \theta} = \sqrt{1 - (1)^2} = 0$$

85. Let $x = 2 \cos \theta$, then $\sqrt{16 - 4x^2} = 2\sqrt{2}$ becomes

$$\sqrt{16 - 4(2 \cos \theta)^2} = 2\sqrt{2}$$

$$\sqrt{16 - 16 \cos^2 \theta} = 2\sqrt{2}$$

$$\sqrt{16(1 - \cos^2 \theta)} = 2\sqrt{2}$$

$$\sqrt{16 \sin^2 \theta} = 2\sqrt{2}$$

$$4 \sin \theta = 2\sqrt{2}$$

$$\sin \theta = \frac{\sqrt{2}}{2}$$

$$\cos \theta = \sqrt{1 - \sin^2 \theta} = \sqrt{1 - \frac{1}{2}} = \sqrt{\frac{1}{2}} = \frac{\sqrt{2}}{2}$$

87. $\sin \theta = \sqrt{1 - \cos^2 \theta}$

Let $y_1 = \sin x$ and $y_2 = \sqrt{1 - \cos^2 x}$, $0 \le x \le 2\pi$.

$y_1 = y_2$ for $0 \le x \le \pi$, so we have

$\sin \theta = \sqrt{1 - \cos^2 \theta}$ for $0 \le \theta \le \pi$.

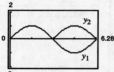

89. $\sec \theta = \sqrt{1 + \tan^2 \theta}$

Let $y_1 = \dfrac{1}{\cos x}$ and $y_2 = \sqrt{1 + \tan^2 x}$, $0 \le x \le 2\pi$.

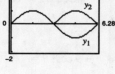

$y_1 = y_2$ for $0 \le x < \dfrac{\pi}{2}$ and $\dfrac{3\pi}{2} < x \le 2\pi$, so we have

$\sec \theta = \sqrt{1 + \tan^2 \theta}$ for $0 \le \theta < \dfrac{\pi}{2}$ and $\dfrac{3\pi}{2} < \theta \le 2\pi$.

91. (a) $\csc^2 132° - \cot^2 132° \approx 1.8107 - 0.8107 = 1$

(b) $\csc^2 \dfrac{2\pi}{7} - \cot^2 \dfrac{2\pi}{7} \approx 1.6360 - 0.6360 = 1$

93. $\cos\left(\dfrac{\pi}{2} - \theta\right) = \sin\theta$

(a) $\theta = 80°$

$\cos(90° - 80°) = \sin 80°$

$0.9848 = 0.9848$

(b) $\theta = 0.8$

$\cos\left(\dfrac{\pi}{2} - 0.8\right) = \sin 0.8$

$0.7174 = 0.7174$

95. $\cot\theta = \dfrac{\cos\theta}{\sin\theta}$

(a) $\theta = 25°$

$\cot\theta = \cot 25° \approx 2.1445$

$\dfrac{\cos\theta}{\sin\theta} = \dfrac{\cos 25°}{\sin 25°} \approx 2.1445$

(b) $\theta = \dfrac{\pi}{8}$

$\cot\theta = \cot\dfrac{\pi}{8} \approx 2.4142$

$\dfrac{\cos\theta}{\sin\theta} = \dfrac{\cos\dfrac{\pi}{8}}{\sin\dfrac{\pi}{8}} \approx 2.4142$

97. $\tan\left(\dfrac{\pi}{2} - \theta\right) = \cot\theta$

(a) $\theta = 5°$

$\tan\left(\dfrac{\pi}{2} - \theta\right) = \tan(90° - 5°) = \tan(85°) \approx 11.4301$

$\cot\theta = \cot 5° = 11.4301$

(b) $\theta = \dfrac{11\pi}{12}$

$\tan\left(\dfrac{\pi}{2} - \theta\right) = \tan\left(\dfrac{\pi}{2} - \dfrac{11\pi}{12}\right) = \tan\left(-\dfrac{5\pi}{12}\right) \approx -3.7321$

$\cot\theta = \cot\dfrac{11\pi}{12} = -3.7321$

99. $\mu W \cos\theta = W\sin\theta$

$\mu = \dfrac{W\sin\theta}{W\cos\theta} = \tan\theta$

101. False. A cofunction identity can be used to transform a tangent function so that it can be represented by a cotangent function.

103. As $x \to 0^{+}$,

$\cos x \to 1$ and $\sec x = \dfrac{1}{\cos x} \to 1.$

105. As $x \to \pi^{+}$,

$\sin x \to 0$ and $\csc x = \dfrac{1}{\sin x} \to -\infty.$

107. The equation is **not** an identity.

$\cot\theta = \pm\sqrt{\csc^2\theta - 1}$

109. The equation is **not** an identity.

$\dfrac{1}{5\cos\theta} = \dfrac{1}{5}\left(\dfrac{1}{\cos\theta}\right) = \dfrac{1}{5}\sec\theta \neq 5\sec\theta$

111. The equation is **not** an identity because the angles may not be the same.

$\sin\theta\csc\phi = \sin\theta\left(\dfrac{1}{\sin\phi}\right) \neq 1$ unless $\theta = \phi$

113. $\cos \theta$

$$\sin \theta = \pm\sqrt{1 - \cos^2 \theta}$$

$$\tan \theta = \frac{\sin \theta}{\cos \theta} = \pm\frac{\sqrt{1 - \cos^2 \theta}}{\cos \theta}$$

$$\csc \theta = \frac{1}{\sin \theta} = \pm\frac{1}{\sqrt{1 - \cos^2 \theta}}$$

$$\sec \theta = \frac{1}{\cos \theta}$$

$$\cot \theta = \frac{1}{\tan \theta} = \pm\frac{\cos \theta}{\sqrt{1 - \cos^2 \theta}}$$

115. $\sqrt{v}\left(\sqrt{20} - \sqrt{5}\right) = \sqrt{20v} - \sqrt{5v}$

$$= 2\sqrt{5v} - \sqrt{5v}$$

$$= \sqrt{5v}$$

117. $\dfrac{50x}{\sqrt{30} - 5} = \dfrac{50x}{\sqrt{30} - 5} \cdot \dfrac{\sqrt{30} + 5}{\sqrt{30} + 5} = \dfrac{50x\left(\sqrt{30} + 5\right)}{30 - 25} = \dfrac{50x\left(\sqrt{30} + 5\right)}{5} = 10x\left(\sqrt{30} + 5\right)$

119. $y = 3 \sec(\pi x - \pi)$

Graph $y = 3 \cos(\pi x - \pi)$ first.

Amplitude: $|3| = 3$

Period: $\dfrac{2\pi}{\pi} = 2$

Shift: $\pi x - \pi = 0 \implies x = 1$ to $\pi x - \pi = 2\pi \implies x = 3$

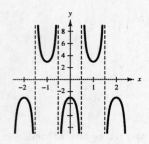

121. $y = \dfrac{1}{4} \sec 2x$

Graph $y = \dfrac{1}{4} \cos 2x$ first.

Amplitude: $\left|\dfrac{1}{4}\right| = \dfrac{1}{4}$

Period: $\dfrac{2\pi}{\pi} = 2$

Shift: $2x = 0 \implies x = 0$ to $2x = 2\pi \implies x = \pi$

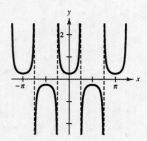

Section 2.2   Verifying Trigonometric Identities

- ■ You should know the difference between an expression, a conditional equation, and an identity.
- ■ You should be able to solve trigonometric identities, using the following techniques.
    - (a) Work with *one* side at a time. Do not "cross" the equal sign.
    - (b) Use algebraic techniques such as combining fractions, factoring expressions, rationalizing denominators, and squaring binomials.
    - (c) Use the fundamental identities.
    - (d) Convert all the terms into sines and cosines.

Solutions to Odd-Numbered Exercises

1. $\sin t \csc t = \sin t \left(\dfrac{1}{\sin t} \right) = 1$

3. $(1 + \sin \alpha)(1 - \sin \alpha) = 1 - \sin^2 \alpha = \cos^2 \alpha$

5. $\cos^2 \beta - \sin^2 \beta = (1 - \sin^2 \beta) - \sin^2 \beta$
$$= 1 - 2 \sin^2 \beta$$

7. $\tan^2 \theta + 4 = (\sec^2 \theta - 1) + 4$
$$= \sec^2 \theta + 3$$

9. $\sin^2 \alpha - \sin^4 \alpha = \sin^2 \alpha (1 - \sin^2 \alpha)$
$$= (1 - \cos^2 \alpha)(\cos^2 \alpha)$$
$$= \cos^2 \alpha - \cos^4 \alpha$$

11. $\dfrac{\csc^2 \theta}{\cot \theta} = \csc^2 \theta \left(\dfrac{1}{\cot \theta} \right) = \csc^2 \theta \tan \theta$
$$= \left(\dfrac{1}{\sin^2 \theta} \right) \left(\dfrac{\sin \theta}{\cos \theta} \right) = \left(\dfrac{1}{\sin \theta} \right) \left(\dfrac{1}{\cos \theta} \right)$$
$$= \csc \theta \sec \theta$$

13. $\dfrac{\cot^2 t}{\csc t} = \dfrac{\cos^2 t}{\sin^2 t} \cdot \sin t$
$$= \dfrac{\cos^2 t}{\sin t}$$
$$= \dfrac{1 - \sin^2 t}{\sin t} = \dfrac{1}{\sin t} - \dfrac{\sin^2 t}{\sin t}$$
$$= \csc t - \sin t$$

15. $\sin^{1/2} x \cos x - \sin^{5/2} x \cos x = \sin^{1/2} x \cos x (1 - \sin^2 x) = \sin^{1/2} x \cos x \cdot \cos^2 x = \cos^3 x \sqrt{\sin x}$

17. $\dfrac{1}{\sec x \tan x} = \cos x \cot x = \cos x \cdot \dfrac{\cos x}{\sin x}$
$$= \dfrac{\cos^2 x}{\sin x}$$
$$= \dfrac{1 - \sin^2 x}{\sin x}$$
$$= \dfrac{1}{\sin x} - \sin x$$
$$= \csc x - \sin x$$

19. $\cot \alpha + \tan \alpha = \dfrac{\cos \alpha}{\sin \alpha} + \dfrac{\sin \alpha}{\cos \alpha}$
$$= \dfrac{\cos^2 \alpha + \sin^2 \alpha}{\sin \alpha \cos \alpha}$$
$$= \dfrac{1}{\sin \alpha \cos \alpha}$$
$$= \dfrac{1}{\sin \alpha} \cdot \dfrac{1}{\cos \alpha}$$
$$= \csc \alpha \sec \alpha$$

21. $\sin x \cos x + \sin^3 x \sec x = \sin x\left[\cos x + \sin^2 x\left(\dfrac{1}{\cos x}\right)\right]$

$$= \sin x\left[\dfrac{\cos^2 x + \sin^2 x}{\cos x}\right]$$

$$= \sin x\left(\dfrac{1}{\cos x}\right)$$

$$= \dfrac{\sin x}{\cos x}$$

$$= \tan x$$

23. $\dfrac{1}{\tan x} + \dfrac{1}{\cot x} = \dfrac{\cot x + \tan x}{\tan x \cot x}$

$$= \dfrac{\cot x + \tan x}{1}$$

$$= \tan x + \cot x$$

25. $\dfrac{\cos\theta \cot\theta}{1 - \sin\theta} - 1 = \dfrac{\cos\theta \cot\theta - (1 - \sin\theta)}{1 - \sin\theta}$

$$= \dfrac{\cos\theta\left(\dfrac{\cos\theta}{\sin\theta}\right) - 1 + \sin\theta}{1 - \sin\theta} \cdot \dfrac{\sin\theta}{\sin\theta}$$

$$= \dfrac{\cos^2\theta - \sin\theta + \sin^2\theta}{\sin\theta(1 - \sin\theta)}$$

$$= \dfrac{1 - \sin\theta}{\sin\theta(1 - \sin\theta)}$$

$$= \dfrac{1}{\sin\theta}$$

$$= \csc\theta$$

27. $\dfrac{1}{\sin x + 1} + \dfrac{1}{\csc x + 1} = \dfrac{\csc x + 1 + \sin x + 1}{(\sin x + 1)(\csc x + 1)}$

$$= \dfrac{\sin x + \csc x + 2}{\sin x \csc x + \sin x + \csc x + 1}$$

$$= \dfrac{\sin x + \csc x + 2}{1 + \sin x + \csc x + 1}$$

$$= \dfrac{\sin x + \csc x + 2}{\sin x + \csc x + 2}$$

$$= 1$$

29. $\tan\left(\dfrac{\pi}{2} - \theta\right)\tan\theta = \cot\theta \tan\theta = \left(\dfrac{1}{\tan\theta}\right)\tan\theta = 1$

31. $\dfrac{\csc(-x)}{\sec(-x)} = \dfrac{\dfrac{1}{\sin(-x)}}{\dfrac{1}{\cos(-x)}}$

$$= \dfrac{\cos(-x)}{\sin(-x)}$$

$$= \dfrac{\cos x}{-\sin x}$$

$$= -\cot x$$

33. $\dfrac{\cos(-\theta)}{1 + \sin(-\theta)} = \dfrac{\cos\theta}{1 - \sin\theta} \cdot \dfrac{1 + \sin\theta}{1 + \sin\theta}$

$$= \dfrac{\cos\theta(1 + \sin\theta)}{1 - \sin^2\theta}$$

$$= \dfrac{\cos\theta(1 + \sin\theta)}{\cos^2\theta}$$

$$= \dfrac{1 + \sin\theta}{\cos\theta}$$

$$= \dfrac{1}{\cos\theta} + \dfrac{\sin\theta}{\cos\theta}$$

$$= \sec\theta + \tan\theta$$

35. $\dfrac{\sin x \cos y + \cos x \sin y}{\cos x \cos y - \sin x \sin y} = \dfrac{\dfrac{\sin x \cos y}{\cos x \cos y} + \dfrac{\cos x \sin y}{\cos x \cos y}}{\dfrac{\cos x \cos y}{\cos x \cos y} - \dfrac{\sin x \sin y}{\cos x \cos y}} = \dfrac{\tan x + \tan y}{1 - \tan x \tan y}$

37. $\dfrac{\tan x + \cot y}{\tan x \cot y} = \dfrac{\dfrac{1}{\cot x} + \dfrac{1}{\tan y}}{\dfrac{1}{\cot x} \cdot \dfrac{1}{\tan y}} \cdot \dfrac{\cot x \tan y}{\cot x \tan y} = \tan y + \cot x$

39. $\sqrt{\dfrac{1 + \sin \theta}{1 - \sin \theta}} = \sqrt{\dfrac{1 + \sin \theta}{1 - \sin \theta} \cdot \dfrac{1 + \sin \theta}{1 + \sin \theta}}$

$= \sqrt{\dfrac{(1 + \sin \theta)^2}{1 - \sin^2 \theta}}$

$= \sqrt{\dfrac{(1 + \sin \theta)^2}{\cos^2 \theta}}$

$= \dfrac{1 + \sin \theta}{|\cos \theta|}$

41. $\cos^2 \beta + \cos^2\left(\dfrac{\pi}{2} - \beta\right) = \cos^2 \beta + \sin^2 \beta = 1$

43. $\sin t \csc\left(\dfrac{\pi}{2} - t\right) = \sin t \sec t = \sin t\left(\dfrac{1}{\cos t}\right)$

$= \dfrac{\sin t}{\cos t} = \tan t$

45. $2 \sec^2 x - 2 \sec^2 x \sin^2 x - \sin^2 x - \cos^2 x = 2 \sec^2 x(1 - \sin^2 x) - (\sin^2 x + \cos^2 x)$

$= 2 \sec^2 x(\cos^2 x) - 1$

$= 2 \cdot \dfrac{1}{\cos^2 x} \cdot \cos^2 x - 1$

$= 2 - 1$

$= 1$

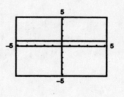

47. $2 + \cos^2 x - 3 \cos^4 x = (1 - \cos^2 x)(2 + 3 \cos^2 x)$

$= \sin^2 x(2 + 3 \cos^2 x)$

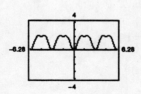

49. $\csc^4 x - 2 \csc^2 x + 1 = (\csc^2 x - 1)^2$

$= (\cot^2 x)^2 = \cot^4 x$

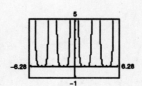

51. $\sec^4 \theta - \tan^4 \theta = (\sec^2 \theta + \tan^2 \theta)(\sec^2 \theta - \tan^2 \theta)$

$\qquad\qquad = (1 + \tan^2 \theta + \tan^2 \theta)(1)$

$\qquad\qquad = 1 + 2\tan^2 \theta$

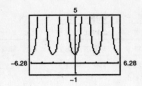

53. $\dfrac{\cos x}{1 + \sin x} = \dfrac{\cos x}{1 + \sin x} \cdot \dfrac{1 - \sin x}{1 - \sin x}$

$\qquad = \dfrac{\cos x(1 - \sin x)}{1 - \sin^2 x}$

$\qquad = \dfrac{\cos x(1 - \sin x)}{\cos^2 x}$

$\qquad = \dfrac{1 - \sin x}{\cos x}$

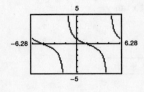

55. $\dfrac{\tan^3 \alpha - 1}{\tan \alpha - 1} = \dfrac{(\tan \alpha - 1)(\tan^2 \alpha + \tan \alpha + 1)}{\tan \alpha - 1} = \tan^2 \alpha + \tan \alpha + 1$

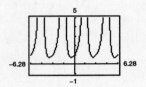

57. $\sin^2 25° + \sin^2 65° = \sin^2 25° + \cos^2(90° - 65°) = \sin^2 25° + \cos^2 25° = 1$

59. $\cos^2 20° + \cos^2 52° + \cos^2 38° + \cos^2 70° = \cos^2 20° + \cos^2 52° + \sin^2(90° - 38°) + \sin^2(90° - 70°)$

$\qquad\qquad\qquad = \cos^2 20° + \cos^2 52° + \sin^2 52° + \sin^2 20°$

$\qquad\qquad\qquad = (\cos^2 20° + \sin^2 20°) + (\cos^2 52° + \sin^2 52°)$

$\qquad\qquad\qquad = 1 + 1$

$\qquad\qquad\qquad = 2$

61. $\cos x - \csc x \cot x = \cos x - \dfrac{1}{\sin x} \dfrac{\cos x}{\sin x}$

$\qquad\qquad = \cos x\left(1 - \dfrac{1}{\sin^2 x}\right)$

$\qquad\qquad = \cos x(1 - \csc^2 x)$

$\qquad\qquad = -\cos x(\csc^2 x - 1)$

$\qquad\qquad = -\cos x \cot^2 x$

63. True. An identity is an equation that is true for all real values in the domain of the variable.

65. $\tan \theta = \sqrt{\sec^2 \theta - 1}$

True identity: $\tan \theta = \pm\sqrt{\sec^2 \theta - 1}$

$\tan \theta = \sqrt{\sec^2 \theta - 1}$ is not true for $\pi/2 < \theta < \pi$ or $3\pi/2 < \theta < 2\pi$. Thus, the equation is not true for $\theta = 3\pi/4$.

67. $\sqrt{\sin^2 x + \cos^2 x} = \sin x + \cos x$

$\sqrt{\sin^2 x + \cos^2 x} \neq \sin x + \cos x$

The left side is 1 for any x, but the right side is not necessarily 1. The equation is not true for $x = \pi/4$.

69. $\sin\left[\dfrac{(12n + 1)\pi}{6}\right] = \sin\left[\dfrac{1}{6}(12n\pi + \pi)\right]$

$\qquad\qquad\qquad = \sin\left(2n\pi + \dfrac{\pi}{6}\right)$

$\qquad\qquad\qquad = \sin\dfrac{\pi}{6} = \dfrac{1}{2}$

Thus, $\sin\left[\dfrac{(12n + 1)\pi}{6}\right] = \dfrac{1}{2}$ for all integers n.

71. $x^2 + 5x - 36 = 0$

$(x + 9)(x - 4) = 0$

$x + 9 = 0 \Longrightarrow x = -9$

$x - 4 = 0 \Longrightarrow x = 4$

$x = 4, -9$

73. $2x^2 - 7x - 15 = 0$

$(2x + 3)(x - 5) = 0$

$2x + 3 = 0 \Longrightarrow x = -\dfrac{3}{2}$

$x - 5 = 0 \Longrightarrow x = 5$

$x = 5, -\dfrac{3}{2}$

75. $\qquad 9x^2 - 1 = 0$

$(3x + 1)(3x - 1) = 0$

$3x + 1 = 0 \Longrightarrow x = -\dfrac{1}{3}$

$3x - 1 = 0 \Longrightarrow x = \dfrac{1}{3}$

$x = \pm\dfrac{1}{3}$

77. $\tan^{-1}\sqrt{3} = \dfrac{\pi}{3} = 60°$

79. $\sin^{-1}\dfrac{1}{2} = \dfrac{\pi}{6} = 30°$

81. $\cos^{-1}\left(-\dfrac{1}{2}\right) = \dfrac{2\pi}{3} = 120°$

Section 2.3   Solving Trigonometric Equations

■   You should be able to identify and solve trigonometric equations.

■   A trigonometric equation is a conditional equation. It is true for a specific set of values.

■   To solve trigonometric equations, use algebraic techniques such as collecting like terms, taking square roots, factoring, squaring, converting to quadratic form, using formulas, and using inverse functions. Study the examples in this section.

Solutions to Odd-Numbered Exercises

1. $2 \cos x - 1 = 0$

(a) $2 \cos \dfrac{\pi}{3} - 1 = 2\left(\dfrac{1}{2}\right) - 1 = 0$

(b) $2 \cos \dfrac{5\pi}{3} - 1 = 2\left(\dfrac{1}{2}\right) - 1 = 0$

3. $3 \tan^2 2x - 1 = 0$

(a) $3\left[\tan 2\left(\dfrac{\pi}{12}\right)\right]^2 - 1 = 3 \tan^2 \dfrac{\pi}{6} - 1$

$\qquad\qquad\qquad = 3\left(\dfrac{1}{\sqrt{3}}\right)^2 - 1$

$\qquad\qquad\qquad = 0$

(b) $3\left[\tan 2\left(\dfrac{5\pi}{12}\right)\right]^2 - 1 = 3 \tan^2 \dfrac{5\pi}{6} - 1$

$\qquad\qquad\qquad = 3\left(-\dfrac{1}{\sqrt{3}}\right)^2 - 1$

$\qquad\qquad\qquad = 0$

5. $2 \sin^2 x - \sin x - 1 = 0$

(a) $2 \sin^2 \dfrac{\pi}{2} - \sin \dfrac{\pi}{2} - 1 = 2(1)^2 - 1 - 1$

$\qquad\qquad\qquad\qquad = 0$

(b) $2 \sin^2 \dfrac{7\pi}{6} - \sin \dfrac{7\pi}{6} - 1 = 2\left(-\dfrac{1}{2}\right)^2 - \left(-\dfrac{1}{2}\right) - 1$

$\qquad\qquad\qquad\qquad = \dfrac{1}{2} + \dfrac{1}{2} - 1$

$\qquad\qquad\qquad\qquad = 0$

7. $2 \cos x + 1 = 0$

$2 \cos x = -1$

$\cos x = -\dfrac{1}{2}$

$x = \dfrac{2\pi}{3} + 2n\pi$

or $x = \dfrac{4\pi}{3} + 2n\pi$

9. $\sqrt{3} \csc x - 2 = 0$

$\sqrt{3} \csc x = 2$

$\csc x = \dfrac{2}{\sqrt{3}}$

$x = \dfrac{\pi}{3} + 2n\pi$

or $x = \dfrac{2\pi}{3} + 2n\pi$

11. $3 \sec^2 x - 4 = 0$

$\sec^2 x = \dfrac{4}{3}$

$\sec x = \pm\dfrac{2}{\sqrt{3}}$

$x = \dfrac{\pi}{6} + n\pi$

or $x = \dfrac{5\pi}{6} + n\pi$

13. $\sin x(\sin x + 1) = 0$

$\sin x = 0 \quad$ or $\quad \sin x = -1$

$x = n\pi \qquad\qquad x = \dfrac{3\pi}{2} + 2n\pi$

15. $4 \cos^2 x - 1 = 0$

$\cos^2 x = \dfrac{1}{4}$

$\cos^2 x = \pm\dfrac{1}{2}$

$x = \dfrac{\pi}{3} + n\pi \quad$ or $\quad x = \dfrac{2\pi}{3} + n\pi$

17. $2 \sin^2 2x = 1$

$\sin 2x = \pm\dfrac{1}{\sqrt{2}} = \pm\dfrac{\sqrt{2}}{2}$

$2x = \dfrac{\pi}{4} + 2n\pi, \; 2x = \dfrac{3\pi}{4} + 2n\pi, \; 2x = \dfrac{5\pi}{4} + 2n\pi, \; 2x = \dfrac{7\pi}{4} + 2n\pi.$

Thus, $x = \dfrac{\pi}{8} + n\pi, \; \dfrac{3\pi}{8} + n\pi, \; \dfrac{5\pi}{8} + n\pi, \; \dfrac{7\pi}{8} + n\pi.$

19. $\tan 3x(\tan x - 1) = 0$

$\tan 3x = 0 \quad$ or $\quad \tan x - 1 = 0$

$3x = n\pi \qquad\qquad \tan x = 1$

$x = \dfrac{n\pi}{3} \qquad\qquad x = \dfrac{\pi}{4} + n\pi$

21. $\cos^3 x = \cos x$

$\cos^3 x - \cos x = 0$

$\cos x(\cos^2 x - 1) = 0$

$\cos x = 0 \qquad$ or $\quad \cos^2 x - 1 = 0$

$x = \dfrac{\pi}{2}, \dfrac{3\pi}{2} \qquad\qquad \cos x = \pm 1$

$x = 0, \; \pi$

23. $3 \tan^3 x - \tan x = 0$

$\tan x(3 \tan^2 x - 1) = 0$

$\tan x = 0 \quad$ or $\quad 3 \tan^2 x - 1 = 0$

$x = 0, \; \pi \qquad\qquad \tan x = \pm\dfrac{\sqrt{3}}{3}$

$x = \dfrac{\pi}{6}, \dfrac{5\pi}{6}, \dfrac{7\pi}{6}, \dfrac{11\pi}{6}$

25. $\sec^2 x - \sec x - 2 = 0$

$(\sec x - 2)(\sec x + 1) = 0$

$\sec x - 2 = 0 \qquad$ or $\quad \sec x + 1 = 0$

$\sec x = 2 \qquad\qquad \sec x = -1$

$x = \dfrac{\pi}{3}, \dfrac{5\pi}{3} \qquad\qquad x = \pi$

27. $2 \sin x + \csc x = 0$

$$2 \sin x + \frac{1}{\sin x} = 0$$

$$2 \sin^2 x + 1 = 0$$

$$\sin^2 x = -\frac{1}{2} \Longrightarrow \text{No solution}$$

29. $2 \cos^2 x + \cos x - 1 = 0$

$$(2 \cos x - 1)(\cos x + 1) = 0$$

$$2 \cos x - 1 = 0 \quad \text{or} \quad \cos x + 1 = 0$$

$$\cos x = \frac{1}{2} \qquad\qquad \cos x = -1$$

$$x = \frac{\pi}{3}, \frac{5\pi}{3} \qquad\qquad x = \pi$$

31.
$$2 \sec^2 x + \tan^2 x - 3 = 0$$

$$2(\tan^2 x + 1) + \tan^2 x - 3 = 0$$

$$3 \tan^2 x - 1 = 0$$

$$\tan x = \pm \frac{\sqrt{3}}{3}$$

$$x = \frac{\pi}{6}, \frac{5\pi}{6}, \frac{7\pi}{6}, \frac{11\pi}{6}$$

33. $\cos 2x = \frac{1}{2}$

$$2x = \frac{\pi}{3} + 2n\pi \quad \text{or} \quad 2x = \frac{5\pi}{3} + 2n\pi$$

$$x = \frac{\pi}{6} + n\pi \qquad\qquad x = \frac{5\pi}{6} + n\pi$$

35. $\tan 3x = 1$

$$3x = \frac{\pi}{4} + 2n\pi \qquad \text{or} \qquad 3x = \frac{5\pi}{4} + 2n\pi$$

$$x = \frac{\pi}{12} + \frac{2n\pi}{3} \qquad\qquad x = \frac{5\pi}{12} + \frac{2n\pi}{3}$$

These can be combined as $x = \frac{\pi}{12} + \frac{n\pi}{3}$.

37. $\cos\left(\frac{x}{2}\right) = \frac{\sqrt{2}}{2}$

$$\frac{x}{2} = \frac{\pi}{4} + 2n\pi \quad \text{or} \quad \frac{x}{2} = \frac{7\pi}{4} + 2n\pi$$

$$x = \frac{\pi}{2} + 4n\pi \qquad\qquad x = \frac{7\pi}{2} + 4n\pi$$

39. $y = \sin \frac{\pi x}{2} + 1$

From the graph in the textbook we see that the curve has x-intercepts at $x = -1$ and at $x = 3$.

41. $y = \tan^2\left(\frac{\pi x}{6}\right) - 3$

From the graph in the textbook we see that the curve has x-intercepts at $x = \pm 2$.

43. $6y^2 - 13y + 6 = 0$

$$(3y - 2)(2y - 3) = 0$$

$$3y - 2 = 0 \quad \text{or} \quad 2y - 3 = 0$$

$$y = \frac{2}{3} \qquad\qquad y = \frac{3}{2}$$

$6 \cos^2 x - 13 \cos x + 6 = 0$

$$(3 \cos x - 2)(2 \cos x - 3) = 0$$

$$3 \cos x - 2 = 0 \qquad \text{or} \quad 2 \cos x - 3 = 0$$

$$\cos x = \frac{2}{3} \qquad\qquad\qquad \cos x = \frac{3}{2} \quad \text{(No solution)}$$

$$x \approx 0.8411 + 2n\pi, \, 5.4421 + 2n\pi$$

45. $2\sin x + \cos x = 0$

$$2\sin x = -\cos x$$

$$2 = -\frac{\cos x}{\sin x}$$

$$2 = -\cot x$$

$$-2 = \cot x$$

$$-\frac{1}{2} = \tan x$$

$$x = \arctan\left(-\frac{1}{2}\right)$$

$$x = \pi - \arctan\left(\frac{1}{2}\right) \approx 2.6779$$

$$\text{or } x = 2\pi - \arctan\left(\frac{1}{2}\right) \approx 5.8195$$

Graph $y_1 = 2\sin x + \cos x$

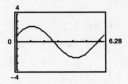

The x-intercepts occur at $x \approx 2.6779$ and $x \approx 5.8195$

47.

$$\frac{1 + \sin x}{\cos x} + \frac{\cos x}{1 + \sin x} = 4$$

$$\frac{(1 + \sin x)^2 + \cos^2 x}{\cos x(1 + \sin x)} = 4$$

$$\frac{1 + 2\sin x + \sin^2 x + \cos^2 x}{\cos x(1 + \sin x)} = 4$$

$$\frac{2 + 2\sin x}{\cos x(1 + \sin x)} = 4$$

$$\frac{2}{\cos x} = 4$$

$$\cos x = \frac{1}{2}$$

$$x = \frac{\pi}{3}, \frac{5\pi}{3}$$

Graph $y_1 = \frac{1 + \sin x}{\cos x} + \frac{\cos x}{1 + \sin x} - 4.$

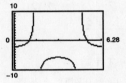

The x-intercepts occur at $x = \frac{\pi}{3} \approx 1.0472$ and $x = \frac{5\pi}{3} \approx 5.2360$.

49. $x\tan x - 1 = 0$

Graph $y_1 = x\tan x - 1$

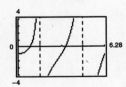

The x-intercepts occur at $x \approx 0.8603$ and $x \approx 3.4256$

51. $\sec^2 x + 0.5\tan x - 1 = 0$

Graph $y_1 = \frac{1}{(\cos x)^2} + 0.5\tan x - 1.$

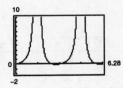

The x-intercepts occur at $x = 0$, $x \approx 2.6779$, $x = \pi \approx 3.1416$, and $x \approx 5.8195$.

53. $2 \tan^2 x + 7 \tan x - 15 = 0$

$(2 \tan x - 3)(\tan x + 5) = 0$

$2 \tan x - 3 = 0$ or $\tan x + 5 = 0$

     $\tan x = 1.5$ $\tan x = -5$

         $x \approx 0.9828, 4.1244$ $x \approx 1.7682, 4.9098$

Graph $y_1 = 2 \tan^2 x + 7 \tan x - 15$.

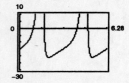

The x-intercepts occur at $x \approx 0.9828$, $x \approx 1.7682$, $x \approx 4.1244$, and $x \approx 4.9098$.

55. $12 \sin^2 x - 13 \sin x + 3 = 0$

$$\sin x = \frac{-(-13) \pm \sqrt{(-13)^2 - 4(12)(3)}}{2(12)}$$

$$= \frac{13 \pm 5}{24}$$

$\sin x = \frac{1}{3}$ or $\sin x = \frac{3}{4}$

   $x \approx 0.3398, 2.8018$ $x \approx 0.8481, 2.2935$

Graph $y_1 = 12 \sin^2 x - 13 \sin x + 3$.

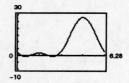

The x-intercepts occur at $x \approx 0.3398$, $x \approx 0.8481$, $x \approx 2.2935$, and $x \approx 2.8018$.

57. $\tan^2 x + 3 \tan x + 1 = 0$

$$\tan x = \frac{-3 \pm \sqrt{3^2 - 4(1)(1)}}{2(1)} = \frac{-3 \pm \sqrt{5}}{2}$$

$\tan x = \dfrac{-3 - \sqrt{5}}{2}$ or $\tan x = \dfrac{-3 + \sqrt{5}}{2}$

   $x \approx 1.9357, 5.0773$ $x \approx 2.7767, 5.9183$

Graph $y_1 = \tan^2 x + 3 \tan x + 1$.

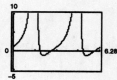

The x-intercepts occur at $x \approx 1.9357$, $x \approx 2.7767$, $x \approx 5.0773$, and $x \approx 5.9183$.

59. $\tan^2 x - 6 \tan x + 5 = 0$

$(\tan x - 1)(\tan x - 5) = 0$

$\tan x - 1 = 0$ or $\tan x - 5 = 0$

$\qquad \tan x = 1 \qquad\qquad \tan x = 5$

$\qquad x = \dfrac{\pi}{4}, \dfrac{5\pi}{4} \qquad x = \arctan 5, \arctan 5 + \pi$

61. $2 \cos^2 x - 5 \cos x + 2 = 0$

$(2 \cos x - 1)(\cos x - 2) = 0$

$2 \cos x - 1 = 0$ or $\cos x - 2 = 0$

$\qquad \cos x = \dfrac{1}{2} \qquad\qquad \cos x = 2$

$\qquad x = \dfrac{\pi}{3}, \dfrac{5\pi}{3} \qquad x = \arccos 2,$

$\qquad\qquad\qquad\qquad\qquad 2\pi - \arccos 2$

63. (a) $f(x) = \sin x + \cos x$

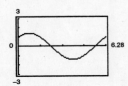

Maximum: $\left(\dfrac{\pi}{4}, \sqrt{2} \right)$

Minimum: $\left(\dfrac{5\pi}{4}, -\sqrt{2} \right)$

(b) $\cos x - \sin x = 0$

$\qquad \cos x = \sin x$

$\qquad 1 = \dfrac{\sin x}{\cos x}$

$\qquad \tan x = 1$

$\qquad x = \dfrac{\pi}{4}, \dfrac{5\pi}{4}$

$f\left(\dfrac{\pi}{4} \right) = \sin \dfrac{\pi}{4} + \cos \dfrac{\pi}{4} = \dfrac{\sqrt{2}}{2} + \dfrac{\sqrt{2}}{2} = \sqrt{2}$

$f\left(\dfrac{5\pi}{4} \right) = \sin \dfrac{5\pi}{4} + \cos \dfrac{5\pi}{4} = -\sin \dfrac{\pi}{4} + \left(-\cos \dfrac{\pi}{4} \right) = -\dfrac{\sqrt{2}}{2} - \dfrac{\sqrt{2}}{2} = -\sqrt{2}$

Therefore, the maximum point in the interval $[0, 2\pi)$ is $\left(\pi/4, \sqrt{2} \right)$ and the minimum point is $\left(5\pi/4, -\sqrt{2} \right)$.

65. $f(x) = \tan \dfrac{\pi x}{4}$

Since $\tan \pi/4 = 1$, $x = 1$ is the smallest nonnegative fixed point.

67. $f(x) = \cos \dfrac{1}{x}$

(a) The domain of $f(x)$ is all real numbers except 0.

(b) The graph has y-axis symmetry and a horizontal asymptote at $y = 1$.

(c) As $x \to 0$, $f(x)$ oscillates between -1 and 1.

(d) There are infinitely many solutions in the interval $[-1, 1]$.

(e) The greatest solution appears to occur at $x \approx 0.6366$.

69.
$$y = \frac{1}{12}(\cos 8t - 3 \sin 8t)$$

$$\frac{1}{12}(\cos 8t - 3 \sin 8t) = 0$$

$$\cos 8t = 3 \sin 8t$$

$$\frac{1}{3} = \tan 8t$$

$$8t \approx 0.32175 + n\pi$$

$$t \approx 0.04 + \frac{n\pi}{8}$$

In the interval $0 \le t \le 1$, $t \approx$ 0.04, 0.43, and 0.83.

71. $S = 74.50 + 43.75 \sin \dfrac{\pi t}{6}$

t	1	2	3	4	5	6	7	8	9	10	11	12
S	96.4	112.4	118.3	112.4	96.4	74.5	52.6	36.6	30.8	36.6	52.6	74.5

Sales exceed 100,000 units during February, March, and April.

73. Range = 1000 yards = 3000 feet

$v_0 = 1200$ feet per second

$f = \frac{1}{32} v_0{}^2 \sin 2\theta$

$3000 = \frac{1}{32}(1200)^2 \sin 2\theta$

$\sin 2\theta \approx 0.066667$

$2\theta \approx 3.8°$

$\theta \approx 1.9°$

75. $f(x) = 3 \sin(0.6x - 2)$

(a) Zero: $\sin(0.6x - 2) = 0$

$0.6x - 2 = 0$

$0.6x = 2$

$x = \dfrac{2}{0.6} = \dfrac{10}{3}$

(c) $-0.45x^2 + 5.52x - 13.70 = 0$

$x = \dfrac{-5.52 \pm \sqrt{(5.52)^2 - 4(-0.45)(-13.70)}}{2(-0.45)}$

$x \approx 3.46, 8.81$

The zero of g on $[0, 6]$ is 3.46. The zero is close to the zero $\frac{10}{3} \approx 3.33$ of f.

(b) $g(x) = -0.45x^2 + 5.52x - 13.70$

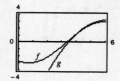

For $3.5 \le x \le 6$ the approximation appears to be good.

77. True. The period of $2 \sin 4t - 1$ is $\dfrac{\pi}{2}$ and the period of $2 \sin t - 1$ is 2π.

In the interval $[0, 2\pi)$ the first equation has four cycles whereas the second equation has only one cycle, thus the first equation has four times the x-intercepts (solutions) as the second equation.

79. $y_1 = 2\sin x$

$y_2 = 3x + 1$

From the graph we see that there is only one point of intersection.

81.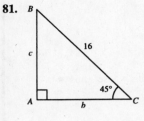

$B = 90° - 45° = 45°$

$\tan 45° = \dfrac{b}{c} = 1 \implies b = c$

$\cos 45° = \dfrac{b}{16}$

$16\left(\dfrac{\sqrt{2}}{2}\right) = b$

$b = 8\sqrt{2} \approx 11.31$

$c \approx 11.31$

83.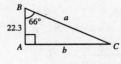

$C = 90° - 66° = 24°$

$\cos 66° = \dfrac{22.3}{a}$

$a \cos 66° = 22.3$

$a = \dfrac{22.3}{\cos 66°} \approx 54.8$

$\tan 66° = \dfrac{b}{22.3}$

$b = 22.3 \tan 66° \approx 50.1$

85. $\theta = 390°$, $\theta' = 390° - 360° = 30°$, θ is in Quadrant I.

$\sin 390° = \sin 30° = \dfrac{1}{2}$

$\cos 390° = \cos 30° = \dfrac{\sqrt{3}}{2}$

$\tan 390° = \tan 30° = \dfrac{1}{\sqrt{3}} = \dfrac{\sqrt{3}}{3}$

87. $\theta = 495°$, $\theta' = 45°$, θ is in Quadrant II.

$\sin 495° = \sin 45° \quad = \dfrac{\sqrt{2}}{2}$

$\cos 495° = -\cos 45° = -\dfrac{\sqrt{2}}{2}$

$\tan 495° = -\tan 45° = -1$

89. $\theta = -1845°$, $\theta' = 45°$, θ is in Quadrant IV.

$\sin(-1845°) = -\sin 45° = -\dfrac{\sqrt{2}}{2}$

$\cos(-1845°) = \cos 45° \quad = \dfrac{\sqrt{2}}{2}$

$\tan(-1845°) = -\tan 45° = -1$

91.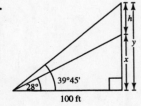

$h = y - x$

$\tan 39.75° = \dfrac{y}{100}$

$100 \tan 39.75° = y$

$\tan 28° = \dfrac{x}{100}$

$100 \tan 28° = x$

$h = 100 \tan 39.75° - 100 \tan 28°$

$h \approx 30$ feet

Section 2.4   Sum and Difference Formulas

- ■ You should know the sum and difference formulas.

  $$\sin(u \pm v) = \sin u \cos v \pm \cos u \sin v$$

  $$\cos(u \pm v) = \cos u \cos v \mp \sin u \sin v$$

  $$\tan(u \pm v) = \frac{\tan u \pm \tan v}{1 \mp \tan u \tan v}$$

- ■ You should be able to use these formulas to find the values of the trigonometric functions of angles whose sums or differences are special angles.

- ■ You should be able to use these formulas to solve trigonometric equations.

Solutions to Odd-Numbered Exercises

1. (a) $\cos\left(\dfrac{\pi}{4} + \dfrac{\pi}{3}\right) = \cos\dfrac{\pi}{4}\cos\dfrac{\pi}{3} - \sin\dfrac{\pi}{4}\sin\dfrac{\pi}{3}$

$$= \frac{\sqrt{2}}{2} \cdot \frac{1}{2} - \frac{\sqrt{2}}{2} \cdot \frac{\sqrt{3}}{2}$$

$$= \frac{\sqrt{2} - \sqrt{6}}{4}$$

(b) $\cos\dfrac{\pi}{4} + \cos\dfrac{\pi}{3} = \dfrac{\sqrt{2}}{2} + \dfrac{1}{2} = \dfrac{\sqrt{2} + 1}{2}$

3. (a) $\sin\left(\dfrac{7\pi}{6} - \dfrac{\pi}{3}\right) = \sin\dfrac{5\pi}{6} = \sin\dfrac{\pi}{6} = \dfrac{1}{2}$

(b) $\sin\dfrac{7\pi}{6} - \sin\dfrac{\pi}{3} = -\dfrac{1}{2} - \dfrac{\sqrt{3}}{2} = \dfrac{-1 - \sqrt{3}}{2}$

5. (a) $\cos(120° + 45°) = \cos 120° \cos 45° - \sin 120° \sin 45°$

$$= \left(-\frac{1}{2}\right)\left(\frac{\sqrt{2}}{2}\right) - \left(\frac{\sqrt{3}}{2}\right)\left(\frac{\sqrt{2}}{2}\right)$$

$$= \frac{-\sqrt{2} - \sqrt{6}}{4}$$

(b) $\cos 120° + \cos 45° = -\dfrac{1}{2} + \dfrac{\sqrt{2}}{2} = \dfrac{-1 + \sqrt{2}}{2}$

7. $\sin 105° = \sin(60° + 45°)$

$$= \sin 60° \cos 45° + \cos 60° \sin 45°$$

$$= \frac{\sqrt{3}}{2} \cdot \frac{\sqrt{2}}{2} + \frac{1}{2} \cdot \frac{\sqrt{2}}{2}$$

$$= \frac{\sqrt{2}}{4}\left(\sqrt{3} + 1\right)$$

$\cos 105° = \cos(60° + 45°)$

$$= \cos 60° \cos 45° - \sin 60° \sin 45°$$

$$= \frac{1}{2} \cdot \frac{\sqrt{2}}{2} - \frac{\sqrt{3}}{2} \cdot \frac{\sqrt{2}}{2}$$

$$= \frac{\sqrt{2}}{4}\left(1 - \sqrt{3}\right)$$

$\tan 105° = \tan(60° + 45°)$

$$= \frac{\tan 60° + \tan 45°}{1 - \tan 60° \tan 45°}$$

$$= \frac{\sqrt{3} + 1}{1 - \sqrt{3}} = \frac{\sqrt{3} + 1}{1 - \sqrt{3}} \cdot \frac{1 + \sqrt{3}}{1 + \sqrt{3}}$$

$$= \frac{4 + 2\sqrt{3}}{-2} = -2 - \sqrt{3}$$

9. $\sin 195° = \sin(225° - 30°)$

$\qquad = \sin 225° \cos 30° - \cos 225° \sin 30°$

$\qquad = -\sin 45° \cos 30° + \cos 45° \sin 30°$

$\qquad = -\dfrac{\sqrt{2}}{2} \cdot \dfrac{\sqrt{3}}{2} + \dfrac{\sqrt{2}}{2} \cdot \dfrac{1}{2}$

$\qquad = \dfrac{\sqrt{2}}{4}\left(1 - \sqrt{3}\right)$

$\cos 195° = \cos(225° - 30°)$

$\qquad = \cos 225° \cos 30° + \sin 225° \sin 30°$

$\qquad = -\cos 45° \cos 30° - \sin 45° \sin 30°$

$\qquad = -\dfrac{\sqrt{2}}{2} \cdot \dfrac{\sqrt{3}}{2} - \dfrac{\sqrt{2}}{2} \cdot \dfrac{1}{2}$

$\qquad = -\dfrac{\sqrt{2}}{4}\left(\sqrt{3} + 1\right)$

$\tan 195° = \tan(225° - 30°)$

$\qquad = \dfrac{\tan 225° - \tan 30°}{1 + \tan 225° \tan 30°}$

$\qquad = \dfrac{\tan 45° - \tan 30°}{1 + \tan 45° \tan 30°}$

$\qquad = \dfrac{1 - \left(\dfrac{\sqrt{3}}{3}\right)}{1 + \left(\dfrac{\sqrt{3}}{3}\right)} = \dfrac{3 - \sqrt{3}}{3 + \sqrt{3}} \cdot \dfrac{3 - \sqrt{3}}{3 - \sqrt{3}}$

$\qquad = \dfrac{12 - 6\sqrt{3}}{6} = 2 - \sqrt{3}$

11. $\sin \dfrac{11\pi}{12} = \sin\left(\dfrac{3\pi}{4} + \dfrac{\pi}{6}\right)$

$\qquad = \sin \dfrac{3\pi}{4} \cos \dfrac{\pi}{6} + \cos \dfrac{3\pi}{4} \sin \dfrac{\pi}{6}$

$\qquad = \dfrac{\sqrt{2}}{2} \cdot \dfrac{\sqrt{3}}{2} + \left(-\dfrac{\sqrt{2}}{2}\right)\dfrac{1}{2}$

$\qquad = \dfrac{\sqrt{2}}{4}\left(\sqrt{3} - 1\right)$

$\cos \dfrac{11\pi}{12} = \cos\left(\dfrac{3\pi}{4} + \dfrac{\pi}{6}\right)$

$\qquad = \cos \dfrac{3\pi}{4} \cos \dfrac{\pi}{6} - \sin \dfrac{3\pi}{4} \sin \dfrac{\pi}{6}$

$\qquad = -\dfrac{\sqrt{2}}{2} \cdot \dfrac{\sqrt{3}}{2} - \dfrac{\sqrt{2}}{2} \cdot \dfrac{1}{2}$

$\qquad = -\dfrac{\sqrt{2}}{4}\left(\sqrt{3} + 1\right)$

$\tan \dfrac{11\pi}{4} = \tan\left(\dfrac{3\pi}{4} + \dfrac{\pi}{6}\right)$

$\qquad = \dfrac{\tan \dfrac{3\pi}{4} + \tan \dfrac{\pi}{6}}{1 - \tan \dfrac{3\pi}{4} \tan \dfrac{\pi}{6}}$

$\qquad = \dfrac{-1 + \dfrac{\sqrt{3}}{3}}{1 - (-1)\dfrac{\sqrt{3}}{3}}$

$\qquad = \dfrac{-3 + \sqrt{3}}{3 + \sqrt{3}} \cdot \dfrac{3 - \sqrt{3}}{3 - \sqrt{3}}$

$\qquad = \dfrac{-12 + 6\sqrt{3}}{6} = -2 + \sqrt{3}$

13. $\sin \dfrac{17\pi}{12} = \sin\left(\dfrac{9\pi}{4} - \dfrac{5\pi}{6}\right)$

$\qquad = \sin \dfrac{9\pi}{4} \cos \dfrac{5\pi}{6} - \cos \dfrac{9\pi}{4} \sin \dfrac{5\pi}{6}$

$\qquad = \dfrac{\sqrt{2}}{2}\left(-\dfrac{\sqrt{3}}{2}\right) - \left(\dfrac{\sqrt{2}}{2}\right)\left(\dfrac{1}{2}\right)$

$\qquad = -\dfrac{\sqrt{2}}{4}\left(\sqrt{3} + 1\right)$

$\cos \dfrac{17\pi}{12} = \cos\left(\dfrac{9\pi}{4} - \dfrac{5\pi}{6}\right)$

$\qquad = \cos \dfrac{9\pi}{4} \cos \dfrac{5\pi}{6} + \sin \dfrac{9\pi}{4} \sin \dfrac{5\pi}{6}$

$\qquad = \dfrac{\sqrt{2}}{2}\left(-\dfrac{\sqrt{3}}{2}\right) + \dfrac{\sqrt{2}}{2}\left(\dfrac{1}{2}\right)$

$\qquad = \dfrac{\sqrt{2}}{4}\left(1 - \sqrt{3}\right)$

$\tan \dfrac{17\pi}{12} = \tan\left(\dfrac{9\pi}{4} - \dfrac{5\pi}{6}\right)$

$\qquad = \dfrac{\tan(9\pi/4) - \tan(5\pi/6)}{1 + \tan(9\pi/4)\tan(5\pi/6)}$

$\qquad = \dfrac{1 - \left(-\sqrt{3}/3\right)}{1 + \left(-\sqrt{3}/3\right)}$

$\qquad = \dfrac{3 + \sqrt{3}}{3 - \sqrt{3}} \cdot \dfrac{3 + \sqrt{3}}{3 + \sqrt{3}}$

$\qquad = \dfrac{12 + 6\sqrt{3}}{6} = 2 + \sqrt{3}$

15. $\qquad 285° = 225° + 60°$

$\sin 285° = \sin(225° + 60°)$

$\qquad = \sin 225° \cos 60° + \cos 225° \sin 60°$

$\qquad = -\dfrac{\sqrt{2}}{2}\left(\dfrac{1}{2}\right) - \dfrac{\sqrt{2}}{2}\left(\dfrac{\sqrt{3}}{2}\right) = -\dfrac{\sqrt{2}}{4}\left(\sqrt{3} + 1\right)$

$\cos 285° = \cos(225° + 60°)$

$\qquad = \cos 225° \cos 60° - \sin 225° \sin 60°$

$\qquad = -\dfrac{\sqrt{2}}{2}\left(\dfrac{1}{2}\right) - \left(-\dfrac{\sqrt{2}}{2}\right)\left(\dfrac{\sqrt{3}}{2}\right) = \dfrac{\sqrt{2}}{4}\left(\sqrt{3} - 1\right)$

$\tan 285° = \tan(225° + 60°)$

$\qquad = \dfrac{\tan 225° + \tan 60°}{1 - \tan 225° \tan 60°} = \dfrac{1 + \sqrt{3}}{1 - \sqrt{3}} \cdot \dfrac{1 + \sqrt{3}}{1 + \sqrt{3}}$

$\qquad = \dfrac{4 + 2\sqrt{3}}{-2} = -2 - \sqrt{3} = -\left(2 + \sqrt{3}\right)$

17. $\qquad -165° = -(120° + 45°)$

$\sin(-165°) = \sin[-(120° + 45°)]$

$\qquad = -\sin(120° + 45°)$

$\qquad = -[\sin 120° \cos 45° + \cos 120° \sin 45°]$

$\qquad = -\left[\dfrac{\sqrt{3}}{2} \cdot \dfrac{\sqrt{2}}{2} - \dfrac{1}{2} \cdot \dfrac{\sqrt{2}}{2}\right]$

$\qquad = -\dfrac{\sqrt{2}}{4}\left(\sqrt{3} - 1\right)$

$\cos(-165°) = \cos[-(120° + 45°)]$

$\qquad = \cos(120° + 45°)$

$\qquad = \cos 120° \cos 45° - \sin 120° \sin 45°$

$\qquad = -\dfrac{1}{2} \cdot \dfrac{\sqrt{2}}{2} - \dfrac{\sqrt{3}}{2} \cdot \dfrac{\sqrt{2}}{2}$

$\qquad = -\dfrac{\sqrt{2}}{4}\left(1 + \sqrt{3}\right)$

$\tan(-165°) = \tan[-(120° + 45°)]$

$\qquad = -\tan(120° + \tan 45°)$

$\qquad = -\dfrac{\tan 120° + \tan 45°}{1 - \tan 120° \tan 45°}$

$\qquad = -\dfrac{-\sqrt{3} + 1}{1 - (-\sqrt{3})(1)}$

$\qquad = -\dfrac{1 - \sqrt{3}}{1 + \sqrt{3}} \cdot \dfrac{1 - \sqrt{3}}{1 - \sqrt{3}}$

$\qquad = -\dfrac{4 - 2\sqrt{3}}{-2}$

$\qquad = 2 - \sqrt{3}$

19. $\dfrac{13\pi}{12} = \dfrac{3\pi}{4} + \dfrac{\pi}{3}$

$\sin\dfrac{13\pi}{12} = \sin\left(\dfrac{3\pi}{4} + \dfrac{\pi}{3}\right)$

$\qquad = \sin\dfrac{3\pi}{4}\cos\dfrac{\pi}{3} + \cos\dfrac{3\pi}{4}\sin\dfrac{\pi}{3}$

$\qquad = \dfrac{\sqrt{2}}{2}\cdot\dfrac{1}{2} + \left(-\dfrac{\sqrt{2}}{2}\right)\left(\dfrac{\sqrt{3}}{2}\right)$

$\qquad = \dfrac{\sqrt{2}}{4}\left(1 - \sqrt{3}\right)$

$\cos\dfrac{13\pi}{12} = \cos\left(\dfrac{3\pi}{4} + \dfrac{\pi}{3}\right)$

$\qquad = \cos\dfrac{3\pi}{4}\cos\dfrac{\pi}{3} - \sin\dfrac{3\pi}{4}\sin\dfrac{\pi}{3}$

$\qquad = -\dfrac{\sqrt{2}}{2}\cdot\dfrac{1}{2} - \dfrac{\sqrt{2}}{2}\cdot\dfrac{\sqrt{3}}{2} = -\dfrac{\sqrt{2}}{4}\left(1 + \sqrt{3}\right)$

$\tan\dfrac{13\pi}{12} = \tan\left(\dfrac{3\pi}{4} + \dfrac{\pi}{3}\right)$

$\qquad = \dfrac{\tan\left(\dfrac{3\pi}{4}\right) + \tan\left(\dfrac{\pi}{3}\right)}{1 - \tan\left(\dfrac{3\pi}{4}\right)\tan\left(\dfrac{\pi}{3}\right)}$

$\qquad = \dfrac{-1 + \sqrt{3}}{1 - (-1)\left(\sqrt{3}\right)}$

$\qquad = -\dfrac{1 - \sqrt{3}}{1 + \sqrt{3}}\cdot\dfrac{1 - \sqrt{3}}{1 - \sqrt{3}}$

$\qquad = -\dfrac{4 - 2\sqrt{3}}{-2}$

$\qquad = 2 - \sqrt{3}$

21. $-\dfrac{13\pi}{12} = -\left(\dfrac{3\pi}{4} + \dfrac{\pi}{3}\right)$

$\sin\left[-\left(\dfrac{3\pi}{4} + \dfrac{\pi}{3}\right)\right] = -\sin\left(\dfrac{3\pi}{4} + \dfrac{\pi}{3}\right)$

$\qquad = -\left[\sin\dfrac{3\pi}{4}\cos\dfrac{\pi}{3} + \cos\dfrac{3\pi}{4}\sin\dfrac{\pi}{3}\right]$

$\qquad = -\left[\dfrac{\sqrt{2}}{2}\left(\dfrac{1}{2}\right) + \left(-\dfrac{\sqrt{2}}{2}\right)\left(\dfrac{\sqrt{3}}{2}\right)\right]$

$\qquad = -\dfrac{\sqrt{2}}{4}\left(1 - \sqrt{3}\right) = \dfrac{\sqrt{2}}{4}\left(\sqrt{3} - 1\right)$

$\cos\left[-\left(\dfrac{3\pi}{4} + \dfrac{\pi}{3}\right)\right] = \cos\left(\dfrac{3\pi}{4} + \dfrac{\pi}{3}\right)$

$\qquad = \cos\dfrac{3\pi}{4}\cos\dfrac{\pi}{3} - \sin\dfrac{3\pi}{4}\sin\dfrac{\pi}{3}$

$\qquad = -\dfrac{\sqrt{2}}{2}\left(\dfrac{1}{2}\right) - \dfrac{\sqrt{2}}{2}\left(\dfrac{\sqrt{3}}{2}\right) = -\dfrac{\sqrt{2}}{4}\left(\sqrt{3} + 1\right)$

$\tan\left[-\left(\dfrac{3\pi}{4} + \dfrac{\pi}{3}\right)\right] = -\tan\left(\dfrac{3\pi}{4} + \dfrac{\pi}{3}\right)$

$\qquad = -\dfrac{\tan\dfrac{3\pi}{4} + \tan\dfrac{\pi}{3}}{1 - \tan\dfrac{3\pi}{4}\tan\dfrac{\pi}{3}} = -\dfrac{-1 + \sqrt{3}}{1 - \left(-\sqrt{3}\right)}$

$\qquad = \dfrac{1 - \sqrt{3}}{1 + \sqrt{3}}\cdot\dfrac{1 - \sqrt{3}}{1 - \sqrt{3}} = \dfrac{4 - 2\sqrt{3}}{-2} = -2 + \sqrt{3}$

23. $\cos 25° \cos 15° - \sin 25° \sin 15° = \cos(25° + 15°) = \cos 40°$

25. $\dfrac{\tan 325° - \tan 86°}{1 + \tan 325° \tan 86°} = \tan(325° - 86°) = \tan 239°$

27. $\sin 3 \cos 1.2 - \cos 3 \sin 1.2 = \sin(3 - 1.2) = \sin 1.8$

29. $\dfrac{\tan 2x + \tan x}{1 - \tan 2x \tan x} = \tan(2x + x) = \tan 3x$

31. $\sin 330° \cos 30° - \cos 330° \sin 30° = \sin(330° - 30°)$

$$= \sin 300°$$
$$= -\dfrac{\sqrt{3}}{2}$$

33. $\sin \dfrac{\pi}{12} \cos \dfrac{\pi}{4} + \cos \dfrac{\pi}{12} \sin \dfrac{\pi}{4} = \sin\left(\dfrac{\pi}{12} + \dfrac{\pi}{4}\right)$

$$= \sin \dfrac{\pi}{3}$$
$$= \dfrac{\sqrt{3}}{2}$$

35. $\dfrac{\tan 25° + \tan 110°}{1 - \tan 25° \tan 110°} = \tan(25° + 110°)$

$$= \tan 135°$$
$$= -1$$

For Exercises 37 – 43, we have:

$$\sin u = \tfrac{5}{13}, \ u \text{ in Quadrant II} \implies \cos u = -\tfrac{12}{13}, \tan u = -\tfrac{5}{12}$$
$$\cos v = -\tfrac{3}{5}, \ v \text{ in Quadrant II} \implies \sin v = \tfrac{4}{5}, \tan v = -\tfrac{4}{3},$$

37. $\sin(u + v) = \sin u \cos v + \cos u \sin v$

$$= \left(\tfrac{5}{13}\right)\left(-\tfrac{3}{5}\right) + \left(-\tfrac{12}{13}\right)\left(\tfrac{4}{5}\right)$$
$$= -\tfrac{63}{65}$$

39. $\cos(u + v) = \cos u \cos v - \sin u \sin v$

$$= \left(-\tfrac{12}{13}\right)\left(-\tfrac{3}{5}\right) - \left(\tfrac{5}{13}\right)\left(\tfrac{4}{5}\right)$$
$$= \tfrac{16}{65}$$

41. $\tan(u + v) = \dfrac{\tan u + \tan v}{1 - \tan u \tan v} = \dfrac{-\tfrac{5}{12} + \left(-\tfrac{4}{3}\right)}{1 - \left(-\tfrac{5}{12}\right)\left(-\tfrac{4}{3}\right)} = \dfrac{-\tfrac{21}{12}}{1 - \tfrac{5}{9}}$

$$= \left(-\tfrac{7}{4}\right)\left(\tfrac{9}{4}\right) = -\dfrac{63}{16}$$

43. $\sec(v - u) = \dfrac{1}{\cos(v - u)} = \dfrac{1}{\cos v \cos u + \sin v \sin u}$

$$= \dfrac{1}{\left(-\tfrac{3}{5}\right)\left(-\tfrac{12}{13}\right) + \left(\tfrac{4}{5}\right)\left(\tfrac{5}{13}\right)} = \dfrac{1}{\left(\tfrac{36}{65}\right) + \left(\tfrac{20}{65}\right)} = \dfrac{1}{\tfrac{56}{65}}$$
$$= \dfrac{65}{56}$$

For Exercises 45–49, we have:

$\sin u = -\frac{7}{25}$, u in Quadrant III $\Longrightarrow \cos u = -\frac{24}{25}$, $\tan u = \frac{7}{24}$

$\cos v = -\frac{4}{5}$, v in Quadrant III $\Longrightarrow \sin v = -\frac{3}{5}$, $\tan v = \frac{3}{4}$

45. $\cos(u + v) = \cos u \cos v - \sin u \sin v$

$$= \left(-\frac{24}{25}\right)\left(-\frac{4}{5}\right) - \left(-\frac{7}{25}\right)\left(-\frac{3}{5}\right)$$

$$= \frac{3}{5}$$

47. $\tan(u - v) = \dfrac{\tan u - \tan v}{1 + \tan u \tan v}$

$$= \frac{\frac{7}{24} - \frac{3}{4}}{1 + \left(\frac{7}{24}\right)\left(\frac{3}{4}\right)} = \frac{-\frac{11}{24}}{\frac{39}{32}} = -\frac{44}{117}$$

49. $\sec(u + v) = \dfrac{1}{\cos(u + v)} = \dfrac{1}{\frac{3}{5}} = \dfrac{5}{3}$

Use Exercise 45 for $\cos(u + v)$.

51. $\sin(\arcsin x + \arccos x) = \sin(\arcsin x)\cos(\arccos x) + \sin(\arccos x)\cos(\arcsin x)$

$$= x \cdot x + \sqrt{1 - x^2} \cdot \sqrt{1 - x^2}$$

$$= x^2 + 1 - x^2$$

$$= 1$$

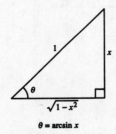

$\theta = \arcsin x$

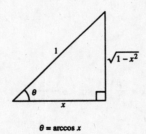

$\theta = \arccos x$

53. $\cos(\arccos x + \arcsin x) = \cos(\arccos x)\cos(\arcsin x) - \sin(\arccos x)\sin(\arcsin x)$

$$= x \cdot \sqrt{1 - x^2} - \sqrt{1 - x^2} \cdot x$$

$$= 0$$

(Use the triangles in Exercise 51.)

55. $\sin(3\pi - x) = \sin 3\pi \cos x - \sin x \cos 3\pi = (0)(\cos x) - (-1)(\sin x) = \sin x$

57. $\sin\left(\frac{\pi}{6} + x\right) = \sin\frac{\pi}{6}\cos x + \cos\frac{\pi}{6}\sin x = \frac{1}{2}\left(\cos x + \sqrt{3}\sin x\right)$

59. $\cos(\pi - \theta) + \sin\left(\frac{\pi}{2} + \theta\right) = \cos\pi\cos\theta + \sin\pi\sin\theta + \sin\frac{\pi}{2}\cos\theta + \cos\frac{\pi}{2}\sin\theta$

$$= (-1)(\cos\theta) + (0)(\sin\theta) + (1)(\cos\theta) + (\sin\theta)(0)$$

$$= -\cos\theta + \cos\theta$$

$$= 0$$

61. $\cos(x + y) \cos(x - y) = (\cos x \cos y - \sin x \sin y)(\cos x \cos y + \sin x \sin y)$

$$= \cos^2 x \cos^2 y - \sin^2 x \sin^2 y$$

$$= \cos^2 x(1 - \sin^2 y) - \sin^2 x \sin^2 y$$

$$= \cos^2 x - \cos^2 x \sin^2 y - \sin^2 x \sin^2 y$$

$$= \cos^2 x - \sin^2 y(\cos^2 x + \sin^2 x)$$

$$= \cos^2 x - \sin^2 y$$

63. $\sin(x + y) + \sin(x - y) = \sin x \cos y + \cos x \sin y + \sin x \cos y - \cos x \sin y$

$$= 2 \sin x \cos y$$

65. $\cos\left(\dfrac{3\pi}{2} - x\right) = \cos\dfrac{3\pi}{2} \cos x + \sin\dfrac{3\pi}{2} \sin x$

$$= (0)(\cos x) + (-1)(\sin x)$$

$$= -\sin x$$

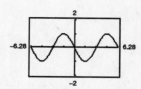

67. $\sin\left(\dfrac{3\pi}{2} - \theta\right) = \sin\dfrac{3\pi}{2} \cos \theta + \cos\dfrac{3\pi}{2} \sin \theta$

$$= (-1)(\cos \theta) + (0)(\sin \theta)$$

$$= -\cos \theta$$

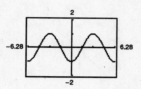

69.
$$\sin\left(x + \dfrac{\pi}{3}\right) + \sin\left(x - \dfrac{\pi}{3}\right) = 1$$

$$\sin x \cos \dfrac{\pi}{3} + \cos x \sin \dfrac{\pi}{3} + \sin x \cos \dfrac{\pi}{3} - \cos x \sin \dfrac{\pi}{3} = 1$$

$$2 \sin x(0.5) = 1$$

$$\sin x = 1$$

$$x = \dfrac{\pi}{2}$$

71.
$$\cos\left(x + \frac{\pi}{4}\right) - \cos\left(x - \frac{\pi}{4}\right) = 1$$

$$\cos x \cos \frac{\pi}{4} - \sin x \sin \frac{\pi}{4} - \left(\cos x \cos \frac{\pi}{4} + \sin x \sin \frac{\pi}{4}\right) = 1$$

$$-2 \sin x \left(\frac{\sqrt{2}}{2}\right) = 1$$

$$-\sqrt{2} \sin x = 1$$

$$\sin x = -\frac{1}{\sqrt{2}}$$

$$\sin x = -\frac{\sqrt{2}}{2}$$

$$x = \frac{5\pi}{4}, \frac{7\pi}{4}$$

73.   Analytically: $\cos\left(x + \frac{\pi}{4}\right) + \cos\left(x - \frac{\pi}{4}\right) = 1$

$$\cos x \cos \frac{\pi}{4} - \sin x \sin \frac{\pi}{4} + \cos x \cos \frac{\pi}{4} + \sin x \sin \frac{\pi}{4} = 1$$

$$2 \cos x \left(\frac{\sqrt{2}}{2}\right) = 1$$

$$\sqrt{2} \cos x = 1$$

$$\cos x = \frac{1}{\sqrt{2}}$$

$$\cos x = \frac{\sqrt{2}}{2}$$

$$x = \frac{\pi}{4}, \frac{7\pi}{4}$$

Graphically: Graph $y_1 = \cos\left(x + \frac{\pi}{4}\right) + \cos\left(x - \frac{\pi}{4}\right)$ and $y_2 = 1$.

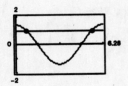

The points of intersection occur at $x = \frac{\pi}{4}$ and $x = \frac{7\pi}{4}$.

75. $y = \dfrac{1}{3} \sin 2t + \dfrac{1}{4} \cos 2t$

(a) $a = \dfrac{1}{3}, \; b = \dfrac{1}{4}, \; B = 2$

$C = \arctan \dfrac{b}{a} = \arctan \dfrac{3}{4} \approx 0.6435$

$y \approx \sqrt{\left(\dfrac{1}{3}\right)^2 + \left(\dfrac{1}{4}\right)^2} \; \sin(2t + 0.6435)$

$= \dfrac{5}{12} \sin(2t + 0.6435)$

(b) Amplitude: $\dfrac{5}{12}$ foot

(c) Frequency: $\dfrac{1}{\text{period}} = \dfrac{B}{2\pi} = \dfrac{2}{2\pi} = \dfrac{1}{\pi}$ cycles per second

77. False. $\sin(u \pm v) = \sin u \cos v \pm \cos u \sin v$.

In Exercises 1–6, parts (a) and (b) are unequal.

79. False. $\cos\left(x - \dfrac{\pi}{2}\right) = \cos x \cos \dfrac{\pi}{2} + \sin x \sin \dfrac{\pi}{2}$

$= (\cos x)(0) + (\sin x)(1)$

$= \sin x$

81. $\cos(n\pi + \theta) = \cos n\pi \cos \theta - \sin n\pi \sin \theta$

$= (-1)^n (\cos \theta) - (0)(\sin \theta)$

$= (-1)^n (\cos \theta)$, where n is an integer.

83. $C = \arctan \dfrac{b}{a} \implies \sin C = \dfrac{b}{\sqrt{a^2 + b^2}}, \cos C = \dfrac{a}{\sqrt{a^2 + b^2}}$

$\sqrt{a^2 + b^2} \sin(B\theta + C) = \sqrt{a^2 + b^2} \left(\sin B\theta \cdot \dfrac{a}{\sqrt{a^2 + b^2}} + \dfrac{b}{\sqrt{a^2 + b^2}} \cdot \cos B\theta \right) = a \sin B\theta + b \cos B\theta$

85. $\sin \theta + \cos \theta$

$a = 1, \; b = 1, \; B = 1$

(a) $C = \arctan \dfrac{b}{a} = \arctan 1 = \dfrac{\pi}{4}$

$\sin \theta + \cos \theta = \sqrt{a^2 + b^2} \sin(B\theta + C)$

$= \sqrt{2} \sin\left(\theta + \dfrac{\pi}{4}\right)$

(b) $C = \arctan \dfrac{a}{b} = \arctan 1 = \dfrac{\pi}{4}$

$\sin \theta + \cos \theta = \sqrt{a^2 + b^2} \cos(B\theta - C)$

$= \sqrt{2} \cos\left(\theta - \dfrac{\pi}{4}\right)$

87. $12 \sin 3\theta + 5 \cos 3\theta$

$a = 12, \; b = 5, \; B = 3$

(a) $C = \arctan \dfrac{b}{a} = \arctan \dfrac{5}{12} \approx 0.3948$

$12 \sin 3\theta + 5 \cos 3\theta = \sqrt{a^2 + b^2} \sin(B\theta + C)$

$\approx 13 \sin(3\theta + 0.3948)$

(b) $C = \arctan \dfrac{a}{b} = \arctan \dfrac{12}{5} \approx 1.1760$

$12 \sin 3\theta + 5 \cos 3\theta = \sqrt{a^2 + b^2} \cos(B\theta - C)$

$\approx 13 \cos(3\theta - 1.1760)$

89. $C = \arctan \dfrac{b}{a} = \dfrac{\pi}{2} \implies a = 0$

$\sqrt{a^2 + b^2} = 2 \implies b = 2$

$B = 1$

$2 \sin\left(\theta + \dfrac{\pi}{2}\right) = (0)(\sin\theta) + (2)(\cos\theta) = 2\cos\theta$

91.

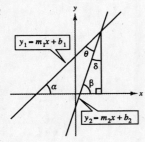

$m_1 = \tan \alpha$ and $m_2 = \tan \beta$

$\beta + \delta = 90° \implies \delta = 90° - \beta$

$\alpha + \theta + \delta = 90° \implies \alpha + \theta + (90° - \beta) = 90° \implies \theta = \beta - \alpha$

Therefore, $\theta = \arctan m_2 - \arctan m_1$

For $y = x$ and $y = \sqrt{3}x$ we have $m_1 = 1$ and $m_2 = \sqrt{3}$

$\theta = \arctan\sqrt{3} - \arctan 1$

$\quad = 60° - 45°$

$\quad = 15°$

93. $\sin^2\left(\theta + \dfrac{\pi}{4}\right) + \sin^2\left(\theta - \dfrac{\pi}{4}\right) = \left[\sin\theta\cos\dfrac{\pi}{4} + \cos\theta\sin\dfrac{\pi}{4}\right]^2 + \left[\sin\theta\cos\dfrac{\pi}{4} - \cos\theta\sin\dfrac{\pi}{4}\right]^2$

$= \left[\dfrac{\sin\theta}{\sqrt{2}} + \dfrac{\cos\theta}{\sqrt{2}}\right]^2 + \left[\dfrac{\sin\theta}{\sqrt{2}} - \dfrac{\cos\theta}{\sqrt{2}}\right]^2$

$= \dfrac{\sin^2\theta}{2} + \sin\theta\cos\theta + \dfrac{\cos^2\theta}{2} + \dfrac{\sin^2\theta}{2} - \sin\theta\cos\theta + \dfrac{\cos^2\theta}{2}$

$= \sin^2\theta + \cos^2\theta$

$= 1$

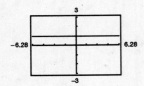

95. To prove the identity for $\sin(u + v)$ we first need to prove the identity for $\cos(u - v)$. Assume $0 < v < u < 2\pi$ and locate u, v, and $u - v$ on the unit circle.

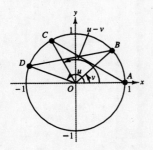

— CONTINUED —

95. — CONTINUED —

The coordinates of the points on the circle are:

$A = (1, 0)$, $B = (\cos v, \sin v)$, $C = (\cos(u - v), \sin(u - v))$, and $D = (\cos u, \sin u)$.

Since $\angle DOB = \angle COA$, chords AC and BD are equal. By the distance formula we have:

$$\sqrt{[\cos(u - v) - 1]^2 + [\sin(u - v) - 0]^2} = \sqrt{(\cos u - \cos v)^2 + (\sin u - \sin v)^2}$$

$$\cos^2(u - v) - 2\cos(u - v) + 1 + \sin^2(u - v) = \cos^2 u - 2\cos u \cos v + \cos^2 v + \sin^2 u - 2\sin u \sin v + \sin^2 v$$

$$[\cos^2(u + v) + \sin^2(u - v)] + 1 - 2\cos(u - v) = (\cos^2 u + \sin^2 u) + (\cos^2 v + \sin^2 v) - 2\cos u \cos v - 2\sin u \sin v$$

$$2 - 2\cos(u - v) = 2 - 2\cos u \cos v - 2\sin u \sin v$$

$$-2\cos(u - v) = -2(\cos u \cos v + \sin u \sin v)$$

$$\cos(u - v) = \cos u \cos v + \sin u \sin v$$

Now, to prove the identity for $\sin(u + v)$, use cofunction identities.

$$\sin(u + v) = \cos\left[\frac{\pi}{2} - (u + v)\right] = \cos\left[\left(\frac{\pi}{2} - u\right) - v\right]$$

$$= \cos\left(\frac{\pi}{2} - u\right)\cos v + \sin\left(\frac{\pi}{2} - u\right)\sin v$$

$$= \sin u \cos v + \cos u \sin v$$

97. $\quad f(x) = 5(x - 3)$

$\qquad y = 5(x - 3)$

$\qquad \dfrac{y}{5} = x - 3$

$\qquad \dfrac{y}{5} + 3 = x$

$\qquad \dfrac{x}{5} + 3 = y$

$\qquad f^{-1}(x) = \dfrac{x + 15}{5}$

$f(f^{-1}(x)) = f\left(\dfrac{x + 15}{5}\right) = 5\left[\dfrac{x + 15}{5} - 3\right]$

$\qquad\qquad = 5\left(\dfrac{x + 15}{5}\right) - 5(3)$

$\qquad\qquad = x + 15 - 15$

$\qquad\qquad = x$

$f^{-1}(f(x)) = f^{-1}(5(x - 3)) = \dfrac{5(x - 3) + 15}{5}$

$\qquad\qquad = \dfrac{5x - 15 + 15}{5}$

$\qquad\qquad = \dfrac{5x}{5}$

$\qquad\qquad = x$

99. $f(x) = x^2 - 8$

f is not one-to-one so f^{-1} does not exist.

Section 2.5   Multiple-Angle and Product-to-Sum Formulas

■ You should know the following double-angle formulas.

(a) $\sin 2u = 2 \sin u \cos u$

(b) $\cos 2u = \cos^2 u - \sin^2 u$

$\qquad = 2 \cos^2 u - 1$

$\qquad = 1 - 2 \sin^2 u$

(c) $\tan 2u = \dfrac{2 \tan u}{1 - \tan^2 u}$

■ You should be able to reduce the power of a trigonometric function.

(a) $\sin^2 u = \dfrac{1 - \cos 2u}{2}$

(b) $\cos^2 u = \dfrac{1 + \cos 2u}{2}$

(c) $\tan^2 u = \dfrac{1 - \cos 2u}{1 + \cos 2u}$

■ You should be able to use the half-angle formulas.

(a) $\sin \dfrac{u}{2} = \pm \sqrt{\dfrac{1 - \cos u}{2}}$

(b) $\cos \dfrac{u}{2} = \pm \sqrt{\dfrac{1 + \cos u}{2}}$

(c) $\tan \dfrac{u}{2} = \dfrac{1 - \cos u}{\sin u} = \dfrac{\sin u}{1 + \cos u}$

■ You should be able to use the product-sum formulas.

(a) $\sin u \sin v = \dfrac{1}{2} \left[\cos(u - v) - \cos(u + v) \right]$

(b) $\cos u \cos v = \dfrac{1}{2} \left[\cos(u - v) + \cos(u + v) \right]$

(c) $\sin u \cos v = \dfrac{1}{2} \left[\sin(u + v) + \sin(u - v) \right]$

(d) $\cos u \sin v = \dfrac{1}{2} \left[\sin(u + v) - \sin(u - v) \right]$

■ You should be able to use the sum-product formulas.

(a) $\sin x + \sin y = 2 \sin \left(\dfrac{x + y}{2} \right) \cos \left(\dfrac{x - y}{2} \right)$

(b) $\sin x - \sin y = 2 \cos \left(\dfrac{x + y}{2} \right) \sin \left(\dfrac{x - y}{2} \right)$

(c) $\cos x + \cos y = 2 \cos \left(\dfrac{x + y}{2} \right) \cos \left(\dfrac{x - y}{2} \right)$

(d) $\cos x - \cos y = -2 \sin \left(\dfrac{x + y}{2} \right) \sin \left(\dfrac{x - y}{2} \right)$

Solutions to Odd-Numbered Exercises

Figure for Exercises 1–7

$$\sin \theta = \frac{\sqrt{17}}{17}$$

$$\cos \theta = \frac{4\sqrt{17}}{17}$$

$$\tan \theta = \frac{1}{4}$$

1. $\sin \theta = \dfrac{\sqrt{17}}{17}$

3. $\cos 2\theta = 2\cos^2 \theta - 1$

$$= 2\left(\frac{4\sqrt{17}}{17}\right)^2 - 1$$

$$= \frac{32}{17} - 1$$

$$= \frac{15}{17}$$

5. $\tan 2\theta = \dfrac{2\tan \theta}{1 - \tan^2 \theta}$

$$= \frac{2\left(\dfrac{1}{4}\right)}{1 - \left(\dfrac{1}{4}\right)^2}$$

$$= \frac{\dfrac{1}{2}}{1 - \dfrac{1}{16}}$$

$$= \frac{1}{2} \cdot \frac{16}{15}$$

$$= \frac{8}{15}$$

7. $\csc 2\theta = \dfrac{1}{\sin 2\theta}$

$$= \frac{1}{2\sin \theta \cos \theta}$$

$$= \frac{1}{2\left(\dfrac{\sqrt{17}}{17}\right)\left(\dfrac{4\sqrt{17}}{17}\right)}$$

$$= \frac{17}{8}$$

9. $\sin 2x - \sin x = 0$

$2\sin x \cos x - \sin x = 0$

$\sin x(2\cos x - 1) = 0$

$\sin x = 0$ or $2\cos x - 1 = 0$

$x = 0, \ \pi$ $\cos x = \dfrac{1}{2}$

$x = \dfrac{\pi}{3}, \dfrac{5\pi}{3}$

$x = 0, \ \dfrac{\pi}{3}, \ \pi, \ \dfrac{5\pi}{3}$

11. $4\sin x \cos x = 1$

$2\sin 2x = 1$

$\sin 2x = \dfrac{1}{2}$

$2x = \dfrac{\pi}{6} + 2n\pi$ or $2x = \dfrac{5\pi}{6} + 2n\pi$

$x = \dfrac{\pi}{12} + n\pi$ $x = \dfrac{5\pi}{12} + n\pi$

$x = \dfrac{\pi}{12}, \dfrac{13\pi}{12}$ $x = \dfrac{5\pi}{12}, \dfrac{17\pi}{12}$

13.
$$\cos 2x = \cos x$$
$$\cos^2 x - \sin^2 x = \cos x$$
$$\cos^2 x - (1 - \cos^2 x) - \cos x = 0$$
$$2 \cos^2 x - \cos x - 1 = 0$$
$$(2 \cos x + 1)(\cos x - 1) = 0$$

$2 \cos x + 1 = 0 \qquad \text{or} \quad \cos x - 1 = 0$

$$\cos x = -\frac{1}{2} \qquad\qquad \cos x = 1$$

$$x = \frac{2\pi}{3}, \frac{4\pi}{3} \qquad\qquad x = 0$$

15.
$$\tan 2x - \cot x = 0$$
$$\frac{2 \tan x}{1 - \tan^2 x} = \cot x$$
$$2 \tan x = \cot x(1 - \tan^2 x)$$
$$2 \tan x = \cot x - \cot x \tan^2 x$$
$$2 \tan x = \cot x - \tan x$$
$$3 \tan x = \cot x$$
$$3 \tan x - \cot x = 0$$
$$3 \tan x - \frac{1}{\tan x} = 0$$
$$\frac{3 \tan^2 x - 1}{\tan x} = 0$$
$$\frac{1}{\tan x}(3 \tan^2 x - 1) = 0$$
$$\cot x(3 \tan^2 x - 1) = 0$$

$\cot x = 0 \qquad \text{or} \quad 3 \tan^2 x - 1 = 0$

$$x = \frac{\pi}{2}, \frac{3\pi}{2} \qquad\qquad \tan^2 x = \frac{1}{3}$$

$$\tan x = \pm\frac{\sqrt{3}}{3}$$

$$x = \frac{\pi}{6}, \frac{5\pi}{6}, \frac{7\pi}{6}, \frac{11\pi}{6}$$

$$x = \frac{\pi}{6}, \frac{\pi}{2}, \frac{5\pi}{6}, \frac{7\pi}{6}, \frac{3\pi}{2}, \frac{11\pi}{6}$$

17.
$$\sin 4x = -2 \sin 2x$$
$$\sin 4x + 2 \sin 2x = 0$$
$$2 \sin 2x \cos 2x + 2 \sin 2x = 0$$
$$2 \sin 2x(\cos 2x + 1) = 0$$

$2 \sin 2x = 0 \qquad\qquad \text{or} \qquad \cos 2x + 1 = 0$

$\sin 2x = 0 \qquad\qquad\qquad \cos 2x = -1$

$2x = n\pi \qquad\qquad\qquad 2x = \pi + 2n\pi$

$x = \dfrac{n}{2}\pi \qquad\qquad\qquad x = \dfrac{\pi}{2} + n\pi$

$x = 0, \dfrac{\pi}{2}, \pi, \dfrac{3\pi}{2} \qquad\qquad x = \dfrac{\pi}{2}, \dfrac{3\pi}{2}$

19. $6 \sin x \cos x = 3(2 \sin x \cos x)$
$$= 3 \sin 2x$$

21. $4 - 8 \sin^2 x = 4(1 - 2 \sin^2 x)$
$$= 4 \cos 2x$$

23. $\sin u = -\dfrac{4}{5}, \pi < u < \dfrac{3\pi}{2} \Rightarrow \cos u = -\dfrac{3}{5}$

$\sin 2u = 2 \sin u \cos u = 2\left(-\dfrac{4}{5}\right)\left(-\dfrac{3}{5}\right) = \dfrac{24}{25}$

$\cos 2u = \cos^2 u - \sin^2 u = \dfrac{9}{25} - \dfrac{16}{25} = -\dfrac{7}{25}$

$\tan 2u = \dfrac{2 \tan u}{1 - \tan^2 u} = \dfrac{2\left(\frac{4}{3}\right)}{1 - \frac{16}{9}} = \dfrac{8}{3}\left(-\dfrac{9}{7}\right) = -\dfrac{24}{7}$

25. $\tan u = \dfrac{3}{4}, 0 < u < \dfrac{\pi}{2} \Rightarrow \sin u = \dfrac{3}{5}$ and $\cos u = \dfrac{4}{5}$

$\sin 2u = 2 \sin u \cos u = 2\left(\dfrac{3}{5}\right)\left(\dfrac{4}{5}\right) = \dfrac{24}{25}$

$\cos 2u = \cos^2 u - \sin^2 u = \dfrac{16}{25} - \dfrac{9}{25} = \dfrac{7}{25}$

$\tan 2u = \dfrac{2 \tan u}{1 - \tan^2 u} = \dfrac{2\left(\frac{3}{4}\right)}{1 - \frac{9}{16}} = \dfrac{3}{2}\left(\dfrac{16}{7}\right) = \dfrac{24}{7}$

27. $\sec u = -\dfrac{5}{2}, \dfrac{\pi}{2} < u < \pi \Rightarrow \sin u = \dfrac{\sqrt{21}}{5}$ and $\cos u = -\dfrac{2}{5}$

$\sin 2u = 2 \sin u \cos u = 2\left(\dfrac{\sqrt{21}}{5}\right)\left(-\dfrac{2}{5}\right) = -\dfrac{4\sqrt{21}}{25}$

$\cos 2u = \cos^2 u - \sin^2 u = \left(-\dfrac{2}{5}\right)^2 - \left(\dfrac{\sqrt{21}}{5}\right)^2 = -\dfrac{17}{25}$

$\tan 2u = \dfrac{2 \tan u}{1 - \tan^2 u} = \dfrac{2\left(-\dfrac{\sqrt{21}}{2}\right)}{1 - \left(-\dfrac{\sqrt{21}}{2}\right)^2}$

$ = \dfrac{-\sqrt{21}}{1 - \frac{21}{4}} = \dfrac{4\sqrt{21}}{17}$

29. $\cos^4 x = (\cos^2 x)(\cos^2 x) = \left(\dfrac{1 + \cos 2x}{2}\right)\left(\dfrac{1 + \cos 2x}{2}\right) = \dfrac{1 + 2\cos 2x + \cos^2 2x}{4}$

$ = \dfrac{1 + 2\cos 2x + \dfrac{1 + \cos 4x}{2}}{4}$

$ = \dfrac{2 + 4\cos 2x + 1 + \cos 4x}{8}$

$ = \dfrac{3 + 4\cos 2x + \cos 4x}{8}$

$ = \dfrac{1}{8}(3 + 4\cos 2x + \cos 4x)$

31. $(\sin^2 x)(\cos^2 x) = \left(\dfrac{1 - \cos 2x}{2}\right)\left(\dfrac{1 + \cos 2x}{2}\right)$

$$= \dfrac{1 - \cos^2 2x}{4}$$

$$= \dfrac{1}{4}\left(1 - \dfrac{1 + \cos 4x}{2}\right)$$

$$= \dfrac{1}{8}(2 - 1 - \cos 4x)$$

$$= \dfrac{1}{8}(1 - \cos 4x)$$

33. $\sin^2 x \cos^4 x = \sin^2 x \cos^2 x \cos^2 x = \left(\dfrac{1 - \cos 2x}{2}\right)\left(\dfrac{1 + \cos 2x}{2}\right)\left(\dfrac{1 + \cos 2x}{2}\right)$

$$= \dfrac{1}{8}(1 - \cos 2x)(1 + \cos 2x)(1 + \cos 2x)$$

$$= \dfrac{1}{8}(1 - \cos^2 2x)(1 + \cos 2x)$$

$$= \dfrac{1}{8}(1 + \cos 2x - \cos^2 2x - \cos^3 2x)$$

$$= \dfrac{1}{8}\left[1 + \cos 2x - \left(\dfrac{1 + \cos 4x}{2}\right) - \cos 2x\left(\dfrac{1 + \cos 4x}{2}\right)\right]$$

$$= \dfrac{1}{16}[2 + 2\cos 2x - 1 - \cos 4x - \cos 2x - \cos 2x \cos 4x]$$

$$= \dfrac{1}{16}\left[1 + \cos 2x - \cos 4x - \left(\dfrac{1}{2}\cos 2x + \dfrac{1}{2}\cos 6x\right)\right]$$

$$= \dfrac{1}{32}(2 + 2\cos 2x - 2\cos 4x - \cos 2x - \cos 6x)$$

$$= \dfrac{1}{32}(2 + \cos 2x - 2\cos 4x - \cos 6x)$$

Figure for Exercises 35 – 39

$\sin \theta = \dfrac{8}{17}$

$\cos \theta = \dfrac{15}{17}$

35. $\cos\dfrac{\theta}{2} = \sqrt{\dfrac{1 + \cos\theta}{2}} = \sqrt{\dfrac{1 + \frac{15}{17}}{2}} = \sqrt{\dfrac{32}{34}} = \sqrt{\dfrac{16}{17}} = \dfrac{4\sqrt{17}}{17}$

37. $\tan\dfrac{\theta}{2} = \dfrac{\sin\theta}{1 + \cos\theta} = \dfrac{\frac{8}{17}}{1 + \frac{15}{17}} = \dfrac{8}{17}\cdot\dfrac{17}{32} = \dfrac{1}{4}$

39. $\csc\dfrac{\theta}{2} = \dfrac{1}{\sin\frac{\theta}{2}} = \dfrac{1}{\sqrt{\dfrac{(1 - \cos\theta)}{2}}} = \dfrac{1}{\sqrt{\dfrac{1 - \frac{15}{17}}{2}}} = \dfrac{1}{\sqrt{\dfrac{1}{17}}} = \sqrt{17}$

41. $\sin 75° = \sin\left(\frac{1}{2} \cdot 150°\right) = \sqrt{\frac{1 - \cos 150°}{2}} = \sqrt{\frac{1 + \frac{\sqrt{3}}{2}}{2}}$

$\qquad = \frac{1}{2}\sqrt{2 + \sqrt{3}}$

$\cos 75° = \cos\left(\frac{1}{2} \cdot 150°\right) = \sqrt{\frac{1 + \cos 150°}{2}} = \sqrt{\frac{1 - \frac{\sqrt{3}}{2}}{2}}$

$\qquad = \frac{1}{2}\sqrt{2 - \sqrt{3}}$

$\tan 75° = \tan\left(\frac{1}{2} \cdot 150°\right) = \frac{\sin 150°}{1 + \cos 150°} = \frac{\frac{1}{2}}{1 - \frac{\sqrt{3}}{2}}$

$\qquad = \frac{1}{2 - \sqrt{3}} \cdot \frac{2 + \sqrt{3}}{2 + \sqrt{3}} = \frac{2 + \sqrt{3}}{4 - 3} = 2 + \sqrt{3}$

43. $\sin 112° 30' = \sin\left(\frac{1}{2} \cdot 225°\right) = \sqrt{\frac{1 - \cos 225°}{2}} = \sqrt{\frac{1 + \frac{\sqrt{2}}{2}}{2}} = \frac{1}{2}\sqrt{2 + \sqrt{2}}$

$\cos 112° 30' = \cos\left(\frac{1}{2} \cdot 225°\right) = -\sqrt{\frac{1 + \cos 225°}{2}} = -\sqrt{\frac{1 - \frac{\sqrt{2}}{2}}{2}} = -\frac{1}{2}\sqrt{2 - \sqrt{2}}$

$\tan 112° 30' = \tan\left(\frac{1}{2} \cdot 225°\right) = \frac{\sin 225°}{1 + \cos 225°} = \frac{-\frac{\sqrt{2}}{2}}{1 - \frac{\sqrt{2}}{2}} = -1 - \sqrt{2}$

45. $\sin\frac{\pi}{8} = \sin\left[\frac{1}{2}\left(\frac{\pi}{4}\right)\right] = \sqrt{\frac{1 - \cos\frac{\pi}{4}}{2}} = \frac{1}{2}\sqrt{2 - \sqrt{2}}$

$\cos\frac{\pi}{8} = \cos\left[\frac{1}{2}\left(\frac{\pi}{4}\right)\right] = \sqrt{\frac{1 + \cos\frac{\pi}{4}}{2}} = \frac{1}{2}\sqrt{2 + \sqrt{2}}$

$\tan\frac{\pi}{8} = \tan\left[\frac{1}{2}\left(\frac{\pi}{4}\right)\right] = \frac{\sin\frac{\pi}{4}}{1 + \cos\frac{\pi}{4}} = \frac{\frac{\sqrt{2}}{2}}{1 + \frac{\sqrt{2}}{2}} = \sqrt{2} - 1$

47. $\sin\frac{3\pi}{8} = \sin\left(\frac{1}{2} \cdot \frac{3\pi}{4}\right) = \sqrt{\frac{1 - \cos\frac{3\pi}{4}}{2}} = \sqrt{\frac{1 + \frac{\sqrt{2}}{2}}{2}} = \frac{1}{2}\sqrt{2 + \sqrt{2}}$

$\cos\frac{3\pi}{8} = \cos\left(\frac{1}{2} \cdot \frac{3\pi}{4}\right) = \sqrt{\frac{1 + \cos\frac{3\pi}{4}}{2}} = \sqrt{\frac{1 - \frac{\sqrt{2}}{2}}{2}} = \frac{1}{2}\sqrt{2 - \sqrt{2}}$

$\tan\frac{3\pi}{8} = \tan\left(\frac{1}{2} \cdot \frac{3\pi}{4}\right) = \frac{\sin\frac{3\pi}{4}}{1 + \cos\frac{3\pi}{4}} = \frac{\frac{\sqrt{2}}{2}}{1 - \frac{\sqrt{2}}{2}} = \frac{\frac{\sqrt{2}}{2}}{\frac{(2 - \sqrt{2})}{2}} = \frac{\sqrt{2}}{2 - \sqrt{2}} = \sqrt{2} + 1$

49. $\sin u = \dfrac{5}{13}, \dfrac{\pi}{2} < u < \pi \implies \cos u = -\dfrac{12}{13}$

$$\sin\left(\frac{u}{2}\right) = \sqrt{\frac{1 - \cos u}{2}} = \sqrt{\frac{1 + \frac{12}{13}}{2}} = \frac{5\sqrt{26}}{26}$$

$$\cos\left(\frac{u}{2}\right) = \sqrt{\frac{1 + \cos u}{2}} = \sqrt{\frac{1 - \frac{12}{13}}{2}} = \frac{\sqrt{26}}{26}$$

$$\tan\left(\frac{u}{2}\right) = \frac{\sin u}{1 + \cos u} = \frac{\frac{5}{13}}{1 - \frac{12}{13}} = 5$$

51. $\tan u = -\dfrac{5}{8}, \dfrac{3\pi}{2} < u < 2\pi \implies \sin u = -\dfrac{5}{\sqrt{89}}$ and $\cos u = \dfrac{8}{\sqrt{89}}$

$$\sin\left(\frac{u}{2}\right) = \sqrt{\frac{1 - \cos u}{2}} = \sqrt{\frac{1 - \frac{8}{\sqrt{89}}}{2}} \sqrt{\frac{\sqrt{89} - 8}{2\sqrt{89}}} = \sqrt{\frac{89 - 8\sqrt{89}}{178}}$$

$$\cos\left(\frac{u}{2}\right) = -\sqrt{\frac{1 + \cos u}{2}} = -\sqrt{\frac{1 + \frac{8}{\sqrt{89}}}{2}} = -\sqrt{\frac{\sqrt{89} + 8}{2\sqrt{89}}} = -\sqrt{\frac{89 + 8\sqrt{89}}{178}}$$

$$\tan\left(\frac{u}{2}\right) = \frac{1 - \cos u}{\sin u} = \frac{1 - \frac{8}{\sqrt{89}}}{-\frac{5}{\sqrt{89}}} = \frac{8 - \sqrt{89}}{5}$$

53. $\csc u = -\dfrac{5}{3}, \pi < u < \dfrac{3\pi}{2} \implies \sin u = -\dfrac{3}{5}$ and $\cos u = -\dfrac{4}{5}$

$$\sin\left(\frac{u}{2}\right) = \sqrt{\frac{1 - \cos u}{2}} = \sqrt{\frac{1 + \frac{4}{5}}{2}} = \frac{3\sqrt{10}}{10}$$

$$\cos\left(\frac{u}{2}\right) = -\sqrt{\frac{1 + \cos u}{2}} = -\sqrt{\frac{1 - \frac{4}{5}}{2}} = -\frac{\sqrt{10}}{10}$$

$$\tan\left(\frac{u}{2}\right) = \frac{1 - \cos u}{\sin u} = \frac{1 + \frac{4}{5}}{-\frac{3}{5}} = -3$$

55. $\sqrt{\dfrac{1 - \cos 6x}{2}} = |\sin 3x|$

57. $-\sqrt{\dfrac{1 - \cos 8x}{1 + \cos 8x}} = -\dfrac{\sqrt{\dfrac{1 - \cos 8x}{2}}}{\sqrt{\dfrac{1 + \cos 8x}{2}}}$

$$= -\left|\frac{\sin 4x}{\cos 4x}\right|$$

$$= -|\tan 4x|$$

59.

$$\sin\frac{x}{2} + \cos x = 0$$

$$\pm\sqrt{\frac{1 - \cos x}{2}} = -\cos x$$

$$\frac{1 - \cos x}{2} = \cos^2 x$$

$$0 = 2\cos^2 x + \cos x - 1$$

$$= (2\cos x - 1)(\cos x + 1)$$

$$\cos x = \frac{1}{2} \quad \text{or} \quad \cos x = -1$$

$$x = \frac{\pi}{3}, \frac{5\pi}{3} \qquad x = \pi$$

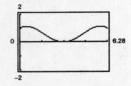

By checking these values in the original equation,
we see that $x = \pi/3$ and $x = 5\pi/3$ are extraneous,
and $x = \pi$ is the only solution.

61.

$$\cos\frac{x}{2} - \sin x = 0$$

$$\pm\sqrt{\frac{1 + \cos x}{2}} = \sin x$$

$$\frac{1 + \cos x}{2} = \sin^2 x$$

$$1 + \cos x = 2\sin^2 x$$

$$1 + \cos x = 2 - 2\cos^2 x$$

$$2\cos^2 x + \cos x - 1 = 0$$

$$(2\cos x - 1)(\cos x + 1) = 0$$

$$2\cos x - 1 = 0 \quad \text{or} \quad \cos x + 1 = 0$$

$$\cos x = \frac{1}{2} \qquad\qquad \cos x = -1$$

$$x = \frac{\pi}{3}, \frac{5\pi}{3} \qquad\qquad x = \pi$$

$$x = \frac{\pi}{3}, \ \pi, \ \frac{5\pi}{3}$$

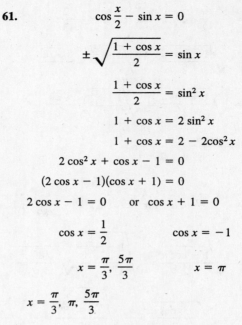

$\pi/3$, π, and $5\pi/3$ are all solutions to the equation.

63. $6\sin\dfrac{\pi}{4}\cos\dfrac{\pi}{4} = 6 \cdot \dfrac{1}{2}\left[\sin\left(\dfrac{\pi}{4} + \dfrac{\pi}{4}\right) + \sin\left(\dfrac{\pi}{4} - \dfrac{\pi}{4}\right)\right] = 3\left(\sin\dfrac{\pi}{2} + \sin 0\right)$

65. $\cos 4\theta \sin 6\theta = \dfrac{1}{2}[\sin(4\theta + 6\theta) - \sin(4\theta - 6\theta)] = \dfrac{1}{2}[\sin 10\theta - \sin(-2\theta)]$

$$= \frac{1}{2}(\sin 10\theta + \sin 2\theta)$$

67. $5\cos(-5\beta)\cos 3\beta = 5 \cdot \dfrac{1}{2}\left[\cos(-5\beta - 3\beta) + \cos(-5\beta + 3\beta)\right] = \dfrac{5}{2}\left[\cos(-8\beta) + \cos(-2\beta)\right]$

$$= \frac{5}{2}(\cos 8\beta + \cos 2\beta)$$

69. $\sin(x + y)\sin(x - y) = \dfrac{1}{2}(\cos 2y - \cos 2x)$

71. $\cos(\theta - \pi)\sin(\theta + \pi) = \dfrac{1}{2}(\sin 2\theta + \sin 2\pi)$

73. $10\cos 75° \cos 15° = 10\left(\dfrac{1}{2}\right)[\cos(75° - 15°) + \cos(75° + 15°)] = 5[\cos 60° + \cos 90°]$

75. $\sin 60° + \sin 30° = 2\sin\left(\dfrac{60° + 30°}{2}\right)\cos\left(\dfrac{60° - 30°}{2}\right) = 2\sin 45° \cos 15°$

77. $\cos\dfrac{3\pi}{4} - \cos\dfrac{\pi}{4} = -2\sin\left(\dfrac{\dfrac{3\pi}{4} + \dfrac{\pi}{4}}{2}\right)\sin\left(\dfrac{\dfrac{3\pi}{4} - \dfrac{\pi}{4}}{2}\right) = -2\sin\dfrac{\pi}{2}\sin\dfrac{\pi}{4}$

79. $\sin 5\theta - \sin 3\theta = 2 \cos\left(\dfrac{5\theta + 3\theta}{2}\right)\sin\left(\dfrac{5\theta - 3\theta}{2}\right) = 2 \cos 4\theta \sin \theta$

81. $\cos 6x + \cos 2x = 2 \cos\left(\dfrac{6x + 2x}{2}\right)\cos\left(\dfrac{6x - 2x}{2}\right) = 2 \cos 4x \cos 2x$

83. $\sin(\alpha + \beta) - \sin(\alpha - \beta) = 2 \cos\left(\dfrac{\alpha + \beta + \alpha - \beta}{2}\right)\sin\left(\dfrac{\alpha + \beta - \alpha + \beta}{2}\right) = 2 \cos \alpha \sin \beta$

85. $\cos\left(\theta + \dfrac{\pi}{2}\right) - \cos\left(\theta - \dfrac{\pi}{2}\right) = -2\sin\left[\dfrac{\left(\theta + \dfrac{\pi}{2}\right) + \left(\theta - \dfrac{\pi}{2}\right)}{2}\right]\sin\left[\dfrac{\left(\theta + \dfrac{\pi}{2}\right) - \left(\theta - \dfrac{\pi}{2}\right)}{2}\right]$

$$= -2\sin\theta \sin\dfrac{\pi}{2}$$

87.
$$\sin 6x + \sin 2x = 0$$
$$2\sin\left(\dfrac{6x + 2x}{2}\right)\cos\left(\dfrac{6x - 2x}{2}\right) = 0$$
$$2(\sin 4x)\cos 2x = 0$$
$$\sin 4x = 0 \quad \text{or} \quad \cos 2x = 0$$
$$4x = n\pi \qquad\qquad 2x = \dfrac{\pi}{2} + n\pi$$
$$x = \dfrac{n\pi}{4} \qquad\qquad x = \dfrac{\pi}{4} + \dfrac{n\pi}{2}$$

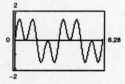

In the interval $[0, 2\pi)$ we have

$$x = 0,\ \dfrac{\pi}{4},\ \dfrac{\pi}{2},\ \dfrac{3\pi}{4},\ \pi,\ \dfrac{5\pi}{4},\ \dfrac{3\pi}{2},\ \dfrac{7\pi}{4}.$$

89.
$$\dfrac{\cos 2x}{\sin 3x - \sin x} - 1 = 0$$
$$\dfrac{\cos 2x}{\sin 3x - \sin x} = 1$$
$$\dfrac{\cos 2x}{2 \cos 2x \sin x} = 1$$
$$2 \sin x = 1$$
$$\sin x = \dfrac{1}{2}$$
$$x = \dfrac{\pi}{6},\ \dfrac{5\pi}{6}$$

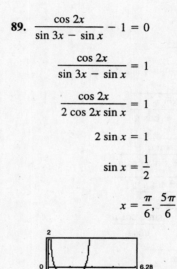

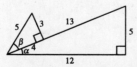

Figure for Exercises 91 and 93

91. $\sin^2 \alpha = \left(\dfrac{5}{13}\right)^2 = \dfrac{25}{169}$

$\sin^2 \alpha = 1 - \cos^2 \alpha = 1 - \left(\dfrac{12}{13}\right)^2$

$\qquad = 1 - \dfrac{144}{169} = \dfrac{25}{169}$

93. $\sin \alpha \cos \beta = \left(\dfrac{5}{13}\right)\left(\dfrac{4}{5}\right) = \dfrac{4}{13}$

$\sin \alpha \cos \beta = \cos\left(\dfrac{\pi}{2} - \alpha\right)\sin\left(\dfrac{\pi}{2} - \beta\right)$

$\qquad = \left(\dfrac{5}{13}\right)\left(\dfrac{4}{5}\right) = \dfrac{4}{13}$

95. $\csc 2\theta = \dfrac{1}{\sin 2\theta}$

$$= \dfrac{1}{2 \sin\theta \cos\theta}$$

$$= \dfrac{1}{\sin\theta} \cdot \dfrac{1}{2\cos\theta}$$

$$= \dfrac{\csc\theta}{2\cos\theta}$$

97. $\cos^2 2\alpha - \sin^2 2\alpha = \cos[2(2\alpha)]$

$$= \cos 4\alpha$$

99. $(\sin x + \cos x)^2 = \sin^2 x + 2\sin x\cos x + \cos^2 x$

$$= (\sin^2 x + \cos^2 x) + 2\sin x\cos x$$

$$= 1 + \sin 2x$$

101. $1 + \cos 10y = 1 + \cos^2 5y - \sin^2 5y$

$$= 1 + \cos^2 5y - (1 - \cos^2 5y)$$

$$= 2\cos^2 5y$$

103. $\sec\dfrac{u}{2} = \dfrac{1}{\cos\dfrac{u}{2}}$

$$= \pm\sqrt{\dfrac{2}{1 + \cos u}}$$

$$= \pm\sqrt{\dfrac{2\sin u}{\sin u(1 + \cos u)}}$$

$$= \pm\sqrt{\dfrac{2\sin u}{\sin u + \sin u\cos u}}$$

$$= \pm\sqrt{\dfrac{\dfrac{2\sin u}{\cos u}}{\dfrac{\sin u}{\cos u} + \dfrac{\sin u\cos u}{\cos u}}}$$

$$= \pm\sqrt{\dfrac{2\tan u}{\tan u + \sin u}}$$

105. $\dfrac{\sin x \pm \sin y}{\cos x + \cos y} = \dfrac{2\sin\left(\dfrac{x \pm y}{2}\right)\cos\left(\dfrac{x \mp y}{2}\right)}{2\cos\left(\dfrac{x + y}{2}\right)\cos\left(\dfrac{x - y}{2}\right)}$

$$= \tan\left(\dfrac{x \pm y}{2}\right)$$

107. $\dfrac{\cos 4x + \cos 2x}{\sin 4x + \sin 2x} = \dfrac{2\cos\left(\dfrac{4x + 2x}{2}\right)\cos\left(\dfrac{4x - 2x}{2}\right)}{2\sin\left(\dfrac{4x + 2x}{2}\right)\cos\left(\dfrac{4x - 2x}{2}\right)}$

$$= \dfrac{2\cos 3x\cos x}{2\sin 3x\cos x}$$

$$= \cot 3x$$

109. $\sin\left(\dfrac{\pi}{6} + x\right) + \sin\left(\dfrac{\pi}{6} - x\right) = 2\sin\dfrac{\pi}{6}\cos x$

$$= 2 \cdot \dfrac{1}{2}\cos x$$

$$= \cos x$$

111. $\cos 3\beta = \cos(2\beta + \beta)$

$= \cos 2\beta \cos\beta - \sin 2\beta \sin \beta$

$= (\cos^2 \beta - \sin^2 \beta)\cos \beta - 2 \sin \beta \cos \beta \sin \beta$

$= \cos^3 \beta - \sin^2 \beta \cos \beta - 2 \sin^2 \beta \cos \beta$

$= \cos^3 \beta - 3 \sin^2 \beta \cos \beta$

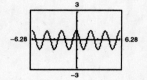

113. $\dfrac{\cos 4x - \cos 2x}{2 \sin 3x} = \dfrac{-2 \sin\left(\dfrac{4x + 2x}{2}\right) \sin\left(\dfrac{4x - 2x}{2}\right)}{2 \sin 3x}$

$= \dfrac{-2 \sin 3x \sin x}{2 \sin 3x}$

$= -\sin x$

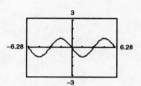

115. $\sin^2 x = \dfrac{1 - \cos 2x}{2} = \dfrac{1}{2} - \dfrac{\cos 2x}{2}$

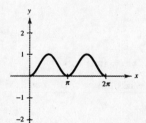

117. $\sin(2 \arcsin x) = 2 \sin(\arcsin x) \cos(\arcsin x) = 2x\sqrt{1 - x^2}$

119. (a) $\qquad A = \dfrac{1}{2}bh$

$\cos \dfrac{\theta}{2} = \dfrac{h}{10} \implies h = 10 \cos \dfrac{\theta}{2}$

$\sin \dfrac{\theta}{2} = \dfrac{(1/2)b}{10} \implies \dfrac{1}{2}b = 10 \sin \dfrac{\theta}{2}$

$A = 10 \sin \dfrac{\theta}{2} 10 \cos \dfrac{\theta}{2} \implies A = 100 \sin \dfrac{\theta}{2} \cos \dfrac{\theta}{2}$

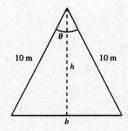

(b) $A = 100 \sin \dfrac{\theta}{2} \cos \dfrac{\theta}{2}$

$A = 50\left(2 \sin \dfrac{\theta}{2} \cos \dfrac{\theta}{2}\right)$

$A = 50 \sin \theta$

When $\theta = \pi/2$, $\sin \theta = 1 \implies$ the area is a maximum.

$A = 50 \sin \dfrac{\pi}{2} = 50(1) = 50$ square feet

121. $\sin\dfrac{\theta}{2} = \dfrac{1}{4.5}$

$\dfrac{\theta}{2} = \arcsin\left(\dfrac{1}{4.5}\right)$

$\theta = 2\arcsin\left(\dfrac{1}{4.5}\right)$

$\theta \approx 0.4482$

123. False. For $u < 0$,

$$\sin 2u = -\sin(-2u)$$
$$= -2\sin(-u)\cos(-u)$$
$$= -2(-\sin u)\cos u$$
$$= 2\sin u\cos u$$

125. (a) $y = 4\sin\dfrac{x}{2} + \cos x$

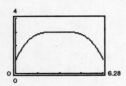

Maximum: $(\pi, 3)$

(b)
$$2\cos\dfrac{x}{2} - \sin x = 0$$

$$2\left(\pm\sqrt{\dfrac{1 + \cos x}{2}}\right) = \sin x$$

$$4\left(\dfrac{1 + \cos x}{2}\right) = \sin^2 x$$

$$2(1 + \cos x) = 1 - \cos^2 x$$

$$\cos^2 x + 2\cos x + 1 = 0$$

$$(\cos x + 1)^2 = 0$$

$$\cos x = -1$$

$$x = \pi$$

127. $f(x) = \sin^4 x + \cos^4 x$

(a) $\sin^4 x + \cos^4 x = (\sin^2 x)^2 + (\cos^2 x)^2$

$\qquad = \left(\dfrac{1 - \cos 2x}{2}\right)^2 + \left(\dfrac{1 + \cos 2x}{2}\right)^2$

$\qquad = \dfrac{1}{4}[(1 - \cos 2x)^2 + (1 + \cos 2x)^2]$

$\qquad = \dfrac{1}{4}(1 - 2\cos 2x + \cos^2 2x + 1 + 2\cos 2x + \cos^2 2x)$

$\qquad = \dfrac{1}{4}(2 + 2\cos^2 2x)$

$\qquad = \dfrac{1}{4}\left[2 + 2\left(\dfrac{1 + \cos 2(2x)}{2}\right)\right]$

$\qquad = \dfrac{1}{4}(3 + \cos 4x)$

(b) $\sin^4 x + \cos^4 x = (\sin^2 x)^2 + \cos^4 x$

$\qquad = (1 - \cos^2 x)^2 + \cos^4 x$

$\qquad = 1 - 2\cos^2 x + \cos^4 x + \cos^4 x$

$\qquad = 2\cos^4 x - 2\cos^2 x + 1$

(c) $\sin^4 x + \cos^4 x = \sin^4 x + 2\sin^2 x\cos^2 x + \cos^4 x - 2\sin^2 x\cos^2 x$

$\qquad = (\sin^2 x + \cos^2 x)^2 - 2\sin^2 x\cos^2 x$

$\qquad = 1 - 2\sin^2 x\cos^2 x$

— CONTINUED —

127. — CONTINUED —

(d) $1 - 2\sin^2 x \cos^2 x = 1 - (2\sin x \cos x)(\sin x \cos x)$

$$= 1 - (\sin 2x)\left(\frac{1}{2}\sin 2x\right)$$

$$= 1 - \frac{1}{2}\sin^2 2x$$

(e) No, it does not mean that one of you is wrong. There is often more than one way to rewrite a trigonometric expression.

129.
$$y = x^2 - 6x + 13$$
$$x^2 - 6x + 13 = y$$
$$x^2 - 6x = y - 13$$
$$x^2 - 6x + 9 = y - 13 + 9$$
$$(x - 3)^2 = (y - 4)$$

Vertex: $(3, 4)$

131.
$$y = 2x^2 - 4x + 3$$
$$2x^2 - 4x + 3 = y$$
$$x^2 - 2x + \frac{3}{2} = \frac{1}{2}y$$
$$x^2 - 2x = \frac{1}{2}y - \frac{3}{2}$$
$$x^2 - 2x + 1 = \frac{1}{2}y - \frac{3}{2} + 1$$
$$(x - 1)^2 = \frac{1}{2}y - \frac{1}{2}$$
$$(x - 1)^2 = \frac{1}{2}(y - 1)$$

Vertex: $(1, 1)$

133. $(x - 5)^2 + y + 8 = 0$

$$(x - 5)^2 = -y - 8$$
$$(x - 5)^2 = -(y + 8)$$

Vertex: $(5, -8)$

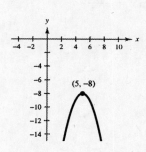

135. Let x = profit for September, then $x + 0.16x$ = profit for October.

$$x + (x + 0.16x) = 507,600$$
$$2.16x = 507,600$$
$$x = 235,000$$
$$x + 0.16x = 272,600$$

Profit for September: $235,000

Profit for October: $272,600

137. Let x = number of gallons of 100% concentrate.

$$0.30(55 - x) + 1.00x = 0.50(55)$$
$$16.50 - 0.30x + x = 27.50$$
$$0.70x = 11$$
$$x \approx 15.7 \text{ gallons}$$

Review Exercises for Chapter 2

Solutions to Odd-Numbered Exercises

1. $\dfrac{1}{\cos x} = \sec x$

3. $\dfrac{1}{\sec x} = \cos x$

5. $\dfrac{\cos x}{\sin x} = \cot x$

7. $\sin x = \dfrac{3}{5}, \ \cos x = \dfrac{4}{5}$

$$\tan x = \frac{\sin x}{\cos x} = \frac{\frac{3}{5}}{\frac{4}{5}} = \frac{3}{4}$$

$$\cot x = \frac{1}{\tan x} = \frac{4}{3}$$

$$\sec x = \frac{1}{\cos x} = \frac{5}{4}$$

$$\csc x = \frac{1}{\sin x} = \frac{5}{3}$$

9. $\sin\left(\dfrac{\pi}{2} - x\right) = \dfrac{\sqrt{2}}{2} \Rightarrow \cos x = \dfrac{1}{\sqrt{2}} = \dfrac{\sqrt{2}}{2}, \ \sin x = -\dfrac{\sqrt{2}}{2}$

$$\tan x = \frac{\sin x}{\cos x} = \frac{-\dfrac{1}{\sqrt{2}}}{\dfrac{1}{\sqrt{2}}} = -1$$

$$\cot x = \frac{1}{\tan x} = -1$$

$$\sec x = \frac{1}{\cos x} = \sqrt{2}$$

$$\csc x = \frac{1}{\sin x} = -\sqrt{2}$$

11. $\dfrac{1}{\cot^2 x + 1} = \dfrac{1}{\csc^2 x} = \sin^2 x$

13. $\tan^2 x(\csc^2 x - 1) = \tan^2 x(\cot^2 x) = \tan^2 x\left(\dfrac{1}{\tan^2 x}\right) = 1$

15. $\dfrac{\sin\left(\dfrac{\pi}{2} - \theta\right)}{\sin \theta} = \dfrac{\cos \theta}{\sin \theta} = \cot \theta$

17. $\cos^2 x + \cos^2 x \cot^2 x = \cos^2 x(1 + \cot^2 x) = \cos^2 x(\csc^2 x)$

$$= \cos^2 x\left(\frac{1}{\sin^2 x}\right) = \frac{\cos^2 x}{\sin^2 x} = \cot^2 x$$

19. $(\tan x + 1)^2\cos x = (\tan^2 x + 2 \tan x + 1)\cos x$

$$= (\sec^2 x + 2 \tan x)\cos x$$

$$= \sec^2 x \cos x + 2\left(\frac{\sin x}{\cos x}\right)\cos x = \sec x + 2 \sin x$$

21. $\dfrac{1}{\csc\theta+1}-\dfrac{1}{\csc\theta-1}=\dfrac{(\csc\theta-1)-(\csc\theta+1)}{(\csc\theta+1)(\csc\theta-1)}$

$$=\dfrac{-2}{\csc^2\theta-1}$$

$$=\dfrac{-2}{\cot^2\theta}$$

$$=-2\tan^2\theta$$

23. $\sin^{-1/2}x\cos x=\dfrac{1}{\sqrt{\sin x}}(\cos x)=\dfrac{\sqrt{\sin x}}{\sin x}(\cos x)$

$$=\sqrt{\sin x}\left(\dfrac{\cos x}{\sin x}\right)=\sqrt{\sin x}\cot x$$

25. $\sec^2 x\cot x-\cot x=\cot x(\sec^2 x-1)=\cot x\tan^2 x$

$$=\left(\dfrac{1}{\tan x}\right)\tan^2 x=\tan x$$

27. $\cot\left(\dfrac{\pi}{2}-x\right)=\tan x$ by the Cofunction Identity

29. $\dfrac{1}{\tan x\csc x\sin x}=\dfrac{1}{(\tan x)\left(\dfrac{1}{\sin x}\right)(\sin x)}=\dfrac{1}{\tan x}$

$$=\cot x$$

31. $\cos^3 x\sin^2 x=\cos x\cos^2 x\sin^2 x$

$$=\cos x(1-\sin^2 x)\sin^2 x$$

$$=\cos x(\sin^2 x-\sin^4 x)$$

$$=(\sin^2 x-\sin^4 x)\cos x$$

33. $4\cos\theta=1+2\cos\theta$

$2\cos\theta=1$

$\cos\theta=\dfrac{1}{2}$

$\theta=\dfrac{\pi}{3}+2n\pi$ or $\dfrac{5\pi}{3}+2n\pi$

35. $\dfrac{1}{2}\sec x-1=0$

$\dfrac{1}{2}\sec x=1$

$\sec x=2$

$\cos x=\dfrac{1}{2}$

$x=\dfrac{\pi}{3}+2n\pi$ or $\dfrac{5\pi}{3}+2n\pi$

37. $4\tan^2 u-1=\tan^2 u$

$3\tan^2 u-1=0$

$\tan^2 u=\dfrac{1}{3}$

$\tan u=\pm\dfrac{1}{\sqrt{3}}=\pm\dfrac{\sqrt{3}}{3}$

$u=\dfrac{\pi}{6}+n\pi$ or $\dfrac{5\pi}{6}+n\pi$

39. $2\sin^2 x-3\sin x=-1$

$2\sin^2 x-3\sin x+1=0$

$(2\sin x-1)(\sin x-1)=0$

$2\sin x-1=0$ or $\sin x-1=0$

$\sin x=\dfrac{1}{2}$ $\sin x=1$

$x=\dfrac{\pi}{6},\dfrac{5\pi}{6}$ $x=\dfrac{\pi}{2}$

41. $\sin^2 x+2\cos x=2$

$1-\cos^2 x+2\cos x=2$

$0=\cos^2 x-2\cos x+1$

$0=(\cos x-1)^2$

$\cos x-1=0$

$\cos x=1$

$x=0$

43. $\sqrt{3}\tan 3x=0$

$\tan 3x=0$

$3x=0,\pi,2\pi,3\pi,4\pi,5\pi$

$x=0,\dfrac{\pi}{3},\dfrac{2\pi}{3},\pi,\dfrac{4\pi}{3},\dfrac{5\pi}{3}$

45. $3 \csc^2 5x = -4$

$$\csc^2 5x = -\frac{4}{3}$$

$$\csc 5x = \pm \sqrt{-\frac{4}{3}}$$

No real solution

47. $2 \cos^2 x + 3 \cos x = 0$

$$\cos x (2 \cos x + 3) = 0$$

$$\cos x = 0 \quad \text{or} \quad 2 \cos x + 3 = 0$$

$$x = \frac{\pi}{2}, \frac{3\pi}{2} \qquad 2 \cos x = -3$$

$$\cos x = -\frac{3}{2}$$

No solution

49. $\sec^2 x - 6 \tan x + 4 = 0$

$$1 + \tan^2 x + 6 \tan x + 4 = 0$$

$$\tan^2 x + 6 \tan x + 5 = 0$$

$$(\tan x + 5)(\tan x + 1) = 0$$

$$\tan x + 5 = 0 \quad \text{or} \qquad \tan x + 1 = 0$$

$$\tan x = -5 \qquad\qquad \tan x = -1$$

$$x = \arctan(-5) + \pi \qquad x = \frac{3\pi}{4}, \frac{7\pi}{4}$$

$$x = \arctan(-5) + 2\pi$$

51. $\sin(345°) = \sin(300° + 45°)$

$$= \sin 300° \cos 45° + \cos 300° \sin 45°$$

$$= -\frac{\sqrt{3}}{2} \cdot \frac{\sqrt{2}}{2} + \frac{1}{2} \cdot \frac{\sqrt{2}}{2}$$

$$= \frac{\sqrt{2}}{4}\left(-\sqrt{3} + 1\right) = \frac{\sqrt{2}}{4}\left(1 - \sqrt{3}\right)$$

$$\cos(345°) = \cos(300° + 45°)$$

$$= \cos 300° \cos 45° - \sin 300° \sin 45°$$

$$= \frac{1}{2} \cdot \frac{\sqrt{2}}{2} - \left(-\frac{\sqrt{3}}{2}\right)\frac{\sqrt{2}}{2}$$

$$= \frac{\sqrt{2}}{4}\left(1 + \sqrt{3}\right)$$

$$\tan(345°) = \tan(300° + 45°)$$

$$= \frac{\tan 300° + \tan 45°}{1 - \tan 300° \tan 45°} = \frac{-\sqrt{3} + 1}{1 + \sqrt{3}(1)} \cdot \frac{1 - \sqrt{3}}{1 - \sqrt{3}}$$

$$= \frac{4 - 2\sqrt{3}}{-2} = -2 + \sqrt{3}$$

53. $\sin\left(\dfrac{19\pi}{12}\right) = \sin\left(\dfrac{11\pi}{6} - \dfrac{\pi}{4}\right)$

$\qquad = \sin\dfrac{11\pi}{6}\cos\dfrac{\pi}{4} - \cos\dfrac{11\pi}{6}\sin\dfrac{\pi}{4}$

$\qquad = -\dfrac{1}{2}\cdot\dfrac{\sqrt{2}}{2} - \dfrac{\sqrt{3}}{2}\cdot\dfrac{\sqrt{2}}{2}$

$\qquad = -\dfrac{\sqrt{2}}{4}\left(1 + \sqrt{3}\right) = -\dfrac{\sqrt{2}}{4}\left(\sqrt{3} + 1\right)$

$\cos\left(\dfrac{19\pi}{12}\right) = \cos\left(\dfrac{11\pi}{6} - \dfrac{\pi}{4}\right)$

$\qquad = \cos\dfrac{11\pi}{6}\cos\dfrac{\pi}{4} + \sin\dfrac{11\pi}{6}\sin\dfrac{\pi}{4}$

$\qquad = \dfrac{\sqrt{3}}{2}\cdot\dfrac{\sqrt{2}}{2} + \left(-\dfrac{1}{2}\right)\dfrac{\sqrt{2}}{2}$

$\qquad = \dfrac{\sqrt{2}}{4}\left(\sqrt{3} - 1\right)$

$\tan\left(\dfrac{19\pi}{12}\right) = \tan\left(\dfrac{11\pi}{6} - \dfrac{\pi}{4}\right)$

$\qquad = \dfrac{\tan\dfrac{11\pi}{6} - \tan\dfrac{\pi}{4}}{1 + \tan\dfrac{11\pi}{6}\tan\dfrac{\pi}{4}}$

$\qquad = \dfrac{-\dfrac{\sqrt{3}}{3} - 1}{1 + \left(-\dfrac{\sqrt{3}}{3}\right)(1)} = \dfrac{-\sqrt{3} - 3}{3 - \sqrt{3}}\cdot\dfrac{3 + \sqrt{3}}{3 + \sqrt{3}}$

$\qquad = \dfrac{-\left(12 + 6\sqrt{3}\right)}{6} = -2 - \sqrt{3}$

55. $\cos 45°\cos 120° - \sin 45°\sin 120° = \cos(45° + 120°) = \cos 165°$

57. $\dfrac{\tan 68° - \tan 115°}{1 + \tan 68°\tan 115°} = \tan(68° - 115°) = \tan(-47°)$

Figures for Exercises 59–63

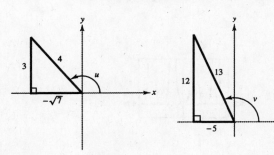

59. $\tan(u + v) = \dfrac{\tan u + \tan v}{1 - \tan u\tan v} = \dfrac{\left(-\dfrac{3}{\sqrt{7}}\right) + \left(-\dfrac{12}{5}\right)}{1 - \left(-\dfrac{3}{\sqrt{7}}\right)\left(-\dfrac{12}{5}\right)}$

$\qquad = \dfrac{15 + 12\sqrt{7}}{36 - 5\sqrt{7}}\cdot\dfrac{36 + 5\sqrt{7}}{36 + 5\sqrt{7}} = \dfrac{960 + 507\sqrt{7}}{1121}$

61. $\sin(u - v) = \sin u\cos v - \cos u\sin v$

$\qquad = \left(\dfrac{3}{4}\right)\left(-\dfrac{5}{13}\right) - \left(-\dfrac{\sqrt{7}}{4}\right)\left(\dfrac{12}{13}\right)$

$\qquad = \dfrac{-15 + 12\sqrt{7}}{52} = \dfrac{12\sqrt{7} - 15}{52}$

63. $\tan(u - v) = \dfrac{\tan u - \tan v}{1 + \tan u\tan v} = \dfrac{\left(-\dfrac{3}{\sqrt{7}}\right) - \left(-\dfrac{12}{5}\right)}{1 + \left(-\dfrac{3}{\sqrt{7}}\right)\left(-\dfrac{12}{5}\right)}$

$\qquad = \dfrac{-15 + 12\sqrt{7}}{36 + 5\sqrt{7}}\cdot\dfrac{36 - 5\sqrt{7}}{36 - 5\sqrt{7}} = \dfrac{-960 + 507\sqrt{7}}{1121}$

65.
$$\cos\left(x + \frac{\pi}{6}\right) - \cos\left(x - \frac{\pi}{6}\right) = 1$$

$$\left(\cos x \cos \frac{\pi}{6} - \sin x \sin \frac{\pi}{6}\right) - \left(\cos x \cos \frac{\pi}{6} + \sin x \sin \frac{\pi}{6}\right) = 1$$

$$-2 \sin x \sin \frac{\pi}{6} = 1$$

$$-2 \sin x \left(\frac{1}{2}\right) = 1$$

$$\sin x = -1$$

$$x = \frac{3\pi}{2}$$

67.
$$\cos\left(x + \frac{3\pi}{4}\right) - \cos\left(x - \frac{3\pi}{4}\right) = 0$$

$$\left(\cos x \cos \frac{3\pi}{4} - \sin x \sin \frac{3\pi}{4}\right) - \left(\cos x \cos \frac{3\pi}{4} + \sin x \sin \frac{3\pi}{4}\right) = 0$$

$$-2 \sin x \sin \frac{3\pi}{4} = 0$$

$$-2 \sin x \left(\frac{\sqrt{2}}{2}\right) = 0$$

$$-\sqrt{2} \sin x = 0$$

$$\sin x = 0$$

$$x = 0, \pi$$

69.
$$\frac{1 - \cos 2x}{1 + \cos 2x} = \frac{1 - (1 - 2 \sin^2 x)}{1 + (2 \cos x^2 - 1)}$$

$$= \frac{2 \sin^2 x}{2 \cos^2 x}$$

$$= \tan^2 x$$

71. $\cos u = -\dfrac{2}{\sqrt{5}}$, $\dfrac{\pi}{2} < u < \pi \Rightarrow \sin u = \dfrac{1}{\sqrt{5}}$ and $\tan u = -\dfrac{1}{2}$

$$\sin 2u = 2 \sin u \cos u = 2\left(\frac{1}{\sqrt{5}}\right)\left(-\frac{2}{\sqrt{5}}\right) = -\frac{4}{5}$$

$$\cos 2u = \cos^2 u - \sin^2 u = \left(-\frac{2}{\sqrt{5}}\right)^2 - \left(\frac{1}{\sqrt{5}}\right)^2 = \frac{3}{5}$$

$$\tan 2u = \frac{2 \tan u}{1 - \tan^2 u} = \frac{2\left(-\frac{1}{2}\right)}{1 - \left(-\frac{1}{2}\right)^2} = \frac{-1}{\frac{3}{4}} = -\frac{4}{3}$$

73. $\tan^2 2x = \dfrac{\sin^2 2x}{\cos^2 2x} = \dfrac{\dfrac{1 - \cos 4x}{2}}{\dfrac{1 + \cos 4x}{2}} = \dfrac{1 - \cos 4x}{1 + \cos 4x}$

75. $\sin^2 x \tan^2 x = \sin^2 x \left(\dfrac{\sin^2 x}{\cos^2 x}\right) = \dfrac{\sin^4 x}{\cos^2 x}$

$\qquad = \dfrac{\left(\dfrac{1 - \cos 2x}{2}\right)^2}{\dfrac{1 + \cos 2x}{2}} = \dfrac{\dfrac{1 - 2\cos 2x + \cos^2 2x}{4}}{\dfrac{1 + \cos 2x}{2}}$

$\qquad = \dfrac{1 - 2\cos 2x + \dfrac{1 + \cos 4x}{2}}{2(1 + \cos 2x)}$

$\qquad = \dfrac{2 - 4\cos 2x + 1 + \cos 4x}{4(1 + \cos 2x)}$

$\qquad = \dfrac{3 - 4\cos 2x + \cos 4x}{4(1 + \cos 2x)}$

77. $\sin(-75°) = -\sqrt{\dfrac{1 - \cos 150°}{2}} = -\sqrt{\dfrac{1 - \left(-\dfrac{\sqrt{3}}{2}\right)}{2}} = -\dfrac{\sqrt{2 + \sqrt{3}}}{2}$

$\qquad = -\dfrac{1}{2}\sqrt{2 + \sqrt{3}}$

$\cos(-75°) = \sqrt{\dfrac{1 + \cos 150°}{2}} = \sqrt{\dfrac{1 + \left(-\dfrac{\sqrt{3}}{2}\right)}{2}} = \dfrac{\sqrt{2 - \sqrt{3}}}{2}$

$\qquad = \dfrac{1}{2}\sqrt{2 - \sqrt{3}}$

$\tan(-75°) = -\left(\dfrac{1 - \cos 150°}{\sin 150°}\right) = -\left(\dfrac{1 - \left(-\dfrac{\sqrt{3}}{2}\right)}{\dfrac{1}{2}}\right) = -\left(2 + \sqrt{3}\right)$

$\qquad = -2 - \sqrt{3}$

79. $\sin\left(\dfrac{19\pi}{12}\right) = -\sqrt{\dfrac{1 - \cos\dfrac{19\pi}{6}}{2}} = -\sqrt{\dfrac{1 - \left(-\dfrac{\sqrt{3}}{2}\right)}{2}} = -\dfrac{\sqrt{2 + \sqrt{3}}}{2}$

$\qquad\qquad\ = -\dfrac{1}{2}\sqrt{2 + \sqrt{3}}$

$\qquad \cos\left(\dfrac{19\pi}{12}\right) = \sqrt{\dfrac{1 + \cos\dfrac{19\pi}{6}}{2}} = \sqrt{\dfrac{1 + \left(-\dfrac{\sqrt{3}}{2}\right)}{2}} = \dfrac{\sqrt{2 + \sqrt{3}}}{2}$

$\qquad\qquad\ = \dfrac{1}{2}\sqrt{2 - \sqrt{3}}$

$\qquad \tan\left(\dfrac{19\pi}{12}\right) = \dfrac{1 - \cos\dfrac{19\pi}{6}}{\sin\dfrac{19\pi}{6}} = \dfrac{1 - \left(-\dfrac{\sqrt{3}}{2}\right)}{-\dfrac{1}{2}} = -2 - \sqrt{3}$

81. $-\sqrt{\dfrac{1 + \cos 10x}{2}} = -\left|\cos\dfrac{10x}{2}\right| = -|\cos 5x|$

83. Volume V of the trough will be the area A of the isosceles triangle times the length l of the trough.

$\qquad V = A \cdot l$

(a) $\qquad A = \dfrac{1}{2}bh$

Not to scale

$\qquad \cos\dfrac{\theta}{2} = \dfrac{h}{0.5} \implies h = 0.5\cos\dfrac{\theta}{2}$

$\qquad \sin\dfrac{\theta}{2} = \dfrac{\dfrac{b}{2}}{0.5} \implies \dfrac{b}{2} = 0.5\sin\dfrac{\theta}{2}$

$\qquad A = 0.5\sin\dfrac{\theta}{2}\,0.5\cos\dfrac{\theta}{2}$

$\qquad\ \ = (0.5)^2\sin\dfrac{\theta}{2}\cos\dfrac{\theta}{2}$

$\qquad\ \ = 0.25\sin\dfrac{\theta}{2}\cos\dfrac{\theta}{2}$ square meters

$\qquad V = (0.25)(4)\sin\dfrac{\theta}{2}\cos\dfrac{\theta}{2}$ cubic meters

$\qquad\ \ = \sin\dfrac{\theta}{2}\cos\dfrac{\theta}{2}$ cubic meters

(b) $V = \sin\dfrac{\theta}{2}\cos\dfrac{\theta}{2}$

$\qquad = \dfrac{1}{2}\left(2\sin\dfrac{\theta}{2}\cos\dfrac{\theta}{2}\right)$

$\qquad = \dfrac{1}{2}\sin\theta$ cubic meters

$\qquad$ Volume is maximum when $\theta = \dfrac{\pi}{2}$.

85. $6\sin 15°\sin 45° = 6\left(\dfrac{1}{2}\right)[\cos(15° - 45°) - \cos(15° + 45°)]$

$\qquad\qquad\qquad\quad\ = 3[\cos(-30°) - \cos 60°]$

$\qquad\qquad\qquad\quad\ = 3(\cos 30° - \cos 60°)$

87. $4\sin 3\alpha\cos 2\alpha = 4\left(\dfrac{1}{2}\right)[\sin(3\alpha + 2\alpha) + \sin(3\alpha - 2\alpha)]$

$\qquad\qquad\qquad\quad\ = 2(\sin 5\alpha + \sin\alpha)$

89. $\cos 3\theta + \cos 2\theta = 2 \cos\left(\dfrac{3\theta + 2\theta}{2}\right) \cos\left(\dfrac{3\theta - 2\theta}{2}\right)$

$$= 2 \cos\dfrac{5\theta}{2} \cos\dfrac{\theta}{2}$$

91. $\sin\left(x + \dfrac{\pi}{4}\right) - \sin\left(x - \dfrac{\pi}{4}\right) = 2 \cos\left[\dfrac{\left(x + \dfrac{\pi}{4}\right) + \left(x - \dfrac{\pi}{4}\right)}{2}\right] \sin\left[\dfrac{\left(x + \dfrac{\pi}{4}\right) - \left(x - \dfrac{\pi}{4}\right)}{2}\right]$

$$= 2 \cos x \sin\dfrac{\pi}{4}$$

93. False. If $\dfrac{\pi}{2} < \theta < \pi$, then $\dfrac{\pi}{4} < \dfrac{\theta}{2} < \dfrac{\pi}{2}$ and $\dfrac{\theta}{2}$ is in Quadrant I. $\cos\dfrac{\theta}{2} > 0$

95. True. $4 \sin(-x)\cos(-x) = 4(-\sin x)\cos x$

$$= -4 \sin x \cos x = -2(2 \sin x \cos x)$$

$$= -2 \sin 2x$$

97. Reciprocal Identities: $\quad \sin\theta = \dfrac{1}{\csc\theta} \qquad \csc\theta = \dfrac{1}{\sin\theta}$

$$\cos\theta = \dfrac{1}{\sec\theta} \qquad \sec\theta = \dfrac{1}{\cos\theta}$$

$$\tan\theta = \dfrac{1}{\cot\theta} \qquad \cot\theta = \dfrac{1}{\tan\theta}$$

Quotient Identities: $\quad \tan\theta = \dfrac{\sin\theta}{\cos\theta} \qquad \cot\theta = \dfrac{\cos\theta}{\sin\theta}$

Pythagorean Identities: $\sin^2\theta + \cos^2\theta = 1$

$$1 + \tan^2\theta = \sec^2\theta$$

$$1 + \cot^2\theta = \csc^2\theta$$

99. No. For an equation to be an identity, the equation must be true for all real numbers. $\sin\theta = \frac{1}{2}$ has an infinite number of solutions but is not an identity.

101. The graph of y_1 is a vertical shift of the graph of y_2 one unit upward so $y_1 = y_2 + 1$.

103. $y = \sqrt{x + 3} + 4 \cos x$

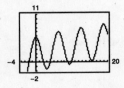

Zeros: $x \approx -1.8431, 2.1758, 3.9903, 8.8935, 9.8820$

Chapter 2   Practice Test

1. Find the value of the other five trigonometric functions, given $\tan x = \frac{4}{11}$, $\sec x < 0$.

2. Simplify $\dfrac{\sec^2 x + \csc^2 x}{\csc^2 x(1 + \tan^2 x)}$.

3. Rewrite as a single logarithm and simplify $\ln|\tan\theta| - \ln|\cot\theta|$.

4. True or false:
$$\cos\left(\frac{\pi}{2} - x\right) = \frac{1}{\csc x}$$

5. Factor and simplify: $\sin^4 x + (\sin^2 x)\cos^2 x$

6. Multiply and simplify: $(\csc x + 1)(\csc x - 1)$

7. Rationalize the denominator and simplify:
$$\frac{\cos^2 x}{1 - \sin x}$$

8. Verify:
$$\frac{1 + \cos\theta}{\sin\theta} + \frac{\sin\theta}{1 + \cos\theta} = 2\csc\theta$$

9. Verify:
$$\tan^4 x + 2\tan^2 x + 1 = \sec^4 x$$

10. Use the sum or difference formulas to determine:
(a) $\sin 105°$ (b) $\tan 15°$

11. Simplify: $(\sin 42°)\cos 38° - (\cos 42°)\sin 38°$

12. Verify $\tan\left(\theta + \dfrac{\pi}{4}\right) = \dfrac{1 + \tan\theta}{1 - \tan\theta}$.

13. Write $\sin(\arcsin x - \arccos x)$ as an algebraic expression in x.

14. Use the double-angle formulas to determine:
(a) $\cos 120°$ (b) $\tan 300°$

15. Use the half-angle formulas to determine:
(a) $\sin 22.5°$ (d) $\tan\dfrac{\pi}{12}$

16. Given $\sin = 4/5$, θ lies in Quadrant II, find $\cos(\theta/2)$.

17. Use the power-reducing identities to write $(\sin^2 x)\cos^2 x$ in terms of the first power of cosine.

18. Rewrite as a sum: $6(\sin 5\theta)\cos 2\theta$.

19. Rewrite as a product:
$\sin(x + \pi) + \sin(x - \pi)$.

20. Verify $\dfrac{\sin 9x + \sin 5x}{\cos 9x - \cos 5x} = -\cot 2x$.

21. Verify:
$(\cos u)\sin v = \frac{1}{2}[\sin(u + v) - \sin(u - v)]$.

22. Find all solutions in the interval $[0, 2\pi)$:
$4\sin^2 x = 1$

23. Find all solutions in the interval $[0, 2\pi)$:
$\tan^2\theta + \left(\sqrt{3} - 1\right)\tan\theta - \sqrt{3} = 0$

24. Find all solutions in the interval $[0, 2\pi)$:
$\sin 2x = \cos x$

25. Use the quadratic formula to find all solutions in the interval $[0, 2\pi)$:
$\tan^2 x - 6\tan x + 4 = 0$

CHAPTER 3
Additional Topics in Trigonometry

195

CHAPTER 3
Additional Topics in Trigonometry

Section 3.1   Law of Sines
Solutions to Odd-Numbered Exercises

■ If ABC is any oblique triangle with sides a, b, and c, then

$$\frac{a}{\sin A} = \frac{b}{\sin B} = \frac{c}{\sin C}.$$

■ You should be able to use the Law of Sines to solve an oblique triangle for the remaining three parts, given:

(a) Two angles and any side (AAS or ASA)

(b) Two sides and an angle opposite one of them (SSA)

   1. If A is acute and $h = b \sin A$:

   (a) $a < h$, no triangle is possible.

   (b) $a = h$ or $a > b$,   one triangle is possible.

   (c) $h < a < b$, two triangles are possible.

   2. If A is obtuse and $h = b \sin A$:

   (a) $a \leq b$, no triangle is possible.

   (b) $a > b$, one triangle is possible.

■ The area of any triangle equals one-half the product of the lengths of two sides times the sine of their included angle.

$$A = \tfrac{1}{2}ab \sin C = \tfrac{1}{2}ac \sin B = \tfrac{1}{2}bc \sin A$$

---

1. Given: $A = 30°$, $B = 45°$, $a = 20$

   $C = 180° - A - B = 105°$

   $b = \dfrac{a}{\sin A}(\sin B) = \dfrac{20 \sin 45°}{\sin 30°} = 20\sqrt{2} \approx 28.28$

   $c = \dfrac{a}{\sin A}(\sin C) = \dfrac{20 \sin 105°}{\sin 30°} \approx 38.64$

3. Given: $A = 25°$, $B = 35°$, $a = 3.5$

   $C = 180° - A - B = 120°$

   $b = \dfrac{a}{\sin A}(\sin B) = \dfrac{3.5}{\sin 25°}(\sin 35°) \approx 4.8$

   $c = \dfrac{a}{\sin A}(\sin C) = \dfrac{3.5}{\sin 25°}(\sin 120°) \approx 7.2$

5. Given: $A = 36°$, $a = 8$, $b = 5$

   $\sin B = \dfrac{b \sin A}{a} = \dfrac{5 \sin 36°}{8} \approx 0.36737 \implies B \approx 21.55°$

   $C = 180° - A - B \approx 180° - 36° - 21.55 = 122.45°$

   $c = \dfrac{a}{\sin A}(\sin C) = \dfrac{8}{\sin 36°}(\sin 122.45°) \approx 11.49$

7. Given: $A = 102.4°$, $C = 16.7°$, $a = 21.6$

   $B = 180° - A - C = 60.9°$

   $b = \dfrac{a}{\sin A}(\sin B) = \dfrac{21.6}{\sin 102.4°}(\sin 60.9°) \approx 19.3$

   $c = \dfrac{a}{\sin A}(\sin C) = \dfrac{21.6}{\sin 102.4°}(\sin 16.7°) \approx 6.4$

9. Given: $A = 83°\ 20'$, $C = 54.6°$, $c = 18.1$

   $B = 180° - A - C = 180° - 83°\ 20' - 54°\ 36' = 42°\ 4'$

   $a = \dfrac{c}{\sin C}\,(\sin A) = \dfrac{18.1}{\sin 54.6°}\,(\sin 83°\ 20') \approx 22.05$

   $b = \dfrac{c}{\sin C}\,(\sin B) = \dfrac{18.1}{\sin 54.6°}\,(\sin 42°\ 4') \approx 14.88$

11. Given: $B = 15°\ 30'$, $a = 4.5$, $b = 6.8$

    $\sin A = \dfrac{a \sin B}{b} = \dfrac{4.5 \sin 15°\ 30'}{6.8} \approx 0.17685 \implies A \approx 10°\ 11'$

    $C = 180° - A - B \approx 180° - 10°\ 11' - 15°\ 30' = 154°\ 19'$

    $c = \dfrac{b}{\sin B}\,(\sin C) = \dfrac{6.8}{\sin 15°\ 30'}\,(\sin 154°\ 19') \approx 11.03$

13. Given: $C = 145°$, $b = 4$, $c = 14$

    $\sin B = \dfrac{b \sin C}{c} = \dfrac{4 \sin 145°}{14} \approx 0.16387 \implies B \approx 9.43°$

    $A = 180° - B - C \approx 180° - 9.43° - 145° = 25.57°$

    $a = \dfrac{c}{\sin C}\,(\sin A) \approx \dfrac{14}{\sin 145°}\,(\sin 25.57°) \approx 10.53$

15. Given: $A = 110°\ 15'$, $a = 48$, $b = 16$

    $\sin B = \dfrac{b \sin A}{a} = \dfrac{16 \sin 110°\ 15'}{48} \approx 0.31273 \implies B \approx 18°\ 13'$

    $C = 180° - A - B \approx 180° - 110°\ 15' - 18°\ 13' = 51°\ 32'$

    $c = \dfrac{a}{\sin A}\,(\sin C) = \dfrac{48}{\sin 110°\ 15'}\,(\sin 51°\ 32') \approx 40.06$

17. Given: $A = 55°$, $B = 42°$, $c = \dfrac{3}{4}$

    $C = 180° - A - B = 83°$

    $a = \dfrac{c}{\sin C}(\sin A) = \dfrac{0.75}{\sin 83°}(\sin 55°) \approx 0.62$

    $b = \dfrac{c}{\sin C}(\sin B) = \dfrac{0.75}{\sin 83°}(\sin 42°) \approx 0.51$

19. Given: $a = 4.5$, $b = 12.8$, $A = 58°$

    $h = 12.8 \sin 58° \approx 10.86$

    Since $a < h$, no triangle is formed.

21. Given: $a = 18$, $b = 20$, $A = 76°$

    $h = 20 \sin 76° \approx 19.41$

    Since $a < h$, no triangle is formed.

23. Given: $a = 125$, $b = 200$, $A = 110°$

    No triangle is formed because A is obtuse and $a < b$.

25. Given: $a = \frac{5}{12}, b = 1\frac{3}{8}, A = 22°$

$h = \left(1\frac{3}{8}\right)\sin 22° \approx 0.52$

Since $a < h$, no triangle is formed.

27. Given: $A = 36°, a = 5$

(a) One solution if $b \le 5$ or $b = \dfrac{5}{\sin 36°}$

(b) Two solutions if $5 < b < \dfrac{5}{\sin 36°}$

(c) No solution if $b > \dfrac{5}{\sin 36°}$

29. Given: $A = 10°, a = 10.8$

(a) One solution if $b \le 10.8$ or $b = \dfrac{10.8}{\sin 10°}$

(b) Two solutions if $10.8 < b < \dfrac{10.8}{\sin 10°}$

(c) No solution if $b > \dfrac{10.8}{\sin 10°}$

31. Area $= \frac{1}{2}ab \sin C = \frac{1}{2}(4)(6) \sin 120° \approx 10.4$

33. Area $= \frac{1}{2}bc \sin A = \frac{1}{2}(57)(85) \sin 43° 45' \approx 1675.2$

35. Area $= \frac{1}{2}ac \sin B = \frac{1}{2}(105)(64)\sin(72°30') \approx 3204.5$

37. $C = 180° - 23° - 94° = 63°$

$h = \dfrac{35}{\sin 63°}(\sin 23°) \approx 15.3$ meters

39. $\dfrac{\sin(42° - \theta)}{10} = \dfrac{\sin 48°}{17}$

$\sin(42° - \theta) \approx 0.43714$

$42° - \theta \approx 25.9°$

$\theta \approx 16.1°$

41. Given: $c = 100$

$A = 74° - 28° = 46°,$

$B = 180° - 41° - 74° = 65°,$

$C = 180° - 46° - 65° = 69°$

$a = \dfrac{c}{\sin C}(\sin A) = \dfrac{100}{\sin 69°}(\sin 46°) \approx 77$ meters

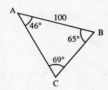

43. (a)

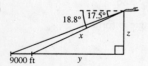

(b) $\dfrac{x}{\sin 17.5°} = \dfrac{9000}{\sin 1.3°}$

$x \approx 119,289.1261$ feet ≈ 22.6 miles

(c) $\dfrac{y}{\sin 71.2°} = \dfrac{x}{\sin 90°}$

$y = x \sin 71.2° \approx 119,289.1261 \sin 71.2°$

$\approx 112,924.963$ feet ≈ 21.4 miles

(d) $z = x \sin 18.8° \approx 119,289.1261 \sin 18.8° \approx 38,443$ feet

45.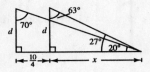

In 15 minutes the boat has traveled

$$(10 \text{ mph})\left(\frac{1}{4} \text{ hr}\right) = \frac{10}{4} \text{ miles}$$

$\tan 63° = \dfrac{x}{d} \qquad \Rightarrow d \tan 63° = x$

$\tan 70° = \dfrac{x + (10/4)}{d} \Rightarrow d \tan 70° = x + \dfrac{10}{4}$

$$\Rightarrow d \tan 70° - \frac{10}{4} = x$$

$d \tan 70° - \dfrac{10}{4} = d \tan 63°$

$d \tan 70° - d \tan 63° = \dfrac{10}{4}$

$d(\tan 70° - \tan 63°) = 2.5$

$d = \dfrac{2.5}{\tan 70° - \tan 63°} \approx 3.2 \text{ miles}$

47. $\alpha = 180 - (\phi + 180 - \theta) = \theta - \phi$

49. False. Two sides and one opposite angle do not necessarily determine a unique triangle.

51. (a) $A = \dfrac{1}{2}(30)(20)\sin\left(\theta + \dfrac{\theta}{2}\right) - \dfrac{1}{2}(8)(20)\sin\dfrac{\theta}{2} - \dfrac{1}{2}(8)(30)\sin\theta$

$= 300 \sin\dfrac{3\theta}{2} - 80 \sin\dfrac{\theta}{2} - 120 \sin\theta$

$= 20\left[15 \sin\dfrac{3\theta}{2} - 4 \sin\dfrac{\theta}{2} - 6 \sin\theta\right]$

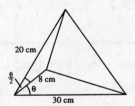

(b)

(c) Domain: $0 \le \theta \le 1.6690$

The domain would increase in length and the area would increase if the 8-centimeter line segment were decreased.

53. $\cos x = \dfrac{1}{5}, \dfrac{3\pi}{2} < x < 2\pi \Rightarrow x$ is in Quadrant IV.

$$\sin^2 x + \left(\dfrac{1}{5}\right)^2 = 1$$

$$\sin^2 x = 1 - \dfrac{1}{25}$$

$$\sin^2 x = \dfrac{24}{25}$$

$$\sin x = -\sqrt{\dfrac{24}{25}} = -\dfrac{2\sqrt{6}}{5}$$

$$\tan x = \dfrac{\sin x}{\cos x} = \dfrac{-\dfrac{2\sqrt{6}}{5}}{\dfrac{1}{5}} = -2\sqrt{6}$$

$$\cot x = \dfrac{1}{\tan x} = \dfrac{1}{-2\sqrt{6}} = -\dfrac{\sqrt{6}}{12}$$

$$\sec x = \dfrac{1}{\cos x} = \dfrac{1}{\dfrac{1}{5}} = 5$$

$$\csc x = \dfrac{1}{\sin x} = \dfrac{1}{-\dfrac{2\sqrt{6}}{5}} = -\dfrac{5}{2\sqrt{6}} = -\dfrac{5\sqrt{6}}{12}$$

55. $\tan x = -5, \dfrac{\pi}{2} < x < \pi \Rightarrow x$ is in Quadrant II.

$$1 + (-5)^2 = \sec^2 x$$

$$26 = \sec^2 x$$

$$\sec x = -\sqrt{26}$$

$$\cos x = \dfrac{1}{\sec x} = \dfrac{1}{-\sqrt{26}} = -\dfrac{\sqrt{26}}{26}$$

$$\dfrac{\sin x}{\cos x} = \tan x \Rightarrow \sin x = \cos x \tan x$$

$$\sin x = \left(-\dfrac{\sqrt{26}}{26}\right)(-5) = \dfrac{5\sqrt{26}}{26}$$

$$\cot x = \dfrac{1}{\tan x} = -\dfrac{1}{5}$$

$$\csc x = \dfrac{1}{\sin x} = \dfrac{1}{\dfrac{5\sqrt{26}}{26}} = \dfrac{26}{5\sqrt{26}} = \dfrac{\sqrt{26}}{5}$$

57. $\tan x \cos x \sec x = \tan x \cos x \dfrac{1}{\cos x} = \tan x$

59. $1 + \cot^2\left(\dfrac{\pi}{2} - x\right) = 1 + \tan^2 x = \sec^2 x$

Section 3.2 Law of Cosines

■ If ABC is any oblique triangle with sides a, b, and c, the following equations are valid.

(a) $a^2 = b^2 + c^2 - 2bc \cos A$ or $\cos A = \dfrac{b^2 + c^2 - a^2}{2bc}$

(b) $b^2 = a^2 + c^2 - 2ac \cos B$ or $\cos B = \dfrac{a^2 + c^2 - b^2}{2ac}$

(c) $c^2 = a^2 + b^2 - 2ab \cos C$ or $\cos C = \dfrac{a^2 + b^2 - c^2}{2ab}$

■ You should be able to use the Law of Cosines to solve an oblique triangle for the remaining three parts, given:

(a) Three sides (SSS)

(b) Two sides and their included angle (SAS)

■ Given any triangle with sides of length a, b, and c, the area of the triangle is

$$\text{Area} = \sqrt{s(s - a)(s - b)(s - c)}, \text{ where } s = \dfrac{a + b + c}{2}. \qquad \text{(Heron's Formula)}$$

Solutions to Odd-Numbered Exercises

1. Given: $a = 7, b = 10, c = 15$

$$\cos C = \frac{a^2 + b^2 - c^2}{2ab} = \frac{49 + 100 - 225}{2(7)(10)} \approx -0.5429 \implies C \approx 122.88°$$

$$\sin B = \frac{b \sin C}{c} = \frac{10 \sin 122.88°}{15} \approx 0.5599 \implies B \approx 34.05°$$

$$A \approx 180° - 34.05° - 122.88° \approx 23.07°$$

3. Given: $A = 30°,\ b = 15,\ c = 30$

$$a^2 = b^2 + c^2 - 2bc \cos A$$
$$= 225 + 900 - 2(15)(30) \cos 30° \approx 345.5771$$
$$a \approx 18.6$$
$$\cos B = \frac{a^2 + c^2 - b^2}{2ac} \approx \frac{(18.6)^2 + 900 - 225}{2(18.6)(30)} \approx 0.9148$$
$$B \approx 23.8°$$
$$C \approx 180° - 30° - 23.8° = 126.2°$$

5. $a = 11, b = 14, c = 20$

$$\cos C = \frac{a^2 + b^2 - c^2}{2ab} = \frac{121 + 196 - 400}{2(11)(14)} \approx -0.2695 \implies C \approx 105.63°$$

$$\sin B = \frac{b \sin C}{c} = \frac{14 \sin 105.63°}{20} \approx 0.6741 \implies B \approx 42.39°$$

$$A \approx 180° - 42.39° - 105.63° \approx 31.98°$$

7. Given: $a = 75.4,\ b = 52,\ c = 52$

$$\cos A = \frac{b^2 + c^2 - a^2}{2bc} = \frac{52^2 + 52^2 - 75.4^2}{2(52)(52)} = -0.05125 \implies A \approx 92.94°$$

$$\sin B = \frac{b \sin A}{a} \approx \frac{52(0.9987)}{75.4} \approx 0.68875 \implies B \approx 43.53°$$

$$C = B \approx 43.53°$$

9. Given: $A = 135°, b = 4, c = 9$

$$a^2 = b^2 + c^2 - 2bc \cos A = 16 + 81 - 2(4)(9)\cos 135° \approx 147.9117 \implies a \approx 12.16$$

$$\sin B = \frac{b \sin A}{a} = \frac{4 \sin 135°}{12.16} \approx 0.2326 \implies B \approx 13.45°$$

$$C \approx 180° - 135° - 13.45° \approx 31.55°$$

11. Given: $B = 10° 35',\ a = 40, c = 30$

$$b^2 = a^2 + c^2 - 2ac \cos B = 1600 + 900 - 2(40)(30)\cos 10° 35' \approx 140.8268 \implies b \approx 11.9$$

$$\sin C = \frac{c \sin B}{b} = \frac{30 \sin 10° 35'}{11.9} \approx 0.4630 \implies C \approx 27.58° \approx 27° 35'$$

$$A \approx 180° - 10° 35' - 27° 35' = 141° 50'$$

13. Given: $C = 125° \, 40'$, $a = 32$, $b = 32$

$c^2 = a^2 + b^2 - 2ab \cos C \approx 32^2 + 32^2 - 2(32)(32)(-0.5831) \approx 3242.1 \implies c \approx 56.9$

$A = B \implies 2A = 180° - 125° \, 40' = 54° \, 20' \implies A = B = 27° \, 10'$

15. $C = 43°, a = \dfrac{4}{9}, b = \dfrac{7}{9}$

$c^2 = a^2 + b^2 - 2ab \cos C = \left(\dfrac{4}{9}\right)^2 + \left(\dfrac{7}{9}\right)^2 - 2\left(\dfrac{4}{9}\right)\left(\dfrac{7}{9}\right)\cos 43° \approx 0.2968 \implies c \approx 0.5448$

$\sin A = \dfrac{a \sin C}{c} = \dfrac{(4/9)\sin 43°}{0.5448} \approx 0.5564 \implies A \approx 33.8°$

$B \approx 180° - 43° - 33.8° \approx 103.2°$

17.

$d^2 = 5^2 + 8^2 - 2(5)(8)\cos 45° \approx 32.4315 \implies d \approx 5.69$

$2\phi = 360° - 2(45°) = 270° \implies \phi = 135°$

$c^2 = 5^2 + 8^2 - 2(5)(8)\cos 135° \approx 145.5685 \implies c \approx 12.07$

19.

$\cos \phi = \dfrac{10^2 + 14^2 - 20^2}{2(10)(14)}$

$\phi \approx 111.8°$

$2\theta \approx 360° - 2(111.8°)$

$\theta = 68.2°$

$d^2 = 10^2 + 14^2 - 2(10)(14) \cos 68.2°$

$d \approx 13.86$

21.

$\cos \alpha = \dfrac{(12.5)^2 + (15)^2 - 10^2}{2(12.5)(15)} = 0.75 \implies \alpha \approx 41.41°$

$\cos \beta = \dfrac{10^2 + 15^2 - (12.5)^2}{2(10)(15)} = 0.5625 \implies \beta \approx 55.77°$

$z = 180° - \alpha - \beta = 82.82°$

— CONTINUED —

21. — CONTINUED —

$$u = 180° - z = 97.18°$$

$$b^2 = 12.5^2 + 10^2 - 2(12.5)(10)\cos 97.18° \approx 287.4967 \Rightarrow b \approx 16.96$$

$$\cos \gamma = \frac{12.5^2 + 16.96^2 - 10^2}{2(12.5)(16.96)} \approx 0.8111 \Rightarrow \gamma \approx 35.80°$$

$$\theta = \alpha + \gamma = 41.41° + 35.80° \approx 77.2°$$

$$2\phi = 360° - 2\theta \Rightarrow \phi = \frac{360° - 2(77.2°)}{2} = 102.8°$$

23. $a = 5$, $b = 7$, $c = 10 \Rightarrow s = \dfrac{a + b + c}{2} = 11$

Area $= \sqrt{s(s - a)(s - b)(s - c)} = \sqrt{11(6)(4)(1)} \approx 16.25$

25. $a = 2.5$, $b = 10.2$, $c = 9 \Rightarrow s = \dfrac{a + b + c}{2} = 10.85$

Area $= \sqrt{s(s - a)(s - b)(s - c)} = \sqrt{10.85(8.35)(0.65)(1.85)} \approx 10.44$

27. $a = 12.32$, $b = 8.46$, $c = 15.05 \Rightarrow s = \dfrac{a + b + c}{2} = 17.915$

Area $= \sqrt{s(s - a)(s - b)(s - c)} = \sqrt{17.915(5.595)(9.455)(2.865)} \approx 52.11$

29.

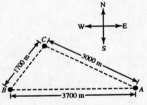

$$\cos B = \frac{1700^2 + 3700^2 - 3000^2}{2(1700)(3700)} \Rightarrow B \approx 52.9°$$

Bearing: $90° - 52.9° = $ N $37.1°$ E

$$\cos C = \frac{1700^2 + 3000^2 - 3700^2}{2(1700)(3000)} \Rightarrow C \approx 100.2°$$

Bearing: $A = 180° - 52.9° - 100.2° = 26.9° \Rightarrow$ S $63.1°$ E

31.

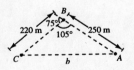

$$b^2 = 220^2 + 250^2 - 2(220)(250)\cos 105° \Rightarrow b \approx 373.3 \text{ meters}$$

33.

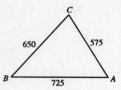

The largest angle is across from the largest side.

$$\cos C = \frac{650^2 + 575^2 - 725^2}{2(650)(575)}$$

$$C \approx 72.3°$$

35. $C = 180° - 53° - 67° = 60°$

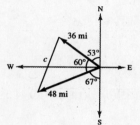

$$c^2 = a^2 + b^2 - 2ab \cos C$$

$$= 36^2 + 48^2 - 2(36)(48)(0.5)$$

$$= 1872$$

$$c \approx 43.3 \text{ mi}$$

37. (a) $\cos \theta = \dfrac{273^2 + 178^2 - 235^2}{2(273)(178)}$

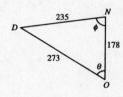

$$\theta \approx 58.4°$$

Bearing: N 58.4° W

(b) $\cos \phi = \dfrac{235^2 + 178^2 - 273^2}{2(235)(178)}$

$$\phi \approx 81.5°$$

Bearing: S 81.5° W

39. $d^2 = 60.5^2 + 90^2 - 2(60.5)(90) \cos 45° \approx 4059.8572 \implies d \approx 63.7 \text{ ft}$

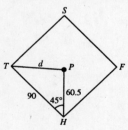

41. $a^2 = 35^2 + 20^2 - 2(35)(20)\cos 42° \implies a \approx 24.2 \text{ miles}$

43. $\overline{RS} = \sqrt{8^2 + 10^2} = \sqrt{164} = 2\sqrt{41} \approx 12.8$ ft

$\overline{PQ} = \frac{1}{2}\sqrt{16^2 + 10^2} = \frac{1}{2}\sqrt{356} = \sqrt{89} \approx 9.4$ ft

$\tan P = \frac{10}{16}$

$P = \arctan \frac{5}{8} \approx 32.0°$

$\overline{QS} \approx \sqrt{8^2 + 9.4^2 - 2(8)(9.4)\cos 32°} \approx \sqrt{24.81} \approx 5.0$ ft

45. $d^2 = 10^2 + 7^2 - 2(10)(7)\cos\theta$

$\theta = \arccos\left[\dfrac{10^2 + 7^2 - d^2}{2(10)(7)}\right]$

$s = \dfrac{360° - \theta}{360°}(2\pi r) = \dfrac{(360° - \theta)\pi}{45°}$

d (inches)	9	10	12	13	14	15	16
θ (degrees)	60.9°	69.5°	88.0°	98.2°	109.6°	122.9°	139.8°
s (inches)	20.88	20.28	18.99	18.28	17.48	16.55	15.37

47. $a = 200, b = 500, c = 600 \implies s = \dfrac{200 + 500 + 600}{2} = 650$

Area $= \sqrt{650(450)(150)(50)} \approx 46{,}837.5$ square feet

49. False. The average of the three sides of a triangle is $\dfrac{a + b + c}{3}$, not $\dfrac{a + b + c}{2}$.

51. (a) Working with $\triangle OBC$, we have $\cos\alpha = \dfrac{\frac{a}{2}}{R}$.

This implies that $2R = \dfrac{a}{\cos\alpha}$.

Since we know that

$$\frac{a}{\sin A} = \frac{b}{\sin B} = \frac{c}{\sin C},$$

we can complete the proof by showing that $\cos\alpha = \sin A$. The solution of the system

$A + B + C = 180°$

$\alpha - C + A = \beta$

$\alpha + \beta = B$

is $\alpha = 90° - A$. Therefore:

$$2R = \frac{a}{\cos\alpha} = \frac{a}{\cos(90° - A)} = \frac{a}{\sin A}.$$

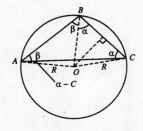

(b) By Heron's Formula, the area of the triangle is

Area $= \sqrt{s(s - a)(s - b)(s - c)}$.

We can also find the area by dividing the area into six triangles and using the fact that the area is $\frac{1}{2}$ the base times the height. Using the figure as given, we have

Area $= \dfrac{1}{2}xr + \dfrac{1}{2}xr + \dfrac{1}{2}yr + \dfrac{1}{2}yr + \dfrac{1}{2}zr + \dfrac{1}{2}zr$

$= r(x + y + z)$

$= rs$.

Therefore: $rs = \sqrt{s(s - a)(s - b)(s - c)} \implies$

$$r = \sqrt{\frac{(s - a)(s - b)(s - c)}{s}}.$$

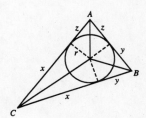

53. Given: $a = 200$ ft, $b = 250$ ft, $c = 325$ ft

$$s = \frac{200 + 250 + 325}{2} \approx 387.5$$

Radius of the inscribed circle: $r = \sqrt{\dfrac{(s-a)(s-b)(s-c)}{s}} = \sqrt{\dfrac{(187.5)(137.5)(62.5)}{387.5}} \approx 64.5$ ft

Circumference of an inscribed circle: $C = 2\pi r \approx 2\pi(64.5) \approx 405.3$ ft

55.
$$\frac{1}{2}bc(1 - \cos A) = \frac{1}{2}bc\left[1 + \frac{a^2 - (b^2 + c^2)}{2bc}\right]$$

$$= \frac{1}{2}bc\left[\frac{2bc + a^2 - b^2 - c^2}{2bc}\right]$$

$$= \frac{a^2 - (b^2 - 2bc + c^2)}{4}$$

$$= \frac{a^2 - (b - c)^2}{4}$$

$$= \left(\frac{a - (b-c)}{2}\right)\left(\frac{a + (b-c)}{2}\right)$$

$$= \frac{a - b + c}{2} \cdot \frac{a + b - c}{2}$$

57. $\arccos 0 = \dfrac{\pi}{2}$

59. $\arctan(-\sqrt{3}) = -\arctan\sqrt{3} = -\dfrac{\pi}{3}$

61. $\arccos\left(-\dfrac{\sqrt{3}}{2}\right) = \pi - \arccos\dfrac{\sqrt{3}}{2} = \pi - \dfrac{\pi}{6} = \dfrac{5\pi}{6}$

63. Let $u = \arccos 3x$

$\cos u = 3x = \dfrac{3x}{1}.$

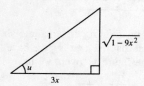

$$\tan(\arccos 3x) = \tan u = \frac{\sqrt{1 - 9x^2}}{3x}$$

65. Let $u = \arcsin\dfrac{x - 1}{2}$

$\sin u = \dfrac{x - 1}{2}.$

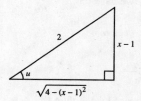

$$\cos\left(\arcsin\frac{x - 1}{2}\right) = \cos u$$

$$= \frac{\sqrt{4 - (x - 1)^2}}{2}$$

67. $x = 2 \cos \theta, -\dfrac{\pi}{2} < \theta < \dfrac{\pi}{2}$

$-\sqrt{2} = \sqrt{4 - x^2}$

$-\sqrt{2} = \sqrt{4 - (2 \cos \theta)^2}$

$-\sqrt{2} = \sqrt{4 - 4 \cos^2 \theta}$

$-\sqrt{2} = \sqrt{4(1 - \cos^2 \theta)}$

$-\sqrt{2} = \sqrt{4 \sin^2 \theta}$

$-\sqrt{2} = 2 \sin \theta$

$-\dfrac{\sqrt{2}}{2} = \sin \theta \implies \cos \theta = \dfrac{\sqrt{2}}{2} \implies x = 2\left(\dfrac{\sqrt{2}}{2}\right) = \sqrt{2}$

$\sec \theta = \dfrac{1}{\cos \theta} = \dfrac{1}{\dfrac{\sqrt{2}}{2}} = \sqrt{2}$

$\csc \theta = \dfrac{1}{\sin \theta} = \dfrac{1}{-\dfrac{\sqrt{2}}{2}} = -\sqrt{2}$

69. $x = 6 \tan \theta, -\dfrac{\pi}{2} < \theta < \dfrac{\pi}{2}$

$12 = \sqrt{36 + x^2}$

$12 = \sqrt{36 + (6 \tan \theta)^2}$

$12 = \sqrt{36 + 36 \tan^2 \theta}$

$12 = \sqrt{36(1 - \tan^2 \theta)}$

$12 = \sqrt{36 \sec^2 \theta}$

$12 = 6 \sec \theta$

$2 = \sec \theta$

$\cos \theta = \dfrac{1}{2}$

$\sin^2 \theta + \left(\dfrac{1}{2}\right)^2 = 1$

$\sin^2 \theta = 1 - \dfrac{1}{4} = \dfrac{3}{4}$

$\sin \theta = \pm\sqrt{\dfrac{3}{4}} = \pm\dfrac{\sqrt{3}}{2}$

$\csc \theta = \dfrac{1}{\sin \theta} = \dfrac{1}{\pm\dfrac{\sqrt{3}}{2}} = \pm\dfrac{2}{\sqrt{3}} = \pm\dfrac{2\sqrt{3}}{3}$

Section 3.3   Vectors in the Plane

- A vector **v** is the collection of all directed line segments that are equivalent to a given directed line segment $\overrightarrow{PQ}$.
- You should be able to *geometrically* perform the operations of vector addition and scalar multiplication.
- The component form of the vector with initial point $P = (p_1, p_2)$ and terminal point $Q = (q_1, q_2)$ is

    $\overrightarrow{PQ} = \langle q_1 - p_1, q_2 - p_2 \rangle = \langle v_1, v_2 \rangle = \mathbf{v}.$
- The magnitude of $\mathbf{v} = \langle v_1, v_2 \rangle$ is given by $\|\mathbf{v}\| = \sqrt{v_1^2 + v_2^2}$.
- If $\|\mathbf{v}\| = 1$, **v** is a unit vector.
- You should be able to perform the operations of scalar multiplication and vector addition in component form.

    (a) $\mathbf{u} + \mathbf{v} = \langle u_1 + v_1, u_2 + v_2 \rangle$ (b) $k\mathbf{u} = \langle ku_1, ku_2 \rangle$
- You should know the following properties of vector addition and scalar multiplication.

    (a) $\mathbf{u} + \mathbf{v} = \mathbf{v} + \mathbf{u}$

    (b) $(\mathbf{u} + \mathbf{v}) + \mathbf{w} = \mathbf{u} + (\mathbf{v} + \mathbf{w})$

    (c) $\mathbf{u} + \mathbf{0} = \mathbf{u}$

    (d) $\mathbf{u} + (-\mathbf{u}) = \mathbf{0}$

    (e) $c(d\mathbf{u}) = (cd)\mathbf{u}$

    (f) $(c + d)\mathbf{u} = c\mathbf{u} + d\mathbf{u}$

    (g) $c(\mathbf{u} + \mathbf{v}) = c\mathbf{u} + c\mathbf{v}$

    (h) $1(\mathbf{u}) = \mathbf{u}, 0\mathbf{u} = \mathbf{0}$

    (i) $\|c\mathbf{v}\| = |c| \, \|\mathbf{v}\|$

— CONTINUED —

— CONTINUED —

■ A unit vector in the direction of **v** is $\mathbf{u} = \dfrac{\mathbf{v}}{\|\mathbf{v}\|}$.

■ The standard unit vectors are $\mathbf{i} = \langle 1, 0 \rangle$ and $\mathbf{j} = \langle 0, 1 \rangle$. $\mathbf{v} = \langle v_1, v_2 \rangle$ can be written as $\mathbf{v} = v_1\mathbf{i} + v_2\mathbf{j}$.

■ A vector **v** with magnitude $\|\mathbf{v}\|$ and direction θ can be written as $\mathbf{v} = a\mathbf{i} + b\mathbf{j} = \|\mathbf{v}\|(\cos\theta)\mathbf{i} + \|\mathbf{v}\|(\sin\theta)\mathbf{j}$ where $\tan\theta = b/a$.

Solutions to Odd-Numbered Exercises

1. Initial point: $(0, 0)$

Terminal point: $(3, 2)$

$\mathbf{v} = \langle 3 - 0, 2 - 0 \rangle = \langle 3, 2 \rangle$

$\|\mathbf{v}\| = \sqrt{3^2 + 2^2} = \sqrt{13}$

3. Initial point: $(2, 2)$

Terminal point: $(-1, 4)$

$\mathbf{v} = \langle -1 - 2, 4 - 2 \rangle = \langle -3, 2 \rangle$

$\|\mathbf{v}\| = \sqrt{(-3)^2 + 2^2} = \sqrt{13}$

5. Initial point: $(3, -2)$

Terminal point: $(3, 3)$

$\mathbf{v} = \langle 3 - 3, 3 - (-2) \rangle = \langle 0, 5 \rangle$

$\|\mathbf{v}\| = \sqrt{0^2 + 5^2} = \sqrt{25} = 5$

7. Initial point: $(-1, 5)$

Terminal point: $(15, 12)$

$\mathbf{v} = \langle 15 - (-1), 12 - 5 \rangle = \langle 16, 7 \rangle$

$\|\mathbf{v}\| = \sqrt{16^2 + 7^2} = \sqrt{305}$

9. Initial point: $(-3, -5)$

Terminal point: $(5, 1)$

$\mathbf{v} = \langle 5 - (-3), 1 - (-5) \rangle = \langle 8, 6 \rangle$

$\|\mathbf{v}\| = \sqrt{8^2 + 6^2} = \sqrt{100} = 10$

11. Initial point: $(1, 3)$

Terminal point: $(-8, -9)$

$\mathbf{v} = \langle -8 - 1, -9 - 3 \rangle = \langle -9, -12 \rangle$

$\|\mathbf{v}\| = \sqrt{(-9)^2 + (-12)^2} = \sqrt{225} = 15$

13.

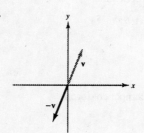

15.

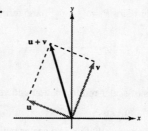

17. u + 2v

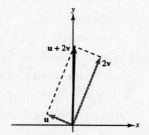

19. $\mathbf{u} = \langle 2, 1 \rangle$, $\mathbf{v} = \langle 1, 3 \rangle$

  (a) $\mathbf{u} + \mathbf{v} = \langle 3, 4 \rangle$

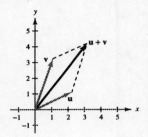

  (b) $\mathbf{u} - \mathbf{v} = \langle 1, -2 \rangle$

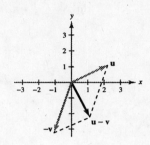

  (c) $2\mathbf{u} - 3\mathbf{v} = \langle 4, 2 \rangle - \langle 3, 9 \rangle = \langle 1, -7 \rangle$

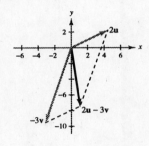

21. $\mathbf{u} = \langle -5, 3 \rangle$, $\mathbf{v} = \langle 0, 0 \rangle$

  (a) $\mathbf{u} + \mathbf{v} = \langle -5, 3 \rangle = \mathbf{u}$

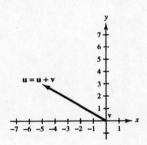

  (b) $\mathbf{u} - \mathbf{v} = \langle -5, 3 \rangle = \mathbf{u}$

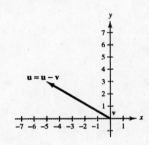

  (c) $2\mathbf{u} - 3\mathbf{v} = 2\mathbf{u} = \langle -10, 6 \rangle$

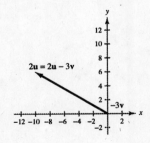

23. u = i + j, v = 2i − 3j

   (a) **u + v = 3i − 2j**

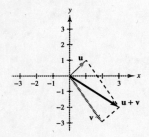

   (b) **u − v = −i + 4j**

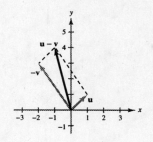

   (c) **2u − 3v = (2i + 2j) − (6i − 9j) = −4i + 11j**

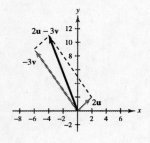

25. u = 2i, v = j

   (a) **u + v = 2i + j**

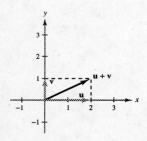

   (b) **u − v = 2i − j**

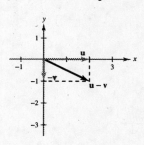

   (c) **2u − 3v = 4i − 3j**

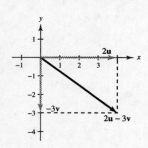

27. $\mathbf{v} = \dfrac{1}{\|\mathbf{u}\|}\mathbf{u} = \dfrac{1}{\sqrt{3^2 + 0^2}}\langle 3, 0 \rangle = \dfrac{1}{3}\langle 3, 0 \rangle = \langle 1, 0 \rangle$

29. $\mathbf{u} = \dfrac{1}{\|\mathbf{v}\|}\mathbf{v} = \dfrac{1}{\sqrt{(-2)^2 + 2^2}}\langle -2, 2 \rangle = \dfrac{1}{2\sqrt{2}}\langle -2, 2 \rangle$

$$= \left\langle -\dfrac{1}{\sqrt{2}}, \dfrac{1}{\sqrt{2}} \right\rangle$$

31. $\mathbf{u} = \dfrac{1}{\|\mathbf{v}\|}\mathbf{v} = \dfrac{1}{\sqrt{6^2 + (-2)^2}}(6\mathbf{i} - 2\mathbf{j}) = \dfrac{1}{\sqrt{40}}(6\mathbf{i} - 2\mathbf{j})$

$\qquad = \dfrac{1}{2\sqrt{10}}(6\mathbf{i} - 2\mathbf{j}) = \dfrac{3}{\sqrt{10}}\mathbf{i} - \dfrac{1}{\sqrt{10}}\mathbf{j}$

33. $\mathbf{u} = \dfrac{1}{\|\mathbf{w}\|}\mathbf{w} = \dfrac{1}{4}(4\mathbf{j}) = \mathbf{j}$

35. $\mathbf{u} = \dfrac{1}{\|\mathbf{w}\|}\mathbf{w} = \dfrac{1}{\sqrt{1^2 + (-2)^2}}(\mathbf{i} - 2\mathbf{j}) = \dfrac{1}{\sqrt{5}}(\mathbf{i} - 2\mathbf{j})$

$\qquad = \dfrac{1}{\sqrt{5}}\mathbf{i} - \dfrac{2}{\sqrt{5}}\mathbf{j}$

37. $5\left(\dfrac{1}{\|\mathbf{u}\|}\mathbf{u}\right) = 5\left(\dfrac{1}{\sqrt{3^2 + 3^2}}\langle 3, 3\rangle\right) = \dfrac{5}{3\sqrt{2}}\langle 3, 3\rangle$

$\qquad = \left\langle \dfrac{5}{\sqrt{2}}, \dfrac{5}{\sqrt{2}}\right\rangle$

39. $9\left(\dfrac{1}{\|\mathbf{u}\|}\mathbf{u}\right) = 9\left(\dfrac{1}{\sqrt{2^2 + 5^2}}\langle 2, 5\rangle\right) = \dfrac{9}{\sqrt{29}}\langle 2, 5\rangle$

$\qquad = \left\langle \dfrac{18}{\sqrt{29}}, \dfrac{45}{\sqrt{29}}\right\rangle$

41. $\mathbf{v} = \tfrac{3}{2}\mathbf{u}$

$\quad = \tfrac{3}{2}(2\mathbf{i} - \mathbf{j})$

$\quad = 3\mathbf{i} - \tfrac{3}{2}\mathbf{j} = \langle 3, -\tfrac{3}{2}\rangle$

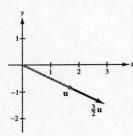

43. $\mathbf{v} = \mathbf{u} + 2\mathbf{w}$

$\quad = (2\mathbf{i} - \mathbf{j}) + 2(\mathbf{i} + 2\mathbf{j})$

$\quad = 4\mathbf{i} + 3\mathbf{j} = \langle 4, 3\rangle$

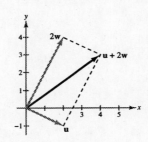

45. $\mathbf{v} = \tfrac{1}{2}(3\mathbf{u} + \mathbf{w})$

$\quad = \tfrac{1}{2}(6\mathbf{i} - 3\mathbf{j} + \mathbf{i} + 2\mathbf{j})$

$\quad = \tfrac{7}{2}\mathbf{i} - \tfrac{1}{2}\mathbf{j} = \langle \tfrac{7}{2}, -\tfrac{1}{2}\rangle$

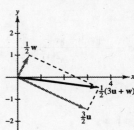

47. $\mathbf{v} = 3(\cos 60°\mathbf{i} + \sin 60°\mathbf{j})$

$\quad \|\mathbf{v}\| = 3, \ \theta = 60°$

49. $\mathbf{v} = 6\mathbf{i} - 6\mathbf{j}$

$\quad \|\mathbf{v}\| = \sqrt{6^2 + (-6)^2} = \sqrt{72} = 6\sqrt{2}$

$\quad \tan \theta = \dfrac{-6}{6} = -1$

Since $\mathbf{v}$ lies in Quadrant IV, $\theta = 315°$.

51. $\mathbf{v} = \langle 3\cos 0°, 3\sin 0°\rangle$

$\quad = \langle 3, 0\rangle$

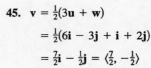

53. $\mathbf{v} = \left\langle \dfrac{7}{2}\cos 150°, \dfrac{7}{2}\sin 150°\right\rangle$

$\quad = \left\langle -\dfrac{7\sqrt{3}}{4}, \dfrac{7}{4}\right\rangle$

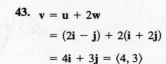

55. $\mathbf{v} = \langle 3\sqrt{2}\cos 150°, \ 3\sqrt{2}\sin 150°\rangle$

$\quad = \left\langle -\dfrac{3\sqrt{6}}{2}, \dfrac{3\sqrt{2}}{2}\right\rangle$

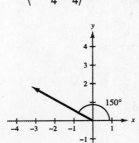

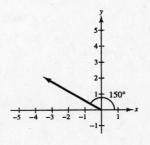

57. $\mathbf{v} = 2\left(\dfrac{1}{\sqrt{1^2 + 3^2}}\right)(\mathbf{i} + 3\mathbf{j})$

$= \dfrac{2}{\sqrt{10}}(\mathbf{i} + 3\mathbf{j})$

$= \dfrac{\sqrt{10}}{5}\mathbf{i} + \dfrac{3\sqrt{10}}{5}\mathbf{j} = \left\langle \dfrac{\sqrt{10}}{5}, \dfrac{3\sqrt{10}}{5} \right\rangle$

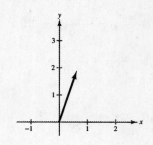

59. $\mathbf{u} = \langle 5 \cos 0°, 5 \sin 0° \rangle = \langle 5, 0 \rangle$

$\mathbf{v} = \langle 5 \cos 90°, 5 \sin 90° \rangle = \langle 0, 5 \rangle$

$\mathbf{u} + \mathbf{v} = \langle 5, 5 \rangle$

61. $\mathbf{u} = \langle 20 \cos 45°, 20 \sin 45° \rangle = \langle 10\sqrt{2}, 10\sqrt{2} \rangle$

$\mathbf{v} = \langle 50 \cos 180°, 50 \sin 180° \rangle = \langle -50, 0 \rangle$

$\mathbf{u} + \mathbf{v} = \langle 10\sqrt{2} - 50, 10\sqrt{2} \rangle$

63. $\mathbf{v} = \mathbf{i} + \mathbf{j}$

$\mathbf{w} = 2\mathbf{i} - 2\mathbf{j}$

$\mathbf{u} = \mathbf{v} - \mathbf{w} = -\mathbf{i} + 3\mathbf{j}$

$\|\mathbf{v}\| = \sqrt{2}$

$\|\mathbf{w}\| = 2\sqrt{2}$

$\|\mathbf{v} - \mathbf{w}\| = \sqrt{10}$

$\cos \alpha = \dfrac{\|\mathbf{v}\|^2 + \|\mathbf{w}\|^2 - \|\mathbf{v} - \mathbf{w}\|^2}{2\|\mathbf{v}\|\,\|\mathbf{w}\|} = \dfrac{2 + 8 - 10}{2\sqrt{2} \cdot 2\sqrt{2}} = 0$

$\alpha = 90°$

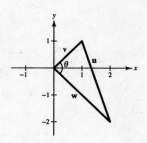

65. $\mathbf{v} = \mathbf{i} + \mathbf{j}$

$\mathbf{w} = 3\mathbf{i} - \mathbf{j}$

$\mathbf{u} = \mathbf{v} - \mathbf{w} = -2\mathbf{i} + 2\mathbf{j}$

$\cos \alpha = \dfrac{\|\mathbf{v}\|^2 + \|\mathbf{w}\|^2 - \|\mathbf{v} - \mathbf{w}\|^2}{2\|\mathbf{v}\|\,\|\mathbf{w}\|} = \dfrac{2 + 10 - 8}{2\sqrt{2}\,\sqrt{10}} \approx 0.4472$

$\alpha = 63.4°$

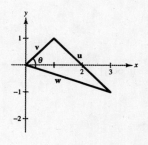

67. Force One: $\mathbf{u} = 45\mathbf{i}$

Force Two: $\mathbf{v} = 60 \cos \theta \mathbf{i} + 60 \sin \theta \mathbf{j}$

Resultant Force: $\mathbf{u} + \mathbf{v} = (45 + 60 \cos \theta)\mathbf{i} + 60 \sin \theta \mathbf{j}$

$\|\mathbf{u} + \mathbf{v}\| = \sqrt{(45 + 60 \cos \theta)^2 + (60 \sin \theta)^2} = 90$

$2025 + 5400 \cos \theta + 3600 = 8100$

$5400 \cos \theta = 2475$

$\cos \theta = \dfrac{2475}{5400} \approx 0.4583$

$\theta \approx 62.7°$

69. $\mathbf{u} = 300\mathbf{i}$

$\mathbf{v} = (125 \cos 45°)\mathbf{i} + (125 \sin 45°)\mathbf{j} = \dfrac{125}{\sqrt{2}}\mathbf{i} + \dfrac{125}{\sqrt{2}}\mathbf{j}$

$\mathbf{R} = \mathbf{u} + \mathbf{v} = \left(300 + \dfrac{125}{\sqrt{2}}\right)\mathbf{i} + \dfrac{125}{\sqrt{2}}\mathbf{j}$

$\|\mathbf{R}\| = \sqrt{\left(300 + \dfrac{125}{\sqrt{2}}\right)^2 + \left(\dfrac{125}{\sqrt{2}}\right)^2} \approx 398.32$ newtons

$\tan \theta = \dfrac{\dfrac{125}{\sqrt{2}}}{300 + \left(\dfrac{125}{\sqrt{2}}\right)} \implies \theta \approx 12.8°$

71. $\mathbf{u} = (75 \cos 30°)\mathbf{i} + (75 \sin 30°)\mathbf{j} \approx 64.95\mathbf{i} + 37.5\mathbf{j}$

$\mathbf{v} = (100 \cos 45°)\mathbf{i} + (100 \sin 45°)\mathbf{j} \approx 70.71\mathbf{i} + 70.71\mathbf{j}$

$\mathbf{w} = (125 \cos 120°)\mathbf{i} + (125 \sin 120°)\mathbf{j} \approx -62.5\mathbf{i} + 108.3\mathbf{j}$

$\mathbf{u} + \mathbf{v} + \mathbf{w} \approx 73.16\mathbf{i} + 216.5\mathbf{j}$

$\|\mathbf{u} + \mathbf{v} + \mathbf{w}\| \approx 228.5$ pounds

$\tan \theta \approx \dfrac{216.5}{73.16} \approx 2.9592$

$\theta \approx 71.3°$

73. Horizontal component of velocity: $70 \cos 35° \approx 57.34$ feet per second

Vertical component of velocity: $70 \sin 35° \approx 40.15$ feet per second

75. Cable $\overrightarrow{AC}$: $\mathbf{u} = \|\mathbf{u}\|(\cos 50°\mathbf{i} - \sin 50°\mathbf{j})$

Cable $\overrightarrow{BC}$: $\mathbf{v} = \|\mathbf{v}\|(\cos 30°\mathbf{i} - \sin 30°\mathbf{j})$

Resultant: $\mathbf{u} + \mathbf{v} = -2000\mathbf{j}$

$\|\mathbf{u}\| \cos 50° - \|\mathbf{v}\| \cos 30° = 0$

$-\|\mathbf{u}\| \sin 50° - \|\mathbf{v}\| \sin 30° = -2000$

Solving this system of equations yields:

$T_{AC} = \|\mathbf{u}\| \approx 1758.8$ pounds

$T_{BC} = \|\mathbf{v}\| \approx 1305.4$ pounds

77. Towline 1: $\mathbf{u} = \|\mathbf{u}\|(\cos 18°\mathbf{i} + \sin 18°\mathbf{j})$

Towline 2: $\mathbf{v} = \|\mathbf{u}\|(\cos 18°\mathbf{i} - \sin 18°\mathbf{j})$

Resultant: $\mathbf{u} + \mathbf{v} = 6000\mathbf{i}$

$\|\mathbf{u}\| \cos 18° + \|\mathbf{u}\| \cos 18° = 6000$

$\|\mathbf{u}\| \approx 3154.4$

Therefore, the tension on each towline is

$\|\mathbf{u}\| \approx 3154.4$ pounds.

79. Airspeed: $\mathbf{u} = (875 \cos 32°)\mathbf{i} - (875 \sin 32°)\mathbf{j}$

Groundspeed: $\mathbf{v} = (800 \cos 40°)\mathbf{i} - (800 \sin 40°)\mathbf{j}$

Wind: $\mathbf{w} = \mathbf{v} - \mathbf{u} = (800 \cos 40° - 875 \cos 32°)\mathbf{i} + (-800 \sin 40° + 875 \sin 32°)\mathbf{j}$

$\approx -129.2065\mathbf{i} - 50.5507\mathbf{j}$

Wind speed: $\|\mathbf{w}\| \approx \sqrt{(-129.2065)^2 + (-50.5507)^2}$

≈ 138.7 kilometers per hour

Wind direction: $\tan \theta \approx \dfrac{-50.5507}{-129.2065}$

$\theta \approx 21.4°$

N $21.4°$ E

81. $W = FD = (100 \cos 50°)(30) = 1928.4$ foot–pounds

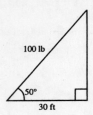

83. True. See Example 1.

85. False, $a = b = 0$.

87. (a) The angle between them is $0°$.

(b) The angle between them is $180°$.

(c) No. At most it can be equal to the sum when the angle between them is $0°$.

89. Let $\mathbf{v} = (\cos \theta)\mathbf{i} + (\sin \theta)\mathbf{j}$.

$\|\mathbf{v}\| = \sqrt{\cos^2 \theta + \sin^2 \theta} = \sqrt{1} = 1$

Therefore, $\mathbf{v}$ is a unit vector for any value of θ.

91. $\mathbf{u} = \langle 5 - 1, 2 - 6 \rangle = \langle 4, -4 \rangle$

$\mathbf{v} = \langle 9 - 4, 4 - 5 \rangle = \langle 5, -1 \rangle$

$\mathbf{u} - \mathbf{v} = \langle -1, -3 \rangle$ or $\mathbf{v} - \mathbf{u} = \langle 1, 3 \rangle$

93. $\sqrt{x^2 - 64} = \sqrt{(8 \sec \theta)^2 - 64}$

$= \sqrt{64(\sec^2 \theta - 1)}$

$= 8\sqrt{\tan^2 \theta}$

$= 8 \tan \theta$ for $0 < \theta < \dfrac{\pi}{2}$

95. $\sqrt{x^2 + 36} = \sqrt{(6 \tan \theta)^2 + 36}$

$= \sqrt{36(\tan^2 \theta + 1)}$

$= 6\sqrt{\sec^2 \theta}$

$= 6 \sec \theta$ for $0 < \theta < \dfrac{\pi}{2}$

97. $\cos x(\cos x + 1) = 0$

$\cos x = 0$ or $\cos x + 1 = 0$

$x = \dfrac{\pi}{2} + n\pi \qquad \cos x = -1$

$x = \pi + 2n\pi$

99. $3 \sec x \sin x - 2\sqrt{3} \sin x = 0$

$\sin x(3 \sec x - 2\sqrt{3}) = 0$

$\sin x = 0$ or $3 \sec x - 2\sqrt{3} = 0$

$x = n\pi \qquad \sec x = \dfrac{2\sqrt{3}}{3}$

$\cos x = \dfrac{3}{2\sqrt{3}} = \dfrac{\sqrt{3}}{2}$

$x = \dfrac{\pi}{6} + 2n\pi$

$x = \dfrac{11\pi}{6} + 2n\pi$

101. $(\sin^2 x - 1)\sin^2 x = 0$

$\sin^2 x - 1 = 0$ or $\sin^2 x = 0$

$\sin^2 x = 1 \qquad \sin x = 0$

$\sin x = \pm 1 \qquad x = n\pi$

$x = \dfrac{\pi}{2} + n\pi$

These solutions can be expressed collectively as $x = \dfrac{n\pi}{2}$.

Section 3.4   Variables and Dot Products

■ Know the definition of the dot product of $\mathbf{u} = \langle u_1, u_2 \rangle$ and $\mathbf{v} = \langle v_1, v_2 \rangle$.

$\mathbf{u} \cdot \mathbf{v} = u_1 v_1 + u_2 v_2$

■ Know the following properties of the dot product:

1. $\mathbf{u} \cdot \mathbf{v} = \mathbf{v} \cdot \mathbf{u}$
2. $\mathbf{0} \cdot \mathbf{v} = 0$
3. $\mathbf{u} \cdot (\mathbf{v} + \mathbf{w}) = \mathbf{u} \cdot \mathbf{v} + \mathbf{u} \cdot \mathbf{w}$
4. $\mathbf{v} \cdot \mathbf{v} = \|\mathbf{v}\|^2$
5. $c(\mathbf{u} \cdot \mathbf{v}) = c\mathbf{u} \cdot \mathbf{v} = \mathbf{u} \cdot c\mathbf{v}$

■ If θ is the angle between two nonzero vectors $\mathbf{u}$ and $\mathbf{v}$, then

$$\cos \theta = \frac{\mathbf{u} \cdot \mathbf{v}}{\|\mathbf{u}\| \, \|\mathbf{v}\|}.$$

■ The vectors $\mathbf{u}$ and $\mathbf{v}$ are orthogonal if $\mathbf{u} \cdot \mathbf{v} = 0$.

■ Know the definition of vector components.

$\mathbf{u} = \mathbf{w}_1 + \mathbf{w}_2$ where $\mathbf{w}_1$ and $\mathbf{w}_2$ are orthogonal, and $\mathbf{w}_1$ is parallel to $\mathbf{v}$. $\mathbf{w}_1$ is called the projection of $\mathbf{u}$ onto $\mathbf{v}$

and is denoted by $\mathbf{w}_1 = \text{proj}_{\mathbf{v}} \mathbf{u} = \left(\dfrac{\mathbf{u} \cdot \mathbf{v}}{\|\mathbf{v}\|^2} \right) \mathbf{v}$. Then we have $\mathbf{w}_2 = \mathbf{u} - \mathbf{w}_1$.

■ Know the definition of work.

1. Projection form: $w = \|\text{proj}_{\overrightarrow{PQ}} \mathbf{F}\| \, \|PQ\|$
2. Dot product form: $w = \mathbf{F} \cdot \overrightarrow{PQ}$

Solutions to Odd-Numbered Exercises

1. $\mathbf{u} = \langle 6, 1 \rangle$, $\mathbf{v} = \langle -2, 3 \rangle$

$\mathbf{u} \cdot \mathbf{v} = 6(-2) + 1(3) = -9$

3. $\mathbf{u} = 4\mathbf{i} - 2\mathbf{j}$, $\mathbf{v} = \mathbf{i} - \mathbf{j}$

$\mathbf{u} \cdot \mathbf{v} = 4(1) + (-2)(-1) = 6$

5. $\mathbf{u} = \langle 2, 2 \rangle$

$\mathbf{u} \cdot \mathbf{u} = 2(2) + 2(2) = 8$

The result is a scalar.

7. $\mathbf{u} = \langle 2, 2 \rangle$, $\mathbf{v} = \langle -3, 4 \rangle$

$(\mathbf{u} \cdot \mathbf{v})\mathbf{v} = [(2)(-3) + 2(4)]\langle -3, 4 \rangle$

$= 2\langle -3, 4 \rangle = \langle -6, 8 \rangle$

The result is a vector.

9. $\mathbf{u} = \langle -5, 12 \rangle$

$\|\mathbf{u}\| = \sqrt{\mathbf{u} \cdot \mathbf{u}} = \sqrt{(-5)^2 + 12^2} = 13$

11. $\mathbf{u} = 20\mathbf{i} + 25\mathbf{j}$

$\|\mathbf{u}\| = \sqrt{(20)^2 + (25)^2} = \sqrt{1025} = 5\sqrt{41}$

13. $\mathbf{u} = 6\mathbf{j}$

$\|\mathbf{u}\| = \sqrt{(0)^2 + (6)^2} = \sqrt{36} = 6$

15. $\mathbf{u} = \langle 1, 0 \rangle$, $\mathbf{v} = \langle 0, -2 \rangle$

$$\cos \theta = \frac{\mathbf{u} \cdot \mathbf{v}}{\|\mathbf{u}\| \, \|\mathbf{v}\|} = \frac{0}{(1)(2)} = 0$$

$$\theta = 90°$$

17. $\mathbf{u} = 3\mathbf{i} + 4\mathbf{j}$, $\mathbf{v} = -2\mathbf{j}$

$$\cos \theta = \frac{\mathbf{u} \cdot \mathbf{v}}{\|\mathbf{u}\| \, \|\mathbf{v}\|} = -\frac{8}{(5)(2)}$$

$$\theta = \arccos\left(-\frac{4}{5}\right)$$

$$\theta \approx 143.13°$$

19. $\mathbf{u} = 2\mathbf{i} - \mathbf{j}$, $\mathbf{v} = 6\mathbf{i} + 4\mathbf{j}$

$$\cos \theta = \frac{\mathbf{u} \cdot \mathbf{v}}{\|\mathbf{u}\| \, \|\mathbf{v}\|} = \frac{8}{\sqrt{5}\sqrt{52}} \implies \theta \approx 60.26°$$

21. $\mathbf{u} = 5\mathbf{i} + 5\mathbf{j}$, $\mathbf{v} = -6\mathbf{i} + 6\mathbf{j}$

$$\cos \theta = \frac{\mathbf{u} \cdot \mathbf{v}}{\|\mathbf{u}\| \, \|\mathbf{v}\|} = 0 \implies \theta = 90°$$

23. $\mathbf{u} = \left(\cos \dfrac{\pi}{3}\right)\mathbf{i} + \left(\sin \dfrac{\pi}{3}\right)\mathbf{j} = \dfrac{1}{2}\mathbf{i} + \dfrac{\sqrt{3}}{2}\mathbf{j}$

$\mathbf{v} = \left(\cos \dfrac{3\pi}{4}\right)\mathbf{i} + \left(\sin \dfrac{3\pi}{4}\right)\mathbf{j} = -\dfrac{\sqrt{2}}{2}\mathbf{i} + \dfrac{\sqrt{2}}{2}\mathbf{j}$

$\|\mathbf{u}\| = \|\mathbf{v}\| = 1$

$$\cos \theta = \frac{\mathbf{u} \cdot \mathbf{v}}{\|\mathbf{u}\| \, \|\mathbf{v}\|} = \mathbf{u} \cdot \mathbf{v} = \left(\frac{1}{2}\right)\left(-\frac{\sqrt{2}}{2}\right) + \left(\frac{\sqrt{3}}{2}\right)\left(\frac{\sqrt{2}}{2}\right) = \frac{-\sqrt{2} + \sqrt{6}}{4}$$

$$\theta = \arccos\left(\frac{-\sqrt{2} + \sqrt{6}}{4}\right) = 75° = \frac{5\pi}{12}$$

25. $P = (1, 2)$, $Q = (3, 4)$, $R = (2, 5)$

$\overrightarrow{PQ} = \langle 2, 2 \rangle$, $\overrightarrow{PR} = \langle 1, 3 \rangle$, $\overrightarrow{QR} = \langle -1, 1 \rangle$

$$\cos \alpha = \frac{\overrightarrow{PQ} \cdot \overrightarrow{PR}}{\|\overrightarrow{PQ}\| \, \|\overrightarrow{PR}\|} = \frac{8}{(2\sqrt{2})(\sqrt{10})} \implies \alpha = \arccos \frac{2}{\sqrt{5}} \approx 26.6°$$

$$\cos \beta = \frac{\overrightarrow{PQ} \cdot \overrightarrow{QR}}{\|\overrightarrow{PQ}\| \, \|\overrightarrow{QR}\|} = 0 \implies \beta = 90°. \quad \text{Thus, } \gamma = 180° - 26.6° - 90° = 63.4°.$$

27. $P = (-3, 0)$, $Q = (2, 2)$, $R = (0, 6)$

$\overrightarrow{QP} = \langle -5, -2 \rangle$, $\overrightarrow{PR} = \langle 3, 6 \rangle$, $\overrightarrow{QR} = \langle -2, 4 \rangle$

$$\cos \alpha = \frac{\overrightarrow{PQ} \cdot \overrightarrow{PR}}{\|\overrightarrow{PQ}\| \, \|\overrightarrow{PR}\|} = \frac{27}{\sqrt{29}\sqrt{45}} \implies \alpha \approx 41.6°$$

$$\cos \beta = \frac{\overrightarrow{QP} \cdot \overrightarrow{QR}}{\|\overrightarrow{QP}\| \, \|\overrightarrow{PR}\|} = \frac{2}{\sqrt{29}\sqrt{20}} \implies \beta \approx 85.2°$$

$$\delta = 180° - 41.6° - 85.2° = 53.2°$$

29. $\mathbf{u} \cdot \mathbf{v} = \|\mathbf{u}\| \, \|\mathbf{v}\| \cos \theta$

$$= (4)(10) \cos \frac{2\pi}{3}$$

$$= 40\left(-\frac{1}{2}\right)$$

$$= -20$$

31. $\mathbf{u} \cdot \mathbf{v} = \|\mathbf{u}\| \, \|\mathbf{v}\| \cos \theta$

$\qquad = (81)(64)\cos \dfrac{\pi}{4}$

$\qquad = 5184\left(\dfrac{\sqrt{2}}{2}\right)$

$\qquad = 2592\sqrt{2}$

33. $\mathbf{u} = \langle -12, 30 \rangle, \ \mathbf{v} = \left\langle \dfrac{1}{2}, \ -\dfrac{5}{4} \right\rangle$

$\mathbf{u} = -24\mathbf{v} \ \Longrightarrow \ \mathbf{u}$ and $\mathbf{v}$ are parallel.

35. $\mathbf{u} = \frac{1}{4}(3\mathbf{i} - \mathbf{j}), \ \mathbf{v} = 5\mathbf{i} + 6\mathbf{j}$

$\mathbf{u} \neq k\mathbf{v} \ \Longrightarrow \ $ Not parallel

$\mathbf{u} \cdot \mathbf{v} \neq 0 \ \Longrightarrow \ $ Not orthogonal

Neither

37. $\mathbf{u} = 2\mathbf{i} - 2\mathbf{j}, \ \mathbf{v} = -\mathbf{i} - \mathbf{j}$

$\mathbf{u} \cdot \mathbf{v} = 0 \ \Longrightarrow \ \mathbf{u}$ and $\mathbf{v}$ are orthogonal.

39. $\mathbf{u} = \langle 2, 2 \rangle, \ \mathbf{v} = \langle 6, 1 \rangle$

$\mathbf{w}_1 = \text{proj}_{\mathbf{v}}\mathbf{u} = \left(\dfrac{\mathbf{u} \cdot \mathbf{v}}{\|\mathbf{v}\|^2}\right)\mathbf{v} = \dfrac{14}{37}\mathbf{v} = \dfrac{14}{37}\langle 6, 1 \rangle$

$\mathbf{w}_2 = \mathbf{u} - \mathbf{w}_1 = \langle 2, 2 \rangle - \dfrac{14}{37}\langle 6, 1 \rangle = \left\langle -\dfrac{10}{37}, \dfrac{60}{37} \right\rangle = \dfrac{10}{37}\langle -1, 6 \rangle$

41. $\mathbf{u} = \langle 0, 3 \rangle, \ \mathbf{v} = \langle 2, 15 \rangle$

$\mathbf{w}_1 = \text{proj}_{\mathbf{v}}\mathbf{u} = \left(\dfrac{\mathbf{u} \cdot \mathbf{v}}{\|\mathbf{v}\|^2}\right)\mathbf{v} = \dfrac{45}{229}\langle 2, 15 \rangle$

$\mathbf{w}_2 = \mathbf{u} - \mathbf{w}_1 = \langle 0, 3 \rangle - \dfrac{45}{229}\langle 2, 15 \rangle = \left\langle -\dfrac{90}{229}, \dfrac{12}{229} \right\rangle = \dfrac{6}{229}\langle -15, 2 \rangle$

43. $\mathbf{u} = \langle 3, 5 \rangle$

For $\mathbf{v}$ to be orthogonal to $\mathbf{u}, \mathbf{u} \cdot \mathbf{v}$ must equal 0.

Two possibilities: $\langle -5, 3 \rangle$ and $\langle 5, -3 \rangle$

45. $\mathbf{u} = \frac{1}{2}\mathbf{i} - \frac{2}{3}\mathbf{j}$

For $\mathbf{u}$ and $\mathbf{v}$ to be orthogonal, $\mathbf{u} \cdot \mathbf{v}$ must equal 0.

Two possibilities: $\frac{2}{3}\mathbf{i} + \frac{1}{2}\mathbf{j}$ and $-\frac{2}{3}\mathbf{i} - \frac{1}{2}\mathbf{j}$

47. $w = \| \text{proj}_{\overrightarrow{PQ}}\, \mathbf{v}\| \|\overrightarrow{PQ}\|$ where $\overrightarrow{PQ} = \langle 4, 7 \rangle$ and $\mathbf{v} = \langle 1, 4 \rangle$.

$\text{proj}_{\overrightarrow{PQ}}\, \mathbf{v} = \left(\dfrac{\mathbf{v} \cdot \overrightarrow{PQ}}{\|\overrightarrow{PQ}\|^2}\right)\overrightarrow{PQ} = \left(\dfrac{32}{65}\right)\langle 4, 7 \rangle$

$w = \| \text{proj}_{\overrightarrow{PQ}}\, \mathbf{v}\| \|\overrightarrow{PQ}\| = \left(\dfrac{32\sqrt{65}}{65}\right)\left(\sqrt{65}\right) = 32$

49. $\mathbf{u} = \langle 1650, 3200 \rangle, \ \mathbf{v} = \langle 15.25, 10.50 \rangle$

$\mathbf{u} \cdot \mathbf{v} = 1650(15.25) + 3200(10.50) = \$58{,}762.50$

This gives the total revenue that can be earned by selling all of the units.

51. (a) $\mathbf{F} = -30{,}000\mathbf{j}$ Gravitational force

$\mathbf{v} = (\cos 5°)\mathbf{i} + (\sin 5°)\mathbf{j}$

$\mathbf{w}_1 = \text{proj}_{\mathbf{v}}\mathbf{F} = \left(\dfrac{\mathbf{F} \cdot \mathbf{v}}{\|\mathbf{v}\|^2}\right)\mathbf{v} = (\mathbf{F} \cdot \mathbf{v})\mathbf{v} \approx -2614.7\mathbf{v}$

The magnitude of this force is 2614.7, therefore a force of 2614.7 pounds is needed to keep the truck from rolling down the hill.

(b) $\mathbf{w}_2 = \mathbf{F} - \mathbf{w}_1 = -30{,}000\mathbf{j} + 2614.7(\cos 5°\mathbf{i} + \sin 5°\mathbf{j})$

$= 2614.7 \cos 5°\mathbf{i} + (2614.7 \sin 5° - 30{,}000)\mathbf{j}$

$\|\mathbf{w}_2\| \approx 29{,}885.8$ pounds

53. $\mathbf{w} = (245)(3) = 735$ Newton-meters

55. $\mathbf{w} = (\cos 30°)(45)(20) \approx 779.4$ foot-pounds

57. False. Work is represented by a scalar.

59. (a) $\mathbf{u} \cdot \mathbf{v} = 0 \implies \mathbf{u}$ and $\mathbf{v}$ are orthogonal and $\theta = \dfrac{\pi}{2}$.

(b) $\mathbf{u} \cdot \mathbf{v} > 0 \implies \cos \theta > 0 \implies 0 \le \theta < \dfrac{\pi}{2}$

(c) $\mathbf{u} \cdot \mathbf{v} < 0 \implies \cos \theta < 0 \implies \dfrac{\pi}{2} < \theta \le \pi$

61. In a rhombus, $\|\mathbf{u}\| = \|\mathbf{v}\|$. The diagonals are $\mathbf{u} + \mathbf{v}$ and $\mathbf{u} - \mathbf{v}$.

$(\mathbf{u} + \mathbf{v}) \cdot (\mathbf{u} - \mathbf{v}) = (\mathbf{u} + \mathbf{v}) \cdot \mathbf{u} - (\mathbf{u} + \mathbf{v}) \cdot \mathbf{v}$

$= \mathbf{u} \cdot \mathbf{u} + \mathbf{v} \cdot \mathbf{u} - \mathbf{u} \cdot \mathbf{v} - \mathbf{v} \cdot \mathbf{v}$

$= \|\mathbf{u}\|^2 - \|\mathbf{v}\|^2 = 0$

Therefore, the diagonals are orthogonal.

63. (a) Let $\mathbf{v} = \langle v_1, v_2 \rangle$.

$\mathbf{0} \cdot \mathbf{v} = 0(v_1) + 0(v_2) = 0$

(b) Let $\mathbf{u} = \langle u_1, u_2 \rangle$, $\mathbf{v} = \langle v_1, v_2 \rangle$ and $\mathbf{w} = \langle w_1, w_2 \rangle$.

$\mathbf{u} \cdot (\mathbf{v} + \mathbf{w}) = \langle u_1, u_2 \rangle \cdot \langle v_1 + w_1, v_2 + w_2 \rangle$

$= u_1(v_1 + w_1) + u_2(v_2 + w_2)$

$= u_1 v_1 + u_1 w_1 + u_2 v_2 + u_2 w_2$

$= (u_1 v_1 + u_2 v_2) + (u_1 w_1 + u_2 w_2)$

$= \mathbf{u} \cdot \mathbf{v} + \mathbf{u} \cdot \mathbf{w}$

(c) Let $\mathbf{u} = \langle u_1, u_2 \rangle$ and $\mathbf{v} = \langle v_1, v_2 \rangle$.

$c(\mathbf{u} \cdot \mathbf{v}) = c(u_1 v_1 + u_2 v_2)$

$= c(u_1 v_1) + c(u_2 v_2)$

$= u_1(c v_1) + u_2(c v_2)$

$= \mathbf{u} \cdot (c\mathbf{v})$

65. $\sin 2x - \sqrt{3} \sin x = 0$

$2 \sin x \cos x - \sqrt{3} \sin x = 0$

$\sin x\left(2 \cos x - \sqrt{3}\right) = 0$

$\sin x = 0$ or $2 \cos x - \sqrt{3} = 0$

$x = 0, \pi$ $\cos x = \dfrac{\sqrt{3}}{2}$

$x = \dfrac{\pi}{6}, \dfrac{11\pi}{6}$

67.

$$2 \tan x = \tan 2x$$

$$2 \tan x = \frac{2 \tan x}{1 - \tan^2 x}$$

$$2 \tan x(1 - \tan^2 x) = 2 \tan x$$

$$2 \tan x(1 - \tan^2 x) - 2 \tan x = 0$$

$$2 \tan x[(1 - \tan^2 x) - 1] = 0$$

$$2 \tan x(-\tan^2 x) = 0$$

$$-2 \tan^3 x = 0$$

$$\tan x = 0$$

$$x = 0, \pi$$

For Exercises 69. and 71.

$$\sin u = -\tfrac{12}{13}, u \text{ in Quadrant IV} \implies \cos u = \tfrac{5}{13}$$

$$\cos v = \tfrac{24}{25}, v \text{ in Quadrant IV} \implies \sin v = -\tfrac{7}{25}$$

69. $\sin(u - v) = \sin u \cos v - \cos u \sin v$

$$= \left(-\tfrac{12}{13}\right)\left(\tfrac{24}{25}\right) - \left(\tfrac{5}{13}\right)\left(-\tfrac{7}{25}\right)$$

$$= -\tfrac{253}{325}$$

71. $\cos(v - u) = \cos v \cos u + \sin v \sin u$

$$= \left(\tfrac{24}{25}\right)\left(\tfrac{5}{13}\right) + \left(-\tfrac{7}{25}\right)\left(-\tfrac{12}{13}\right)$$

$$= \tfrac{204}{325}$$

Review Exercises for Chapter 3

1. Given: $A = 35°, B = 71°, a = 8$

$C = 180° - 35° - 71° = 74°$

$b = \dfrac{a \sin B}{\sin A} = \dfrac{8 \sin 71°}{\sin 35°} \approx 13.19$

$c = \dfrac{a \sin C}{\sin A} = \dfrac{8 \sin 74°}{\sin 35°} \approx 13.41$

3. Given: $B = 72°, C = 82°, b = 54$

$A = 180° - 72° - 82° = 26°$

$a = \dfrac{b \sin A}{\sin B} = \dfrac{54 \sin 26°}{\sin 72°} \approx 24.89$

$c = \dfrac{b \sin C}{\sin B} = \dfrac{54 \sin 82°}{\sin 72°} \approx 56.23$

5. Given: $A = 16°, B = 98°, c = 8.4$

$C = 180° - 16° - 98° = 66°$

$a = \dfrac{c \sin A}{\sin C} = \dfrac{8.4 \sin 16°}{\sin 66°} \approx 2.53$

$b = \dfrac{c \sin B}{\sin C} = \dfrac{8.4 \sin 98°}{\sin 66°} \approx 9.11$

7. Given: $A = 24°, C = 48°, b = 27.5$

$B = 180° - 24° - 48° = 108°$

$a = \dfrac{b \sin A}{\sin B} = \dfrac{27.5 \sin 24°}{\sin 108°} \approx 11.76$

$c = \dfrac{b \sin C}{\sin B} = \dfrac{27.5 \sin 48°}{\sin 108°} \approx 21.49$

9. Given: $B = 150°, b = 30, c = 10$

$\sin C = \dfrac{c \sin B}{b} = \dfrac{10 \sin 150°}{30} \approx 0.1667 \Rightarrow C \approx 9.59°$

$A \approx 180° - 150° - 9.59° = 20.41°$

$a = \dfrac{b \sin A}{\sin B} = \dfrac{30 \sin 20.41°}{\sin 150°} \approx 20.92$

11. $A = 75°, a = 51.2, b = 33.7$

$\sin B = \dfrac{b \sin A}{a} = \dfrac{33.7 \sin 75°}{51.2} \approx 0.6358 \Rightarrow B \approx 39.48°$

$C \approx 180° - 75° - 39.48° = 65.52°$

$c = \dfrac{a \sin C}{\sin A} = \dfrac{51.2 \sin 65.52°}{\sin 75°} \approx 48.24$

13. Area $= \dfrac{1}{2} bc \sin A = \dfrac{1}{2}(5)(7)\sin 27° \approx 7.945$

15. Area $= \dfrac{1}{2} ab \sin C = \dfrac{1}{2}(16)(5)\sin 123° \approx 33.547$

17. $\tan 17° = \dfrac{h}{x + 50} \Rightarrow h = (x + 50)\tan 17°$

$h = x \tan 17° + 50 \tan 17°$

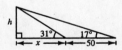

$\tan 31° = \dfrac{h}{x} \Rightarrow h = x \tan 31°$

$x \tan 17° + 50 \tan 17° = x \tan 31°$

$50 \tan 17° = x(\tan 31° - \tan 17°)$

$\dfrac{50 \tan 17°}{\tan 31° - \tan 17°} = x$

$x \approx 51.7959$

$h = x \tan 31°$

$\approx 51.7959 \tan 31°$

≈ 31.1 meters

19.

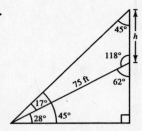

$$\frac{h}{\sin 17°} = \frac{75}{\sin 45°}$$

$$h = \frac{75 \sin 17°}{\sin 45°}$$

$$h \approx 31.01 \text{ feet}$$

21. Given: $a = 80, b = 60, c = 100$

$$\cos C = \frac{a^2 + b^2 - c^2}{2ab} = \frac{6400 + 3600 - 10{,}000}{2(80)(60)}$$

$$= 0 \Longrightarrow C = 90°$$

$$\sin A = \frac{80}{100} = 0.8 \Longrightarrow A \approx 53.13°$$

$$\sin B = \frac{60}{100} = 0.6 \Longrightarrow B \approx 36.87°$$

23. Given: $a = 16.4, b = 8.8, c = 12.2$

$$\cos A = \frac{b^2 + c^2 - a^2}{2bc} = \frac{8.8^2 + 12.2^2 - 16.4^2}{2(8.8)(12.2)} \approx -0.1988 \Longrightarrow A \approx 101.47°$$

$$\sin B \approx \frac{b \sin A}{a} = \frac{8.8 \sin 101.47°}{16.4} \approx 0.5259 \Longrightarrow B \approx 31.73°$$

$$C \approx 180° - 101.47° - 31.73° = 46.80°$$

25. Given: $B = 150°, a = 10, c = 20$

$$b^2 = 10^2 + 20^2 - 2(10)(20)\cos 150° \Longrightarrow b \approx 29.09$$

$$\sin A = \frac{a \sin B}{b} \approx \frac{10 \sin 150°}{29.09} \Longrightarrow A \approx 9.90°$$

$$C \approx 180° - 150° - 9.90° = 20.10°$$

27. Given: $A = 62°, b = 11.34, c = 19.52$

$$a^2 = 11.34^2 + 19.52^2 - 2(11.34)(19.52)\cos 62° \Longrightarrow a \approx 17.37$$

$$\sin B = \frac{b \sin A}{a} \approx \frac{11.34 \sin 62°}{17.37} \Longrightarrow B \approx 35.20°$$

$$C \approx 180° - 62° - 35.20° = 82.80°$$

29. $d^2 = 850^2 + 1060^2 - 2(850)(1060)\cos 72°$

$\approx 1{,}289{,}251$

$d \approx 1135$ miles

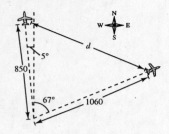

31. $a = 15, b = 8, c = 10$

$s = \dfrac{15 + 8 + 10}{2} = 16.5$

Area $= \sqrt{16.5(1.5)(8.5)(6.5)} \approx 36.979$

33. $a = 38.1, b = 26.7, c = 19.4$

$s = \dfrac{38.1 + 26.7 + 19.4}{2} = 42.1$

Area $= \sqrt{42.1(4)(15.4)(22.7)} \approx 242.630$

35. Initial point: $(3, 4)$

Terminal point: $(-5, -7)$

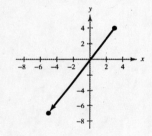

37. Initial point: $(-6, -8)$

Terminal point: $(8, 3)$

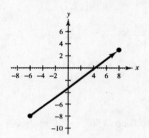

39. Initial point: $(0, 1)$

Terminal point: $\left(6, \frac{7}{2}\right)$

$\mathbf{v} = \left\langle 6 - 0, \frac{7}{2} - 1 \right\rangle = \left\langle 6, \frac{5}{2} \right\rangle$

41. Initial point: $(1, 5)$

Terminal point: $(15, 9)$

$\mathbf{v} = \langle 15 - 1, 9 - 5 \rangle = \langle 14, 4 \rangle$

43. $\|\mathbf{v}\| = \frac{1}{2}, \theta = 225°$

$\left\langle \frac{1}{2}\cos 225°, \frac{1}{2}\sin 225° \right\rangle = \left\langle -\frac{\sqrt{2}}{4}, -\frac{\sqrt{2}}{4} \right\rangle$

45. $\mathbf{u} = 6\mathbf{i} - 5\mathbf{j}, \ \mathbf{v} = 10\mathbf{i} + 3\mathbf{j}$

$4\mathbf{u} - 5\mathbf{v} = (24\mathbf{i} - 20\mathbf{j}) - (50\mathbf{i} + 15\mathbf{j}) = -26\mathbf{i} - 35\mathbf{j}$

$= \langle -26, -35 \rangle$

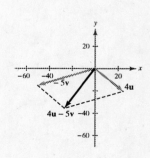

47. $\mathbf{v} = 10\mathbf{i} + 3\mathbf{j}$

$\frac{1}{2}\mathbf{v} = 5\mathbf{i} + \frac{3}{2}\mathbf{j} = \left\langle 5, \frac{3}{2} \right\rangle$

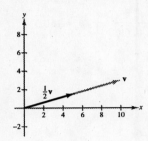

49. $\mathbf{u} = \langle -6, -8 \rangle = -6\mathbf{i} - 8\mathbf{j}$

51. Initial point: $(-2, 7)$

Terminal point: $(5, -9)$

$\mathbf{u} = \langle 5 - (-2), -9 - 7 \rangle = \langle 7, -16 \rangle = 7\mathbf{i} - 16\mathbf{j}$

53. $\mathbf{v} = 4\mathbf{i} - \mathbf{j}$

$\|\mathbf{v}\| = \sqrt{4^2 + (-1)^2} = \sqrt{17}$

$\tan \theta = \dfrac{-1}{4}$, θ in Quadrant IV $\implies \theta \approx 346°$

$\mathbf{v} \approx \sqrt{17}(\mathbf{i} \cos 346° + \mathbf{j} \sin 346°)$

55. $\mathbf{v} = 3(\cos 150°\mathbf{i} + \sin 150° \mathbf{j})$

$\|\mathbf{v}\| = 3$, $\theta = 150°$

57. $\mathbf{v} = -4\mathbf{i} + 7\mathbf{j}$

$\|\mathbf{v}\| = \sqrt{(-4)^2 + 7^2} = \sqrt{65}$

$\tan \theta = \dfrac{7}{-4}$, θ in Quadrant II $\implies \theta \approx 119.7°$

59. $\mathbf{v} = 8\mathbf{i} - \mathbf{j}$

$\|\mathbf{v}\| = \sqrt{8^2 + (-1)^2} = \sqrt{65}$

$\tan \theta = \dfrac{-1}{8}$, θ in Quadrant IV $\implies \theta \approx 352.9°$

61. Rope One: $\mathbf{u} = \|\mathbf{u}\|(\cos 30°\mathbf{i} - \sin 30°\mathbf{j}) = \|\mathbf{u}\|\left(\dfrac{\sqrt{3}}{2}\mathbf{i} - \dfrac{1}{2}\mathbf{j} \right)$

Rope Two: $\mathbf{v} = \|\mathbf{u}\|(-\cos 30°\mathbf{i} - \sin 30°\mathbf{j}) = \|\mathbf{u}\|\left(-\dfrac{\sqrt{3}}{2}\mathbf{i} - \dfrac{1}{2}\mathbf{j} \right)$

Resultant: $\mathbf{u} + \mathbf{v} = -\|\mathbf{u}\|\mathbf{j} = -180\mathbf{j}$

$\|\mathbf{u}\| = 180$

Therefore, the tension on each rope is $\|\mathbf{u}\| = 180$ lb.

63. $\mathbf{u} = \langle 6, 7 \rangle$

$\mathbf{v} = \langle -3, 9 \rangle$

$\mathbf{u} \cdot \mathbf{v} = 6(-3) + 7(9) = 45$

65. $\mathbf{u} = 3\mathbf{i} + 7\mathbf{j}$

$\mathbf{v} = 11\mathbf{i} - 5\mathbf{j}$

$\mathbf{u} \cdot \mathbf{v} = 3(11) + 7(-5) = -2$

67. $\mathbf{u} = \langle -3, 4 \rangle$

$2\mathbf{u} = \langle -6, 8 \rangle$

$2\mathbf{u} \cdot \mathbf{u} = (-6)(-3) + 8(4) = 50$

The result is a scalar.

69. $\mathbf{u} = \langle -3, 4 \rangle$, $\mathbf{v} = \langle 2, 1 \rangle$

$\mathbf{u} \cdot \mathbf{v} = (-3)(2) + 4(1) = -2$

$\mathbf{u}(\mathbf{u} \cdot \mathbf{v}) = \mathbf{u}(-2) = -2\mathbf{u} = \langle 6, -8 \rangle$

The result is a vector.

71. $\mathbf{u} = \cos\dfrac{7\pi}{4}\mathbf{i} + \sin\dfrac{7\pi}{4}\mathbf{j} = \left\langle \dfrac{1}{\sqrt{2}}, -\dfrac{1}{\sqrt{2}} \right\rangle$

$\mathbf{v} = \cos\dfrac{5\pi}{6}\mathbf{i} + \sin\dfrac{5\pi}{6}\mathbf{j} = \left\langle -\dfrac{\sqrt{3}}{2}, \dfrac{1}{2} \right\rangle$

$\cos\theta = \dfrac{\mathbf{u}\cdot\mathbf{v}}{\|\mathbf{u}\|\,\|\mathbf{v}\|} = \dfrac{-\sqrt{3}-1}{2\sqrt{2}} \Rightarrow \theta = \dfrac{11\pi}{12}$

73. $\mathbf{u} = \langle 2\sqrt{2}, -4 \rangle, \mathbf{v} = \langle -\sqrt{2}, 1 \rangle$

$\cos\theta = \dfrac{\mathbf{u}\cdot\mathbf{v}}{\|\mathbf{u}\|\,\|\mathbf{v}\|} = \dfrac{-8}{\left(\sqrt{24}\right)\left(\sqrt{3}\right)} \Rightarrow \theta \approx 160.5°$

75. $\mathbf{u} = \langle -3, 8 \rangle$

$\mathbf{v} = \langle 8, 3 \rangle$

$\mathbf{u}\cdot\mathbf{v} = -3(8) + 8(3) = 0$

$\mathbf{u}$ and $\mathbf{v}$ are orthogonal.

77. $\mathbf{u} = -\mathbf{i}$

$\mathbf{v} = \mathbf{i} + 2\mathbf{j}$

$\mathbf{u}\cdot\mathbf{v} \neq 0 \Rightarrow$ Not orthogonal

$\mathbf{v} \neq k\mathbf{u} \Rightarrow$ Not parallel

Neither

79. $\mathbf{u} = \langle -4, 3 \rangle, \mathbf{v} = \langle -8, -2 \rangle$

$\mathbf{w}_1 = \text{proj}_{\mathbf{v}}\mathbf{u} = \left(\dfrac{\mathbf{u}\cdot\mathbf{v}}{\|\mathbf{v}\|^2}\right)\mathbf{v} = \left(\dfrac{26}{68}\right)\langle -8, -2 \rangle$

$\qquad = -\dfrac{13}{17}\langle 4, 1 \rangle$

$\mathbf{w}_2 = \mathbf{u} - \mathbf{w}_1 = \langle -4, 3 \rangle - \left(-\dfrac{13}{17}\right)\langle 4, 1 \rangle$

$\qquad = \dfrac{16}{17}\langle -1, 4 \rangle$

81. $\mathbf{u} = \langle 2, 7 \rangle, \mathbf{v} = \langle 1, -1 \rangle$

$\mathbf{w}_1 = \text{proj}_{\mathbf{v}}\mathbf{u} = \left(\dfrac{\mathbf{u}\cdot\mathbf{v}}{\|\mathbf{v}\|^2}\right)\mathbf{v} = -\dfrac{5}{2}\langle 1, -1 \rangle$

$\qquad = \dfrac{5}{2}\langle -1, 1 \rangle$

$\mathbf{w}_2 = \mathbf{u} - \mathbf{w}_1 = \langle 2, 7 \rangle - \left(\dfrac{5}{2}\right)\langle -1, 1 \rangle$

$\qquad = \dfrac{9}{2}\langle 1, 1 \rangle$

83. $P = (5, 3), Q = (8, 9) \Rightarrow \overrightarrow{PQ} = \langle 3, 6 \rangle$

$W = F\cdot PQ = \langle 2, 7 \rangle \cdot \langle 3, 6 \rangle = 48$

85. True. $\sin 90°$ is defined in the Law of Sines.

87. $\dfrac{a}{\sin A} = \dfrac{b}{\sin B} = \dfrac{c}{\sin C}$ or $\dfrac{\sin A}{a} = \dfrac{\sin B}{b} = \dfrac{\sin C}{c}$

89. Since $\cos 90° = 0$, $c^2 = a^2 + b^2 - 2ab\cos 90°$ becomes $c^2 = a^2 + b^2$.

This is the Pythagorean Theorem.

91. *A* and *C* appear to have the same magnitude and direction.

93. If $k > 0$, the direction of $k\mathbf{u}$ is the same, and the magnitude is $k\|\mathbf{u}\|$.

If $k < 0$, the direction of $k\mathbf{u}$ is the opposite direction of $\mathbf{u}$ and the magnitude is $|k|\|\mathbf{u}\|$.

Chapter 3   Practice Test

For Exercises 1 and 2, use the Law of Sines to find the remaining sides and angles of the triangle.

1. $A = 40°$, $B = 12°$, $b = 100$
2. $C = 150°$, $a = 5$, $c = 20$

3. Find the area of the triangle: $a = 3$, $b = 6$, $C = 130°$.

4. Determine the number of solutions to the triangle: $a = 10$, $b = 35$, $A = 22.5°$.

For Exercises 5 and 6, use the Law of Cosines to find the remaining sides and angles of the triangle.

5. $a = 49$, $b = 53$, $c = 38$
6. $C = 29°$, $a = 100$, $b = 300$

7. Use Heron's Formula to find the area of the triangle: $a = 4.1$, $b = 6.8$, $c = 5.5$.

8. A ship travels 40 miles due east, then adjusts its course 12° southward. After traveling 70 miles in that direction, how far is the ship from its point of departure?

9. $\mathbf{w} = 4\mathbf{u} - 7\mathbf{v}$ where $\mathbf{u} = 3\mathbf{i} + \mathbf{j}$ and $\mathbf{v} = -\mathbf{i} + 2\mathbf{j}$. Find $\mathbf{w}$.

10. Find a unit vector in the direction of $\mathbf{v} = 5\mathbf{i} - 3\mathbf{j}$.

11. Find the dot product and the angle between $\mathbf{u} = 6\mathbf{i} + 5\mathbf{j}$ and $\mathbf{v} = 2\mathbf{i} - 3\mathbf{j}$.

12. $\mathbf{v}$ is a vector of magnitude 4 making an angle of 30° with the positive x-axis. Find $\mathbf{v}$ in component form.

13. Find the projection of $\mathbf{u}$ onto $\mathbf{v}$ given $\mathbf{u} = \langle 3, -1 \rangle$ and $\mathbf{v} = \langle -2, 4 \rangle$.

14. Given $\|\mathbf{u}\| = 7$, $\theta\mathbf{u} = 35°$
    $\|\mathbf{v}\| = 4$, $\theta\mathbf{v} = 123°$
    Find the component form of $\mathbf{u} + \mathbf{v}$.

15. Find two vectors orthogonal to $\langle -3, 10 \rangle$.

C H A P T E R   4
Complex Numbers

CHAPTER 4
Complex Numbers

Section 4.1   Complex Numbers

■ Standard form: $a + bi$.

If $b = 0$, then $a + bi$ is a real number.

If $a = 0$ and $b \neq 0$, then $a + bi$ is a pure imaginary number.

■ Equality of Complex Numbers: $a + bi = c + di$ if and only if $a = c$ and $b = d$

■ Operations on complex numbers

(a) Addition: $(a + bi) + (c + di) = (a + c) + (b + d)i$

(b) Subtraction: $(a + bi) - (c + di) = (a - c) + (b - d)i$

(c) Multiplication: $(a + bi)(c + di) = (ac - bd) + (ad + bc)i$

(d) Division: $\dfrac{a + bi}{c + di} = \dfrac{a + bi}{c + di} \cdot \dfrac{c - di}{c - di} = \dfrac{ac + bd}{c^2 + d^2} + \dfrac{bc - ad}{c^2 + d^2}i$

■ The complex conjugate of $a + bi$ is $a - bi$:

$(a + bi)(a - bi) = a^2 + b^2$

■ The additive inverse of $a + bi$ is $-a - bi$.

■ The multiplicative inverse of $a + bi$ is

$\dfrac{a - bi}{a^2 + b^2}.$

■ $\sqrt{-a} = \sqrt{a}\, i$ for $a > 0$.

Solutions to Odd-Numbered Exercises

1. $a + bi = -10 + 6i$

   $a = -10$

   $b = 6$

3. $(a - 1) + (b + 3)i = 5 + 8i$

   $a - 1 = 5 \implies a = 6$

   $b + 3 = 8 \implies b = 5$

5. $4 + \sqrt{-9} = 4 + 3i$

7. $2 - \sqrt{-27} = 2 - \sqrt{27}i = 2 - 3\sqrt{3}i$

9. $\sqrt{-75} = \sqrt{75}i = 5\sqrt{3}i$

11. $8 = 8 + 0i = 8$

13. $-6i + i^2 = -6i - 1 = -1 - 6i$

15. $\sqrt{-0.09} = \sqrt{0.09}i = 0.3i$

17. $(5 + i) + (6 - 2i) = 11 - i$

19. $(8 - i) - (4 - i) = 8 - i - 4 + i = 4$

21. $\left(-2 + \sqrt{-8}\right) + \left(5 - \sqrt{-50}\right) = -2 + 2\sqrt{2}i + 5 - 5\sqrt{2}i = 3 - 3\sqrt{2}i$

23. $13i - (14 - 7i) = 13i - 14 + 7i = -14 + 20i$

25. $-\left(\frac{3}{2} + \frac{5}{2}i\right) + \left(\frac{5}{3} + \frac{11}{3}i\right) = -\frac{3}{2} - \frac{5}{2}i + \frac{5}{3} + \frac{11}{3}i$

    $= -\frac{9}{6} - \frac{15}{6}i + \frac{10}{6} + \frac{22}{6}i$

    $= \frac{1}{6} + \frac{7}{6}i$

27. $\sqrt{-6} \cdot \sqrt{-2} = \left(\sqrt{6}i\right)\left(\sqrt{2}i\right) = \sqrt{12}i^2 = \left(2\sqrt{3}\right)(-1)$

    $= -2\sqrt{3}$

29. $\left(\sqrt{-10}\right)^2 = \left(\sqrt{10}i\right)^2 = 10i^2 = -10$

31. $(1 + i)(3 - 2i) = 3 - 2i + 3i - 2i^2 = 3 + i + 2 = 5 + i$

33. $6i(5 - 2i) = 30i - 12i^2 = 30i + 12 = 12 + 30i$

35. $\left(\sqrt{14} + \sqrt{10}i\right)\left(\sqrt{14} - \sqrt{10}i\right) = 14 - 10i^2 = 14 + 10 = 24$

37. $(4 + 5i)^2 = 16 + 40i + 25i^2$

    $= 16 + 40i - 25$

    $= -9 + 40i$

39. $(2 + 3i)^2 + (2 - 3i)^2 = 4 + 12i + 9i^2 + 4 - 12i + 9i^2$

    $= 4 + 12i - 9 + 4 - 12i - 9$

    $= -10$

41. The complex conjugate of $6 + 3i$ is $6 - 3i$.

    $(6 + 3i)(6 - 3i) = 36 - (3i)^2 = 36 + 9 = 45$

43. The complex conjugate of $-1 - \sqrt{5}i$ is $-1 + \sqrt{5}i$.

    $\left(-1 - \sqrt{5}i\right)\left(-1 + \sqrt{5}i\right) = (-1)^2 - \left(\sqrt{5}i\right)^2$

    $= 1 + 5 = 6$

45. The complex conjugate of $\sqrt{-20} = 2\sqrt{5}i$ is $-2\sqrt{5}i$.

    $\left(2\sqrt{5}i\right)\left(-2\sqrt{5}i\right) = -20i^2 = 20$

47. The complex conjugate of $\sqrt{8}$ is $\sqrt{8}$.

    $\left(\sqrt{8}\right)\left(\sqrt{8}\right) = 8$

49. $\dfrac{5}{i} = \dfrac{5}{i} \cdot \dfrac{-i}{-i} = \dfrac{-5i}{1} = -5i$

51. $\dfrac{2}{4 - 5i} = \dfrac{2}{4 - 5i} \cdot \dfrac{4 + 5i}{4 + 5i} = \dfrac{2(4 + 5i)}{16 + 25} = \dfrac{8 + 10i}{41} = \dfrac{8}{41} + \dfrac{10}{41}i$

53. $\dfrac{3+i}{3-i} = \dfrac{3+i}{3-i} \cdot \dfrac{3+i}{3+i}$

$= \dfrac{9+6i+i^2}{9+1}$

$= \dfrac{8+6i}{10}$

$= \dfrac{4}{5} + \dfrac{3}{5}i$

55. $\dfrac{6-5i}{i} = \dfrac{6-5i}{i} \cdot \dfrac{-i}{-i}$

$= \dfrac{-6i+5i^2}{1}$

$= -5 - 6i$

57. $\dfrac{3i}{(4-5i)^2} = \dfrac{3i}{16-40i+25i^2} = \dfrac{3i}{-9-40i} \cdot \dfrac{-9+40i}{-9+40i}$

$= \dfrac{-27i+120i^2}{81+1600} = \dfrac{-120-27i}{1681}$

$= -\dfrac{120}{1681} - \dfrac{27}{1681}i$

59. $\dfrac{2}{1+i} - \dfrac{3}{1-i} = \dfrac{2(1-i) - 3(1+i)}{(1+i)(1-i)}$

$= \dfrac{2-2i-3-3i}{1+1}$

$= \dfrac{-1-5i}{2}$

$= -\dfrac{1}{2} - \dfrac{5}{2}i$

61. $\dfrac{i}{3-2i} + \dfrac{2i}{3+8i} = \dfrac{i(3+8i) + 2i(3-2i)}{(3-2i)(3+8i)}$

$= \dfrac{3i+8i^2+6i-4i^2}{9+24i-6i-16i^2}$

$= \dfrac{4i^2+9i}{9+18i+16}$

$= \dfrac{-4+9i}{25+18i} \cdot \dfrac{25-18i}{25-18i}$

$= \dfrac{-100+72i+225i-162i^2}{625+324}$

$= \dfrac{-100+297i+162}{949}$

$= \dfrac{62+297i}{949}$

$= \dfrac{62}{949} + \dfrac{297}{949}i$

63. $x^2 - 2x + 2 = 0; \ a = 1, \ b = -2, \ c = 2$

$x = \dfrac{-(-2) \pm \sqrt{(-2)^2 - 4(1)(2)}}{2(1)}$

$= \dfrac{2 \pm \sqrt{-4}}{2}$

$= \dfrac{2 \pm 2i}{2}$

$= 1 \pm i$

65. $4x^2 + 16x + 17 = 0; \ a = 4, \ b = 16, \ c = 17$

$x = \dfrac{-16 \pm \sqrt{(16)^2 - 4(4)(17)}}{2(4)} = \dfrac{-16 \pm \sqrt{-16}}{8} = \dfrac{-16 \pm 4i}{8} = -2 \pm \dfrac{1}{2}i$

67. $4x^2 + 16x + 15 = 0; \ a = 4, \ b = 16, \ c = 15$

$x = \dfrac{-16 \pm \sqrt{(16)^2 - 4(4)(15)}}{2(4)} = \dfrac{-16 \pm \sqrt{16}}{8} = \dfrac{-16 \pm 4}{8}$

$x = -\dfrac{12}{8} = -\dfrac{3}{2} \quad \text{or} \quad x = -\dfrac{20}{8} = -\dfrac{5}{2}$

69. $\frac{3}{2}x^2 - 6x + 9 = 0$ Multiply both sides by 2.

$3x^2 - 12x + 18 = 0$

$x = \dfrac{-(-12) \pm \sqrt{(-12)^2 - 4(3)(18)}}{2(3)} = \dfrac{12 \pm \sqrt{-72}}{6} = \dfrac{12 \pm 6\sqrt{2}i}{6} = 2 \pm \sqrt{2}i$

71. $1.4x^2 - 2x - 10 = 0$ Multiply both sides by 5

$7x^2 - 10x - 50 = 0$

$x = \dfrac{-(-10) \pm \sqrt{(-10)^2 - 4(7)(-50)}}{2(7)} = \dfrac{10 \pm \sqrt{1500}}{14} = \dfrac{10 \pm 10\sqrt{15}}{14} = \dfrac{5 \pm 5\sqrt{15}}{7} = \dfrac{5}{7} \pm \dfrac{5\sqrt{15}}{7}$

73. (a) $i^{40} = i^4 \cdot i^4 \cdot i^4 \cdot i^4 \cdot i^4 \cdot i^4 \cdot i^4 \cdot i^4 \cdot i^4 \cdot i^4$

  $= 1 \cdot 1 \cdot 1 \cdot 1 \cdot 1 \cdot 1 \cdot 1 \cdot 1 \cdot 1 \cdot 1$

  $= 1$

 (c) $i^{50} = i^{25} \cdot i^{25} = i \cdot i = i^2 = -1$

 (b) $i^{25} = i^4 \cdot i^4 \cdot i^4 \cdot i^4 \cdot i^4 \cdot i^4 \cdot i$

  $= 1 \cdot 1 \cdot 1 \cdot 1 \cdot 1 \cdot 1 \cdot i$

  $= i$

 (d) $i^{67} = i^{50} \cdot i^{17} = -1 \cdot i^4 \cdot i^4 \cdot i^4 \cdot i^4 \cdot i = -i$

75. $4i^2 - 2i^3 = -4 + 2i$

77. $(-i)^3 = (-1)(i^3) = (-1)(-i) = i$

79. $\left(\sqrt{-2}\right)^6 = \left(\sqrt{2}i\right)^6 = 8i^6 = 8i^4i^2 = -8$

81. $\dfrac{1}{(2i)^3} = \dfrac{1}{8i^3} = \dfrac{1}{-8i} \cdot \dfrac{8i}{8i} = \dfrac{8i}{-64i^2} = \dfrac{1}{8}i$

83. (a) $2^4 = 16,$

 (b) $(-2)^4 = 16$

 (c) $(2i)^4 = 2^4i^4 = 16(1) = 16$

 (d) $(-2i)^4 = (-2)^4i^4 = 16(1) = 16$

85. False, if $b = 0$ then $a + bi = a - bi = a$. That is, if the complex number is real, the number equals its conjugate.

87. False.

$i^{44} + i^{150} - i^{74} - i^{109} + i^{61} = (i^4)^{11} + (i^4)^{37}(i^2) - (i^4)^{18}(i^2) - (i^4)^{27}(i) + (i^4)^{15}(i)$

$= (1)^{11} + (1)^{37}(-1) - (1)^{18}(-1) - (1)^{27}(i) + (1)^{15}(i)$

$= 1 + (-1) + 1 - i + i = 1$

89. $(a_1 + b_1i)(a_2 + b_2i) = a_1a_2 + a_1b_2i + a_2b_1i + b_1b_2i^2$

  $= (a_1a_2 - b_1b_2) + (a_1b_2 + a_2b_1)i$

The conjugate of this product is $(a_1a_2 - b_1b_2) - (a_1b_2 + a_2b_1)i$.

The product of the conjugates is:

  $(a_1 - b_1i)(a_2 - b_2i) = a_1a_2 - a_1b_2i - a_2b_1i + b_1b_2i$

  $= (a_1a_2 - b_1b_2) - (a_1b_2 + a_2b_1)i$

Thus, the conjugate of the product of two complex numbers is the product of their conjugates.

91. $(4 + 3x) + (8 - 6x - x^2) = -x^2 - 3x + 12$

93. $\left(3x - \frac{1}{2}\right)(x + 4) = 3x^2 + 12x - \frac{1}{2}x - 2 = 3x^2 + \frac{23}{2}x - 2$

95. $-x - 12 = 19$

$-x = 31$

$x = -31$

97. $4(5x - 6) - 3(6x + 1) = 0$

$20x - 24 - 18x - 3 = 0$

$2x - 27 = 0$

$2x = 27$

$x = \dfrac{27}{2}$

99. $V = \dfrac{4}{3}\pi a^2 b$

$3V = 4\pi a^2 b$

$\dfrac{3V}{4\pi b} = a^2$

$\sqrt{\dfrac{3V}{4\pi b}} = a$

$a = \dfrac{1}{2}\sqrt{\dfrac{3V}{\pi b}}$

101. Let $x = $ # liters withdrawn and replaced.

$0.50(5 - x) + 1.00x = 0.60(5)$

$2.50 - 0.50x + 1.00x = 3.00$

$0.50x = 0.50$

$x = 1$ liter

Section 4.2   Complex Solutions of Equations

- If f is a polynomial with real coefficients of degree $n > 0$, then f has exactly n solutions (zeros, roots) in the complex number system.
- Given the quadratic equation, $ax^2 + bx + c = 0$, the discriminant can be used to determine the types of solutions.

     If $b^2 - 4ac < 0$, then both solutions are complex.

     If $b^2 - 4ac = 0$, then there is one repeating real solution.

     If $b^2 - 4ac > 0$, then both solutions are real.

- If $a + bi$ is a complex solution to a polynomial with real coefficients, then so is $a - bi$.

Solutions to Odd-Numbered Exercises

1. $2x^3 + 3x + 1 = 0$ has degree three so there are three solutions in the complex number system.

3. $50 - 2x^4 = 0$ has degree four so there are four solutions in the complex number system.

5. $2x^2 - 5x + 5 = 0$

$b^2 - 4ac = (-5)^2 - 4(2)(5) = -15 < 0$

Both solutions are complex. There are no real solutions.

7. $\frac{1}{5}x^2 + \frac{6}{5}x - 8 = 0$

$b^2 - 4ac = \left(\frac{6}{5}\right)^2 - 4\left(\frac{1}{5}\right)(-8) = \frac{196}{25} > 0$

Both solutions are real.

9. $2x^2 - x - 15 = 0$

$b^2 - 4ac = (-1)^2 - 4(2)(-15) = 121 > 0$

Both solutions are real.

11. $x^2 + 2x + 10 = 0$

$b^2 - 4ac = (2)^2 - 4(1)(10) = -36 < 0$

Both solutions are complex. There are no real solutions.

13. $x^2 - 5 = 0$

$x^2 = 5$

$x = \pm\sqrt{5}$

15. $(x + 5)^2 - 6 = 0$

$(x + 5)^2 = 6$

$x + 5 = \pm\sqrt{6}$

$x = -5 \pm \sqrt{6}$

17. $x^2 - 8x + 16 = 0$

$(x - 4)^2 = 0$

$x = 4$

19. $x^2 + 2x + 5 = 0$

$x = \dfrac{-2 \pm \sqrt{2^2 - 4(1)(5)}}{2(1)}$

$= \dfrac{-2 \pm \sqrt{-16}}{2}$

$= \dfrac{-2 \pm 4i}{2}$

$= -1 \pm 2i$

21. $4x^2 - 4x + 5 = 0$

$x = \dfrac{-(-4) \pm \sqrt{(-4)^2 - 4(4)(5)}}{2(4)} = \dfrac{4 \pm \sqrt{-64}}{8} = \dfrac{4 \pm 8i}{8} = \dfrac{1}{2} \pm i$

23. $230 + 20x - 0.5x^2 = 0$

$$x = \frac{-20 \pm \sqrt{(20)^2 - 4(-0.5)(230)}}{2(-0.5)} = \frac{-20 \pm \sqrt{860}}{-1} = 20 \pm 2\sqrt{215}$$

25. $8 + (x + 3)^2 = 0$

$$(x + 3)^2 = -8$$

$$x + 3 = \pm\sqrt{-8}$$

$$x = -3 \pm 2\sqrt{2}\,i$$

27. $f(x) = x^3 - 4x^2 + x - 4 = x^2(x - 4) + 1(x - 4)$

$$= (x - 4)(x^2 + 1)$$

The zeros are $x = 4, \pm i$. The only real zero of $f(x)$ is $x = 4$. This corresponds to the x-intercept of $(4, 0)$ on the graph.

29. $f(x) = x^4 + 4x^2 + 4 = (x^2 + 2)^2$

The zeros are $x = \pm\sqrt{2}\,i$. $f(x)$ has no real zeros and the graph of $f(x)$ has no x-intercepts.

31. $f(x) = x^2 + 25$

$$= (x + 5i)(x - 5i)$$

The zeros of $f(x)$ are $x = \pm 5i$.

33. $h(x) = x^2 - 4x + 1$

h has no rational zeros.

By the Quadratic Formula, the zeros are $x = \dfrac{4 \pm \sqrt{16 - 4}}{2} = 2 \pm \sqrt{3}$.

$$h(x) = \left[x - \left(2 + \sqrt{3}\right)\right]\left[x - \left(2 - \sqrt{3}\right)\right] = (x - 2 - \sqrt{3})(x - 2 + \sqrt{3})$$

35. $f(x) = x^4 - 81$

$$= (x^2 - 9)(x^2 + 9)$$

$$= (x + 3)(x - 3)(x + 3i)(x - 3i)$$

The zeros of $f(x)$ are $x = \pm 3$ and $x = \pm 3i$.

37. $g(x) = x^3 - 6x^2 + 13x - 10$

Possible rational zeros: ± 1, ± 2, ± 5, ± 10

$$
\begin{array}{r|rrrr}
2 & 1 & -6 & 13 & -10 \\
  &   & 2 & -8 & 10 \\
\hline
  & 1 & -4 & 5 & 0
\end{array}
$$

By the Quadratic Formula, the zeros of $x^2 - 4x + 5$ are $x = \dfrac{4 \pm \sqrt{16 - 20}}{2} = 2 \pm i$.

The zeros of $g(x)$ are $x = 2$ and $x = 2 \pm i$.

$$g(x) = (x - 2)[x - (2 + i)][x - (2 - i)]$$

$$= (x - 2)(x - 2 - i)(x - 2 + i)$$

39. $h(x) = x^3 - x + 6$

Possible rational zeros: ± 1, ± 2, ± 3, ± 6

$$
\begin{array}{r|rrrr}
-2 & 1 & 0 & -1 & 6 \\
   &   & -2 & 4 & -6 \\
\hline
   & 1 & -2 & 3 & 0
\end{array}
$$

By the Quadratic Formula, the zeros of $x^2 - 2x + 3$ are $x = \dfrac{2 \pm \sqrt{4 - 12}}{2} = 1 \pm \sqrt{2}\,i$.

The zeros of $h(x)$ are $x = -2$ and $x = 1 \pm \sqrt{2}\,i$.

$$h(x) = [x - (-2)]\left[x - \left(1 + \sqrt{2}\,i\right)\right]\left[x - \left(1 - \sqrt{2}\,i\right)\right]$$

$$= (x + 2)(x - 1 - \sqrt{2}\,i)(x - 1 + \sqrt{2}\,i)$$

41. $f(x) = 5x^3 - 9x^2 + 28x + 6$

Possible rational zeros: $\pm 1, \pm 2, \pm 3, \pm 6, \pm\frac{1}{5}, \pm\frac{2}{5}, \pm\frac{3}{5}, \pm\frac{6}{5}$

$$
-\tfrac{1}{5} \ \big|
\begin{array}{rrrr}
5 & -9 & 28 & 6 \\
  & -1 & 2 & -6 \\
\hline
5 & -10 & 30 & 0
\end{array}
$$

By the Quadratic Formula, the zeros of
$5x^2 - 10x + 30 = 5(x^2 - 2x + 6)$ are

$$x = \frac{2 \pm \sqrt{4 - 24}}{2} = 1 \pm \sqrt{5}\,i.$$

The zeros of $f(x)$ are $x = -\frac{1}{5}$ and $x = 1 \pm \sqrt{5}\,i.$

$$
\begin{aligned}
f(x) &= \left[x - \left(-\tfrac{1}{5}\right)\right](5)\left[x - \left(1 + \sqrt{5}\,i\right)\right]\left[x - \left(1 - \sqrt{5}\,i\right)\right] \\
&= (5x + 1)\left(x - 1 - \sqrt{5}\,i\right)\left(x - 1 + \sqrt{5}\,i\right)
\end{aligned}
$$

43. $g(x) = x^4 - 4x^3 + 8x^2 - 16x + 16$

Possible rational zeros: $\pm 1, \pm 2, \pm 4, \pm 8, \pm 16$

$$
2 \ \big|
\begin{array}{rrrrr}
1 & -4 & 8 & -16 & 16 \\
  & 2 & -4 & 8 & -16 \\
\hline
1 & -2 & 4 & -8 & 0
\end{array}
$$

$$
2 \ \big|
\begin{array}{rrrr}
1 & -2 & 4 & -8 \\
  & 2 & 0 & 8 \\
\hline
1 & 0 & 4 & 0
\end{array}
$$

$g(x) = (x - 2)(x - 2)(x^2 + 4) = (x - 2)^2(x + 2i)(x - 2i)$

The zeros of $g(x)$ are $x = 2$ and $x = \pm 2i.$

45. $f(x) = x^4 + 10x^2 + 9$

$\qquad = (x^2 + 1)(x^2 + 9)$

$\qquad = (x + i)(x - i)(x + 3i)(x - 3i)$

The zeros of $f(x)$ are $x = \pm i$ and $x = \pm 3i.$

47. $f(x) = 2x^3 + 3x^2 + 50x + 75$

Since $5i$ is a zero, so is $-5i.$

$$
5i \ \big|
\begin{array}{rrrr}
2 & 3 & 50 & 75 \\
  & 10i & -50 + 15i & -75 \\
\hline
2 & 3 + 10i & 15i & 0
\end{array}
$$

$$
-5i \ \big|
\begin{array}{rrr}
2 & 3 + 10i & 15i \\
  & -10i & -15i \\
\hline
2 & 3 & 0
\end{array}
$$

The zero of $2x + 3$ is $x = -\frac{3}{2}.$

The zeros of $f(x)$ are $x = -\frac{3}{2}$ and $x = \pm 5i.$

Alternate Solution

Since $x = \pm 5i$ are zeros of $f(x)$,
$(x + 5i)(x - 5i) = x^2 + 25$
is a factor of $f(x)$. By long division we have:

$$
\begin{array}{r}
2x + 3 \\
x^2 + 0x + 25 \overline{)\, 2x^3 + 3x^2 + 50x + 75} \\
\underline{2x^3 + 0x^2 + 50x} \\
3x^2 + 0x + 75 \\
\underline{3x^2 + 0x + 75} \\
0
\end{array}
$$

Thus, $f(x) = (x^2 + 25)(2x + 3)$ and the zeros of $f(x)$ are
$x = \pm 5i$ and $x = -\frac{3}{2}.$

49. $f(x) = 2x^4 - x^3 + 7x^2 - 4x - 4$

Since $2i$ is a zero, so is $-2i.$

$$
2i \ \big|
\begin{array}{rrrrr}
2 & -1 & 7 & -4 & -4 \\
  & 4i & -8 - 2i & 4 - 2i & 4 \\
\hline
2 & -1 + 4i & -1 - 2i & -2i & 0
\end{array}
$$

$$
-2i \ \big|
\begin{array}{rrrr}
2 & -1 + 4i & -1 - 2i & -2i \\
  & -4i & 2i & 2i \\
\hline
2 & -1 & -1 & 0
\end{array}
$$

The zeros of $2x^2 - x - 1 = (2x + 1)(x - 1)$ are $x = -\frac{1}{2}$
and $x = 1.$

The zeros of $f(x)$ are $x = \pm 2i$, $x = -\frac{1}{2}$, and $x = 1.$

Alternate Solution

Since $x = \pm 2i$ are zeros of $f(x)$,
$(x + 2i)(x - 2i) = x^2 + 4$ is a factor of $f(x)$.
By long division we have:

$$
\begin{array}{r}
2x^2 - x - 1 \\
x^2 + 0x + 4 \overline{)\, 2x^4 - x^3 + 7x^2 - 4x - 4} \\
\underline{2x^4 + 0x^3 + 8x^2} \\
-x^3 - x^2 - 4x \\
\underline{-x^3 + 0x^2 - 4x} \\
-x^2 + 0x - 4 \\
\underline{-x^2 + 0x - 4} \\
0
\end{array}
$$

Thus, $f(x) = (x^2 + 4)(2x^2 - x - 1)$

$\qquad = (x + 2i)(x - 2i)(2x + 1)(x - 1)$
and the zeros of $f(x)$ are $x = \pm 2i$, $x = -\frac{1}{2}$, and $x = 1.$

51. $g(x) = 4x^3 + 23x^2 + 34x - 10$

Since $-3 + i$ is a zero, so is $-3 - i$.

$$
\begin{array}{r|rrrr}
-3 + i & 4 & 23 & 34 & -10 \\
       &   & -12 + 4i & -37 - i & 10 \\
\hline
       & 4 & 11 + 4i & -3 - i & 0
\end{array}
$$

$$
\begin{array}{r|rrr}
-3 - i & 4 & 11 + 4i & -3 - i \\
       &   & -12 - 4i & 3 + i \\
\hline
       & 4 & -1 & 0
\end{array}
$$

The zero of $4x - 1$ is $x = \frac{1}{4}$. The zeros of $g(x)$ are $x = -3 \pm i$ and $x = \frac{1}{4}$.

<u>Alternate Solution</u>

Since $-3 \pm i$ are zeros of $g(x)$,
$$[x - (-3 + i)][x - (-3 - i)] = [(x + 3) - i][(x + 3) + i]$$
$$= (x + 3)^2 - i^2$$
$$= x^2 + 6x + 10$$
is a factor of $g(x)$. By long division we have:

$$
\begin{array}{r}
4x - 1 \\
x^2 + 6x + 10 \overline{\smash{\big)}\ 4x^3 + 23x^2 + 34x - 10} \\
\underline{4x^3 + 24x^2 + 40x} \\
-x^2 - 6x - 10 \\
\underline{-x^2 - 6x - 10} \\
0
\end{array}
$$

Thus, $g(x) = (x^2 + 6x + 10)(4x - 1)$ and the zeros of $g(x)$ are $x = -3 \pm i$ and $x = \frac{1}{4}$.

53. $f(x) = x^4 + 3x^3 - 5x^2 - 21x + 22$

Since $-3 + \sqrt{2}\,i$ is a zero, so is $-3 - \sqrt{2}\,i$, and

$$\left[x - \left(-3 + \sqrt{2}\,i\right)\right]\left[x - \left(-3 - \sqrt{2}\,i\right)\right]$$
$$= \left[(x + 3) - \sqrt{2}\,i\right]\left[(x + 3) + \sqrt{2}\,i\right]$$
$$= (x + 3)^2 - \left(\sqrt{2}\,i\right)^2$$
$$= x^2 + 6x + 11$$

is a factor of $f(x)$. By long division, we have:

$$
\begin{array}{r}
x^2 - 3x + 2 \\
x^2 + 6x + 11 \overline{\smash{\big)}\ x^4 + 3x^3 - 5x^2 - 21x + 22} \\
\underline{x^4 + 6x^3 + 11x^2} \\
-3x^3 - 16x^2 - 21x \\
\underline{-3x^3 - 18x^2 - 33x} \\
2x^2 + 12x + 22 \\
\underline{2x^2 + 12x + 22} \\
0
\end{array}
$$

Thus, $f(x) = (x^2 + 6x + 11)(x^2 - 3x + 2)$
$$= (x^2 + 6x + 11)(x - 1)(x - 2)$$

and the zeros of $f(x)$ are $x = -3 \pm \sqrt{2}\,i$, $x = 1$, and $x = 2$.

55. $f(x) = (x - 1)(x - 5i)(x + 5i)$

$= (x - 1)(x^2 + 25)$

$= x^3 - x^2 + 25x - 25$

Note: $f(x) = a(x^3 - x^2 + 25x - 25)$.
where a is any nonzero real number, has
the zeros 1 and $\pm 5i$.

57. $f(x) = (x - 6)[x - (-5 + 2i)][x - (-5 - 2i)]$

$= (x - 6)[(x + 5) - 2i][(x + 5) + 2i]$

$= (x - 6)[(x + 5)^2 - (2i)^2]$

$= (x - 6)(x^2 + 10x + 25 + 4)$

$= (x - 6)(x^2 + 10x + 29)$

$= x^3 + 4x^2 - 31x - 174$

Note: $f(x) = a(x^3 + 4x^2 - 31x - 174)$, where
a is any nonzero real number, has the zeros 6, and
$-5 \pm 2i$.

59. If $3 + \sqrt{2}i$ is a zero, so is its conjugate, $3 - \sqrt{2}i$.

$f(x) = (3x - 2)(x + 1)[x - (3 + \sqrt{2}i)][x - (3 - \sqrt{2}i)]$

$= (3x - 2)(x + 1)[(x - 3) - \sqrt{2}i][(x - 3) + \sqrt{2}i]$

$= (3x^2 + x - 2)[(x - 3)^2 - (\sqrt{2}i)^2]$

$= (3x^2 + x - 2)(x^2 - 6x + 9 + 2)$

$= (3x^2 + x - 2)(x^2 - 6x + 11)$

$= 3x^4 - 17x^3 + 25x^2 + 23x - 22$

Note: $f(x) = a(3x^4 - 17x^3 + 25x^2 + 23x - 22)$,
where a is any nonzero real number, has the zeros $\frac{2}{3}$, -1,
and $3 \pm \sqrt{2}i$.

61. $h = -16t^2 + 48t$, $0 \le t \le 3$

$= -16(t^2 - 3t)$

$= -16\left(t^2 - 3t + \frac{9}{4} - \frac{9}{4}\right)$

$= -16\left[\left(t - \frac{3}{2}\right)^2 - \frac{9}{4}\right]$

$= -16\left(t - \frac{3}{2}\right)^2 + 36$

The maximum height that the baseball reaches is 36 feet
when $t = 1.5$ seconds. No, it is not possible for the ball
to reach a height of 64 feet.

Alternate Solution

Let $h = 64$ and solve for t.

$$64 = -16t^2 + 48t$$

$$16t^2 - 48t + 64 = 0$$

$$16(t^2 - 3t + 4) = 0$$

$$t^2 - 3t + 4 = 0$$

$$t = \frac{3 \pm \sqrt{9 - 16}}{2} = \frac{3 \pm \sqrt{7}i}{2}$$

No, it is not possible since solving this equation yields
only imaginary roots.

63. False. The most nonreal complex zeros it can have is
two and the Linear Factorization Theorem guarantees that
there are 3 linear factors, so one zero must be real.

65. $g(x) = -f(x)$. This function would have the same
zeros as $f(x)$ so r_1, r_2, and r_3 are also zeros of $g(x)$.

67. $g(x) = f(x - 5)$. The graph of $g(x)$ is a horizontal
shift of the graph of $f(x)$ five units to the right so
the zeros of $g(x)$ are $5 + r_1$, $5 + r_2$, and $5 + r_3$.

69. $g(x) = 3 + f(x)$. Since $g(x)$ is a vertical shift of the graph
of $f(x)$, the zeros of $g(x)$ cannot be determined.

71. $f(x) = x^4 - 4x^2 + k$

$$x^2 = \frac{-(-4) \pm \sqrt{(-4)^2 - 4(1)(k)}}{2(1)} = \frac{4 \pm 2\sqrt{4-k}}{2} = 2 \pm \sqrt{4-k}$$

$$x = \pm\sqrt{2 \pm \sqrt{4-k}}$$

(a) For there to be four distinct real roots, both $4 - k$ and $2 \pm \sqrt{4-k}$ must be positive. This occurs when $0 < k < 4$. Thus, some possible k-values are $k = 1, k = 2, k = 3, k = \frac{1}{2}, k = \sqrt{2}$, etc.

(b) For there to be two real roots, each of multiplicity 2, $4 - k$ must equal zero. Thus, $k = 4$.

(c) For there to be two real zeros and two complex zeros, $2 + \sqrt{4-k}$ must be positive and $2 - \sqrt{4-k}$ must be negative. This occurs when $k < 0$. Thus, some possible k-values are $k = -1, k = -2, k = -\frac{1}{2}$, etc.

(d) For there to be four complex zeros, $2 \pm \sqrt{4-k}$ must be nonreal. This occurs when $k > 4$. Some possible k-values are $k = 5, k = 6, k = 7.4$, etc.

73. (a) $f(x) = (x - \sqrt{b}i)(x + \sqrt{b}i) = x^2 + b$

(b) $f(x) = [x - (a + bi)][x - (a - bi)]$

$$= [(x - a) - bi][(x - a) + bi]$$

$$= (x - a)^2 - (bi)^2$$

$$= x^2 - 2ax + a^2 + b^2$$

75. $(-3 + 6i) - (8 - 3i) = -3 + 6i - 8 + 3i = -11 + 9i$

77. $(6 - 2i)(1 + 7i) = 6 + 42i - 2i - 14i^2 = 20 + 40i$

79. $\dfrac{1+i}{1-i} = \dfrac{1+i}{1-i} \cdot \dfrac{1+i}{1+i} = \dfrac{1 + 2i + i^2}{1 + 1} = \dfrac{2i}{2} = i$

81. $g(x) = f(x - 2)$

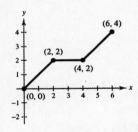

Horizontal shift two units
to the right

83. $g(x) = 2f(x)$

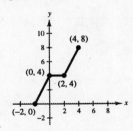

Vertical stretch

85. $g(x) = f(2x)$

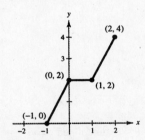

Horizontal shrink

Section 4.3   Trigonometric Form of a Complex Number

- You should be able to graphically represent complex numbers and know the following facts about them.
- The absolute value of the complex numbers $z = a + bi$ is $|z| = \sqrt{a^2 + b^2}$.
- The trigonometric form of the complex number $z = a + bi$ is $z = r(\cos \theta + i \sin \theta)$ where
  - (a) $a = r \cos \theta$
  - (b) $b = r \sin \theta$
  - (c) $r = \sqrt{a^2 + b^2}$; r is called the modulus of z.
  - (d) $\tan \theta = \dfrac{b}{a}$; θ is called the argument of z.
- Given $z_1 = r_1(\cos \theta_1 + i \sin \theta_1)$ and $z_2 = r_2(\cos \theta_2 + i \sin \theta_2)$:
  - (a) $z_1 z_2 = r_1 r_2 [\cos(\theta_1 + \theta_2) + i \sin(\theta_1 + \theta_2)]$
  - (b) $\dfrac{z_1}{z_2} = \dfrac{r_1}{r_2}[\cos(\theta_1 - \theta_2) + i \sin(\theta_1 - \theta_2)]$, $z_2 \neq 0$

Solutions to Odd-Numbered Exercises

1. $|-7i| = \sqrt{0^2 + (-7)^2}$
   $= \sqrt{49} = 7$

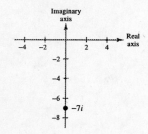

3. $|-4 + 4i| = \sqrt{(-4)^2 + (4)^2}$
   $= \sqrt{32} = 4\sqrt{2}$

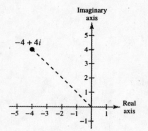

5. $|6 - 7i| = \sqrt{6^2 + (-7)^2}$
   $= \sqrt{85}$

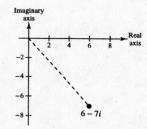

7. $z = 3i$

$r = \sqrt{0^2 + 3^2} = \sqrt{9} = 3$

$\tan \theta = \dfrac{3}{0}$, undefined $\implies \theta = \dfrac{\pi}{2}$

$z = 3\left(\cos \dfrac{\pi}{2} + i \sin \dfrac{\pi}{2}\right)$

9. $z = 3 - i$

$r = \sqrt{(3)^2 + (-1)^2} = \sqrt{10}$

$\tan \theta = -\dfrac{1}{3}$, θ is in Quadrant IV.

$\theta \approx 5.96$ radians

$z \approx \sqrt{10}(\cos 5.96 + i \sin 5.96)$

11. $z = 3 - 3i$

$r = \sqrt{3^2 + (-3)^2} = \sqrt{18} = 3\sqrt{2}$

$\tan \theta = \dfrac{-3}{3} = -1, \theta$ is in Quadrant IV $\Rightarrow \theta = \dfrac{7\pi}{4}$.

$z = 3\sqrt{2}\left(\cos \dfrac{7\pi}{4} + i \sin \dfrac{7\pi}{4}\right)$

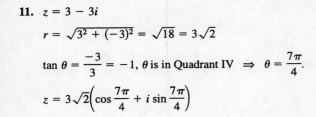

13. $z = \sqrt{3} + i$

$r = \sqrt{\left(\sqrt{3}\right)^2 + 1^2} = \sqrt{4} = 2$

$\tan \theta = \dfrac{1}{\sqrt{3}} = \dfrac{\sqrt{3}}{3} \Rightarrow \theta = \dfrac{\pi}{6}$

$z = 2\left(\cos \dfrac{\pi}{6} + i \sin \dfrac{\pi}{6}\right)$

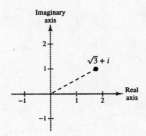

15. $z = -2\left(1 + \sqrt{3}i\right)$

$r = \sqrt{(-2)^2 + (-2\sqrt{3})^2} = \sqrt{16} = 4$

$\tan \theta = \dfrac{\sqrt{3}}{1} = \sqrt{3}, \theta$ is in Quadrant III $\Rightarrow \theta = \dfrac{4\pi}{3}$.

$z = 4\left(\cos \dfrac{4\pi}{3} + i \sin \dfrac{4\pi}{3}\right)$

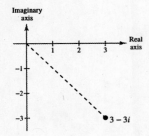

17. $z = -5i$

$r = \sqrt{0^2 + (-5)^2} = \sqrt{25} = 5$

$\tan \theta = \dfrac{-5}{0}$, undefined $\Rightarrow \theta = \dfrac{3\pi}{2}$

$z = 5\left(\cos \dfrac{3\pi}{2} + i \sin \dfrac{3\pi}{2}\right)$

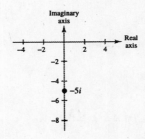

19. $z = -7 + 4i$

$r = \sqrt{(-7)^2 + (4)^2} = \sqrt{65}$

$\tan \theta = \dfrac{4}{-7}, \theta$ is in Quadrant II $\Rightarrow \theta \approx 2.62$.

$z \approx \sqrt{65}\left(\cos 2.62 + i \sin 2.62\right)$

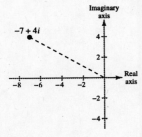

21. $z = 7 + 0i$

$r = \sqrt{(7)^2 + (0)^2} = \sqrt{49} = 7$

$\tan \theta = \dfrac{0}{7} = 0 \Rightarrow \theta = 0$

$z = 7(\cos 0 + i \sin 0)$

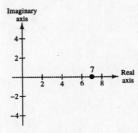

23. $z = 3 + \sqrt{3}i$

$r = \sqrt{(3)^2 + \left(\sqrt{3}\right)^2} = \sqrt{12}$

$\qquad\qquad\qquad\quad = 2\sqrt{3}$

$\tan \theta = \dfrac{\sqrt{3}}{3} \implies \theta = \dfrac{\pi}{6}$

$z = 2\sqrt{3}\left(\cos \dfrac{\pi}{6} + i \sin \dfrac{\pi}{6}\right)$

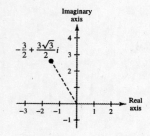

25. $z = -3 - i$

$r = \sqrt{(-3)^2 + (-1)^2} = \sqrt{10}$

$\tan \theta = \dfrac{-1}{-3} = \dfrac{1}{3}$, θ is in Quadrant III $\implies$ $\theta \approx 3.46$.

$z \approx \sqrt{10} \,(\cos 3.46 + i \sin 3.46)$

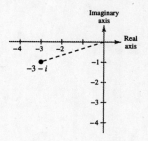

27. $z = 5 + 2i$

$r \approx 5.39$

$\theta \approx 0.38$

$z \approx 5.39(\cos 0.38 + i \sin 0.38)$

29. $z = -3 + i$

$r \approx 3.16$

$\theta \approx 2.82$

$z \approx 3.16(\cos 2.82 + i \sin 2.82)$

31. $z = 3\sqrt{2} - 7i$

$r \approx 8.19$

$\theta = \tan^{-1}\left(-\dfrac{7}{3\sqrt{2}}\right) + 2\pi \approx 5.26$

$z \approx 8.19(\cos 5.26 + i \sin 5.26)$

33. $z = -8 - 5\sqrt{3}i$

$r \approx 11.79$

$\theta \approx 3.97$

$z \approx 11.79(\cos 3.97 + i \sin 3.97)$

35. $3(\cos 120° + i \sin 120°) = 3\left(-\dfrac{1}{2} + \dfrac{\sqrt{3}}{2}i\right)$

$\qquad\qquad\qquad\qquad\qquad = -\dfrac{3}{2} + \dfrac{3\sqrt{3}}{2}i$

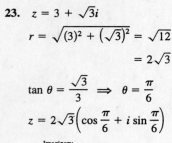

37. $\dfrac{3}{2}(\cos 300° + i \sin 300°) = \dfrac{3}{2}\left[\dfrac{1}{2} + i\left(-\dfrac{\sqrt{3}}{2}\right)\right]$

$\qquad\qquad\qquad\qquad\qquad = \dfrac{3}{4} - \dfrac{3\sqrt{3}}{4}i$

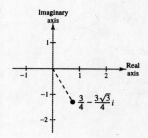

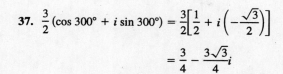

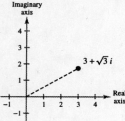

39. $3.75\left(\cos\frac{3\pi}{4} + i\sin\frac{3\pi}{4}\right) = -\frac{15\sqrt{2}}{8} + \frac{15\sqrt{2}}{8}i$

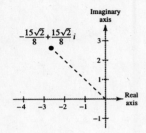

41. $8\left(\cos\frac{\pi}{2} + i\sin\frac{\pi}{2}\right) = 8(0 + i) = 8i$

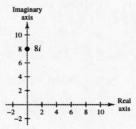

43. $3[\cos(18°45') + i\sin(18°45')] \approx 2.8408 + 0.9643i$

45. $5\left(\cos\frac{\pi}{9} + i\sin\frac{\pi}{9}\right) \approx 4.70 + 1.71i$

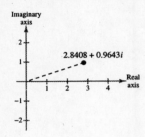

47. $3(\cos 165.5° + i\sin 165.5°) \approx -2.90 + 0.75i$

49. $\left[2\left(\cos\frac{\pi}{4} + i\sin\frac{\pi}{4}\right)\right]\left[6\left(\cos\frac{\pi}{12} + i\sin\frac{\pi}{12}\right)\right] = (2)(6)\left[\cos\left(\frac{\pi}{4} + \frac{\pi}{12}\right) + i\sin\left(\frac{\pi}{4} + \frac{\pi}{12}\right)\right]$

$$= 12\left(\cos\frac{\pi}{3} + i\sin\frac{\pi}{3}\right)$$

51. $\left[\frac{5}{3}(\cos 140° + i\sin 140°)\right]\left[\frac{2}{3}(\cos 60° + i\sin 60°)\right] = \left(\frac{5}{3}\right)\left(\frac{2}{3}\right)[\cos(140° + 60°) + i\sin(140° + 60°)]$

$$= \frac{10}{9}(\cos 200° + i\sin 200°)$$

53. $[0.45(\cos 310° + i\sin 310°)][0.60(\cos 200° + i\sin 200°)] = (0.45)(0.60)[\cos(310° + 200°) + i\sin(310° + 200°)]$

$$= 0.27(\cos 510° + i\sin 510°)$$

$$= 0.27(\cos 150° + i\sin 150°)$$

55. $\dfrac{\cos 50° + i\sin 50°}{\cos 20° + i\sin 20°} = \cos(50° - 20°) + i\sin(50° - 20°)$

$$= \cos 30° + i\sin 30°$$

57. $\dfrac{\cos\frac{5\pi}{3} + i\sin\frac{5\pi}{3}}{\cos\pi + i\sin\pi} = \cos\left(\frac{5\pi}{3} - \pi\right) + i\sin\left(\frac{5\pi}{3} - \pi\right) = \cos\left(\frac{2\pi}{3}\right) + i\sin\left(\frac{2\pi}{3}\right)$

59. $\dfrac{12(\cos 52° + i\sin 52°)}{3(\cos 110° + i\sin 110°)} = \dfrac{12}{3}[\cos(52° - 110°) + i\sin(52° - 110°)]$

$$= 4[\cos(-58°) + i\sin(-58°)]$$

61. (a) $2 + 2i = 2\sqrt{2}\left(\cos\frac{\pi}{4} + i\sin\frac{\pi}{4}\right)$

$1 - i = \sqrt{2}\left[\cos\left(\frac{7\pi}{4}\right) + i\sin\left(\frac{7\pi}{4}\right)\right]$

(b) $(2 + 2i)(1 - i) = \left[2\sqrt{2}\left(\cos\frac{\pi}{4} + i\sin\frac{\pi}{4}\right)\right]\left[\sqrt{2}\left(\cos\left(\frac{7\pi}{4}\right) + i\sin\left(\frac{7\pi}{4}\right)\right)\right]$

$= 4(\cos 2\pi + i\sin 2\pi) = 4(\cos 0 + i\sin 0) = 4$

(c) $(2 + 2i)(1 - i) = 2 - 2i + 2i - 2i^2 = 2 + 2 = 4$

63. (a) $-2i = 2\left[\cos\left(\frac{3\pi}{2}\right) + i\sin\left(\frac{3\pi}{2}\right)\right]$

$1 + i = \sqrt{2}\left(\cos\frac{\pi}{4} + i\sin\frac{\pi}{4}\right)$

(b) $-2i(1 + i) = 2\left[\cos\left(\frac{3\pi}{2}\right) + i\sin\left(\frac{3\pi}{2}\right)\right]\left[\sqrt{2}\left(\cos\frac{\pi}{4} + i\sin\frac{\pi}{4}\right)\right]$

$= 2\sqrt{2}\left[\cos\left(\frac{7\pi}{4}\right) + i\sin\left(\frac{7\pi}{4}\right)\right]$

$= 2\sqrt{2}\left[\frac{1}{\sqrt{2}} - \frac{1}{\sqrt{2}}i\right] = 2 - 2i$

(c) $-2i(1 + i) = -2i - 2i^2 = -2i + 2 = 2 - 2i$

65. (a) $3 + 4i = 5(\cos 0.93 + i\sin 0.93)$

$1 - \sqrt{3}i = 2\left(\cos\frac{5\pi}{3} + i\sin\frac{5\pi}{3}\right)$

(b) $\frac{3 + 4i}{1 - \sqrt{3}i} = \frac{5(\cos 0.93 + i\sin 0.93)}{2\left(\cos\frac{5\pi}{3} + i\sin\frac{5\pi}{3}\right)}$

$\approx 2.5[\cos(-4.31) + i\sin(-4.31)]$

$\approx 2.5(\cos 1.97 + i\sin 1.97)$

$\approx -0.982 + 2.299i$

(c) $\frac{3 + 4i}{1 - \sqrt{3}i} = \frac{3 + 4i}{1 - \sqrt{3}i} \cdot \frac{1 + \sqrt{3}i}{1 + \sqrt{3}i}$

$= \frac{3 + \left(4 + 3\sqrt{3}\right)i + 4\sqrt{3}i^2}{1 + 3}$

$= \frac{3 - 4\sqrt{3}}{4} + \frac{4 + 3\sqrt{3}}{4}i$

$\approx -0.982 + 2.299i$

67. (a) $5 = 5(\cos 0 + i\sin 0)$

$2 + 3i \approx \sqrt{13}(\cos 0.98 + i\sin 0.98)$

(b) $\frac{5}{2 + 3i} \approx \frac{5(\cos 0 + i\sin 0)}{\sqrt{13}(\cos 0.98 + i\sin 0.98)} = \frac{5}{\sqrt{13}}[\cos(-0.98) + i\sin(-0.98)]$

$\approx \frac{5}{\sqrt{13}}(\cos 5.30 + i\sin 5.30) \approx 0.769 - 1.154i$

(c) $\frac{5}{2 + 3i} = \frac{5}{2 + 3i} \cdot \frac{2 - 3i}{2 - 3i} = \frac{10 - 15i}{13} = \frac{10}{13} - \frac{15}{13}i \approx 0.769 - 1.154i$

69. Let $z = x + iy$ such that:

$$|z| = 2 \implies 2 = \sqrt{x^2 + y^2}$$

$$\implies 4 = x^2 + y^2: \text{circle with radius of 2}$$

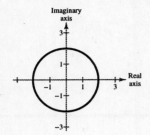

71. Let $\theta = \dfrac{\pi}{6}$.

Let $z = x + iy$ such that:

$$\tan \frac{\pi}{6} = \frac{y}{x}$$

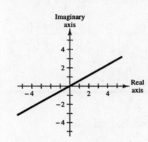

73. True, by the definition of the absolute value of a complex number

75.
$$\frac{z_1}{z_2} = \frac{r_1(\cos \theta_1 + i \sin \theta_1)}{r_2(\cos \theta_2 + i \sin \theta_2)} \cdot \frac{(\cos \theta_2 + i \sin \theta_2)}{(\cos \theta_2 + i \sin \theta_2)}$$

$$= \frac{r_1}{r_2(\cos^2 \theta_2 + \sin^2 \theta_2)} [\cos \theta_1 \cos \theta_2 + \sin \theta_1 \sin \theta_2 + i(\sin \theta_1 \cos \theta_2 - \sin \theta_2 \cos \theta_1)]$$

$$= \frac{r_1}{r_2}[\cos(\theta_1 - \theta_2) + i \sin(\theta_1 - \theta_2)]$$

77. (a) $z\bar{z} = [r(\cos \theta + i \sin \theta)][r(\cos(-\theta) + i \sin(-\theta))]$

$\qquad = r^2[\cos(\theta - \theta) + i \sin(\theta - \theta)]$

$\qquad = r^2[\cos 0 + i \sin 0]$

$\qquad = r^2$

(b) $\dfrac{z}{\bar{z}} = \dfrac{r(\cos \theta + i \sin\theta)}{r[\cos(-\theta) + i \sin(-\theta)]}$

$\qquad = \dfrac{r}{r}[\cos(\theta - (-\theta)) + i \sin(\theta - (-\theta))]$

$\qquad = \cos 2\theta + i \sin 2\theta$

79. $A = 22°, a = 8$

$B = 90° - A = 68°$

$\tan 22° = \dfrac{8}{b} \implies b = \dfrac{8}{\tan 22°} \approx 19.80$

$\sin 22° = \dfrac{8}{c} \implies c = \dfrac{8}{\sin 22°} \approx 21.36$

81. $A = 30°, b = 112.6$

$B = 90° - A = 60°$

$\tan 30° = \dfrac{a}{112.6} \implies a = 112.6 \tan 30° \approx 65.01$

$\cos 30° = \dfrac{112.6}{c} \implies c = \dfrac{112.6}{\cos 30°} \approx 130.02$

83. $A = 42°15' = 42.25°, c = 11.2$

$B = 90° - A = 47°45'$

$\sin 42.25° = \dfrac{a}{11.2} \implies a = 11.2 \sin 42.25° \approx 7.53$

$\cos 42.25° = \dfrac{b}{11.2} \implies b = 11.2 \cos 42.25° \approx 8.29$

85. $d = 16 \cos \dfrac{\pi}{4}t$

Maximum displacement: $|16| = 16$

$16 \cos \dfrac{\pi}{4}t = 0 \implies \dfrac{\pi}{4}t = \dfrac{\pi}{2} \implies t = 2$

87. $d = \dfrac{1}{16} \sin \dfrac{5}{4}\pi t$

Maximum displacement: $\left|\dfrac{1}{16}\right| = \dfrac{1}{16}$

$\dfrac{1}{16} \sin \dfrac{5}{4}\pi t = 0 \implies \dfrac{5}{4}\pi t = 0 \implies t = 0$

Section 4.4   DeMoivre's Theorem

■ You should know DeMoivre's Theorem: If $z = r(\cos \theta + i \sin \theta)$, then for any positive integer n,

$z^n = r^n (\cos n\theta + i \sin n\theta)$.

■ You should know that for any positive integer n, $z = r(\cos \theta + i \sin \theta)$ has n distinct nth roots given by

$$\sqrt[n]{r}\left[\cos\left(\frac{\theta + 2\pi k}{n}\right) + i \sin\left(\frac{\theta + 2\pi k}{n}\right)\right]$$

where $k = 0, 1, 2, \ldots, n - 1$.

Solutions to Odd-Numbered Exercises

1. $(1 + i)^5 = \left[\sqrt{2}\left(\cos \frac{\pi}{4} + i \sin \frac{\pi}{4}\right)\right]^5$

$= (\sqrt{2})^5\left(\cos \frac{5\pi}{4} + i \sin \frac{5\pi}{4}\right)$

$= 4\sqrt{2}\left(-\frac{\sqrt{2}}{2} - \frac{\sqrt{2}}{2}i\right)$

$= -4 - 4i$

3. $(-1 + i)^{10} = \left[\sqrt{2}\left(\cos \frac{3\pi}{4} + i \sin \frac{3\pi}{4}\right)\right]^{10}$

$= (\sqrt{2})^{10}\left(\cos \frac{30\pi}{4} + i \sin \frac{30\pi}{4}\right)$

$= 32\left[\cos\left(\frac{3\pi}{2} + 6\pi\right) + i \sin\left(\frac{3\pi}{2} + 6\pi\right)\right]$

$= 32\left(\cos \frac{3\pi}{2} + i \sin \frac{3\pi}{2}\right)$

$= 32[0 + i(-1)]$

$= -32i$

5. $2(\sqrt{3} + i)^7 = 2\left[2\left(\cos \frac{\pi}{6} + i \sin \frac{\pi}{6}\right)\right]^7$

$= 2\left[2^7\left(\cos \frac{7\pi}{6} + i \sin \frac{7\pi}{6}\right)\right]$

$= 256\left(-\frac{\sqrt{3}}{2} - \frac{1}{2}i\right)$

$= -128\sqrt{3} - 128i$

7. $[5(\cos 20° + i \sin 20°)]^3 = 5^3(\cos 60° + i \sin 60°) = \frac{125}{2} + \frac{125\sqrt{3}}{2}i$

9. $\left(\cos \frac{\pi}{4} + i \sin \frac{\pi}{4}\right)^{12} = \cos \frac{12\pi}{4} + i \sin \frac{12\pi}{4}$

$= \cos 3\pi + i \sin 3\pi$

$= -1$

11. $[5(\cos 3.2 + i \sin 3.2)]^4 = 5^4(\cos 12.8 + i \sin 12.8)$

$\approx 608.02 + 144.69i$

13. $(3 - 2i)^5 \approx \{3.6056[\cos(5.695) + i \sin(5.695)]\}^5$

$\approx (3.6056)^5[\cos(28.475) + i \sin(28.475)]$

$\approx -597 - 122i$

15. $(\sqrt{5} - 4i)^3 \approx \{4.5826[\cos(5.222) + i \sin(5.222)]\}^3$

$\approx (4.5826)^3[\cos(15.666) + i \sin(15.666)]$

$\approx -96.15 + 4.04i$

The exact answer is $-43\sqrt{5} + 4i$

17. $[3(\cos 15° + i \sin 15°)]^4 = 81(\cos 60° + i \sin 60°)$

$$= \frac{81}{2} + \frac{8\sqrt{3}}{2}i$$

19. $[5(\cos 95° + i \sin 95°)]^3 = 125(\cos 285° + i \sin 285°)$

$$\approx 32.3524 - 120.7407i$$

21. $\left[2\left(\cos \dfrac{\pi}{10} + i \sin \dfrac{\pi}{10}\right)\right]^5 = 2^5\left(\cos \dfrac{\pi}{2} + i \sin \dfrac{\pi}{2}\right)$

$$= 32i$$

23. $\left[3\left(\cos \dfrac{2\pi}{3} + i \sin \dfrac{2\pi}{3}\right)\right]^3 = 27(\cos 2\pi + i \sin 2\pi)$

$$= 27$$

25. (a) Square roots of $5(\cos 120° + i \sin 120°)$:

$$\sqrt{5}\left[\cos\left(\frac{120° + 360°k}{2}\right) + i \sin\left(\frac{120° + 360°k}{2}\right)\right], \ k = 0, 1$$

$k = 0$: $\sqrt{5}(\cos 60° + i \sin 60°)$

$k = 1$: $\sqrt{5}(\cos 240° + i \sin 240°)$

(c) $\dfrac{\sqrt{5}}{2} + \dfrac{\sqrt{15}}{2}i, \ -\dfrac{\sqrt{5}}{2} - \dfrac{\sqrt{15}}{2}i$

(b)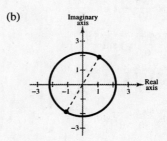

27. (a) Cube roots of $8\left(\cos \dfrac{2\pi}{3} + i \sin \dfrac{2\pi}{3}\right)$:

$$\sqrt[3]{8}\left[\cos\left(\frac{\frac{2\pi}{3} + 2\pi k}{3}\right) + i \sin\left(\frac{\frac{2\pi}{3} + 2\pi k}{3}\right)\right], \ k = 0, 1, 2$$

$k = 0$: $2\left(\cos \dfrac{2\pi}{9} + i \sin \dfrac{2\pi}{9}\right)$

$k = 1$: $2\left(\cos \dfrac{8\pi}{9} + i \sin \dfrac{8\pi}{9}\right)$

$k = 2$: $2\left(\cos \dfrac{14\pi}{9} + i \sin \dfrac{14\pi}{9}\right)$

(c) $1.5321 + 1.2856i$

$-1.8794 + 0.6840i$

$0.3473 - 1.9696i$

(b)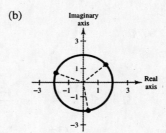

29. (a) Fifth roots of $243\left(\cos\dfrac{\pi}{6} + i\sin\dfrac{\pi}{6}\right)$

$$\sqrt[5]{243}\left[\cos\left(\dfrac{\dfrac{\pi}{6} + 2\pi k}{5}\right) + i\sin\left(\dfrac{\dfrac{\pi}{6} + 2\pi k}{5}\right)\right],$$

$k = 0, 1, 2, 3, 4$

$k = 0:\ 3\left(\cos\dfrac{\pi}{30} + i\sin\dfrac{\pi}{30}\right)$

$k = 1:\ 3\left(\cos\dfrac{13\pi}{30} + i\sin\dfrac{13\pi}{30}\right)$

$k = 2:\ 3\left(\cos\dfrac{5\pi}{6} + i\sin\dfrac{5\pi}{6}\right)$

$k = 3:\ 3\left(\cos\dfrac{37\pi}{30} + i\sin\dfrac{37\pi}{30}\right)$

$k = 4:\ 3\left(\cos\dfrac{49\pi}{30} + i\sin\dfrac{49\pi}{30}\right)$

(b)

(c) $2.9836 + 0.3136i$

$0.6237 + 2.9344i$

$-2.5981 + 1.5i$

$-2.2294 - 2.0074i$

$1.2202 - 2.7406i$

31. (a) Square roots of $-25i = 25\left(\cos\dfrac{3\pi}{2} + i\sin\dfrac{3\pi}{2}\right)$:

$$\sqrt{25}\left[\cos\left(\dfrac{\dfrac{3\pi}{2} + 2k\pi}{2}\right) + i\sin\left(\dfrac{\dfrac{3\pi}{2} + 2k\pi}{2}\right)\right],\ k = 0, 1$$

$k = 0:\ 5\left(\cos\dfrac{3\pi}{4} + i\sin\dfrac{3\pi}{4}\right)$

$k = 1:\ 5\left(\cos\dfrac{7\pi}{4} + i\sin\dfrac{7\pi}{4}\right)$

(b)

(c) $-\dfrac{5\sqrt{2}}{2} + \dfrac{5\sqrt{2}}{2}i,\ \dfrac{5\sqrt{2}}{2} - \dfrac{5\sqrt{2}}{2}i$

33. (a) Fourth roots of $81i = 81\left(\cos\dfrac{\pi}{2} + i\sin\dfrac{\pi}{2}\right)$

$$\sqrt[4]{81}\left[\cos\left(\dfrac{\dfrac{\pi}{2} + 2\pi k}{4}\right) + i\sin\left(\dfrac{\dfrac{\pi}{2} + 2\pi k}{4}\right)\right]$$

$k = 0, 1, 2, 3$

$k = 0$: $3\left(\cos\dfrac{\pi}{8} + i\sin\dfrac{\pi}{8}\right)$

$k = 1$: $3\left(\cos\dfrac{5\pi}{8} + i\sin\dfrac{5\pi}{8}\right)$

$k = 2$: $3\left(\cos\dfrac{9\pi}{8} + i\sin\dfrac{9\pi}{8}\right)$

$k = 3$: $3\left(\cos\dfrac{13\pi}{8} + i\sin\dfrac{13\pi}{8}\right)$

(c) $2.7716 + 1.1481i$

$-1.1481 + 2.7716i$

$-2.7716 - 1.1481i$

$1.1481 - 2.7716i$

(b)

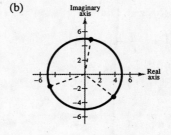

35. (a) Cube roots of $-\dfrac{125}{2}(1 + \sqrt{3}i) = 125\left(\cos\dfrac{4\pi}{3} + i\sin\dfrac{4\pi}{3}\right)$:

$$\sqrt[3]{125}\left[\cos\left(\dfrac{\dfrac{4\pi}{3} + 2k\pi}{3}\right) + i\sin\left(\dfrac{\dfrac{4\pi}{3} + 2k\pi}{3}\right)\right],\ k = 0, 1, 2$$

$k = 0$: $5\left(\cos\dfrac{4\pi}{9} + i\sin\dfrac{4\pi}{9}\right)$

$k = 1$: $5\left(\cos\dfrac{10\pi}{9} + i\sin\dfrac{10\pi}{9}\right)$

$k = 2$: $5\left(\cos\dfrac{16\pi}{9} + i\sin\dfrac{16\pi}{9}\right)$

(c) $0.8682 + 4.924i,\ -4.6985 - 1.7101i,\ 3.8302 - 3.214i$

(b)

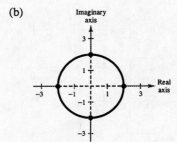

37. (a) Fourth roots of $16 = 16(\cos 0 + i\sin 0)$

$$\sqrt[4]{16}\left[\cos\left(\dfrac{0 + 2\pi k}{4}\right) + i\sin\left(\dfrac{0 + 2\pi k}{4}\right)\right], k = 0, 1, 2, 3$$

$k = 0$: $2(\cos 0 + i\sin 0)$

$k = 1$: $2\left(\cos\dfrac{\pi}{2} + i\sin\dfrac{\pi}{2}\right)$

$k = 2$: $2(\cos \pi + i\sin \pi)$

$k = 3$: $2\left(\cos\dfrac{3\pi}{2} + i\sin\dfrac{3\pi}{2}\right)$

(c) $2, 2i, -2, -2i$

39. (a) Fifth roots of $1 = \cos 0 + i \sin 0$:

$$\cos\left(\frac{2k\pi}{5}\right) + i \sin\left(\frac{2k\pi}{5}\right), k = 0, 1, 2, 3, 4$$

$k = 0$: $\cos 0 + i \sin 0$

$k = 1$: $\cos \dfrac{2\pi}{5} + i \sin \dfrac{2\pi}{5}$

$k = 2$: $\cos \dfrac{4\pi}{5} + i \sin \dfrac{4\pi}{5}$

$k = 3$: $\cos \dfrac{6\pi}{5} + i \sin \dfrac{6\pi}{5}$

$k = 4$: $\cos \dfrac{8\pi}{5} + i \sin \dfrac{8\pi}{5}$

(b)

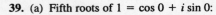

(c) $1, 0.3090 + 0.9511i, -0.8090 + 0.5878i, -0.8090 - 0.5878i, 0.3090 - 0.9511i$

41. (a) The cube roots of $-125 = 125(\cos \pi + i \sin \pi)$ are:

$$\sqrt[3]{125}\left[\cos\left(\frac{\pi + 2\pi k}{3}\right) + i \sin\left(\frac{\pi + 2\pi k}{3}\right)\right], k = 0, 1, 2$$

$k = 0$: $5\left(\cos \dfrac{\pi}{3} + i \sin \dfrac{\pi}{3}\right)$

$k = 1$: $5(\cos \pi + i \sin \pi)$

$k = 2$: $5\left(\cos \dfrac{5\pi}{3} + i \sin \dfrac{5\pi}{3}\right)$

(b)

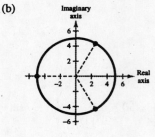

(c) $\dfrac{5}{2} + \dfrac{5\sqrt{3}}{2}i, -5, \dfrac{5}{2} - \dfrac{5\sqrt{3}}{2}i$

43. (a) The fifth roots of

$$128(-1 + i) = 128\sqrt{2}\left(\cos \frac{3\pi}{4} + i \sin \frac{3\pi}{4}\right) \text{ are:}$$

$$\sqrt[5]{128\sqrt{2}}\left[\cos\left(\frac{\frac{3\pi}{4} + 2\pi k}{5}\right) + i \sin\left(\frac{\frac{3\pi}{4} + 2\pi k}{5}\right)\right],$$

$k = 0, 1, 2, 3, 4$

$k = 0$: $2\sqrt[5]{4\sqrt{2}}\left(\cos \dfrac{3\pi}{20} + i \sin \dfrac{3\pi}{20}\right)$

$k = 1$: $2\sqrt[5]{4\sqrt{2}}\left(\cos \dfrac{11\pi}{20} + i \sin \dfrac{11\pi}{20}\right)$

$k = 2$: $2\sqrt[5]{4\sqrt{2}}\left(\cos \dfrac{19\pi}{20} + i \sin \dfrac{19\pi}{20}\right)$

$k = 3$: $2\sqrt[5]{4\sqrt{2}}\left(\cos \dfrac{27\pi}{20} + i \sin \dfrac{27\pi}{20}\right)$

$k = 4$: $2\sqrt[5]{4\sqrt{2}}\left(\cos \dfrac{7\pi}{4} + i \sin \dfrac{7\pi}{4}\right)$

(b)

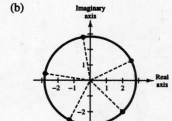

(c) $2.5201 + 1.2841i, -0.4425 + 2.7936i, -2.7936 + 0.4425i, -1.2841 - 2.5201i, 2 - 2i$

45. $x^4 - i = 0$

$x^4 = i$

The solutions are the fourth roots of $i = \cos\dfrac{\pi}{2} + i\sin\dfrac{\pi}{2}$:

$$\sqrt[4]{1}\left[\cos\left(\frac{\dfrac{\pi}{2} + 2k\pi}{4}\right) + i\sin\left(\frac{\dfrac{\pi}{2} + 2k\pi}{4}\right)\right], \; k = 0, 1, 2, 3$$

$k = 0$: $\cos\dfrac{\pi}{8} + i\sin\dfrac{\pi}{8}$

$k = 1$: $\cos\dfrac{5\pi}{8} + i\sin\dfrac{5\pi}{8}$

$k = 2$: $\cos\dfrac{9\pi}{8} + i\sin\dfrac{9\pi}{8}$

$k = 3$: $\cos\dfrac{13\pi}{8} + i\sin\dfrac{13\pi}{8}$

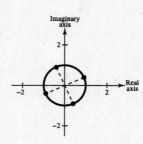

47. $x^6 + 1 = 0$

$x^6 = -1$

The solutions are the sixth roots of $-1 = \cos\pi + i\sin\pi$:

$$\sqrt[6]{1}\left[\cos\left(\frac{\pi + 2\pi k}{6}\right) + i\sin\left(\frac{\pi + 2\pi k}{6}\right)\right],$$

$k = 0, 1, 2, 3, 4, 5$

$k = 0$: $\cos\dfrac{\pi}{6} + i\sin\dfrac{\pi}{6}$

$k = 1$: $\cos\dfrac{\pi}{2} + i\sin\dfrac{\pi}{2}$

$k = 2$: $\cos\dfrac{5\pi}{6} + i\sin\dfrac{5\pi}{6}$

$k = 3$: $\cos\dfrac{7\pi}{6} + i\sin\dfrac{7\pi}{6}$

$k = 4$: $\cos\dfrac{3\pi}{2} + i\sin\dfrac{3\pi}{2}$

$k = 5$: $\cos\dfrac{11\pi}{6} + i\sin\dfrac{11\pi}{6}$

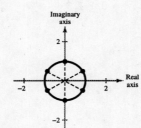

49. $x^5 + 243 = 0$

$$x^5 = -243$$

The solutions are the fifth roots of $-243 = 243(\cos \pi + i \sin \pi)$:

$$\sqrt[5]{243}\left[\cos\left(\frac{\pi + 2k\pi}{5}\right) + i \sin\left(\frac{\pi + 2k\pi}{5}\right)\right], \ k = 0, 1, 2, 3, 4$$

$k = 0: \ 3\left(\cos \dfrac{\pi}{5} + i \sin \dfrac{\pi}{5}\right)$

$k = 1: \ 3\left(\cos \dfrac{3\pi}{5} + i \sin \dfrac{3\pi}{5}\right)$

$k = 2: \ 3(\cos \pi + i \sin \pi) = -3$

$k = 3: \ 3\left(\cos \dfrac{7\pi}{5} + i \sin \dfrac{7\pi}{5}\right)$

$k = 4: \ 3\left(\cos \dfrac{9\pi}{5} + i \sin \dfrac{9\pi}{5}\right)$

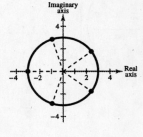

51. $x^5 - 32 = 0$

$$x^5 = 32$$

The solutions are the fifth roots of $32 = 32(\cos 0 + i \sin 0)$:

$$\sqrt[5]{32}\left[\cos\left(\frac{2\pi k}{5}\right) + i \sin\left(\frac{2\pi k}{5}\right)\right], k = 0, 1, 2, 3, 4$$

$k = 0: \ 2(\cos 0 + i \sin 0)$

$k = 1: \ 2\left(\cos \dfrac{2\pi}{5} + i \sin \dfrac{2\pi}{5}\right)$

$k = 2: \ 2\left(\cos \dfrac{4\pi}{5} + i \sin \dfrac{4\pi}{5}\right)$

$k = 3: \ 2\left(\cos \dfrac{6\pi}{5} + i \sin \dfrac{6\pi}{5}\right)$

$k = 4: \ 2\left(\cos \dfrac{8\pi}{5} + i \sin \dfrac{8\pi}{5}\right)$

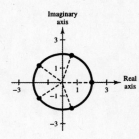

53. $x^4 + 16i = 0$

$$x^4 = -16i$$

The solutions are the fourth roots of $-16i = 16\left(\cos \dfrac{3\pi}{2} + i \sin \dfrac{3\pi}{2}\right)$:

$$\sqrt[4]{16}\left[\cos\left(\frac{\frac{3\pi}{2} + 2\pi k}{4}\right) + i \sin\left(\frac{\frac{3\pi}{2} + 2\pi k}{4}\right)\right], k = 0, 1, 2, 3$$

$k = 0: \ 2\left(\cos \dfrac{3\pi}{8} + i \sin \dfrac{3\pi}{8}\right)$

$k = 1: \ 2\left(\cos \dfrac{7\pi}{8} + i \sin \dfrac{7\pi}{8}\right)$

$k = 2: \ 2\left(\cos \dfrac{11\pi}{8} + i \sin \dfrac{11\pi}{8}\right)$

$k = 3: \ 2\left(\cos \dfrac{15\pi}{8} + i \sin \dfrac{15\pi}{8}\right)$

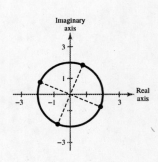

55. $x^4 - 16i = 0$

$$x^4 = 16i$$

The solutions are the fourth roots of $16i = 16\left(\cos\dfrac{\pi}{2} + i\sin\dfrac{\pi}{2}\right)$:

$$\sqrt[4]{16}\left[\cos\left(\dfrac{\dfrac{\pi}{2} + 2\pi k}{4}\right) + i\sin\left(\dfrac{\dfrac{\pi}{2} + 2\pi k}{4}\right)\right], k = 0, 1, 2, 3$$

$k = 0$: $2\left(\cos\dfrac{\pi}{8} + i\sin\dfrac{\pi}{8}\right)$

$k = 1$: $2\left(\cos\dfrac{5\pi}{8} + i\sin\dfrac{5\pi}{8}\right)$

$k = 2$: $2\left(\cos\dfrac{9\pi}{8} + i\sin\dfrac{9\pi}{8}\right)$

$k = 3$: $2\left(\cos\dfrac{13\pi}{8} + i\sin\dfrac{13\pi}{8}\right)$

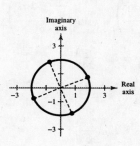

57. $x^3 - (1 - i) = 0$

$$x^3 = 1 - i = \sqrt{2}\left(\cos\dfrac{7\pi}{4} + i\sin\dfrac{7\pi}{4}\right)$$

The solutions are the cube roots of $1 - i$:

$$\sqrt[3]{\sqrt{2}}\left[\cos\left(\dfrac{\dfrac{7\pi}{4} + 2\pi k}{3}\right) + i\sin\left(\dfrac{\dfrac{7\pi}{4} + 2\pi k}{3}\right)\right], k = 0, 1, 2$$

$k = 0$: $\sqrt[6]{2}\left(\cos\dfrac{7\pi}{12} + i\sin\dfrac{7\pi}{12}\right)$

$k = 1$: $\sqrt[6]{2}\left(\cos\dfrac{5\pi}{4} + i\sin\dfrac{5\pi}{4}\right)$

$k = 2$: $\sqrt[6]{2}\left(\cos\dfrac{23\pi}{12} + i\sin\dfrac{23\pi}{12}\right)$

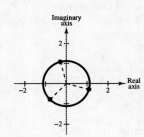

59. $x^6 + (1 + i) = 0$

$$x^6 = -(1 + i) = -1 - i$$

The solutions are the sixth roots of $-1 - i = \sqrt{2}\left(\cos\dfrac{5\pi}{4} + i\sin\dfrac{5\pi}{4}\right)$

$$\sqrt[6]{\sqrt{2}}\left[\cos\left(\dfrac{\dfrac{5\pi}{4} + 2\pi k}{6}\right) + i\sin\left(\dfrac{\dfrac{5\pi}{4} + 2\pi k}{6}\right)\right], k = 0, 1, 2, 3, 4, 5$$

$k = 0$: $\quad\sqrt[12]{2}\left(\cos\dfrac{5\pi}{24} + i\sin\dfrac{5\pi}{24}\right)$

$k = 1$: $\quad\sqrt[12]{2}\left(\cos\dfrac{13\pi}{24} + i\sin\dfrac{13\pi}{24}\right)$

$k = 2$: $\quad\sqrt[12]{2}\left(\cos\dfrac{7\pi}{8} + i\sin\dfrac{7\pi}{8}\right)$

$k = 3$: $\quad\sqrt[12]{2}\left(\cos\dfrac{29\pi}{24} + i\sin\dfrac{29\pi}{24}\right)$

$k = 4$: $\quad\sqrt[12]{2}\left(\cos\dfrac{37\pi}{24} + i\sin\dfrac{37\pi}{24}\right)$

$k = 5$: $\quad\sqrt[12]{2}\left(\cos\dfrac{15\pi}{8} + i\sin\dfrac{15\pi}{8}\right)$

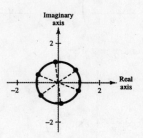

61. False. They are equally spaced along the circle centered at the origin with radius $\sqrt[n]{r}$.

63. $\quad -\dfrac{1}{2}(1 + \sqrt{3}i) = \cos\dfrac{4\pi}{3} + i\sin\dfrac{4\pi}{3}$

$$\left[-\dfrac{1}{2}(1 + \sqrt{3}i)\right]^6 = \left[\cos\dfrac{4\pi}{3} + i\sin\dfrac{4\pi}{3}\right]^6$$

$$= \cos 8\pi + i\sin 8\pi$$

$$= 1$$

65. (a) In trigonometric form we have:

$2(\cos 30° + i\sin 30°)$

$2(\cos 150° + i\sin 150°)$

$2(\cos 270° + i\sin 270°)$

(b) There are three roots evenly spaced around a circle of radius 2. Therefore, they represent the cube roots of some number of modulus 8. Cubing them shows that they are all cube roots of $8i$.

(c) $[2(\cos 30° + i\sin 30°)]^3 = 8i$

$[2(\cos 150° + i\sin 150°)]^3 = 8i$

$[2(\cos 270° + i\sin 270°)]^3 = 8i$

For 67-75, use the following figure.

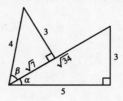

67. $\cos(\alpha + \beta) = \cos\alpha\cos\beta - \sin\alpha\sin\beta$

$$= \frac{5}{\sqrt{34}} \cdot \frac{\sqrt{7}}{4} - \frac{3}{\sqrt{34}} \cdot \frac{3}{4}$$

$$= \frac{5\sqrt{7} - 9}{4\sqrt{34}} = \frac{\sqrt{34}}{136}(5\sqrt{7} - 9)$$

69. $\sin(\alpha - \beta) = \sin\alpha\cos\beta - \cos\alpha\sin\beta$

$$= \frac{3}{\sqrt{34}} \cdot \frac{\sqrt{7}}{4} - \frac{5}{\sqrt{34}} \cdot \frac{3}{4}$$

$$= \frac{3\sqrt{7} - 15}{4\sqrt{34}} = \frac{3\sqrt{34}}{136}(\sqrt{7} - 5)$$

71. $\tan(\alpha + \beta) = \dfrac{\tan\alpha + \tan\beta}{1 - \tan\alpha\tan\beta}$

$$= \frac{\dfrac{3}{5} + \dfrac{3}{\sqrt{7}}}{1 - \left(\dfrac{3}{5}\right)\left(\dfrac{3}{\sqrt{7}}\right)} \cdot \frac{5\sqrt{7}}{5\sqrt{7}}$$

$$= \frac{3\sqrt{7} + 15}{5\sqrt{7} - 9} \cdot \frac{5\sqrt{7} + 9}{5\sqrt{7} + 9}$$

$$= \frac{240 + 102\sqrt{7}}{94} = \frac{120 + 51\sqrt{7}}{47}$$

73. $\tan 2\alpha = \dfrac{2\tan\alpha}{1 - \tan^2\alpha}$

$$= \frac{2(3/5)}{1 - (3/5)^2}$$

$$= \frac{6}{5} \cdot \frac{25}{16} = \frac{15}{8}$$

75. $\sin\dfrac{\beta}{2} = \sqrt{\dfrac{1 - \cos\beta}{2}} = \sqrt{\dfrac{1 - \sqrt{7}/4}{2}} = \sqrt{\dfrac{4 - \sqrt{7}}{8}} = \dfrac{1}{2}\sqrt{\dfrac{4 - \sqrt{7}}{2}}$

77. $\|\mathbf{u}\|^2 = \mathbf{u} \cdot \mathbf{u} = \langle -3, 4 \rangle \cdot \langle -3, 4 \rangle = 25$

$\|\mathbf{u}\| = \sqrt{25} = 5$

79. $\|\mathbf{u}\|^2 = \mathbf{u} \cdot \mathbf{u} = \langle -9, 40 \rangle \cdot \langle -9, 40 \rangle = 1681$

$\|\mathbf{u}\| = \sqrt{1681} = 41$

81. $\|\mathbf{u}\|^2 = \mathbf{u} \cdot \mathbf{u} = (22\mathbf{i} - 3\mathbf{j}) \cdot (22\mathbf{i} - 3\mathbf{j}) = 493$

$\|\mathbf{u}\| = \sqrt{493}$

83. $\|\mathbf{u}\|^2 = \mathbf{u} \cdot \mathbf{u} = (-13\mathbf{i} + 6\mathbf{j}) \cdot (-13\mathbf{i} + 6\mathbf{j}) = 205$

$\|\mathbf{u}\| = \sqrt{205}$

Review Exercises for Chapter 4

Solutions to Odd-Numbered Exercises

1. $6 + \sqrt{-4} = 6 + 2i$

3. $i^2 + 3i = -1 + 3i$

5. $(7 + 5i) + (-4 + 2i) = (7 - 4) + (5i + 2i) = 3 + 7i$

7. $5i(13 - 8i) = 65i - 40i^2 = 40 + 65i$

9. $(10 - 8i)(2 - 3i) = 20 - 30i - 16i + 24i^2 = -4 - 46i$

11. $\dfrac{6 + i}{4 - i} = \dfrac{6 + i}{4 - i} \cdot \dfrac{4 + i}{4 + i} = \dfrac{24 + 10i + i^2}{16 + 1} = \dfrac{23 + 10i}{17}$

$\qquad\qquad = \dfrac{23}{17} + \dfrac{10}{17}i$

13. $\dfrac{4}{2 - 3i} + \dfrac{2}{1 + i} = \dfrac{4}{2 - 3i} \cdot \dfrac{2 + 3i}{2 + 3i} + \dfrac{2}{1 + i} \cdot \dfrac{1 - i}{1 - i}$

$\qquad\qquad = \dfrac{8 + 12i}{4 + 9} + \dfrac{2 - 2i}{1 + 1}$

$\qquad\qquad = \dfrac{8}{13} + \dfrac{12}{13}i + 1 - i$

$\qquad\qquad = \left(\dfrac{8}{13} + 1\right) + \left(\dfrac{12}{13}i - i\right)$

$\qquad\qquad = \dfrac{21}{13} - \dfrac{1}{13}i$

15. $3x^2 + 1 = 0$

$\qquad 3x^2 = -1$

$\qquad x^2 = -\dfrac{1}{3}$

$\qquad x = \pm\sqrt{-\dfrac{1}{3}}$

$\qquad\; = \pm\sqrt{\dfrac{1}{3}}\,i$

17. $x^2 - 2x + 10 = 0$

$\qquad x^2 - 2x + 1 = -10 + 1$

$\qquad (x - 1)^2 = -9$

$\qquad x - 1 = \pm\sqrt{-9}$

$\qquad x = 1 \pm 3i$

19. $x^5 - 2x^4 + 3x^2 - 5 = 0$ is a fifth degree polynomial equation, so it has **five** zeros in the complex number system.

21. $\frac{1}{2}x^4 + \frac{2}{3}x^3 - x^2 + \frac{3}{10} = 0$ is a fourth degree polynomial equation, so it has **four** zeros in the complex number system.

23. $6x^2 + x - 2 = 0$

$\qquad b^2 - 4ac = 1^2 - 4(6)(-2) = 49 > 0$

Two real solutions

25. $0.13x^2 - 0.45x + 0.65 = 0$

$\qquad b^2 - 4ac = (-0.45)^2 - 4(0.13)(0.65) = -0.1355 < 0$

No real solutions

27. $x^2 - 2x = 0$

$\qquad x(x - 2) = 0$

$\qquad x = 0, x = 2$

29. $x^2 + 8x + 10 = 0$

$$x^2 + 8x = -10$$

$$x^2 + 8x + 16 = -10 + 16$$

$$(x + 4)^2 = 6$$

$$x + 4 = \pm\sqrt{6}$$

$$x = -4 \pm \sqrt{6}$$

31. $2x^2 + 2x + 3 = 0$

$$x = \frac{-2 \pm \sqrt{2^2 - 4(2)(3)}}{2(2)}$$

$$= \frac{-2 \pm \sqrt{-20}}{4}$$

$$= \frac{-2 \pm 2\sqrt{5}i}{4}$$

$$= -\frac{1}{2} \pm \frac{\sqrt{5}}{2}i$$

33. $f(x) = 4x^3 - 11x^2 + 10x - 3$

Possible rational zeros: $\pm 1, \pm 3, \pm\frac{1}{2}, \pm\frac{3}{2}, \pm\frac{1}{4}, \pm\frac{3}{4}$

$$
\begin{array}{r|rrrr}
1 & 4 & -11 & 10 & -3 \\
  &   & 4   & -7 & 3 \\
\hline
  & 4 & -7 & 3 & 0
\end{array}
$$

$4x^3 - 11x^2 + 10x - 3 = (x - 1)(4x^2 - 7x + 3) = (x - 1)^2(4x - 3)$

Thus, the zeros of $f(x)$ are $x = 1$ and $x = \frac{3}{4}$.

35. $f(x) = 6x^4 - 25x^3 + 14x^2 + 27x - 18$

Possible rational zeros: $\pm 1, \pm 2, \pm 3, \pm 6, \pm 9, \pm 18, \pm\frac{1}{2}, \pm\frac{3}{2}, \pm\frac{9}{2}, \pm\frac{1}{3}, \pm\frac{2}{3}, \pm\frac{1}{6}$

$$
\begin{array}{r|rrrrr}
-1 & 6 & -25 & 14 & 27 & -18 \\
   &   & -6 & 31 & -45 & 18 \\
\hline
   & 6 & -31 & 45 & -18 & 0
\end{array}
$$

$$
\begin{array}{r|rrrr}
3 & 6 & -31 & 45 & -18 \\
  &   & 18 & -39 & 18 \\
\hline
  & 6 & -13 & 6 & 0
\end{array}
$$

$6x^4 - 25x^3 + 14x^2 + 27x - 18 = (x + 1)(x - 3)(6x^2 - 13x + 6)$

$$= (x + 1)(x - 3)(3x - 2)(2x - 3)$$

Thus, the zeros of $f(x)$ are $x = -1, x = 3, x = \frac{2}{3}$, and $x = \frac{3}{2}$.

37. $f(x) = x^3 + 3x^2 - 24x + 28$

Zero: $2 \Rightarrow x - 2$ is a factor of $f(x)$.

$$
\begin{array}{r}
x^2 + 5x - 14 \\
x - 2 \overline{)\, x^3 + 3x^2 - 24x + 28} \\
\underline{x^3 - 2x^2} \\
5x^2 - 24x \\
\underline{5x^2 - 10x} \\
-14x + 28 \\
\underline{-14x + 28} \\
0
\end{array}
$$

Thus, $f(x) = (x - 2)(x^2 + 5x - 14)$

$\qquad = (x - 2)(x - 2)(x + 7)$

$\qquad = (x - 2)^2(x + 7)$

The zeros of $f(x)$ are $x = 2$ and $x = -7$.

39. $f(x) = x^3 + 3x^2 - 5x + 25$

Zero: $-5 \Rightarrow x + 5$ is a factor of $f(x)$.

$$
\begin{array}{r}
x^2 - 2x + 5 \\
x + 5 \overline{)\, x^3 + 3x^2 - 5x + 25} \\
\underline{x^3 + 5x^2} \\
-2x^2 - 5x \\
\underline{-2x^2 - 10x} \\
5x + 25 \\
\underline{5x + 25} \\
0
\end{array}
$$

Thus, $f(x) = (x + 5)(x^2 - 2x + 5)$ and by the Quadratic Formula, the zeros of $x^2 - 2x + 5$ are $1 \pm 2i$. The zeros of $f(x)$ are $x = -5$ and $x = 1 \pm 2i$.

$f(x) = (x + 5)[x - (1 + 2i)][x - (1 - 2i)]$

$\qquad = (x + 5)(x - 1 - 2i)(x - 1 + 2i)$

41. $h(x) = 2x^3 - 19x^2 + 58x + 34$

Zero: $5 + 3i \Rightarrow 5 - 3i$ is also a zero and $(x - 5 - 3i)(x - 5 + 3i) = x^2 - 10x + 34$ is a factor of $h(x)$.

$$
\begin{array}{r}
2x + 1 \\
x^2 - 10x + 34 \overline{)\, 2x^3 - 19x^2 + 58x + 34} \\
\underline{2x^3 - 20x^2 + 68x} \\
x^2 - 10x + 34 \\
\underline{x^2 - 10x + 34} \\
0
\end{array}
$$

Thus, $h(x) = (2x + 1)(x^2 - 10x + 34)$

$\qquad = (2x + 1)(x - 5 - 3i)(x - 5 + 3i)$

The zeros of $h(x)$ are $x = -\frac{1}{2}$ and $x = 5 \pm 3i$.

43. $f(x) = x^4 + 5x^3 + 2x^2 - 50x - 84$

Zero: $-3 + \sqrt{5}i \Rightarrow -3 - \sqrt{5}i$ is also a zero and $\left(x + 3 - \sqrt{5}i\right)\left(x + 3 + \sqrt{5}i\right) = x^2 + 6x + 14$ is a factor of $f(x)$.

$$
\begin{array}{r}
x^2 - x - 6 \\
x^2 + 6x + 14 \overline{)\, x^4 + 5x^3 + 2x^2 - 50x - 84} \\
\underline{x^4 + 6x^3 + 14x^2} \\
-x^3 - 12x^2 - 50x \\
\underline{-x^3 - 6x^2 - 14x} \\
-6x^2 - 36x - 84 \\
\underline{-6x^2 - 36x - 84} \\
0
\end{array}
$$

Thus, $f(x) = (x^2 - x - 6)(x^2 + 6x + 14)$

$\qquad = (x + 2)(x - 3)\left(x + 3 - \sqrt{5}i\right)\left(x + 3 + \sqrt{5}i\right)$

The zeros of $f(x)$ are $x = -2$, $x = 3$, and $x = -3 \pm \sqrt{5}i$.

45. Zeros: $1, 1, \frac{1}{4}, -\frac{2}{3}$

$f(x) = (x - 1)(x - 1)(4x - 1)(3x + 2)$

$\qquad = (x^2 - 2x + 1)(12x^2 + 5x - 2)$

$\qquad = 12x^4 - 19x^3 + 9x - 2$

47. Zeros: $3, 2 - \sqrt{3}, 2 + \sqrt{3}$

$f(x) = (x - 3)\left(x - 2 + \sqrt{3}\right)\left(x - 2 - \sqrt{3}\right)$

$\qquad = (x - 3)(x^2 - 4x + 1)$

$\qquad = x^3 - 7x^2 + 13x - 3$

49. Zeros: $\frac{2}{3}, 4, \sqrt{3}i, -\sqrt{3}i$

$$f(x) = (3x - 2)(x - 4)(x - \sqrt{3}i)(x + \sqrt{3}i)$$
$$= (3x^2 - 14x + 8)(x^2 + 3)$$
$$= 3x^4 - 14x^3 + 17x^2 - 42x + 24$$

51. Zeros: $-\sqrt{2}i, \sqrt{2}i, -5i, 5i$

$$f(x) = (x + \sqrt{2}i)(x - \sqrt{2}i)(x + 5i)(x - 5i)$$
$$= (x^2 + 2)(x^2 + 25)$$
$$= x^4 + 27x^2 + 50$$

53. $P = xp - C$

$9,000,000 = x(140 - 0.0001x) - (75x + 100,000)$

$9,000,000 = 140x - 0.0001x^2 - 75x - 100,000$

$0.0001x^2 - 65x + 9,100,000 = 0$

$b^2 - 4ac = (-65)^2 - 4(0.0001)(9,100,000) = 585 > 0$

There are two real solutions. By the Quadratic Formula we have:

$x \approx 445,934$ units or $x \approx 204,066$ units

$p \approx \$95.41$ \qquad $p \approx \$119.59$

There are two possible values for p that will yield a profit of approximately 9 million.

55. $|-6i| = \sqrt{0^2 + (-6)^2} = \sqrt{36} = 6$

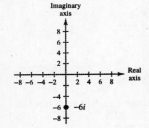

57. $|-10 - 4i| = \sqrt{(-10)^2 + (-4)^2} = \sqrt{116} = 2\sqrt{29}$

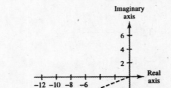

59. $z = 5 + 12i$

$r = \sqrt{5^2 + 12^2} = \sqrt{169} = 13$

$\tan \theta = \frac{12}{5} \Longrightarrow \theta \approx 1.176$

$z \approx 13(\cos 1.176 + i \sin 1.176)$

61. $z = -7$

$r = |-7| = 7$

$\theta = \pi$

$z = 7(\cos \pi + i \sin \pi)$

63. (a) $z_1 = -3(1 + i) = -3 - 3i$

$r = \sqrt{(-3)^2 + (-3)^2} = \sqrt{18} = 3\sqrt{2}$

$\tan \theta = \frac{-3}{-3} = 1$ and z_1 is in Quadrant III. $\Longrightarrow \theta = \frac{5\pi}{4}$

$z_1 = 3\sqrt{2}\left(\cos \frac{5\pi}{4} + i \sin \frac{5\pi}{4}\right)$

$z_2 = 2(\sqrt{3} + i) = 2\sqrt{3} + 2i$

$r = \sqrt{(2\sqrt{3})^2 + 2^2} = \sqrt{12 + 4} = \sqrt{16} = 4$

$\tan \theta = \frac{2}{2\sqrt{3}} = \frac{1}{\sqrt{3}}$ and z_2 is in Quadrant I. $\Longrightarrow \theta = \frac{\pi}{6}$

$z_2 = 4\left(\cos \frac{\pi}{6} + i \sin \frac{\pi}{6}\right)$

—CONTINUED—

63. —CONTINUED—

(b) $z_1 z_2 = \left[3\sqrt{2}\left(\cos\dfrac{5\pi}{4} + i\sin\dfrac{5\pi}{4}\right)\right]\left[4\left(\cos\dfrac{\pi}{6} + i\sin\dfrac{\pi}{6}\right)\right]$

$\qquad = (3\sqrt{2})(4)\left[\cos\left(\dfrac{5\pi}{4} + \dfrac{\pi}{6}\right) + i\sin\left(\dfrac{5\pi}{4} + \dfrac{\pi}{6}\right)\right]$

$\qquad = 12\sqrt{2}\left(\cos\dfrac{17\pi}{12} + i\sin\dfrac{17\pi}{12}\right)$

$\dfrac{z_1}{z_2} = \dfrac{3\sqrt{2}\left(\cos\dfrac{5\pi}{4} + i\sin\dfrac{5\pi}{4}\right)}{4\left(\cos\dfrac{\pi}{6} + i\sin\dfrac{\pi}{6}\right)}$

$\qquad = \dfrac{3\sqrt{2}}{4}\left[\cos\left(\dfrac{5\pi}{4} - \dfrac{\pi}{6}\right) + i\sin\left(\dfrac{5\pi}{4} - \dfrac{\pi}{6}\right)\right]$

$\qquad = \dfrac{3\sqrt{2}}{4}\left(\cos\dfrac{13\pi}{12} + i\sin\dfrac{13\pi}{12}\right)$

65. $\left[2\left(\cos\dfrac{4\pi}{15} + i\sin\dfrac{4\pi}{15}\right)\right]^5 = 2^5\left[\cos 5\left(\dfrac{4\pi}{15}\right) + i\sin 5\left(\dfrac{4\pi}{15}\right)\right]$

$\qquad\qquad = 32\left(\cos\dfrac{4\pi}{3} + i\sin\dfrac{4\pi}{3}\right)$

$\qquad\qquad = 32\left(-\dfrac{1}{2} - \dfrac{\sqrt{3}}{2}i\right)$

$\qquad\qquad = -16 - 16\sqrt{3}i$

67. $(1 - i)^8 = \left[\sqrt{2}\left(\cos\dfrac{7\pi}{4} + i\sin\dfrac{7\pi}{4}\right)\right]^8$

$\qquad = (\sqrt{2})^8\left(\cos 8\left(\dfrac{7\pi}{4}\right) + i\sin 8\left(\dfrac{7\pi}{4}\right)\right)$

$\qquad = 16(\cos 14\pi + i\sin 14\pi)$

$\qquad = 16(1 + 0i)$

$\qquad = 16$

69. (a) There are four roots shown and they each have an absolute value of 4.

$4(\cos 60° + i\sin 60°)$

$4(\cos 150° + i\sin 150°)$

$4(\cos 240° + i\sin 240°)$

$4(\cos 330° + i\sin 330°)$

(b) $[4(\cos 60° + i\sin 60°)]^4 = 256(\cos 240° + i\sin 240°)$

$\qquad\qquad = 256\left(-\dfrac{1}{2} - \dfrac{\sqrt{3}}{2}i\right)$

$\qquad\qquad = -128 - 128\sqrt{3}i$

(c) By using a calculator, show that each of the roots raised to the fourth power yields $-128 - 128\sqrt{3}i$.

71. Fourth roots of 256

$256 = 256(\cos 0 + i\sin 0)$

$\sqrt[4]{256} = \sqrt[4]{256}\left(\cos\dfrac{2\pi k}{4} + i\sin\dfrac{2\pi k}{4}\right), k = 0, 1, 2, 3$

$k = 0$: $4(\cos 0 + i\sin 0) = 4$

$k = 1$: $4\left(\cos\dfrac{\pi}{2} + i\sin\dfrac{\pi}{2}\right) = 4i$

$k = 2$: $4(\cos\pi + i\sin\pi) = -4$

$k = 3$: $4\left(\cos\dfrac{3\pi}{2} + i\sin\dfrac{3\pi}{2}\right) = -4i$

73. $x^5 - 32 = 0$

$\qquad x^5 = 32$

The solutions of the equation are the fifth roots of 32.

$\qquad 32 = 32(\cos 0 + i \sin 0)$

$\qquad \sqrt[5]{32} = \sqrt[5]{32}\left(\cos \dfrac{2\pi k}{5} + i \sin \dfrac{2\pi k}{5}\right), k = 0, 1, 2, 3, 4.$

$k = 0$: $2(\cos 0 + i \sin 0) = 2$

$k = 1$: $2\left(\cos \dfrac{2\pi}{5} + i \sin \dfrac{2\pi}{5}\right) \approx 0.6180 + 1.9021i$

$k = 2$: $2\left(\cos \dfrac{4\pi}{5} + i \sin \dfrac{4\pi}{5}\right) \approx -1.6180 + 1.1756i$

$k = 3$: $2\left(\cos \dfrac{6\pi}{5} + i \sin \dfrac{6\pi}{5}\right) \approx -1.6180 - 1.1756i$

$k = 4$: $2\left(\cos \dfrac{8\pi}{5} + i \sin \dfrac{8\pi}{5}\right) \approx 0.6180 - 1.9021i$

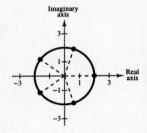

75. $(x^3 - 1)(x^2 + 1) = 0$

$\qquad x^3 - 1 = 0 \text{ or } x^2 + 1 = 0$

$\qquad\quad x^3 = 1 \text{ or } \qquad x^2 = -1$

The solutions of the equation are the cube roots of 1 or the square roots of -1.

$\qquad 1 = \cos 0 + i \sin 0$

$\qquad \sqrt[3]{1} = \cos\left(\dfrac{2\pi k}{3}\right) + i \sin\left(\dfrac{2\pi k}{3}\right), k = 0, 1, 2$

$k = 0$: $\cos 0 + i \sin 0 = 1$

$k = 1$: $\cos \dfrac{2\pi}{3} + i \sin \dfrac{2\pi}{3} = -\dfrac{1}{2} + \dfrac{\sqrt{3}}{2}i$

$k = 2$: $\cos \dfrac{4\pi}{3} + i \sin \dfrac{4\pi}{3} = -\dfrac{1}{2} - \dfrac{\sqrt{3}}{2}i$

$\qquad -1 = \cos \pi + i \sin \pi$

$\qquad \sqrt{-1} = \cos\left(\dfrac{\pi + 2\pi k}{2}\right) + i \sin\left(\dfrac{\pi + 2\pi k}{2}\right), k = 0, 1$

$k = 0$: $\cos \dfrac{\pi}{2} + i \sin \dfrac{\pi}{2} = i$

$k = 1$: $\cos \dfrac{3\pi}{2} + i \sin \dfrac{3\pi}{2} = -i$

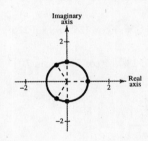

77. False. The equation $325x^2 - 717x + 398 = 0$ has no *real* solution. By the Quadratic Formula, the complex solutions are:

$\qquad x = \dfrac{717}{650} \pm \dfrac{\sqrt{3311}}{650}i.$

79. Answers will vary.

(a) Two distinct real solutions. $\Rightarrow b^2 - 4ac > 0$

$\qquad x^2 - 5x + 6 = 0$

$\qquad 3x^2 + 8x - 40 = 0$

(b) Two complex solutions. **All** quadratic equations have two complex solutions since the degree of any quadratic equation is 2.

(c) No real solutions. $\Rightarrow b^2 - 4ac < 0$

$\qquad x^2 + 5x + 14 = 0$

$\qquad 6x^2 - 3x + 11 = 0$

81. (a) There are 4 fourth roots of z in the complex number system, so 3 roots are not shown.

(b) The four roots are equally spaced around the circle, so they are 90° apart. The 4 roots are:

$\qquad 2(\cos 30° + i \sin 30°)$

$\qquad 2(\cos 120° + i \sin 120°)$

$\qquad 2(\cos 210° + i \sin 210°)$

$\qquad 2(\cos 300° + i \sin 300°)$

Chapter 4   Practice Test

1. Write $4 + \sqrt{-81} - 3i^2$ in standard form.

2. Write the result in standard form: $\dfrac{3 + i}{5 - 4i}$

3. Use the Quadratic Formula to solve $x^2 - 4x + 7 = 0$.

4. True or false: $\sqrt{-6}\sqrt{-6} = \sqrt{36} = 6$

5. Use the discriminant to determine the type of solutions of $3x^2 - 8x + 7 = 0$.

6. Find all the zeros of $f(x) = x^4 + 13x^2 + 36$.

7. Find a polynomial function that has the following zeros: $3, -1 \pm 4i$

8. Use the zero $x = 4 + i$ to find all the zeros of $f(x) = x^3 - 10x^2 + 33x - 34$.

9. Give the trigonometric form of $z = 5 - 5i$.

10. Give the standard form of $z = 6(\cos 225° + i \sin 225°)$.

11. Multiply $[7(\cos 23° + i \sin 23°)][4(\cos 7° + i \sin 7°)]$.

12. Divide $\dfrac{9\left(\cos \dfrac{5\pi}{4} + i \sin \dfrac{5\pi}{4}\right)}{3(\cos \pi + i \sin \pi)}$

13. Find $(2 + 2i)^8$.

14. Find the cube roots of $8\left(\cos \dfrac{\pi}{3} + i \sin \dfrac{\pi}{3}\right)$.

15. Find all the solutions to $x^4 + i = 0$.

C H A P T E R 5
Exponential and Logarithmic Functions

CHAPTER 5
Exponential and Logarithmic Functions

Section 5.1   Exponential Functions and Their Graphs

Solutions to Odd-Numbered Exercises

- You should know that a function of the form $f(x) = a^x$, where $a > 0$, $a \neq 1$, is called an exponential function with base a.
- You should be able to graph exponential functions.
- You should know formulas for compound interest.

    (a) For n compoundings per year: $A = P\left(1 + \dfrac{r}{n}\right)^{nt}$.

    (b) For continuous compoundings: $A = Pe^{rt}$.

---

1. $(3.4)^{5.6} \approx 946.852$

3. $(1.005)^{400} \approx 7.352$

5. $5^{-\pi} \approx 0.006$

7. $100^{\sqrt{2}} \approx 673.639$

9. $e^{-3/4} \approx 0.472$

11. $f(x) = 2^x$

Increasing

Asymptote: $y = 0$

Intercept: $(0, 1)$

Matches graph (d).

13. $f(x) = 2^{-x}$

Decreasing

Asymptote: $y = 0$

Intercept: $(0, 1)$

Matches graph (a).

15. $f(x) = 3^x$

$g(x) = 3^{x-4}$

Because $g(x) = f(x - 4)$, the graph of g can be obtained by shifting the graph of f four units to the right.

17. $f(x) = -2^x$

$g(x) = 5 - 2^x$

Because $g(x) = 5 + f(x)$, the graph of g can be obtained by shifting the graph of f five units upward.

19. $f(x) = \left(\frac{3}{5}\right)^x$

$g(x) = -\left(\frac{3}{5}\right)^{x+4}$

Because $g(x) = -f(x + 4)$, the graph of g can be obtained by reflecting the graph of f in the x-axis and shifting f four units to the left.

21. $f(x) = 0.3^x$

$g(x) = -0.3^x + 5$

Because $g(x) = -f(x) + 5$, the graph of g can be obtained by reflecting the graph of f in the x-axis and shifting f five units upward.

23. $f(x) = \left(\frac{1}{2}\right)^x$

x	-2	-1	0	1	2
$f(x)$	4	2	1	0.5	0.25

Asymptote: $y = 0$

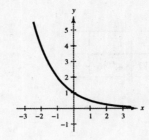

25. $f(x) = \left(\frac{1}{2}\right)^{-x} = 2^x$

x	-2	-1	0	1	2
$f(x)$	0.25	0.5	1	2	4

Asymptote: $y = 0$

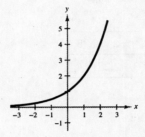

27. $f(x) = 2^{x-1}$

x	-2	-1	0	1	2
$f(x)$	0.125	0.25	0.5	1	2

Asymptote: $y = 0$

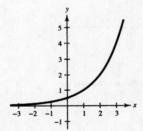

29. $f(x) = e^x$

x	-2	-1	0	1	2
$f(x)$	0.135	0.368	1	2.718	7.389

Asymptote: $y = 0$

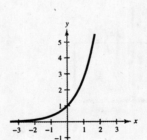

31. $f(x) = 3e^{x+4}$

x	-8	-7	-6	-5	-4
$f(x)$	0.055	0.149	0.406	1.104	3

Asymptote: $y = 0$

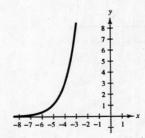

33. $f(x) = 2e^{x-2} + 4$

x	-2	-1	0	1	2
$f(x)$	4.037	4.100	4.271	4.736	6

Asymptote: $y = 4$

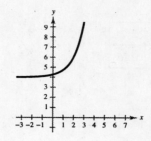

35. $f(x) = 4^{x-3} + 3$

x	-1	0	1	2	3
$f(x)$	3.003	3.016	3.063	3.25	4

Asymptote: $y = 3$

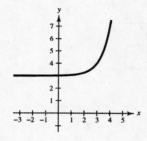

37. $g(x) = 5^x$

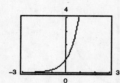

39. $f(x) = \left(\dfrac{1}{5}\right)^x = 5^{-x}$

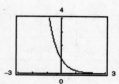

41. $h(x) = 5^{x-2}$

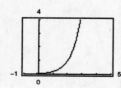

43. $g(x) = 5^{-x} - 3$

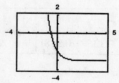

45. $y = 2^{-x^2}$

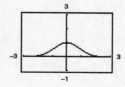

47. $f(x) = 3^{x-2} + 1$

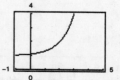

49. $y = 1.08^{-5x}$

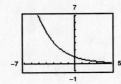

51. $s(t) = 2e^{0.12t}$

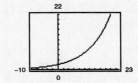

53. $g(x) = 1 + e^{-x}$

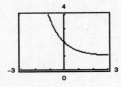

55. $P = \$2500$, $r = 8\%$, $t = 10$ years

Compounded n times per year: $A = P\left(1 + \dfrac{r}{n}\right)^{nt} = 2500\left(1 + \dfrac{0.08}{n}\right)^{10n}$

Compounded continuously: $A = Pe^{rt} = 2500e^{0.08(10)}$

n	1	2	4	12	365	Continuous Compounding
A	$5397.31	$5477.81	$5520.10	$5549.10	$5563.36	$5563.85

57. $P = \$2500$, $r = 8\%$, $t = 20$ years

Compounded n times per year: $A = P\left(1 + \dfrac{r}{n}\right)^{nt} = 2500\left(1 + \dfrac{0.08}{n}\right)^{20n}$

Compounded continuously: $A = Pe^{rt} = 2500e^{0.08(20)}$

n	1	2	4	12	365	Continuous Compounding
A	$11,652.39	$12,002.55	$12,188.60	$12,317.01	$12,380.41	$12,382.58

59. $A = Pe^{rt}$

$A = 12000e^{0.08t}$

t	1	10	20	30	40	50
A	$12,999.44	$26,706.49	$59,436.39	$132,278.12	$294,390.36	$655,177.80

61. $A = Pe^{rt}$

$A = 12000e^{0.065t}$

t	1	10	20	30	40	50
A	$12,805.91	$22,986.49	$44,031.56	$84,344.25	$161,564.86	$309,484.08

63. $A = 25{,}000e^{(0.0875)(25)} \approx \$222{,}822.57$

65. (a) The steeper curve represents the investment earning
compound interest, because compound interest earns
more than simple interest. With simple interest there
is no compounding so the growth is linear.

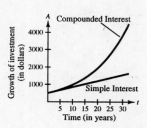

(b) Compound interest formula: $A = 500\left(1 + \dfrac{0.07}{1}\right)^{(1)t}$

$$= 500(1.07)^t$$

Simple interest formula: $A = Prt + P$

$$= 500(0.07)t + 500$$

67. $C(10) = 23.95(1.04)^{10} \approx \$35.45$

69. $P(t) = 100e^{0.2197t}$

(a) $P(0) = 100$

(b) $P(5) \approx 300$

(c) $P(10) \approx 900$

71. $Q = 25\left(\frac{1}{2}\right)^{t/1620}$

(a) When $t = 0$, $Q = 25\left(\frac{1}{2}\right)^{0/1620} = 25(1) = 25$ units.

(b) When $t = 1000$, $Q = 25\left(\frac{1}{2}\right)^{1000/1620} \approx 16.30$ units.

(c)
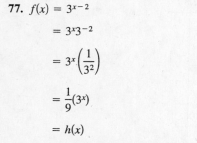

73. $P = 102{,}303e^{-0.137h}$

(a)
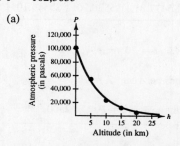

(b)

h	0	5	10	15	20
P	102,303	51,570	25,996	13,104	6606

(c) $P(8) \approx 34{,}190$ Pascals

(d) $21{,}000 = 102{,}303e^{-0.137h}$ when $h \approx 11.6$ km

75. True. As $x \to -\infty$, $f(x) \to 0$ but never reaches zero.

77. $f(x) = 3^{x-2}$

$$= 3^x 3^{-2}$$

$$= 3^x\left(\frac{1}{3^2}\right)$$

$$= \frac{1}{9}(3^x)$$

$$= h(x)$$

Thus, $f(x) \neq g(x)$, but $f(x) = h(x)$.

79.

$f(x) = 16(4^{-x})$ and $f(x) = 16(4^{-x})$

$\qquad = 4^2(4^{-x})$ $\qquad\qquad\qquad = 16(2^2)^{-x}$

$\qquad = 4^{2-x}$ $\qquad\qquad\qquad\quad = 16(2^{-2x})$

$\qquad = \left(\frac{1}{4}\right)^{-(2-x)}$ $\qquad\qquad = h(x)$

$\qquad = \left(\frac{1}{4}\right)^{x-2}$

$\qquad = g(x)$

Thus, $f(x) = g(x) = h(x)$.

81. $y = 3^x$ and $y = 4^x$

x	-2	-1	0	1	2
3^x	$\frac{1}{9}$	$\frac{1}{3}$	1	3	9
4^x	$\frac{1}{16}$	$\frac{1}{4}$	1	4	16

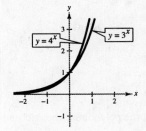

(a) $4^x < 3^x$ when $x < 0$.

(b) $4^x > 3^x$ when $x > 0$.

83. (a) $f(x) = \dfrac{8}{1 + e^{-0.5x}}$

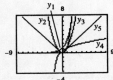

Horizontal asymptotes: $y = 0$ and $y = 8$

(b) $g(x) = \dfrac{8}{1 + e^{-0.5/x}}$

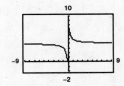

Horizontal asymptote: $y = 4$

Vertical asymptote: $x = 0$

85.

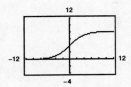

$y_1 = e^x$

$y_2 = x^2$

$y_3 = x^3$

$y_4 = \sqrt{x}$

$y_5 = |x|$

The function that increases at the fastest rate for "large" values of x is $y_1 = e^x$. (Note: One of the intersection points of $y = e^x$ and $y = x^3$ is approximately $(4.536, 93)$ and past this point $e^x > x^3$. This is not shown on the graph above.)

87. It usually implies rapid growth.

89. In Exercise 88 $f(x) = \left[1 + \dfrac{0.5}{x}\right]^x$ appears to approach $g(x) = e^{0.5}$ as x increases without bound. Therefore,

the value of $\left[1 + \dfrac{r}{x}\right]^x$ approaches e^r as x increases without bound.

x	1	10	100	200	500	1100	10,000
$\left[1 + \left(\dfrac{1}{x}\right)\right]^x$	2	2.5937	2.7048	2.7115	2.7156	2.7170	2.718

$e^1 \approx 2.718281828\ldots$

91. Since $\sqrt{2} \approx 1.414$ we know that $1 < \sqrt{2} < 2$.

Thus, $2^1 < 2^{\sqrt{2}} < 2^2$

$2 < 2^{\sqrt{2}} < 4$.

93. $y_4 = 1 + \dfrac{x}{1!} + \dfrac{x^2}{2!} + \dfrac{x^3}{3!} + \dfrac{x^4}{4!}$

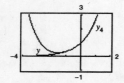

As more terms are added, the polynomial approaches e^x.

$$e^x = 1 + \frac{x}{1!} + \frac{x^2}{2!} + \frac{x^3}{3!} + \frac{x^4}{4!} + \frac{x^5}{5!} + \cdots$$

95. $2x - 7y + 14 = 0$

$\quad\quad 2x + 14 = 7y$

$\quad\quad \frac{1}{7}(2x + 14) = y$

97. $x^2 + y^2 = 25$

$\quad\quad y^2 = 25 - x^2$

$\quad\quad y = \pm\sqrt{25 - x^2}$

99. $1, 8i, -8i$

$\quad f(x) = (x - 1)(x - 8i)(x - (-8i))$

$\quad f(x) = (x - 1)(x^2 + 64)$

$\quad f(x) = x^3 - x^2 + 64x - 64$

101. $4, 6 + i, 6 - i$

$\quad f(x) = (x - 4)(x - 6 - i)(x - 6 + i)$

$\quad f(x) = (x - 4)(x^2 - 12x + 37)$

$\quad f(x) = x^3 - 16x^2 + 85x - 148$

Section 5.2   Logarithmic Functions and Their Graphs

- ■ You should know that a function of the form $y = \log_a x$, where $a > 0$, $a \neq 1$, and $x > 0$, is called a logarithm of x to base a.

- ■ You should be able to convert from logarithmic form to exponential form and vice versa.

    $\quad y = \log_a x \iff a^y = x$

- ■ You should know the following properties of logarithms.

    (a) $\log_a 1 = 0$ since $a^0 = 1$.

    (b) $\log_a a = 1$ since $a^1 = a$.

    (c) $\log_a a^x = x$ since $a^x = a^x$.

    (d) If $\log_a x = \log_a y$, then $x = y$.

- ■ You should know the definition of the natural logarithmic function.

    $\quad \log_e x = \ln x, \; x > 0$

- ■ You should know the properties of the natural logarithmic function.

    (a) $\ln 1 = 0$ since $e^0 = 1$.

    (b) $\ln e = 1$ since $e^1 = e$.

    (c) $\ln e^x = x$ since $e^x = e^x$.

    (d) If $\ln x = \ln y$, then $x = y$.

- ■ You should be able to graph logarithmic functions.

Solutions to Odd-Numbered Exercises

1. $\log_4 64 = 3 \implies 4^3 = 64$

3. $\log_7 \frac{1}{49} = -2 \implies 7^{-2} = \frac{1}{49}$

5. $\log_{32} 4 = \frac{2}{5} \implies 32^{2/5} = 4$

7. $\ln 1 = 0 \implies e^0 = 1$

9. $5^3 = 125 \implies \log_5 125 = 3$

11. $81^{1/4} = 3 \implies \log_{81} 3 = \frac{1}{4}$

13. $6^{-2} = \frac{1}{36} \implies \log_6 \frac{1}{36} = -2$

15. $e^3 = 20.0855\ldots \implies \ln 20.0855\ldots = 3$

17. $e^0 = 1 \implies \ln 1 = 0$

19. $\log_2 16 = \log_2 2^4 = 4$

21. $\log_{16} 4 = \log_{16} 16^{1/2} = \frac{1}{2}$

23. $\log_7 1 = \log_7 7^0 = 0$

25. $\log_{10} 0.01 = \log_{10} 10^{-2} = -2$

27. $\log_8 32 = \log_8 8^{5/3} = \frac{5}{3}$

29. $\ln e^3 = 3$

31. $\log_a a^2 = 2$

33. $\log_{10} 345 \approx 2.538$

35. $\log_{10} \frac{4}{5} \approx -0.097$

37. $\ln 18.42 \approx 2.913$

39. $3 \ln 0.32 \approx -3.418$

41. $\ln\left(1 + \sqrt{3}\right) \approx 1.005$

43. $\ln \frac{2}{3} \approx -0.405$

45. $f(x) = \log_3 x + 2$

Asymptote: $x = 0$

Point on graph: $(1, 2)$

Matches graph (c).

47. $f(x) = -\log_3(x + 2)$

Asymptote: $x = -2$

Point on graph: $(-1, 0)$

Matches graph (d).

49. $f(x) = \log_3(1 - x)$

Asymptote: $x = 1$

Point on graph: $(0, 0)$

Matches graph (b).

51. $f(x) = \log_4 x$

Domain: $x > 0 \implies$ The domain is $(0, \infty)$.

x-intercept: $(1, 0)$

Vertical asymptote: $x = 0$

$y = \log_4 x \implies 4^y = x$

x	$\frac{1}{4}$	1	4	2
$f(x)$	-1	0	1	$\frac{1}{2}$

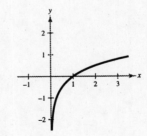

53. $y = -\log_3 x + 2$

Domain: $(0, \infty)$

x-intercept:

$$-\log_3 x + 2 = 0$$
$$2 = \log_3 x$$
$$3^2 = x$$
$$9 = x$$

The x-intercept is $(9, 0)$.

Vertical asymptote: $x = 0$

$y = -\log_3 x + 2$

$\log_3 x = 2 - y \implies 3^{2-y} = x$

x	27	9	3	1	$\frac{1}{3}$
y	-1	0	1	2	3

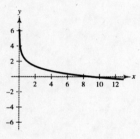

55. $f(x) = -\log_6(x + 2)$

Domain: $x + 2 > 0 \implies x > -2$

The domain is $(-2, \infty)$.

x-intercept:

$$0 = -\log_6(x + 2)$$
$$0 = \log_6(x + 2)$$
$$6^0 = x + 2$$
$$1 = x + 2$$
$$-1 = x$$

The x-intercept is $(-1, 0)$.

Vertical asymptote: $x + 2 = 0 \implies x = -2$

$$y = -\log_6(x + 2)$$
$$-y = \log_6(x + 2)$$

$6^{-y} - 2 = x$

x	4	-1	$-1\frac{5}{6}$	$-1\frac{35}{36}$
$f(x)$	-1	0	1	2

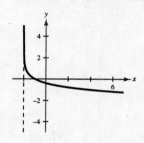

57. $y = \log_{10}\left(\dfrac{x}{5}\right)$

Domain: $\dfrac{x}{5} > 0 \implies x > 0$

The domain is $(0, \infty)$.

x-intercept:

$$\log_{10}\left(\frac{x}{5}\right) = 0$$
$$\frac{x}{5} = 10^0$$
$$\frac{x}{5} = 1 \implies x = 5$$

The x-intercept is $(5, 0)$.

Vertical asymptote: $\dfrac{x}{5} = 0 \implies x = 0$

The vertical asymptote is the y-axis.

x	1	2	3	4	5	6	7
y	-0.70	-0.40	-0.22	-0.10	0	0.08	0.15

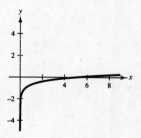

61. $g(x) = \ln(-x)$

Domain: $-x > 0 \implies x < 0$

The domain is $(-\infty, 0)$.

x-intercept:

$$0 = \ln(-x)$$
$$e^0 = -x$$
$$-1 = x$$

The x-intercept is $(-1, 0)$.

Vertical asymptote: $-x = 0 \implies x = 0$

x	-0.5	-1	-2	-3
$g(x)$	-0.69	0	0.69	1.10

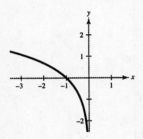

59. $f(x) = \ln(x - 2)$

Domain: $x - 2 > 0 \implies x > 2$

The domain is $(2, \infty)$.

x-intercept:

$$0 = \ln(x - 2)$$
$$e^0 = x - 2$$
$$3 = x$$

The x-intercept is $(3, 0)$.

Vertical asymptote: $x - 2 = 0 \implies x = 2$

x	2.5	3	4	5
$f(x)$	-0.69	0	0.69	1.10

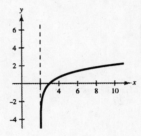

63. $y_1 = \log(x + 1)$

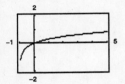

65. $y_1 = \ln(x - 1)$

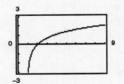

67. $y = \ln x + 2$

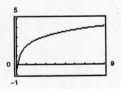

69. $f(t) = 80 - 17 \log_{10}(t + 1), \ 0 \le t \le 12$

(a) $f(0) = 80 - 17 \log_{10} 1 = 80.0$

(b) $f(4) = 80 - 17 \log_{10} 5 \approx 68.1$

(c) $f(10) = 80 - 17 \log_{10} 11 \approx 62.3$

71. $t = \dfrac{\ln 2}{r}$

(a)

r	0.005	0.01	0.015	0.02	0.025	0.03
t	138.6	69.3	46.2	34.7	27.7	23.1

(b) Answers will vary.

73. $y = 80.4 - 11 \ln x$

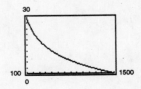

$$y(300) = 80.4 - 11 \ln 300 \approx 17.66 \text{ ft}^3/\text{min}$$

75. $W = 19,440(\ln 9 - \ln 3) \approx 21,357$ ft-lb

77. $t = 12.542 \ln\left(\dfrac{1100.65}{1100.65 - 1000}\right) \approx 30$ years

79. Total amount $= (1100.65)(12)(30) = \$396,234$

Interest $= 396,234 - 150,000 = \$246,234$

81. $f(x) = \dfrac{\ln x}{x}$

(a)

x	1	5	10	10^2	10^4	10^6
$f(x)$	0	0.322	0.230	0.046	0.00092	0.0000138

(b) As $x \rightarrow \infty$, $f(x) \rightarrow 0$.

(c)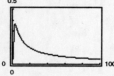

83. True, $\log_3 27 = 3 \implies 3^3 = 27$.

85. $f(x) = 5^x$, $g(x) = \log_5 x$

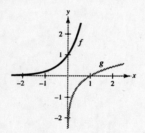

f and g are inverses. Their graphs are reflected about the line $y = x$.

87. $f(x) = 10^x$, $g(x) = \log_{10} x$

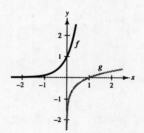

f and g are inverses. Their graphs are reflected about the line $y = x$.

89. (a) False. If y were an exponential function of x, then $y = a^x$, but $a^1 = a$, not 0. Because one point is $(1, 0)$, y is not an exponential function of x.

(c) True. $x = a^y$

For $a = 2$, $x = 2^y$.

$y = 0$, $2^0 = 1$

$y = 1$, $2^1 = 2$

$y = 3$, $2^3 = 8$

(b) True. $y = \log_a x$

For $a = 2$, $y = \log_2 x$.

$x = 1$, $\log_2 1 = 0$

$x = 2$, $\log_2 2 = 1$

$x = 8$, $\log_2 8 = 3$

(d) False. If y were a linear function of x, the slope between $(1, 0)$ and $(2, 1)$ and the slope between $(2, 1)$ and $(8, 3)$ would be the same.

However,

$$m_1 = \frac{1 - 0}{2 - 1} = 1 \text{ and } m_2 = \frac{3 - 1}{8 - 2} = \frac{2}{6} = \frac{1}{3}.$$

Therefore, y is not a linear function of x.

91. $y_4 = (x - 1) - \frac{1}{2}(x - 1)^2 + \frac{1}{3}(x - 1)^3 - \frac{1}{4}(x - 1)^4$

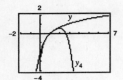

The pattern implies that $\ln x = (x - 1) - \frac{1}{2}(x - 1)^2 + \frac{1}{3}(x - 1)^3 - \frac{1}{4}(x - 1)^4 + \cdots$.

93. $f(x) = |\ln x|$

(a)

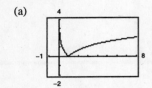

(b) Increasing on $(1, \infty)$

Decreasing on $(0, 1)$

(c) Relative minimum: $(1, 0)$

95. $f(x) = \dfrac{x}{2} - \ln \dfrac{x}{4}$

(a)

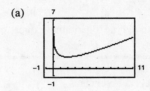

(b) Increasing on $(2, \infty)$

Decreasing on $(0, 2)$

(c) Relative minimum: $\left(2, 1 - \ln \frac{1}{2}\right)$

97. $8n - 3$

99. $83.95 + 37.50t$ Parts and labor

101. $e^6 \approx 403.429$

103. $e^{-4} \approx 0.018$

105. $f(x) = 2^x + 3$

Horizontal asymptote: $y = 3$

x	-2	-1	0	1	2
$f(x)$	3.25	3.5	4	5	7

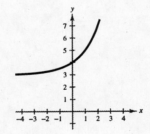

107. $f(x) = \left(\frac{3}{2}\right)^{x+1}$

Horizontal asymptote: $y = 0$

x	-2	-1	0	1	2
$f(x)$	0.667	1	1.5	2.25	3.375

Section 5.3   Properties of Logarithms

■ You should know the following properties of logarithms.

(a) $\log_a x = \dfrac{\log_b x}{\log_b a}$ $\qquad \log_a x = \dfrac{\log_{10} x}{\log_{10} a}$ $\qquad \log_a x = \dfrac{\ln x}{\ln a}$

(b) $\log_a(uv) = \log_a u + \log_a v$ $\qquad \ln(uv) = \ln u + \ln v$

(c) $\log_a(u/v) = \log_a u - \log_a v$ $\qquad \ln(u/v) = \ln u - \ln v$

(d) $\log_a u^n = n \log_a u$ $\qquad\qquad\quad \ln u^n = n \ln u$

■ You should be able to rewrite logarithmic expressions using these properties.

Solutions to Odd-Numbered Exercises

1. $\log_3 7 = \dfrac{\log_{10} 7}{\log_{10} 3} = \dfrac{\ln 7}{\ln 3} \approx 1.771$

3. $\log_{1/2} 4 = \dfrac{\log_{10} 4}{\log_{10}(1/2)} = \dfrac{\ln 4}{\ln(1/2)} = -2.000$

5. $\log_9(0.4) = \dfrac{\log_{10} 0.4}{\log_{10} 9} = \dfrac{\ln 0.4}{\ln 9} \approx -0.417$

7. $\log_{15} 1250 = \dfrac{\log_{10} 1250}{\log_{10} 15} = \dfrac{\ln 1250}{\ln 15} \approx 2.633$

9. (a) $\log_5 x = \dfrac{\log_{10} x}{\log_{10} 5}$

(b) $\log_5 x = \dfrac{\ln x}{\ln 5}$

11. (a) $\log_{\frac{1}{5}} x = \dfrac{\log_{10} x}{\log_{10}\left(\frac{1}{5}\right)}$

(b) $\log_{\frac{1}{5}} x = \dfrac{\ln x}{\ln\left(\frac{1}{5}\right)}$

13. (a) $\log_x \dfrac{3}{10} = \dfrac{\log_{10}\left(\frac{3}{10}\right)}{\log_{10} x}$

(b) $\log_x \dfrac{3}{10} = \dfrac{\ln\left(\frac{3}{10}\right)}{\ln x}$

15. (a) $\log_{2.6} x = \dfrac{\log_{10} x}{\log_{10} 2.6}$

(b) $\log_{2.6} x = \dfrac{\ln x}{\ln 2.6}$

17. $f(x) = \log_2 x = \dfrac{\log_{10} x}{\log_{10} 2} = \dfrac{\ln x}{\ln 2}$

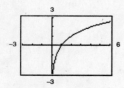

19. $f(x) = \log_{\frac{1}{2}} x = \dfrac{\log_{10} x}{\log_{10} \frac{1}{2}} = \dfrac{\ln x}{\ln\left(\frac{1}{2}\right)}$

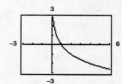

21. $f(x) = \log_{11.8} x = \dfrac{\log_{10} x}{\log_{10} 11.8} = \dfrac{\ln x}{\ln 11.8}$

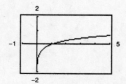

23. $\log_{10} 5x = \log_{10} 5 + \log_{10} x$

25. $\log_{10} \dfrac{5}{x} = \log_{10} 5 - \log_{10} x$

27. $\log_8 x^4 = 4 \log_8 x$

29. $\ln \sqrt{z} = \ln z^{1/2} = \frac{1}{2} \ln z$

31. $\ln xyz = \ln x + \ln y + \ln z$

33. $\ln \sqrt{a-1} = \frac{1}{2} \ln(a-1), \; a > 1$

35. $\ln z(z-1)^2 = \ln z + \ln(z-1)^2$
$$= \ln z + 2 \ln(z-1), \; z > 1$$

37. $\ln \sqrt[3]{\dfrac{x}{y}} = \dfrac{1}{3} \ln \dfrac{x}{y}$
$$= \frac{1}{3} [\ln x - \ln y]$$
$$= \frac{1}{3} \ln x - \frac{1}{3} \ln y$$

39. $\ln \left(\dfrac{x^4 \sqrt{y}}{z^5} \right) = \ln x^4 \sqrt{y} - \ln z^5$
$$= \ln x^4 + \ln \sqrt{y} - \ln z^5$$
$$= 4 \ln x + \frac{1}{2} \ln y - 5 \ln z$$

41. $\log_b \left(\dfrac{x^2}{y^2 z^3} \right) = \log_b x^2 - \log_b y^2 z^3$
$$= \log_b x^2 - [\log_b y^2 + \log_b z^3]$$
$$= 2 \log_b x - 2 \log_b y - 3 \log_b z$$

43. $\ln x + \ln 3 = \ln 3x$

45. $\log_4 z - \log_4 y = \log_4 \dfrac{z}{y}$

47. $2 \log_2(x+4) = \log_2(x+4)^2$

49. $\frac{1}{4} \log_3 5x = \log_3 (5x)^{1/4} = \log_3 \sqrt[4]{5x}$

51. $\ln x - 3 \ln(x+1) = \ln x - \ln(x+1)^3$
$$= \ln \frac{x}{(x+1)^3}$$

53. $\ln(x-2) - \ln(x+2) = \ln \left(\dfrac{x-2}{x+2} \right)$

55. $\ln x - 4[\ln(x+2) + \ln(x-2)] = \ln x - 4 \ln(x+2)(x-2)$
$$= \ln x - 4 \ln(x^2 - 4)$$
$$= \ln x - \ln(x^2 - 4)^4$$
$$= \ln \frac{x}{(x^2 - 4)^4}$$

57. $\frac{1}{3}[2\ln(x+3) + \ln x - \ln(x^2 - 1)] = \frac{1}{3}[\ln(x+3)^2 + \ln x - \ln(x^2 - 1)]$

$$= \frac{1}{3}[\ln x(x+3)^2 - \ln(x^2 - 1)]$$

$$= \frac{1}{3}\ln\frac{x(x+3)^2}{x^2 - 1}$$

$$= \ln\sqrt[3]{\frac{x(x+3)^2}{x^2 - 1}}$$

59. $\frac{1}{3}[\ln y + 2\ln(y+4)] - \ln(y-1) = \frac{1}{3}[\ln y + \ln(y+4)^2] - \ln(y-1)$

$$= \frac{1}{3}\ln y(y+4)^2 - \ln(y-1)$$

$$= \ln\sqrt[3]{y(y+4)^2} - \ln(y-1)$$

$$= \ln\frac{\sqrt[3]{y(y+4)^2}}{y-1}$$

61. $2\ln 3 - \frac{1}{2}\ln(x^2 + 1) = \ln 3^2 - \ln\sqrt{x^2 + 1}$

$$= \ln\frac{9}{\sqrt{x^2 + 1}}$$

63. $\log_2\frac{32}{4} = \log_2 32 - \log_2 4 \neq \frac{\log_2 32}{\log_2 4}$

The first two expressions are equal by Property 2.

65. $\log_3 9 = 2\log_3 3 = 2$

67. $\log_4 16^{1.2} = 1.2(\log_4 16) = 1.2\log_4 4^2 = 1.2(2) = 2.4$

69. $\log_3(-9)$ is undefined. -9 is not in the domain of $\log_3 x$.

71. $\log_5 75 - \log_5 3 = \log_5 \frac{75}{3} = \log_5 25 = \log_5 5^2 = 2\log_5 5 = 2$

73. $\ln e^2 - \ln e^5 = 2 - 5 = -3$

75. $\log_{10} 0$ is undefined. 0 is not in the domain of $\log_{10} x$.

77. $\ln e^{4.5} = 4.5$

79. $\log_4 8 = \frac{\log_2 8}{\log_2 4} = \frac{\log_2 2^3}{\log_2 2^2} = \frac{3}{2}$

81. $\log_5 \frac{1}{250} = \log_5\left(\frac{1}{125}\cdot\frac{1}{2}\right) = \log_5\frac{1}{125} + \log_5\frac{1}{2}$

$$= \log_5 5^{-3} + \log_5 2^{-1}$$

$$= -3 - \log_5 2$$

83. $\ln(5e^6) = \ln 5 + \ln e^6 = \ln 5 + 6 = 6 + \ln 5$

85. $f(t) = 90 - 15\log_{10}(t+1), \quad 0 \le t \le 12$

(a) $f(0) = 90$

(b) $f(6) \approx 77$

(c) $f(12) \approx 73$

(d) $\quad 75 = 90 - 15\log_{10}(t+1)$

$\quad -15 = -15\log_{10}(t+1)$

$\quad 1 = \log_{10}(t+1)$

$\quad 10^1 = t + 1$

$\quad t = 9$ months

(e) $f(t) = 90 - \log_{10}(t+1)^{15}$

(f)

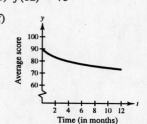

87. $f(x) = \ln x$

False, $f(0) \neq 0$ since 0 is not in the domain of $f(x)$. $f(1) = \ln 1 = 0$

89. False. $f(x) - f(2) = \ln x - \ln 2 = \ln \dfrac{x}{2} \neq \ln(x - 2)$

91. False. $f(u) = 2f(v) \implies \ln u = 2 \ln v \implies \ln u = \ln v^2 \implies u = v^2$

93. Let $x = \log_b u$ and $y = \log_b v$, then $b^x = u$ and $b^y = v$.

$$\frac{u}{v} = \frac{b^x}{b^y} = b^{x-y}$$

Then $\log_b\left(\dfrac{u}{v}\right) = \log_b(b^{x-y}) = x - y = \log_b u - \log_b v$

95. $f(x) = \log_{10} x$

$g(x) = \dfrac{\ln x}{\ln 10}$

$f(x) = g(x)$

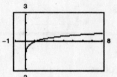

97. $f(x) = \ln \dfrac{x}{2}$, $g(x) = \dfrac{\ln x}{\ln 2}$, $h(x) = \ln x - \ln 2$

$f(x) = h(x)$ by Property 2.

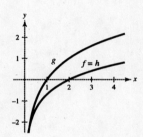

99. $\dfrac{24xy^{-2}}{16x^{-3}y} = \dfrac{24xx^3}{16yy^2} = \dfrac{3x^4}{2y^3}, x \neq 0$

101. $(18x^3y^4)^{-3}(18x^3y^4)^3 = \dfrac{(18x^3y^4)^3}{(18x^3y^4)^3} = 1$ if $x \neq 0, y \neq 0$.

103. $(2.8)^{7.6} \approx 2502.655$

105. $7^{-\pi} \approx 0.002$

107. $\sqrt[4]{350} \approx 4.325$

109. $\log_{10} 26 \approx 1.415$

111. $\ln 10.6 \approx 2.361$

Section 5.4   Exponential and Logarithmic Equations

■ To solve an exponential equation, isolate the exponential expression, then take the logarithm of both sides. Then solve for the variable.

     1. $\log_a a^x = x$

     2. $\ln e^x = x$

■ To solve a logarithmic equation, rewrite it in exponential form. Then solve for the variable.

     1. $a^{\log_a x} = x$

     2. $e^{\ln x} = x$

■ If $a > 0$ and $a \neq 1$ we have the following:

     1. $\log_a x = \log_a y \iff x = y$

     2. $a^x = a^y \iff x = y$

■ Check for extraneous solutions.

Solutions to Odd-Numbered Exercises

1. $4^{2x-7} = 64$

   (a) $x = 5$

     $4^{2(5)-7} = 4^3 = 64$

     Yes, $x = 5$ is a solution.

   (b) $x = 2$

     $4^{2(2)-7} = 4^{-3} = \frac{1}{64} \neq 64$

     No, $x = 2$ is not a solution.

3. $3e^{x+2} = 75$

   (a) $x = -2 + e^{25}$

     $3e^{(-2+e^{25})+2} = 3e^{e^{25}} \neq 75$

     No, $x = -2 + e^{25}$ is not a solution.

   (b) $x = -2 + \ln 25$

     $3e^{(-2+\ln 25)+2} = 3e^{\ln 25} = 3(25) = 75$

     Yes, $x = -2 + \ln 25$ is a solution.

   (c) $x \approx 1.2189$

     $3e^{1.2189+2} = 3e^{3.2189} \approx 75$

     Yes, $x \approx 1.2189$ is a solution.

5. $\log_4(3x) = 3 \implies 3x = 4^3 \implies 3x = 64$

   (a) $x \approx 20.3560$

     $3(20.3560) = 61.0680 \neq 64$

     No, $x \approx 20.3560$ is not a solution.

   (b) $x = -4$

     $3(-4) = -12 \neq 64$

     No, $x = -4$ is not a solution.

   (c) $x = \frac{64}{3}$

     $3\left(\frac{64}{3}\right) = 64$

     Yes, $x = \frac{64}{3}$ is a solution.

7. $4^x = 16$

   $4^x = 4^2$

   $x = 2$

9. $5^x = 625$

   $5^x = 5^4$

   $x = 4$

11. $7^x = \frac{1}{49}$

   $7^x = 7^{-2}$

   $x = -2$

13. $\left(\frac{1}{2}\right)^x = 32$

$2^{-x} = 2^5$

$-x = 5$

$x = -5$

15. $\left(\frac{3}{4}\right)^x = \frac{27}{64}$

$\left(\frac{3}{4}\right)^x = \left(\frac{3}{4}\right)^3$

$x = 3$

17. $3^{x-1} = 27$

$3^{x-1} = 3^3$

$x - 1 = 3$

$x = 4$

19. $\ln x - \ln 2 = 0$

$\ln x = \ln 2$

$x = 2$

21. $e^x = 2$

$\ln e^x = \ln 2$

$x = \ln 2$

$x \approx 0.693$

23. $\ln x = -1$

$e^{\ln x} = e^{-1}$

$x = e^{-1}$

$x \approx 0.368$

25. $\log_4 x = 3$

$4^{\log_4 x} = 4^3$

$x = 4^3$

$x = 64$

27. $\log_{10} x - 2 = 0$

$\log_{10} x = 2$

$10^{\log_{10} x} = 10^2$

$x = 10^2$

$x = 100$

29. $\log_{10} x = -1$

$10^{\log_{10} x} = 10^{-1}$

$x = 10^{-1}$

$x = \frac{1}{10}$

31. $f(x) = g(x)$

$2^x = 8$

$2^x = 2^3$

$x = 3$

Point of intersection: $(3, 8)$

33. $f(x) = g(x)$

$\log_3 x = 2$

$x = 3^2$

$x = 9$

Point of intersection: $(9, 2)$

35. $\log_{10} 10^{x^2} = x^2$

37. $8^{\log_8(x-2)} = x - 2$

39. $\ln e^{7x+2} = 7x + 2$

41. $e^{\ln(5x+2)} = 5x + 2$

43. $-1 + \ln e^{2x} = -1 + 2x = 2x - 1$

45. $e^x = 10$

$x = \ln 10 \approx 2.303$

47. $7 - 2e^x = 5$

$-2e^x = -2$

$e^x = 1$

$x = \ln 1 = 0$

49. $e^{3x} = 12$

$3x = \ln 12$

$x = \frac{\ln 12}{3} \approx 0.828$

51. $500e^{-x} = 300$

$e^{-x} = \frac{3}{5}$

$-x = \ln \frac{3}{5}$

$x = -\ln \frac{3}{5} = \ln \frac{5}{3} \approx 0.511$

53. $e^{2x} - 4e^x - 5 = 0$

$(e^x + 1)(e^x - 5) = 0$

$e^x = -1$ or $e^x = 5$
(No solution) $x = \ln 5 \approx 1.609$

55. $20(100 - e^{x/2}) = 500$

$100 - e^{x/2} = 25$

$-e^{x/2} = -75$

$e^{x/2} = 75$

$\dfrac{x}{2} = \ln 75$

$x = 2 \ln 75 \approx 8.635$

57. $10^x = 42$

$x = \log_{10} 42 \approx 1.623$

59. $3^{2x} = 80$

$\ln 3^{2x} = \ln 80$

$2x \ln 3 = \ln 80$

$x = \dfrac{\ln 80}{2 \ln 3} \approx 1.994$

61. $5^{-t/2} = 0.20$

$5^{-t/2} = \dfrac{1}{5}$

$5^{-t/2} = 5^{-1}$

$-\dfrac{t}{2} = -1$

$t = 2$

63. $2^{3-x} = 565$

$\ln 2^{3-x} = \ln 565$

$(3 - x) \ln 2 = \ln 565$

$3 \ln 2 - x \ln 2 = \ln 565$

$-x \ln 2 = \ln 565 - \ln 2^3$

$x \ln 2 = \ln 8 - \ln 565$

$x = \dfrac{\ln 8 - \ln 565}{\ln 2} \approx -6.142$

65. $g(x) = 6e^{1-x} - 25$

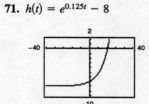

The zero is $x \approx -0.427$.

67. $f(x) = 3e^{3x/2} - 962$

The zero is $x \approx 3.847$.

69. $g(t) = e^{0.09t} - 3$

The zero is $x \approx 12.207$.

71. $h(t) = e^{0.125t} - 8$

The zero is $x \approx 16.636$.

73.
$$8(10^{3x}) = 12$$
$$10^{3x} = \frac{12}{8}$$
$$\log_{10} 10^{3x} = \log_{10}\left(\frac{3}{2}\right)$$
$$3x = \log_{10}\left(\frac{3}{2}\right)$$
$$x = \frac{1}{3}\log_{10}\left(\frac{3}{2}\right) \approx 0.059$$

75.
$$3(5^{x-1}) = 21$$
$$5^{x-1} = 7$$
$$\ln 5^{x-1} = \ln 7$$
$$(x - 1)\ln 5 = \ln 7$$
$$x - 1 = \frac{\ln 7}{\ln 5}$$
$$x = 1 + \frac{\ln 7}{\ln 5} \approx 2.209$$

77.
$$\left(1 + \frac{0.065}{365}\right)^{365t} = 4$$
$$\ln\left(1 + \frac{0.065}{365}\right)^{365t} = \ln 4$$
$$365t \ln\left(1 + \frac{0.065}{365}\right) = \ln 4$$
$$t = \frac{\ln 4}{365 \ln\left(1 + \frac{0.065}{365}\right)} \approx 21.330$$

79.
$$\left(1 + \frac{0.10}{12}\right)^{12t} = 2$$
$$\ln\left(1 + \frac{0.10}{12}\right)^{12t} = \ln 2$$
$$12t \ln\left(1 + \frac{0.10}{12}\right) = \ln 2$$
$$t = \frac{\ln 2}{12 \ln\left(1 + \frac{0.10}{12}\right)} \approx 6.960$$

81.
$$\frac{3000}{2 + e^{2x}} = 2$$
$$3000 = 2(2 + e^{2x})$$
$$1500 = 2 + e^{2x}$$
$$1498 = e^{2x}$$
$$\ln 1498 = 2x$$
$$x = \frac{\ln 1498}{2} \approx 3.656$$

83.
$$\ln x = -3$$
$$x = e^{-3} \approx 0.050$$

85.
$$\ln 2x = 2.4$$
$$2x = e^{2.4}$$
$$x = \frac{e^{2.4}}{2} \approx 5.512$$

87.
$$3 \ln 5x = 10$$
$$\ln 5x = \frac{10}{3}$$
$$5x = e^{10/3}$$
$$x = \frac{e^{10/3}}{5} \approx 5.606$$

89.
$$\ln \sqrt{x + 2} = 1$$
$$\sqrt{x + 2} = e^1$$
$$x + 2 = e^2$$
$$x = e^2 - 2 \approx 5.389$$

91. $\ln(x + 1)^2 = 2$

$2 \ln|x + 1| = 2$

$\ln|x + 1| = 1$

$|x + 1| = e^1$

$x + 1 = \pm e$

$x = \pm e - 1$

$x \approx -3.718$ or $x \approx 1.718$

93. $\ln x + \ln(x - 2) = 1$

$\ln[x(x - 2)] = 1$

$x(x - 2) = e^1$

$x^2 - 2x - e = 0$

$x = \dfrac{2 \pm \sqrt{4 + 4e}}{2}$

$= \dfrac{2 \pm 2\sqrt{1 + e}}{2}$

$= 1 \pm \sqrt{1 + e}$

The negative value is extraneous. The only solution is

$x = 1 + \sqrt{1 + e} \approx 2.928$.

95. $\ln(x + 5) = \ln(x - 1) - \ln(x + 1)$

$\ln(x + 5) = \ln\left(\dfrac{x - 1}{x + 1}\right)$

$x + 5 = \dfrac{x - 1}{x + 1}$

$(x + 5)(x + 1) = x - 1$

$x^2 + 6x + 5 = x - 1$

$x^2 + 5x + 6 = 0$

$(x + 2)(x + 3) = 0$

$x = -2$ or $x = -3$

Both of these solutions are extraneous, so the equation has no solution.

97. $\log_{10}(z - 3) = 2$

$10^{\log_{10}(z - 3)} = 10^2$

$z - 3 = 10^2$

$z = 10^2 + 3 = 103$

99. $6 \log_3(0.5x) = 11$

$\log_3(0.5x) = \dfrac{11}{6}$

$3^{\log_3(0.5x)} = 3^{11/6}$

$0.5x = 3^{11/6}$

$x = 2(3^{11/6}) \approx 14.988$

101. $\log_{10}(x + 4) - \log_{10} x = \log_{10}(x + 2)$

$\log_{10}\left(\dfrac{x + 4}{x}\right) = \log_{10}(x + 2)$

$\dfrac{x + 4}{x} = x + 2$

$x + 4 = x^2 + 2x$

$0 = x^2 + x - 4$

$x = \dfrac{-1 \pm \sqrt{17}}{2}$ Quadratic Formula

Choosing the positive value of x (the negative value is extraneous), we have

$x = \dfrac{-1 + \sqrt{17}}{2} \approx 1.562$.

103. $\log_4 x - \log_4(x - 1) = \dfrac{1}{2}$

$\log_4\left(\dfrac{x}{x - 1}\right) = \dfrac{1}{2}$

$4^{\log_4\left(\frac{x}{x-1}\right)} = 4^{1/2}$

$\dfrac{x}{x - 1} = 4^{1/2}$

$x = 2(x - 1)$

$x = 2x - 2$

$-x = -2$

$x = 2$

105. $\log_{10} 8x - \log_{10}\left(1 + \sqrt{x}\right) = 2$

$$\log_{10} \frac{8x}{1 + \sqrt{x}} = 2$$

$$\frac{8x}{1 + \sqrt{x}} = 10^2$$

$$8x = 100\left(1 + \sqrt{x}\right)$$

$$2x = 25\left(1 + \sqrt{x}\right)$$

$$2x = 25 + 25\sqrt{x}$$

$$2x - 25 = 25\sqrt{x}$$

$$(2x - 25)^2 = \left(25\sqrt{x}\right)^2$$

$$4x^2 - 100x + 625 = 625x$$

$$4x^2 - 725x + 625 = 0$$

$$x = \frac{725 \pm \sqrt{725^2 - 4(4)(625)}}{2(4)}$$

$$x = \frac{725 \pm \sqrt{515625}}{8}$$

$$x = \frac{25\left(29 \pm 5\sqrt{33}\right)}{8}$$

$$x \approx 0.866 \text{ (extraneous)} \quad \text{or} \quad x \approx 180.384$$

The only solution is $x = \dfrac{25\left(29 + 5\sqrt{33}\right)}{8} \approx 180.384$

107. $y_1 = 7$

$y_2 = 2^x$

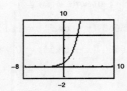

From the graph we have $x \approx 2.807$ when $y = 7$.

The point of intersection is approximately $(2.807, 7)$.

109. $y_1 = 3$

$y_2 = \ln x$

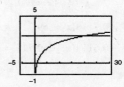

From the graph we have $x \approx 20.806$ when $y = 3$.

The point of intersection is approximately $(20.086, 3)$.

111. $A = Pe^{rt}$

$$2000 = 1000e^{0.085t}$$

$$2 = e^{0.085t}$$

$$\ln 2 = 0.085t$$

$$\frac{\ln 2}{0.085} = t$$

$$t \approx 8.2 \text{ years}$$

113. $A = Pe^{rt}$

$$3000 = 1000e^{0.085t}$$

$$3 = e^{0.085t}$$

$$\ln 3 = 0.085t$$

$$\frac{\ln 3}{0.085} = t$$

$$t \approx 12.9 \text{ years}$$

115. $p = 500 - 0.5(e^{0.004x})$

(a)
$$p = 350$$
$$350 = 500 - 0.5(e^{0.004x})$$
$$300 = e^{0.004x}$$
$$0.004x = \ln 300$$
$$x \approx 1426 \text{ units}$$

(b)
$$p = 300$$
$$300 = 500 - 0.5(e^{0.004x})$$
$$400 = e^{0.004x}$$
$$0.004x = \ln 400$$
$$x \approx 1498 \text{ units}$$

117. $V = 6.7e^{-48.1/t}$, $t \geq 0$

(a)
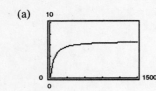

(b) As $t \to \infty$, $V \to 6.7$.

Horizontal asymptote: $V = 6.7$

The yield will approach 6.7 million cubic feet per acre.

(c)
$$1.3 = 6.7e^{-48.1/t}$$
$$\frac{1.3}{6.7} = e^{-48.1/t}$$
$$\ln\left(\frac{13}{67}\right) = \frac{-48.1}{t}$$
$$t = \frac{-48.1}{\ln(13/67)} \approx 29.3 \text{ years}$$

119. (a) From the graph shown in the textbook, we see horizontal asymptotes at $y = 0$ and $y = 100$. These represent the lower and upper percent bounds; the range falls between 0% and 100%.

(b) Males

$$50 = \frac{100}{1 + e^{-0.6114(x - 69.71)}}$$
$$1 + e^{-0.6114(x - 69.71)} = 2$$
$$e^{-0.6114(x - 69.71)} = 1$$
$$-0.6114(x - 69.71) = \ln 1$$
$$-0.6114(x - 69.71) = 0$$
$$x = 69.71 \text{ inches}$$

Females

$$50 = \frac{100}{1 + e^{-0.66607(x - 64.51)}}$$
$$1 + e^{-0.66607(x - 64.51)} = 2$$
$$e^{-0.66607(x - 64.51)} = 1$$
$$-0.66607(x - 64.51) = \ln 1$$
$$-0.66607(x - 64.51) = 0$$
$$x = 64.51 \text{ inches}$$

121. $T = 20[1 + 7(2^{-h})]$

(a) From the graph in the textbook we see a horizontal asymptote at $T = 20$. This represents the room temperature.

(b)
$$100 = 20[1 + 7(2^{-h})]$$
$$5 = 1 + 7(2^{-h})$$
$$4 = 7(2^{-h})$$
$$\frac{4}{7} = 2^{-h}$$
$$\ln\left(\frac{4}{7}\right) = \ln 2^{-h}$$
$$\ln\left(\frac{4}{7}\right) = -h \ln 2$$
$$\frac{\ln\left(\frac{4}{7}\right)}{-\ln 2} = h$$
$$h \approx 0.81 \text{ hour}$$

123. $\log_a(uv) = \log_a u + \log_a v$

True by Property 1 in Section 5.3.

125. $\log_a(u - v) = \log_a u - \log_a v$

False.

$1.95 \approx \log_{10}(100 - 10) \neq \log_{10} 100 - \log_{10} 10 = 1$

127. $A = Pe^{rt}$

(a) $A = (2P)e^{rt} = 2(Pe^{rt})$ This doubles your money.

(b) $A = Pe^{(2r)t} = Pe^{rt}e^{rt} = e^{rt}(Pe^{rt})$

(c) $A = Pe^{r(2t)} = Pe^{rt}e^{rt} = e^{rt}(Pe^{rt})$

Doubling the interest rate yields the same result as doubling the number of years.

If $2 > e^{rt}$ (i.e., $rt < \ln 2$), then doubling your investment would yield the most money. If $rt > \ln 2$, then doubling either the interest rate or the number of years would yield more money.

129. No. Doubling time does not depend on the amount of the investment, but depends on the interest rate, r.

$$2P = Pe^{rt}$$
$$2 = e^{rt}$$
$$\ln 2 = rt$$
$$\frac{\ln 2}{r} = t$$

131. $\sqrt{32} - 2\sqrt{25} = \sqrt{16 \cdot 2} - 2(5)$
$$= 4\sqrt{2} - 10$$

133. $\dfrac{3}{\sqrt{10} - 2} = \dfrac{3}{\sqrt{10} - 2} \cdot \dfrac{\sqrt{10} + 2}{\sqrt{10} + 2}$
$$= \frac{3(\sqrt{10} + 2)}{10 - 4}$$
$$= \frac{3(\sqrt{10} + 2)}{6}$$
$$= \frac{\sqrt{10} + 2}{2}$$
$$= \frac{1}{2}\sqrt{10} + 1$$

135. $t = \dfrac{k}{s^3}$

137. $x = \dfrac{k}{b - 3}$

139. $\log_6 9 = \dfrac{\log_{10} 9}{\log_{10} 6} = \dfrac{\ln 9}{\ln 6} \approx 1.226$

141. $\log_{3/4} 5 = \dfrac{\log_{10} 5}{\log_{10}\left(\dfrac{3}{4}\right)} = \dfrac{\ln 5}{\ln\left(\dfrac{3}{4}\right)} \approx -5.595$

Section 5.5 Exponential and Logarithmic Models

■ You should be able to solve growth and decay problems.

(a) Exponential growth if $b0$ and $y = ae^{bx}$.

(b) Exponential decay if $b > 0$ and $y = ae^{-bx}$.

■ You should be able to use the Gaussian model
$$y = ae^{-(x-b)^2/c}.$$

■ You should be able to use the logistics growth model

$$y = \frac{a}{1 + be^{-rt}}.$$

■ You should be able to use the logarithmic models
$$y = a + b \ln x, \quad y = a + b \log_{10} x.$$

Solutions to Odd-Numbered Exercises

1. $y = 2e^{x/4}$

This is an exponential growth model. Matches graph (c)

3. $y = 6 + \log_{10}(x + 2)$

This is a logarithmic function shifted up 6 units and left 2 units. Matches graph (b)

5. $y = \ln(x + 1)$

This is a logarithmic model. Matches graph (d)

7. Since $A = 1000e^{0.12t}$, the time to double is given by $2000 = 1000e^{0.12t}$ and we have

$2000 = 1000e^{0.12t}$

$2 = e^{0.12t}$

$\ln 2 = \ln e^{0.12t}$

$\ln 2 = 0.12t$

$t = \dfrac{\ln 2}{0.12} \approx 5.78$ years.

Amount after 10 years: $A = 1000e^{1.2} \approx \$3320.12$

9. Since $A = 750e^{rt}$ and $A = 1500$ when $t = 7.75$, we have the following.

$1500 = 750e^{7.75r}$

$2 = e^{7.75r}$

$\ln 2 = \ln e^{7.75r}$

$\ln 2 = 7.75r$

$r = \dfrac{\ln 2}{7.75} \approx 0.089438 = 8.9438\%$

Amount after 10 years: $A = 750e^{0.089438(10)} \approx \$1834.37$

11. Since $A = 500e^{rt}$ and $A = \$1505.00$ when $t = 10$, we have the following.

$$1505.00 = 500e^{10r}$$

$$r = \frac{\ln(1505.00/500)}{10} \approx 0.110 = 11.0\%$$

The time to double is given by

$$1000 = 500e^{0.110t}$$

$$t = \frac{\ln 2}{0.110} \approx 6.3 \text{ years.}$$

13. Since $A = Pe^{0.045t}$ and $A = 10,000.00$ when $t = 10$, we have the following.

$$10,000.00 = Pe^{0.045(10)}$$

$$\frac{10,000.00}{e^{0.045(10)}} = P \approx \$6376.28$$

The time to double is given by

$$t = \frac{\ln 2}{0.045} \approx 15.40 \text{ years.}$$

15. $500,000 = P\left(1 + \dfrac{0.075}{12}\right)^{12(20)}$

$$P = \frac{500,000}{\left(1 + \dfrac{0.075}{12}\right)^{12(20)}} = \frac{500,000}{1.00625^{240}} \approx \$112,087.09$$

17. $P = 1000, r = 11\%$

(a) $\qquad n = 1$

$$(1 + 0.11)^t = 2$$

$$t \ln 1.11 = \ln 2$$

$$t = \frac{\ln 2}{\ln 1.11} \approx 6.642 \text{ years}$$

(b) $\qquad n = 12$

$$\left(1 + \frac{0.11}{12}\right)^{12t} = 2$$

$$12t \ln\left(1 + \frac{0.11}{12}\right) = \ln 2$$

$$t = \frac{\ln 2}{12 \ln\left(1 + \frac{0.11}{12}\right)} \approx 6.330 \text{ years}$$

(c) $\qquad n = 365$

$$\left(1 + \frac{0.11}{365}\right)^{365t} = 2$$

$$365t \ln\left(1 + \frac{0.11}{365}\right) = \ln 2$$

$$t = \frac{\ln 2}{365 \ln\left(1 + \frac{0.11}{365}\right)} \approx 6.302 \text{ years}$$

(d) Continuously

$$e^{0.11t} = 2$$

$$0.11t = \ln 2$$

$$t = \frac{\ln 2}{0.11} \approx 6.301 \text{ years}$$

19. $\qquad 3P = Pe^{rt}$

$$3 = e^{rt}$$

$$\ln 3 = rt$$

$$\frac{\ln 3}{r} = t$$

r	2%	4%	6%	8%	10%	12%
$t = \dfrac{\ln 3}{r}$ (years)	54.93	27.47	18.31	13.73	10.99	9.16

21. $\qquad 3P = P(1 + r)^t$

$$3 = (1 + r)^t$$

$$\ln 3 = \ln(1 + r)^t$$

$$\ln 3 = t \ln(1 + r)$$

$$\frac{\ln 3}{\ln(1 + r)} = t$$

r	2%	4%	6%	8%	10%	12%
$t = \dfrac{\ln 3}{\ln(1 + r)}$ (years)	55.48	18.01	18.85	14.27	11.53	9.69

23. Continuous compounding results in faster growth.

$A = 1 + 0.075[\![t]\!]$ and $A = e^{0.07t}$

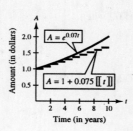

25.
$$\frac{1}{2}C = Ce^{k(1620)}$$
$$0.5 = e^{k(1620)}$$
$$\ln 0.5 = \ln e^{k(1620)}$$
$$\ln 0.5 = k(1620)$$
$$k = \frac{\ln 0.5}{1620}$$

Given $C = 10$ grams after 1000, years we have
$$y = 10e^{[(\ln 0.5)/1620](1000)}$$
$$\approx 6.52 \text{ grams.}$$

27.
$$\frac{1}{2}C = Ce^{k(5730)}$$
$$0.5 = e^{k(5730)}$$
$$\ln 0.5 = \ln e^{k(5730)}$$
$$\ln 0.5 = k(5730)$$
$$k = \frac{\ln 0.5}{5730}$$

Given $y = 2$ grams after 1000 years, we have
$$2 = Ce^{[(\ln 0.5)/5730](1000)}$$
$$C \approx 2.26 \text{ grams.}$$

29.
$$\frac{1}{2}C = Ce^{k(24,360)}$$
$$0.5 = e^{k(24,360)}$$
$$\ln 0.5 = \ln e^{k(24,360)}$$
$$\ln 0.5 = k(24,360)$$
$$k = \frac{\ln 0.5}{24,360}$$

Given $y = 2.1$ grams after 1000 years, we have
$$2.1 = Ce^{[(\ln 0.5)/24,360](1000)}$$
$$C \approx 2.16 \text{ grams.}$$

31.
$$y = ae^{bx}$$
$$1 = ae^{b(0)} \implies 1 = a$$
$$10 = e^{b(3)}$$
$$\ln 10 = 3b$$
$$\frac{\ln 10}{3} = b \implies b \approx 0.7675$$

Thus, $y = e^{0.7675x}$.

33.
$$y = ae^{bx}$$
$$5 = ae^{b(0)} \implies 5 = a$$
$$1 = 5e^{b(4)}$$
$$\frac{1}{5} = e^{4b}$$
$$\ln\left(\frac{1}{5}\right) = 4b$$
$$\frac{\ln\left(\frac{1}{5}\right)}{4} = b \implies b \approx -0.4024$$

Thus, $y = 5e^{-0.4024x}$.

35.
$$P = 105,300e^{0.015t}$$
$$150,000 = 105,300e^{0.015t}$$
$$\ln \tfrac{1500}{1053} = 0.015t$$
$$t \approx 23.59$$

The population will reach 150,000 during 2023.
[Note: 2000 + 23.59]

37. $P = 2500e^{kt}$

For 1945, use $t = -55$

$$1350 = 2500e^{k(-55)}$$
$$\ln\left(\frac{1350}{2500}\right) = -55k \implies k \approx 0.0112$$

For 2010, use $t = 10$

$$P = 2500e^{0.0112(10)} \approx 2796 \text{ people}$$

39.

Country	1997	2020
Croatia	5.0	4.8
Mali	9.9	20.4
Singapore	3.5	4.3
Sweden	8.9	9.5

(a) Croatia: $a = 5.0$

$4.8 = 5.0e^{b(23)}$

$\ln\left(\dfrac{4.8}{5.0}\right) = 23b \implies b \approx -0.0018$

For 2030, use $t = 33$

$y = 5.0e^{-0.0018(33)} \approx 4.7$ million

Mali: $a = 9.9$

$20.4 = 9.9e^{b(23)}$

$\ln\left(\dfrac{20.4}{9.9}\right) = 23b \implies b \approx 0.0314$

For 2030, use $t = 33$

$y = 9.9e^{0.0314(33)} \approx 27.9$ million

Singapore: $a = 3.5$

$4.3 = 3.5e^{b(23)}$

$\ln\left(\dfrac{4.3}{3.5}\right) = 23b \implies b \approx 0.0090$

For 2030, use $t = 33$

$y = 3.5e^{0.0090(33)} \approx 4.7$ million

Sweden: $a = 8.9$

$9.5 = 8.9e^{b(23)}$

$\ln\left(\dfrac{9.5}{8.9}\right) = 23b \implies b \approx 0.0028$

For 2030, use $t = 33$

$y = 8.9e^{0.0028(33)} \approx 9.8$ million

(b) The constant b determines the growth rates.
The greater the rate of growth, the greater
the value of b.

(c) The constant b determines whether the population
is increasing ($b > 0$) or decreasing ($b < 0$).

41. $N = 250e^{kt}$

$280 = 250e^{k(10)}$

$1.12 = e^{10k}$

$k = \dfrac{\ln 1.12}{10}$

$N = 250e^{[(\ln 1.12)/10]t}$

$500 = 250e^{[(\ln 1.12)/10]t}$

$2 = e^{[(\ln 1.12)/10]t}$

$\ln 2 = [(\ln 1.12)/10]t$

$t = \dfrac{\ln 2}{(\ln 1.12)/10} \approx 61.16$ hours

43. $y = Ce^{kt}$

$$\frac{1}{2}C = Ce^{5730k}$$

$$\ln\frac{1}{2} = 5730k$$

$$k = \frac{\ln(1/2)}{5730}$$

The ancient charcoal has only 15% as much radioactive carbon.

$$0.15C = Ce^{[(\ln 0.5)/5730]t}$$

$$\ln 0.15 = \frac{\ln 0.5}{5730}t$$

$$t = \frac{5730\ln 0.15}{\ln 0.5} \approx 15,683 \text{ years}$$

45. $(0, 2000), (2, 500)$

(a) $m = \dfrac{500 - 2000}{2 - 0} = -750$

$$V = -750t + 2000$$

(b) $500 = 2000e^{k(2)}$

$$\ln\frac{1}{4} = 2k \implies k \approx -0.6931$$

$$V = 2000e^{-0.6931t}$$

(c)

The exponential model depreciates faster in the first 2 years.

(d)

t	1	3
$V = -750t + 2000$	\$1250	$-\$250$
$V = 2000e^{-0.6931t}$	\$1000	\$250

(e) The slope of the linear model means that the computer depreciates \$750 per year.

47. $S = \dfrac{500,000}{1 + 0.6e^{kt}}$

(a) $300,000 = \dfrac{500,000}{1 + 0.6e^{2k}}$

$$1 + 0.6e^{2k} = \frac{5}{3}$$

$$0.6e^{2k} = \frac{2}{3}$$

$$e^{2k} = \frac{10}{9}$$

$$2k = \ln\left(\frac{10}{9}\right)$$

$$k = \frac{1}{2}\ln\left(\frac{10}{9}\right) \approx 0.053$$

$$S = \frac{500,000}{1 + 0.6e^{0.053t}}$$

(b) When $t = 5$:

$$S = \frac{500,000}{1 + 0.6e^{[0.5\ln(10/9)](5)}} \approx 280,771 \text{ units}$$

49. $y = ae^{bt}$

$$632,000 = 742,000e^{b(2)}$$

$$\frac{632}{742} = e^{2b}$$

$$b = \frac{1}{2}\ln\left(\frac{632}{742}\right)$$

$$y = 742,000e^{0.5[\ln(632/742)](3)} \approx \$583,275$$

51. $p(t) = \dfrac{1000}{1 + 9e^{-0.1656t}}$

(a)

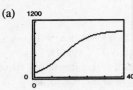

The horizontal asymptotes are $p = 0$ and $p = 1000$. The asymptote with the larger p-value, $p = 1000$, indicates that the population size will approach 1000 as time increases.

(b) $p(5) = \dfrac{1000}{1 + 9e^{-0.1656(5)}} \approx 203$ animals

(c) $\qquad 500 = \dfrac{1000}{1 + 9e^{-0.1656t}}$

$1 + 9e^{-0.1656t} = 2$

$9e^{-0.1656t} = 1$

$e^{-0.1656t} = \dfrac{1}{9}$

$t = -\dfrac{\ln(1/9)}{0.1656} \approx 13$ months

53. $R = \log_{10} \dfrac{I}{I_0} = \log_{10} I$ since $I_0 = 1$.

(a) $8.6 = \log_{10} I$

$10^{8.6} = I \approx 398{,}107{,}171$

(b) $6.7 = \log_{10} I$

$10^{6.7} = I \approx 5{,}011{,}872$

(c) $7.7 = \log_{10} I$

$10^{7.7} = I \approx 50{,}118{,}723$

55. $\beta(I) = 10 \log_{10} \dfrac{I}{I_0}$ where $I_0 = 10^{-12}$ watt/m^2

(a) $\beta(10^{-9}) = 10 \log_{10} \dfrac{10^{-9}}{10^{-12}} = 10 \log_{10} 10^3 = 30$ decibels

(b) $\beta(10^{-3.5}) = 10 \log_{10} \dfrac{10^{-3.5}}{10^{-12}} = 10 \log_{10} 10^{8.5} = 85$ decibels

(c) $\beta(10^{-3}) = 10 \log_{10} \dfrac{10^{-3}}{10^{-12}} = 10 \log_{10} 10^9 = 90$ decibels

(d) $\beta(10^{-0.5}) = 10 \log_{10} \dfrac{10^{-0.5}}{10^{-12}} = 10 \log_{10} 10^{11.5} = 115$ decibels

57. $\beta = 10 \log_{10} \dfrac{I}{I_0}$

$10^{\beta/10} = \dfrac{I}{I_0}$

$I = I_0 10^{\beta/10}$

% decrease $= \dfrac{I_0 10^{8.8} - I_0 10^{7.2}}{I_0 10^{8.8}} \times 100 \approx 97\%$

59. $\text{pH} = -\log_{10}[\text{H}^+] = -\log_{10}[11.3 \times 10^{-6}] \approx 4.95$

61. $3.2 = -\log_{10}[\text{H}^+]$

$10^{-3.2} = [\text{H}^+]$

$[\text{H}^+] \approx 6.3 \times 10^{-4}$ moles per liter

63. $\text{pH} - 1 = -\log_{10}[\text{H}^+]$

$-(\text{pH} - 1) = \log_{10}[\text{H}^+]$

$10^{-(\text{pH}-1)} = [\text{H}^+]$

$10^{-\text{pH}+1} = [\text{H}^+]$

$10^{-\text{pH}} \cdot 10 = [\text{H}^+]$

The hydrogen ion concentration is increased by a factor of 10.

65. $u = 120{,}000 \left[\dfrac{0.075t}{1 - \left(\dfrac{1}{1 + 0.075/12} \right)^{12t}} - 1 \right]$

(a)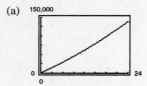

(b) From the graph, $u = \$120{,}000$ when $x \approx 21$ years. It would take approximately 37.6 years to pay \$240,000 in interest. Yes, it is possible to pay twice as much in interest charges as the size of the mortgage. It is especially likely when the interest rates are higher.

67. $t = -2.5 \ln \dfrac{T - 70}{98.6 - 70}$

At 9:00 A.M. we have:

$t = -2.5 \ln \dfrac{85.7 - 70}{98.6 - 70} \approx 1.5$ hours

From this you can conclude that the person died at 7:30 A.M.

69. False. A logistics growth function never has an x-intercept.

71. (a) Logarithmic

(b) Logistic

(c) Exponential (decay)

(d) Linear

(e) None of the above (appears to be a combination of a linear and a quadratic)

(f) Exponential (growth)

73. Answers will vary.

75. $y = -4x - 1$

Line

Slope: $m = -4$

y-intercept: $(0, -1)$

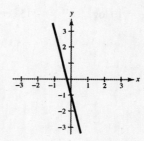

77. $y = 2x^2 - 7x - 30$

$\quad = (2x + 5)(x - 6)$

$\quad = 2\left(x - \frac{7}{4}\right)^2 - \frac{289}{8}$

Parabola

Vertex: $\left(\frac{7}{4}, -\frac{289}{8}\right)$

x-intercepts: $\left(-\frac{5}{2}, 0\right), (6, 0)$

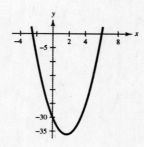

79. $-x^2 - 8y = 0$

$\qquad x^2 = -8y$

Parabola

Vertex: $(0, 0)$

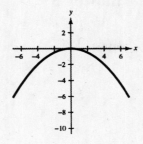

81. $y = \dfrac{x^2}{-x - 2} = -x + 2 + \dfrac{4}{-x - 2}$

Vertical asymptote: $x = -2$

Slant asymptote: $y = -x + 2$

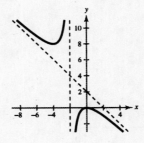

83. $(x - 4)^2 + (y + 7) = 4$

$\qquad (x - 4)^2 = -y - 7 + 4$

$\qquad (x - 4)^2 = -(y + 3)$

Parabola

Vertex: $(4, -3)$

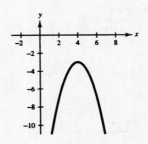

85. $f(x) = -2^{-x-1} - 1$

Horizontal asymptote: $y = -1$

x	-2	-1	0	1	2
$f(x)$	-3	-2	$-\frac{3}{2}$	$-\frac{5}{4}$	$-\frac{9}{8}$

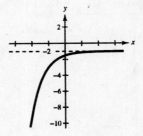

87. $f(x) = -3^x + 4$

Horizontal asymptote: $y = 4$

x	-2	-1	0	1	2
$f(x)$	$3\frac{8}{9}$	$3\frac{2}{3}$	3	1	-5

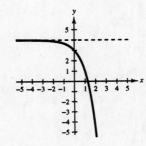

Review Exercises for Chapter 5

Solutions to Odd-Numbered Exercises

1. $(6.1)^{2.4} \approx 76.699$

3. $2^{-0.5\pi} \approx 0.337$

5. $60^{\sqrt{3}} \approx 1201.845$

7. $f(x) = 4^x$

Intercept: $(0, 1)$

Horizontal asymptote: x-axis

Increasing on: $(-\infty, \infty)$

Matches graph (c)

9. $f(x) = -4^x$

Intercept: $(0, -1)$

Horizontal asymptote: x-axis

Decreasing on: $(-\infty, \infty)$

Matches graph (a)

11. $f(x) = 4^{-x} + 4$

Horizontal asymptote: $y = 4$

x	-1	0	1	2	3
$f(x)$	8	5	4.25	4.0625	4.016

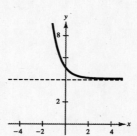

13. $f(x) = -2.65^{x+1}$

Horizontal asymptote: $y = 0$

x	-2	-1	0	1	2
$f(x)$	-0.377	-1	-2.65	-7.023	-18.61

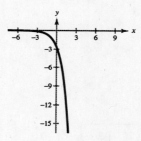

15. $f(x) = 5^{x-2} + 4$

Horizontal asymptote: $y = 4$

x	-1	0	1	2	3
$f(x)$	4.008	4.04	4.2	5	9

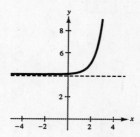

17. $f(x) = \left(\frac{1}{2}\right)^{-x} + 3 = 2^x + 3$

Horizontal asymptote: $y = 3$

x	-2	-1	0	1	2
$f(x)$	3.25	3.5	4	5	7

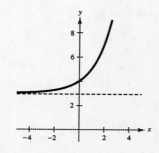

19. $e^8 \approx 2980.958$

21. $e^{-1.7} \approx 0.183$

23. $h(x) = e^{-x/2}$

x	-2	-1	0	1	2
$h(x)$	2.72	1.65	1	0.61	0.37

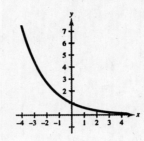

25. $f(x) = e^{x+2}$

x	-3	-2	-1	0	1
$f(x)$	0.37	1	2.72	7.39	20.09

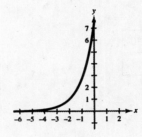

27. $A = 3500\left(1 + \dfrac{0.065}{n}\right)^{10n}$ or $A = 3500e^{(0.065)(10)}$

n	1	2	4	12	365	Continuous Compounding
A	\$6569.98	\$6635.43	\$6669.46	\$6692.64	\$6704.00	\$6704.39

29. $200{,}000 = Pe^{0.08t}$

$$P = \frac{200{,}000}{e^{0.08t}}$$

t	1	10	20	30	40	50
P	\$184,623.27	\$89,865.79	\$40,379.30	\$18,143.59	\$8,152.44	\$3,663.13

31. $F(t) = 1 - e^{-t/3}$

(a) $F\left(\tfrac{1}{2}\right) \approx 0.154$

(b) $F(2) \approx 0.487$

(c) $F(5) \approx 0.811$

33. (a) $A = 50{,}000e^{(0.0875)(35)} \approx \$1{,}069{,}047.14$

(b) The doubling time is

$$\frac{\ln 2}{0.0875} \approx 7.9 \text{ years.}$$

35. $4^3 = 64$

$\log_4 64 = 3$

37. $\log_{10} 1000 = \log_{10} 10^3 = 3$

39. $\log_2 \tfrac{1}{8} = \log_2 2^{-3} = -3$

41. $g(x) = \log_7 x \implies x = 7^y$

Vertical asymptote: $x = 0$

x	$\frac{1}{7}$	1	7	49
$g(x)$	-1	0	1	2

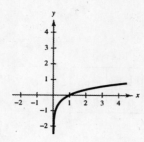

43. $f(x) = \log_{10}\left(\dfrac{x}{3}\right) \implies \dfrac{x}{3} = 10^y \implies x = 3(10^y)$

Vertical asymptote: $x = 0$

x	0.03	0.3	3	30
$f(x)$	-2	-1	0	1

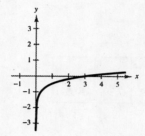

45. $f(x) = 4 - \log_{10}(x + 5)$

Vertical asymptote: $x = -5$

x	-4	-3	-2	-1	0	1
$f(x)$	4	3.70	3.52	3.40	3.30	3.22

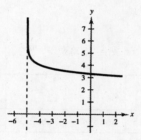

47. $\ln 22.6 \approx 3.118$

49. $\ln e^{-12} = -12$

51. $\ln\left(\sqrt{7} + 5\right) \approx 2.034$

53. $f(x) = \ln x + 3$

Domain: $(0, \infty)$

Vertical asymptote: $x = 0$

x	1	2	3	$\frac{1}{2}$	$\frac{1}{4}$
$f(x)$	3	3.69	4.10	2.31	1.61

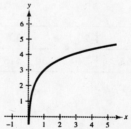

55. $h(x) = \ln(x^2) = 2\ln|x|$

Vertical asymptote: $x = 0$

x	± 0.5	± 1	± 2	± 3	± 4
y	-1.39	0	1.39	2.20	2.77

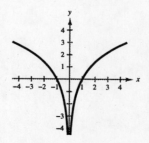

57. $s = 25 - \dfrac{13 \ln(10/12)}{\ln 3} \approx 27.16$ miles

59. $\log_{12} 200 = \dfrac{\log_{10} 200}{\log_{10} 12} \approx 2.132$

$\log_{12} 200 = \dfrac{\ln 200}{\ln 12} \approx 2.132$

61. $\log_3 0.28 = \dfrac{\log_{10} 0.28}{\log_{10} 3} \approx -1.159$

$\log_3 0.28 = \dfrac{\ln 0.28}{\ln 3} \approx -1.159$

63. $-\ln\left(\frac{1}{12}\right) = -[\ln 1 - \ln 12] = -[0 - \ln 12] = \ln 12$

65. $\log_8\left(\dfrac{\sqrt{x}}{y^3}\right) = \log_8 \sqrt{x} - \log_8 y^3 = \dfrac{1}{2}\log_8 x - 3\log_8 y$

67. $\log_5 5x^2 = \log_5 5 + \log_5 x^2$

$= 1 + 2\log_5 |x|$

69. $\log_{10} \dfrac{5\sqrt{y}}{x^2} = \log_{10} 5\sqrt{y} - \log_{10} x^2$

$= \log_{10} 5 + \log_{10} \sqrt{y} - \log_{10} x^2$

$= \log_{10} 5 + \dfrac{1}{2}\log_{10} y - 2\log_{10}|x|$

71. $\log_2 5 + \log_2 x = \log_2 5x$

73. $\dfrac{1}{2}\ln|2x - 1| - 2\ln|x + 1| = \ln\sqrt{|2x - 1|} - \ln|x + 1|^2$

$= \ln\dfrac{\sqrt{|2x - 1|}}{(x + 1)^2}$

75. $t = 50\log_{10}\dfrac{18,000}{18,000 - h}$

(a) Domain: $0 \le h < 18,000$

(b)

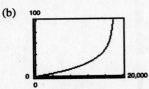

Vertical asymptote: $h = 18,000$

(c) As the plane approaches its absolute ceiling, it climbs at a slower rate, so the time required increases.

(d) $50\log_{10}\dfrac{18,000}{18,000 - 4000} \approx 5.46$ minutes

77. $3^x = 729$

$3^x = 3^6$

$x = 6$

79. $6^{x-2} = 1296$

$6^{x-2} = 6^4$

$x - 2 = 4$

$x = 6$

81. $\log_x 243 = 5$

$x^5 = 243$

$x^5 = 3^5$

$x = 3$

83. $e^{3x} = 25$

$\ln e^{3x} = \ln 25$

$3x = \ln 25$

$x = \dfrac{\ln 25}{3} \approx 1.073$

85. $14e^{3x+2} = 560$

$e^{3x+2} = 40$

$\ln e^{3x+2} = \ln 40$

$3x + 2 = \ln 40$

$x = \dfrac{(\ln 40) - 2}{3} \approx 0.563$

87. $e^x - 28 = -8$

$e^x = 20$

$x = \ln 20 \approx 2.996$

89. $2(12^x) = 190$

$12^x = 95$

$\ln 12^x = \ln 95$

$x \ln 12 = \ln 95$

$x = \dfrac{\ln 95}{\ln 12} \approx 1.833$

91. $e^{2x} - 6e^x + 8 = 0$

$(e^x - 2)(e^x - 4) = 0$

$e^x = 2$ or $e^x = 4$

$x = \ln 2$ $x = \ln 4$

$x \approx 0.693$ $x \approx 1.386$

93. $4^{-0.2x} + x = 0$

Graph $y_1 = 4^{-0.2x} + x$.

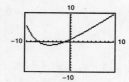

The x-intercepts are at $x \approx -7.04$ and $x \approx -1.53$.

95. $4e^{1.2x} = 9$

Graph $y_1 = 4e^{1.2x}$ and $y_2 = 9$.

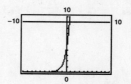

The intersection is at $x \approx 0.68$.

97. $\ln 5x = 7.2$

$5x = e^{7.2}$

$x = \dfrac{e^{7.2}}{5} \approx 267.886$

99. $4 \ln 3x = 15$

$\ln 3x = \dfrac{15}{4}$

$3x = e^{15/4}$

$x = \dfrac{e^{15/4}}{3} \approx 14.174$

101. $\ln \sqrt{x + 8} = 3$

$\frac{1}{2} \ln(x + 8) = 3$

$\ln(x + 8) = 6$

$x + 8 = e^6$

$x = e^6 - 8 \approx 395.429$

103. $\ln x - \ln 5 = 4$

$\ln \dfrac{x}{5} = 4$

$\dfrac{x}{5} = e^4$

$x = 5e^4 \approx 272.991$

105. $\log_{10}(x + 2) - \log_{10} x = \log_{10}(x + 5)$

$\log_{10}\left(\dfrac{x + 2}{x}\right) = \log_{10}(x + 5)$

$\dfrac{x + 2}{x} = x + 5$

$x + 2 = x^2 + 5x$

$0 = x^2 + 4x - 2$

$x = -2 \pm \sqrt{6}$ Quadratic Formula

Only $x = -2 + \sqrt{6} \approx 0.449$ is a valid solution.

107. $\log_{10}(-x - 4) = 2$

$-x - 4 = 10^2$

$-x = 100 + 4$

$x = -104$

109. $6 \log_{10}(x^2 + 1) - x = 0$

Graph

$$y_1 = 6 \log_{10}(x^2 + 1) - x.$$

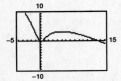

Zoom in to see the behavior near the origin.

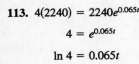

The solutions of the equation occur at the x-intercepts, which are at $x = 0$, $x \approx 0.42$, and $x \approx 13.63$.

111. $x - 2 \log_{10}(x + 4) = 0$

Graph

$$y_1 = x - 2 \log_{10}(x + 4)$$

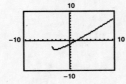

Note that $x = -4$ is a vertical asymptote, but the graph's behavior near $x = -4$ is not visible in most typical viewing windows. Zoom in to see the behavior near $x = -4$.

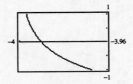

The solutions of the equation occur at the x-intercepts, which are at $x \approx -3.99$ and $x \approx 1.48$.

113. $4(2240) = 2240e^{0.065t}$

$$4 = e^{0.065t}$$

$$\ln 4 = 0.065t$$

$$\frac{\ln 4}{0.065} = t$$

$$t \approx 21.3 \text{ years}$$

115. $y = e^{-2x/3}$

Exponential decay model

Matches graph (e)

117. $y = \ln(x + 3)$

Logarithmic model

Vertical asymptote: $x = -3$

Graph includes $(-2, 0)$

Matches graph (f)

119. $y = 2e^{-(x+4)^2/3}$

Gaussian model

Matches graph (a)

121.
$$17000 = 12620e^{0.0118t}$$

$$\frac{17000}{12620} = e^{0.0118t}$$

$$\ln\left(\frac{17000}{12620}\right) = 0.0118t$$

$$\frac{\ln\left(\dfrac{17000}{12620}\right)}{0.0118} = t$$

$$t \approx 25.25 \text{ years}$$

This corresponds to the year 2025.

123. (a) $20{,}000 = 10{,}000e^{r(5)}$

$$2 = e^{5r}$$

$$\ln 2 = 5r$$

$$\frac{\ln 2}{5} = r$$

$$r \approx 0.138629 = 13.8629\%$$

(b) $A = 10{,}000e^{0.138629}$

$$\approx \$11{,}486.98$$

125. $y = ae^{bx}$

$$\frac{1}{2} = ae^{b(0)} \implies a = \frac{1}{2}$$

$$5 = \frac{1}{2}e^{b(5)}$$

$$10 = e^{5b}$$

$$\ln 10 = 5b$$

$$\frac{\ln 10}{5} = b$$

$$b \approx 0.4605$$

$$y = \frac{1}{2}e^{0.4605x}$$

127. $N = \dfrac{157}{1 + 5.4e^{-0.12t}}$

(a) When $N = 50$:

$$50 = \frac{157}{1 + 5.4e^{-0.12t}}$$

$$1 + 5.4e^{-0.12t} = \frac{157}{50}$$

$$5.4e^{-0.12t} = \frac{107}{50}$$

$$e^{-0.12t} = \frac{107}{270}$$

$$-0.12t = \ln\frac{107}{270}$$

$$t = \frac{\ln(107/270)}{-0.12} \approx 7.7 \text{ weeks}$$

(b) When $N = 75$:

$$75 = \frac{157}{1 + 5.4e^{-0.12t}}$$

$$1 + 5.4e^{-0.12t} = \frac{157}{75}$$

$$5.4e^{-0.12t} = \frac{82}{75}$$

$$e^{-0.12t} = \frac{82}{405}$$

$$-0.12t = \ln\frac{82}{405}$$

$$t = \frac{\ln(82/405)}{-0.12} \approx 13.3 \text{ weeks}$$

129. $R = \log_{10} I$ since $I_0 = 1$.

(a) $\log_{10} I = 8.4$

$I = 10^{8.4} \approx 251,188,643$

(b) $\log_{10} I = 6.85$

$I = 10^{6.85} \approx 7,079,458$

(c) $\log_{10} I = 9.1$

$I = 10^{9.1} \approx 1,258,925,412$

131. True by properties of exponents.

$$e^{x-1} = e^x \cdot e^{-1} = \frac{e^x}{e}$$

133. False.

$$\ln(x \cdot y) = \ln x + \ln y \neq \ln(x + y)$$

135. False. The domain of $f(x) = \ln x$ is $(0, \infty)$.

Chapter 5   Practice Test

1. Solve for x: $x^{3/5} = 8$.

2. Solve for x: $3^{x-1} = \frac{1}{81}$.

3. Graph $f(x) = 2^{-x}$.

4. Graph $g(x) = e^x + 1$.

5. If $5000 is invested at 9% interest, find the amount after three years if the interest is compounded

   (a) monthly.   (b) quarterly.   (c) continuously.

6. Write the equation in logarithmic form: $7^{-2} = \frac{1}{49}$.

7. Solve for x: $x - 4 = \log_2 \frac{1}{64}$.

8. Given $\log_b 2 = 0.3562$ and $\log_b 5 = 0.8271$, evaluate $\log_b \sqrt[4]{8/25}$.

9. Write $5 \ln x - \frac{1}{2} \ln y + 6 \ln z$ as a single logarithm.

10. Using your calculator and the change of base formula, evaluate $\log_9 28$.

11. Use your calculator to solve for N: $\log_{10} N = 0.6646$

12. Graph $y = \log_4 x$.

13. Determine the domain of $f(x) = \log_3(x^2 - 9)$.

14. Graph $y = \ln(x - 2)$.

15. True or false: $\dfrac{\ln x}{\ln y} = \ln(x - y)$

16. Solve for x: $5^x = 41$

17. Solve for x: $x - x^2 = \log_5 \frac{1}{25}$

18. Solve for x: $\log_2 x + \log_2(x - 3) = 2$

19. Solve for x: $\dfrac{e^x + e^{-x}}{3} = 4$

20. Six thousand dollars is deposited into a fund at an annual interest rate of 13%. Find the time required for the investment to double if the interest is compounded continuously.

C H A P T E R 6
Topics in Analytic Geometry

CHAPTER 6
Topics in Analytic Geometry

Section 6.1   Lines

Solutions to Odd-Numbered Exercises

- The **inclination** of a nonhorizontal line is the positive angle θ ($\theta < 180°$) measured counterclockwise from the x-axis to the line. A horizontal line has an inclination of zero.

- If a nonvertical line has inclination of θ and slope m, then $m = \tan \theta$.

- If two nonperpendicular lines have slopes m_1 and m_2, then the angle between the lines is given by
  $$\tan \theta = \left| \frac{m_2 - m_1}{1 + m_1 m_2} \right|.$$

- The distance between a point (x_1, y_1) and a line $Ax + By + C = 0$ is given by
  $$d = \frac{|Ax_1 + By_1 + C|}{\sqrt{A^2 + B^2}}.$$

---

1. $m = \tan \dfrac{\pi}{6} = \dfrac{\sqrt{3}}{3}$

3. $m = \tan \dfrac{3\pi}{4} = -1$

5. $m = \tan \dfrac{\pi}{3} = \sqrt{3}$

7. $m = \tan 1.27 \approx 3.2236$

9. $m = -1$

$-1 = \tan \theta$

$\theta = 180° + \arctan(-1) = \dfrac{3\pi}{4}$ radians $= 135°$

11. $m = 1$

$1 = \tan \theta$

$\theta = \dfrac{\pi}{4}$ radians $= 45°$

13. $m = \dfrac{3}{4}$

$\dfrac{3}{4} = \tan \theta$

$\theta = \arctan\left(\dfrac{3}{4}\right) \approx 0.6435$ radian $\approx 36.9°$

15. $(6, 1)$, $(10, 8)$

$m = \dfrac{8 - 1}{10 - 6} = \dfrac{7}{4}$

$\dfrac{7}{4} = \tan \theta$

$\theta = \arctan\left(\dfrac{7}{4}\right) \approx 1.0517$ radians $\approx 60.3°$

17. $(-2, 20)$, $(10, 0)$

$m = \dfrac{0 - 20}{10 - (-2)} = -\dfrac{20}{12} = -\dfrac{5}{3}$

$-\dfrac{5}{3} = \tan \theta$

$\theta = \pi + \arctan\left(-\dfrac{5}{3}\right) \approx 2.1112$ radians $\approx 121.0°$

19. $6x - 2y + 8 = 0$

$y = 3x + 4 \Rightarrow m = 3$

$3 = \tan \theta$

$\theta = \arctan 3 \approx 1.2490$ radians $\approx 71.6°$

21. $5x + 3y = 0$

$$y = -\frac{5}{3}x \implies m = -\frac{5}{3}$$

$$-\frac{5}{3} = \tan\theta$$

$$\theta = \pi + \arctan\left(-\frac{5}{3}\right) \approx 2.1112 \text{ radians} \approx 121.0°$$

23. $3x + y = 3 \implies y = -3x + 3 \implies m_1 = -3$

$$x - y = 2 \implies y = x - 2 \quad \implies m_2 = 1$$

$$\tan\theta = \left|\frac{1 - (-3)}{1 + (-3)(1)}\right| = 2$$

$$\theta = \arctan 2 \approx 1.1071 \text{ radians} \approx 63.4°$$

25. $x - y = 0 \implies y = x \quad \implies m_1 = 1$

$$3x - 2y = -1 \implies y = \frac{3}{2}x + \frac{1}{2} \implies m_2 = \frac{3}{2}$$

$$\tan\theta = \left|\frac{\frac{3}{2} - 1}{1 + \left(\frac{3}{2}\right)(1)}\right| = \frac{1}{5}$$

$$\theta = \arctan\frac{1}{5} \approx 0.1974 \text{ radian} \approx 11.3°$$

27. $x - 2y = 7 \implies y = \frac{1}{2}x - \frac{7}{2} \implies m_1 = \frac{1}{2}$

$$6x + 2y = 5 \implies y = -3x + \frac{5}{2} \implies m_2 = -3$$

$$\tan\theta = \left|\frac{-3 - \frac{1}{2}}{1 + \left(\frac{1}{2}\right)(-3)}\right| = 7$$

$$\theta = \arctan 7 \approx 1.4289 \text{ radians} \approx 81.9°$$

29. $x + 2y = 8 \implies y = -\frac{1}{2}x + 4 \implies m_1 = -\frac{1}{2}$

$$x - 2y = 2 \implies y = \frac{1}{2}x - 1 \quad \implies m_2 = \frac{1}{2}$$

$$\tan\theta = \left|\frac{\frac{1}{2} - \left(-\frac{1}{2}\right)}{1 + \left(-\frac{1}{2}\right)\left(\frac{1}{2}\right)}\right| = \frac{4}{3}$$

$$\theta = \arctan\left(\frac{4}{3}\right) \approx 0.9273 \text{ radian} \approx 53.1°$$

31. $0.05x - 0.03y = 0.21 \implies y = \frac{5}{3}x - 7 \quad \implies m_1 = \frac{5}{3}$

$$0.07x + 0.02y = 0.16 \implies y = -\frac{7}{2}x + 8 \implies m_2 = -\frac{7}{2}$$

$$\tan\theta = \left|\frac{\left(-\frac{7}{2}\right) - \left(\frac{5}{3}\right)}{1 + \left(\frac{5}{3}\right)\left(-\frac{7}{2}\right)}\right| = \frac{31}{29}$$

$$\theta = \arctan\left(\frac{31}{29}\right) \approx 0.8187 \text{ radian} \approx 46.9°$$

33. Let $A = (2, 1)$, $B = (4, 4)$, and $C = (6, 2)$.

Slope of AB: $m_1 = \dfrac{1 - 4}{2 - 4} = \dfrac{3}{2}$

Slope of BC: $m_2 = \dfrac{4 - 2}{4 - 6} = -1$

Slope of AC: $m_3 = \dfrac{1 - 2}{2 - 6} = \dfrac{1}{4}$

$$\tan A = \left|\frac{\frac{1}{4} - \frac{3}{2}}{1 + \left(\frac{3}{2}\right)\left(\frac{1}{4}\right)}\right| = \frac{\frac{5}{4}}{\frac{11}{8}} = \frac{10}{11}$$

$$A = \arctan\left(\frac{10}{11}\right) \approx 42.3°$$

$$\tan B = \left|\frac{\frac{3}{2} - (-1)}{1 + (-1)\left(\frac{3}{2}\right)}\right| = \frac{\frac{5}{2}}{\frac{1}{2}} = 5$$

$$B = \arctan 5 \approx 78.7°$$

$$\tan C = \left|\frac{-1 - \frac{1}{4}}{1 + \left(\frac{1}{4}\right)(-1)}\right| = \frac{\frac{5}{4}}{\frac{3}{4}} = \frac{5}{3}$$

$$C = \arctan\left(\frac{5}{3}\right) \approx 59.0°$$

35. Let $A = (-4, -1)$, $B = (3, 2)$, and $C = (1, 0)$.

Slope of AB: $m_1 = \dfrac{-1 - 2}{-4 - 3} = \dfrac{3}{7}$

Slope of BC: $m_2 = \dfrac{2 - 0}{3 - 1} = 1$

Slope of AC: $m_3 = \dfrac{-1 - 0}{-4 - 1} = \dfrac{1}{5}$

$$\tan A = \left|\frac{\frac{1}{5} - \frac{3}{7}}{1 + \left(\frac{3}{7}\right)\left(\frac{1}{5}\right)}\right| = \frac{\frac{8}{35}}{\frac{38}{35}} = \frac{4}{19}$$

$$A = \arctan\left(\frac{4}{19}\right) \approx 11.9°$$

$$\tan B = \left|\frac{1 - \frac{3}{7}}{1 + \left(\frac{3}{7}\right)(1)}\right| = \frac{\frac{4}{7}}{\frac{10}{7}} = \frac{2}{5}$$

$$B = \arctan\left(\frac{2}{5}\right) \approx 21.8°$$

$$C = 180° - A - B$$

$$\approx 180° - 11.9° - 21.8° = 146.3°$$

37. $(0, 0) \Rightarrow x_1 = 0$ and $y_1 = 0$

$4x + 3y = 0 \Rightarrow A = 4, B = 3,$ and $C = 0$

$d = \dfrac{|4(0) + 3(0) + 0|}{\sqrt{4^2 + 3^2}} = \dfrac{0}{5} = 0$

Note: The point is *on* the line.

39. $(2, 3) \Rightarrow x_1 = 2$ and $y_1 = 3$

$4x + 3y - 10 = 0 \Rightarrow A = 4, B = 3,$ and $C = -10$

$d = \dfrac{|4(2) + 3(3) + (-10)|}{\sqrt{4^2 + 3^2}} = \dfrac{7}{5}$

41. $(6, 2) \Rightarrow x_1 = 6$ and $y_1 = 2$

$x + 1 = 0 \Rightarrow A = 1, B = 0,$ and $C = 1$

$d = \dfrac{|1(6) + 0(2) + 1|}{\sqrt{1^2 + 0^2}} = 7$

43. $(0, 8) \Rightarrow x_1 = 0$ and $y_1 = 8$

$6x - y = 0 \Rightarrow A = 6, B = -1,$ and $C = 0$

$d = \dfrac{|6(0) + (-1)(8) + 0|}{\sqrt{6^2 + (-1)^2}}$

$= \dfrac{8}{\sqrt{37}} = \dfrac{8\sqrt{37}}{37} \approx 1.3152$

45. $A = (0, 0), B = (1, 4), C = (4, 0)$

(a) The slope the line through AC is $m = \dfrac{0 - 0}{4 - 0} = 0$.

The equation of the line through AC is $y = 0$.

The distance between the line and $B = (1, 4)$ is $d = \dfrac{|0(1) + (1)(4) + 0|}{\sqrt{0^2 + 1^2}} = 4$.

(b) The distance between A and C is 4.

$A = \dfrac{1}{2}(4)(4) = 8$ square units.

47. $A = \left(-\dfrac{1}{2}, \dfrac{1}{2}\right), B = (2, 3), C = \left(\dfrac{5}{2}, 0\right)$

(a) The slope of the line through AC is $m = \dfrac{\frac{1}{2} - 0}{\left(-\frac{1}{2}\right) - \frac{5}{2}} = -\dfrac{1}{6}$.

The equation of the line through AC is $y - 0 = -\dfrac{1}{6}\left(x - \dfrac{5}{2}\right) \Rightarrow 2x + 12y - 5 = 0$.

The distance between the line and $B = (2, 3)$ is $d = \dfrac{|2(2) + 12(3) + (-5)|}{\sqrt{2^2 + 12^2}} = \dfrac{35}{\sqrt{148}} = \dfrac{35\sqrt{37}}{74}$.

(b) The distance between A and C is $d = \sqrt{\left[\left(-\dfrac{1}{2}\right) - \left(\dfrac{5}{2}\right)\right]^2 + \left[\left(\dfrac{1}{2}\right) - 0\right]^2} = \dfrac{\sqrt{37}}{2}$.

$A = \dfrac{1}{2}\left(\dfrac{\sqrt{37}}{2}\right)\left(\dfrac{35\sqrt{37}}{74}\right) = \dfrac{35}{8}$ square units

49. $x + y = 1 \Rightarrow (0, 1)$ is a point on the line $\Rightarrow x_1 = 0$ and $y_1 = 1$

$x + y = 5 \Rightarrow A = 1, B = 1,$ and $C = -5$

$d = \dfrac{|1(0) + 1(1) + (-5)|}{\sqrt{1^2 + 1^2}} = \dfrac{4}{\sqrt{2}} = 2\sqrt{2}$

51. Slope: $m = \tan 0.1 \approx 0.1003$

Change in elevation: $\tan 0.1 = \dfrac{x}{2(5280)}$

$x \approx 1059$ feet

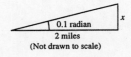

0.1 radian
2 miles
(Not drawn to scale)

53. (a) $\tan \theta = \dfrac{1}{3}$

$\theta = \arctan\left(\dfrac{1}{3}\right) \approx 18.4°$

(b) $\dfrac{5}{x} = \sin 18.4°$

$x = \dfrac{5}{\sin 18.4°}$

$x \approx 15.8$ meters

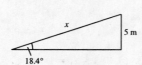

55. $\tan \gamma = \dfrac{6}{9}$

$\gamma = \arctan\left(\dfrac{2}{3}\right) \approx 33.69°$

$\beta = 90 - \gamma \approx 56.31°$

Also, since the right triangles containing α and β are equal, $\alpha = \gamma \approx 33.69°$

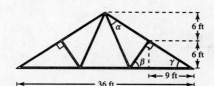

57. True. The inclination of a line is related to its slope by $m = \tan \theta$.

If the angle is greater than $\dfrac{\pi}{2}$ but less than π, then the angle is in the second quadrant where the tangent function is negative.

59. (a) $(0, 0) \Rightarrow x_1 = 0$ and $y_1 = 0$

$y = mx + 4 \Rightarrow 0 = mx - y + 4$

$d = \dfrac{|m(0) + (-1)(0) + 4|}{\sqrt{m^2 + (-1)^2}} = \dfrac{4}{\sqrt{m^2 + 1}}$

(c) The maximum distance of 4 occurs when the slope m is 0 and the line through $(0, 4)$ is horizontal.

(d) The graph has a horizontal asymptote at $d = 0$. As the slope becomes larger, the distance between the origin and the line, $y = mx + 4$, becomes smaller and approaches 0.

(b)

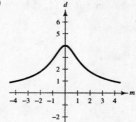

61. $f(x) = (x - 7)^2$

x-intercept: $0 = (x - 7)^2 \Rightarrow x = 7$

$(7, 0)$

y-intercept: $y = (0 - 7)^2 = 49$

$(0, 49)$

63. $f(x) = (x - 5)^2 - 5$

x-intercepts: $0 = (x - 5)^2 - 5$

$5 = (x - 5)^2$

$\pm\sqrt{5} = x - 5$

$5 \pm \sqrt{5} = x$

$\left(5 \pm \sqrt{5}, 0\right)$

y-intercept: $y = (0 - 5)^2 - 5 = 20$

$(0, 20)$

65. $f(x) = x^2 - 7x - 1$

x-intercepts: $0 = x^2 - 7x - 1$

$$x = \frac{7 \pm \sqrt{53}}{2} \text{ by the Quadratic Formula}$$

$$\left(\frac{7 \pm \sqrt{53}}{2}, 0\right)$$

y-intercept: $y = 0^2 - 7(0) - 1 = -1$

$$(0, -1)$$

67. $f(x) = 3x^2 + 2x - 16$

$$= 3\left(x^2 + \frac{2}{3}x\right) - 16$$

$$= 3\left(x^2 + \frac{2}{3}x + \frac{1}{9}\right) - \frac{1}{3} - 16$$

$$= 3\left(x + \frac{1}{3}\right)^2 - \frac{49}{3}$$

Vertex: $\left(-\frac{1}{3}, -\frac{49}{3}\right)$

69. $f(x) = 5x^2 + 34x - 7$

$$= 5\left(x^2 + \frac{34}{5}x\right) - 7$$

$$= 5\left(x^2 + \frac{34}{5}x + \frac{289}{25}\right) - \frac{289}{5} - 7$$

$$= 5\left(x + \frac{17}{5}\right)^2 - \frac{324}{5}$$

Vertex: $\left(-\frac{17}{5}, -\frac{324}{5}\right)$

71. $f(x) = 6x^2 - x - 12$

$$= 6\left(x^2 - \frac{1}{6}x\right) - 12$$

$$= 6\left(x^2 - \frac{1}{6}x + \frac{1}{144}\right) - \frac{1}{24} - 12$$

$$= 6\left(x - \frac{1}{12}\right)^2 - \frac{289}{24}$$

Vertex: $\left(\frac{1}{12}, -\frac{289}{24}\right)$

Section 6.2   Introduction to Conics: Parabolas

■ A **parabola** is the set of all points *(x, y)* that are equidistant from a fixed line (**directrix**) and a fixed point (**focus**) not on the line.

■ The standard equation of a parabola with vertex *(h, k)* and:
   (a) Vertical axis $x = h$ and directrix $y = k - p$ is:
       $(x - h)^2 = 4p(y - k), p \neq 0$
   (b) Horizontal axis $y = k$ and directrix $x = h - p$ is:
       $(y - k)^2 = 4p(x - h), p \neq 0$

■ The tangent line to a parabola at a point *P* makes **equal angles** with:
   (a) the line through *P* and the focus.
   (b) the axis of the parabola.

Solutions to Odd-Numbered Exercises

1. A circle is formed when a plane intersects the top or bottom half of a double-napped cone and is parallel to the vertex.

3. A parabola is formed when a plane intersects the top or bottom half of a double-napped cone, is parallel to the side of the cone, and does not intersect the vertex.

5. $y^2 = -4x$

Vertex: $(0, 0)$

Opens to the left since *p* is negative.

Matches graph (e).

7. $x^2 = -8y$

Vertex: $(0, 0)$

Opens downward since *p* is negative.

Matches graph (d).

9. $(y - 1)^2 = 4(x - 3)$

Vertex: $(3, 1)$

Opens to the right since *p* is positive.

Matches graph (a).

11. $y = \frac{1}{2}x^2$

$x^2 = 2y$

$x^2 = 4\left(\frac{1}{2}\right)y \Rightarrow h = 0, k = 0, p = \frac{1}{2}$

Vertex: $(0, 0)$

Focus: $\left(\frac{1}{2}, 0\right)$

Directrix: $y = -\frac{1}{2}$

13. $y^2 = -6x$

$y^2 = 4\left(-\frac{3}{2}\right)x \Rightarrow h = 0, k = 0, p = -\frac{3}{2}$

Vertex: $(0, 0)$

Focus: $\left(-\frac{3}{2}, 0\right)$

Directrix: $x = \frac{3}{2}$

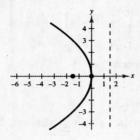

15. $x^2 + 6y = 0$

$x^2 = -6y = 4\left(-\frac{3}{2}\right)y \Rightarrow h = 0, k = 0, p = -\frac{3}{2}$

Vertex: $(0, 0)$

Focus: $\left(0, -\frac{3}{2}\right)$

Directrix: $y = \frac{3}{2}$

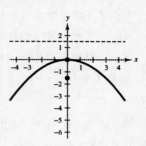

17. $(x - 1)^2 + 8(y + 2) = 0$

$\qquad (x - 1)^2 = 4(-2)(y + 2)$

$h = 1, k = -2, p = -2$

Vertex: $(1, -2)$

Focus: $(1, -4)$

Directrix: $y = 0$

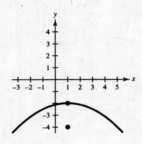

19. $\left(x + \frac{3}{2}\right)^2 = 4(y - 2)$

$\left(x + \frac{3}{2}\right)^2 = 4(1)(y - 2)$

$h = -\frac{3}{2}, k = 2, p = 1$

Vertex: $\left(-\frac{3}{2}, 2\right)$

Focus: $\left(-\frac{3}{2}, 3\right)$

Directrix: $y = 1$

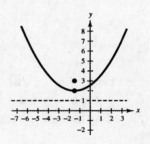

21. $\qquad y = \frac{1}{4}(x^2 - 2x + 5)$

$\qquad 4y = x^2 - 2x + 5$

$4y - 5 + 1 = x^2 - 2x + 1$

$\qquad 4y - 4 = (x - 1)^2$

$\qquad (x - 1)^2 = 4(1)(y - 1)$

$h = 1, k = 1, p = 1$

Vertex: $(1, 1)$

Focus: $(1, 2)$

Directrix: $y = 0$

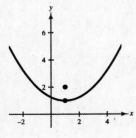

23. $y^2 + 6y + 8x + 25 = 0$

$\qquad y^2 + 6y + 9 = -8x - 25 + 9$

$\qquad (y + 3)^2 = 4(-2)(x + 2)$

$h = -2, k = -3, p = -2$

Vertex: $(-2, -3)$

Focus: $(-4, -3)$

Directrix: $x = 0$

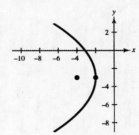

25. $x^2 + 4x + 6y - 2 = 0$

$x^2 + 4x = -6y + 2$

$x^2 + 4x + 4 = -6y + 2 + 4$

$(x + 2)^2 = -6(y - 1)$

$(x + 2)^2 = 4\left(-\frac{3}{2}\right)(y - 1)$

$h = -2, k = 1, p = -\frac{3}{2}$

Vertex: $(-2, 1)$

Focus: $\left(-2, -\frac{1}{2}\right)$

Directrix: $y = \frac{5}{2}$

On the graphing calculator, enter:

$y_1 = \frac{1}{6}(x^2 + 4x - 2)$

27. $y^2 + x + y = 0$

$y^2 + y + \frac{1}{4} = -x + \frac{1}{4}$

$\left(y + \frac{1}{2}\right)^2 = 4\left(-\frac{1}{4}\right)\left(x - \frac{1}{4}\right)$

$h = \frac{1}{4}, k = -\frac{1}{2}, p = -\frac{1}{4}$

Vertex: $\left(\frac{1}{4}, -\frac{1}{2}\right)$

Focus: $\left(0, -\frac{1}{2}\right)$

Directrix: $x = \frac{1}{2}$

To use a graphing calculator, enter:

$y_1 = -\frac{1}{2} + \sqrt{\frac{1}{4} - x}$

$y_2 = -\frac{1}{2} - \sqrt{\frac{1}{4} - x}$

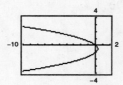

29. $y^2 - 8x = 0 \Rightarrow y = \pm\sqrt{8x}$

$x - y + 2 = 0 \Rightarrow y = x + 2$

The point of tangency is $(2, 4)$.

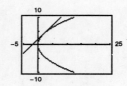

31. Vertex: $(0, 0) \Rightarrow h = 0, k = 0$

Graph opens upward.

$x^2 = 4py$

Point on graph: $(3, 6)$

$3^2 = 4p(6)$

$9 = 24p$

$\frac{3}{8} = p$

Thus, $x^2 = 4\left(\frac{3}{8}\right)y \Rightarrow x^2 = \frac{3}{2}y$

33. Vertex: $(0, 0) \Rightarrow h = 0, k = 0$

Focus: $\left(0, -\frac{3}{2}\right) \Rightarrow p = -\frac{3}{2}$

$x^2 = 4py$

$x^2 = 4\left(-\frac{3}{2}\right)y$

$x^2 = -6y$

35. Vertex: $(0, 0) \Rightarrow h = 0, k = 0$

Focus: $(-2, 0) \Rightarrow p = -2$

$y^2 = 4px$

$y^2 = 4(-2)x$

$y^2 = -8x$

37. Vertex: $(0, 0) \Rightarrow h = 0, k = 0$

Directrix: $y = -1 \Rightarrow p = 1$

$x^2 = 4py$

$x^2 = 4(1)y$

$x^2 = 4y$

39. Vertex: $(0, 0) \Rightarrow h = 0, k = 0$

Directrix: $y = 2 \Rightarrow p = -2$

$x^2 = 4py$

$x^2 = 4(-2)y$

$x^2 = -8y$

41. Vertex: $(0, 0) \Rightarrow h = 0, k = 0$

Horizontal axis and passes through the point $(4, 6)$

$y^2 = 4px$

$6^2 = 4p(4)$

$36 = 16p \Rightarrow p = \frac{9}{4}$

$y^2 = 4\left(\frac{9}{4}\right)x$

$y^2 = 9x$

43. Vertex: $(3, 1)$ and opens downward. Passes through $(2, 0)$ and $(4, 0)$.

$y = -(x - 2)(x - 4)$

$\quad = -x^2 + 6x - 8$

$\quad = -(x - 3)^2 + 1$

$(x - 3)^2 = -(y - 1)$

45. Vertex: $(-4, 0)$ and opens to the right. Passes through $(0, 4)$.

$(y - 0)^2 = 4p(x + 4)$

$4^2 = 4p(0 + 4)$

$16 = 16p$

$1 = p$

$y^2 = 4(x + 4)$

47. Vertex: $(5, 2)$

Focus: $(3, 2)$

Horizontal axis

$p = 3 - 5 = -2$

$(y - 2)^2 = 4(-2)(x - 5)$

$(y - 2)^2 = -8(x - 5)$

49. Vertex: $(0, 4)$

Directrix: $y = 2$

Vertical axis

$p = 4 - 2 = 2$

$(x - 0)^2 = 4(2)(y - 4)$

$\quad x^2 = 8(y - 4)$

51. Focus: $(2, 2)$

Directrix: $x = -2$

Horizontal axis

Vertex: $(0, 2)$

$p = 2 - 0 = 2$

$(y - 2)^2 = 4(2)(x - 0)$

$(y - 2)^2 = 8x$

53. $(y - 3)^2 = 6(x + 1)$

For the upper half of the parabola:

$y - 3 = +\sqrt{6(x + 1)}$

$y = \sqrt{6(x + 1)} + 3$

55. $x^2 = 2y \Rightarrow p = \frac{1}{2}$

Point: $(x_1, y_1) = (4, 8)$

Use: $y - y_1 = \frac{x_1}{2p}(x - x_1)$

$y - 8 = \frac{4}{2(1/2)}(x - 4)$

$y - 8 = 4x - 16$

$y = 4x - 8 \Rightarrow 0 = 4x - y - 8$

x-intercept: $(2, 0)$

57. $y = -2x^2 \Rightarrow x^2 = -\frac{1}{2}y \Rightarrow p = -\frac{1}{8}$

Point: $(x_1, y_1) = (-1, -2)$

Use: $y - y_1 = \frac{x_1}{2p}(x - x_1)$

$y + 2 = \frac{-1}{2(-1/8)}(x + 1)$

$y + 2 = 4x(x + 1)$

$y = 4x + 2 \Rightarrow 0 = 4x - y + 2$

x-intercept: $\left(-\frac{1}{2}, 0\right)$

59. $R = 265x - \frac{5}{4}x^2$

The revenue is maximum when $x = 106$ units.

61. Vertex: $(0, 0) \Rightarrow h = 0, k = 0$

Focus: $(0, 4.5) \Rightarrow p = 4.5$

$(x - h)^2 = 4p(y - k)$

$(x - 0)^2 = 4(4.5)(y - 0)$

$\quad x^2 = 18y$ or $y = \frac{1}{18}x^2$

63. (a) Vertex: $(0, 0) \Rightarrow h = 0, k = 0$

Points on the parabola: $(\pm 16, -0.4)$

$$x^2 = 4py$$

$$(\pm 16)^2 = 4p(-0.4)$$

$$256 = -1.6p$$

$$-160 = p$$

$$x^2 = 4(-160y)$$

$$x^2 = -640y$$

$$y = -\frac{1}{640}x^2$$

(b) When $y = 0.1$ we have, $0.1 = -\frac{1}{640}x^2$

$$64 = x^2$$

$$\pm 8 = x$$

Thus, 8 feet away from the center of the road, the road surface is 0.1 foot lower than in the middle.

65. (a) $V = 17,500\sqrt{2}$ mi/hr

$\approx 24,750$ mi/hr

(b) $p = -4100, (h, k) = (0, 4100)$

$$(x - 0)^2 = 4(-4100)(y - 4100)$$

$$x^2 = -16,400(y - 4100)$$

67. The position of the target is on the x-axis, so first, let $y = 0$.

$$0 = 30,000 - \frac{x^2}{39,204}$$

$$\frac{x^2}{39,204} = 30,000$$

$$x^2 = 30,000(39,204)$$

$$x = \sqrt{30,000(39,204)}$$

$$x \approx 34,294.606 \text{ feet}$$

Since the bomber is flying at 792 feet per second,

$$\frac{34,294.606 \text{ feet}}{792 \text{ feet per second}} \approx 43.3 \text{ seconds}$$

The bomb should be released 43.3 seconds prior to being over the target.

69. True. If the axis (line connecting the vertex and focus) is horizontal, then the directrix must be vertical.

71. (a) $A = \frac{8}{3}(2)^{1/2}(4)^{3/2} = \frac{8}{3}\left(\sqrt{2}\right)(8) = \frac{64\sqrt{2}}{3}$ square units

(b) As p approaches zero, the parabola becomes narrower and narrower, thus the area becomes smaller and smaller.

73. $f(x) = (x - 3)[x - (2 + i)][x - (2 - i)]$

$= (x - 3)[(x - 2) - i][(x - 2) + i]$

$= (x - 3)(x^2 - 4x + 5)$

$= x^3 - 7x^2 + 17x - 15$

75. $g(x) = 6x^4 + 7x^3 - 29x^2 - 28x + 20$

Possible rational roots: $\pm 1, \pm 2, \pm 4, \pm 5, \pm 10, \pm 20,$
$\pm\frac{1}{2}, \pm\frac{5}{2}, \pm\frac{1}{3}, \pm\frac{2}{3}, \pm\frac{4}{3}, \pm\frac{5}{3}, \pm\frac{10}{3}, \pm\frac{20}{3}, \pm\frac{1}{6}, \pm\frac{5}{6}$

$x = \pm 2$ are both solutions.

$$
\begin{array}{r|rrrrr}
2 & 6 & 7 & -29 & -28 & 20 \\
  &   & 12 & 38 & 18 & -20 \\
\hline
  & 6 & 19 & 9 & -10 & 0 \\
\end{array}
$$

$$
\begin{array}{r|rrrr}
-2 & 6 & 19 & 9 & -10 \\
   &   & -12 & -14 & 10 \\
\hline
   & 6 & 7 & -5 & 0 \\
\end{array}
$$

$g(x) = (x - 2)(x + 2)(6x^2 + 7x - 5)$

$= (x - 2)(x + 2)(2x - 1)(3x + 5)$

The zeros of $g(x)$ are $x = \pm 2, x = \frac{1}{2}, x = -\frac{5}{3}$.

Section 6.3   Ellipses

- An **ellipse** is the set of all points *(x, y)* the sum of whose distances from two distinct fixed points (**foci**) is constant.
- The standard equation of an ellipse with center *(h, k)* and major and minor axes of lengths 2*a* and 2*b* is:

    (a) $\dfrac{(x - h)^2}{a^2} + \dfrac{(y - k)^2}{b^2} = 1$ if the major axis is horizontal.

    (b) $\dfrac{(x - h)^2}{b^2} + \dfrac{(y - k)^2}{a^2} = 1$ if the major axis is vertical.

- $c^2 = a^2 - b^2$ where *c* is the distance from the center to a focus.
- The eccentricity of an ellipse is $e = \dfrac{c}{a}$.

Solutions to Odd-Numbered Exercises

1. $\dfrac{x^2}{4} + \dfrac{y^2}{9} = 1$

Center: $(0, 0)$

$a = 3, b = 2$

Vertical major axis

Matches graph (b).

3. $\dfrac{x^2}{4} + \dfrac{y^2}{25} = 1$

Center: $(0, 0)$

$a = 5, b = 2$

Vertical major axis

Matches graph (d).

5. $\dfrac{(x - 2)^2}{16} + (y + 1)^2 = 1$

Center: $(2, -1)$

$a = 4, b = 1$

Horizontal major axis

Matches graph (a).

7. $\dfrac{x^2}{25} + \dfrac{y^2}{16} = 1$

Center: $(0, 0)$

$a = 5, b = 4, c = 3$

Foci: $(\pm 3, 0)$

Vertices: $(\pm 5, 0)$

$e = \dfrac{3}{5}$

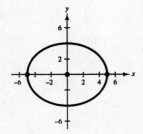

9. $\dfrac{x^2}{5} + \dfrac{y^2}{9} = 1$

$a = 3, b = \sqrt{5}, c = 2$

Center: $(0, 0)$

Foci: $(0, \pm 2)$

Vertices: $(0, \pm 3)$

$e = \dfrac{\sqrt{5}}{3}$

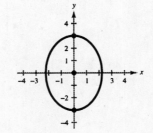

11. $\dfrac{(x + 3)^2}{16} + \dfrac{(y - 5)^2}{25} = 1$

Center: $(-3, 5)$

$a = 5, b = 4, c = 3$

Foci: $(-3, 8)(-3, 2)$

Vertices: $(-3, 10)(-3, 0)$

$e = \dfrac{3}{5}$

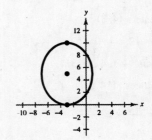

13. $\dfrac{(x+5)^2}{\frac{9}{4}} + (y-1)^2 = 1$

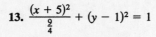

Center: $(-5, 1)$

$a = \dfrac{3}{2}, b = 1, c = \dfrac{\sqrt{5}}{2}$

Foci: $\left(-5 + \dfrac{\sqrt{5}}{2}, 1\right), \left(-5 - \dfrac{\sqrt{5}}{2}, 1\right)$

Vertices: $\left(-\dfrac{7}{2}, 1\right), \left(-\dfrac{13}{2}, 1\right)$

$e = \dfrac{\frac{\sqrt{5}}{2}}{\frac{3}{2}} = \dfrac{\sqrt{5}}{3}$

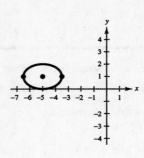

15. $\qquad 9x^2 + 4y^2 + 36x - 24y + 36 = 0$

$9(x^2 + 4x + 4) + 4(y^2 - 6y + 9) = -36 + 36 + 36$

$\qquad\qquad 9(x+2)^2 + 4(y-3) = 36$

$\qquad\qquad \dfrac{(x+2)^2}{4} + \dfrac{(y-3)^2}{9} = 1$

$a = 3, b = 2, c = \sqrt{5}$

Center: $(-2, 3)$

Foci: $\left(-2, 3 \pm \sqrt{5}\right)$

Vertices: $(-2, 6), (-2, 0)$

$e = \dfrac{\sqrt{5}}{3}$

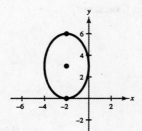

17. $\qquad x^2 + 5y^2 - 8x - 30y - 39 = 0$

$(x^2 - 8x + 16) + 5(y^2 - 6y + 9) = 39 + 16 + 45$

$\qquad\qquad (x-4)^2 + 5(y-3)^2 = 100$

$\qquad\qquad \dfrac{(x-4)^2}{100} + \dfrac{(y-3)^2}{20} = 1$

Center: $(4, 3)$

$a = 10, b = 2\sqrt{5}, c = 4\sqrt{5}$

Foci: $\left(4 \pm 4\sqrt{5}, 3\right)$

Vertices: $(14, 3), (-6, 3)$

$e = \dfrac{4\sqrt{5}}{10} = \dfrac{2\sqrt{5}}{5}$

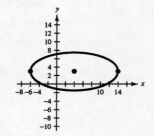

19.
$$6x^2 + 2y^2 + 18x - 10y + 2 = 0$$

$$6\left(x^2 + 3x + \frac{9}{4}\right) + 2\left(y^2 - 5y + \frac{25}{4}\right) = -2 + \frac{27}{2} + \frac{25}{2}$$

$$6\left(x + \frac{3}{2}\right)^2 + 2\left(y - \frac{5}{2}\right)^2 = 24$$

$$\frac{\left(x + \frac{3}{2}\right)^2}{4} + \frac{\left(y - \frac{5}{2}\right)^2}{12} = 1$$

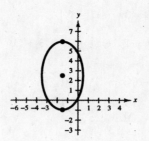

Center: $\left(-\frac{3}{2}, \frac{5}{2}\right)$

$a = 2\sqrt{3}, b = 2, c = 2\sqrt{2}$

Foci: $\left(-\frac{3}{2}, \frac{5}{2} \pm 2\sqrt{2}\right)$

Vertices: $\left(-\frac{3}{2}, \frac{5}{2} \pm 2\sqrt{3}\right)$

$e = \frac{2\sqrt{2}}{2\sqrt{3}} = \frac{\sqrt{6}}{3}$

21.
$$16x^2 + 25y^2 - 32x + 50y + 16 = 0$$

$$16(x^2 - 2x + 1) + 25(y^2 + 2y + 1) = -16 + 16 + 25$$

$$16(x - 1)^2 + 25(y + 1)^2 = 25$$

$$\frac{(x - 1)^2}{\frac{25}{16}} + (y + 1)^2 = 1$$

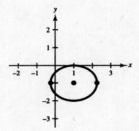

$a = \frac{5}{4}, b = 1, c = \frac{3}{4}$

Center: $(1, -1)$

Foci: $\left(\frac{7}{4}, -1\right), \left(\frac{1}{4}, -1\right)$

Vertices: $\left(\frac{9}{4}, -1\right), \left(-\frac{1}{4}, -1\right)$

$e = \frac{3}{5}$

23. $5x^2 + 3y^2 = 15$

$\frac{x^2}{3} + \frac{y^2}{5} = 1$

Center: $(0, 0)$

$a = \sqrt{5}, b = \sqrt{3}, c = \sqrt{2}$

Foci: $\left(0, \pm\sqrt{2}\right)$

Vertices: $\left(0, \pm\sqrt{5}\right)$

To graph, solve for y.

$y^2 = \frac{15 - 5x^2}{3}$

$y_1 = \sqrt{\frac{15 - 5x^2}{3}}$

$y_2 = -\sqrt{\frac{15 - 5x^2}{3}}$

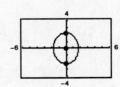

25. $12x^2 + 20y^2 - 12x + 40y - 37 = 0$

$$12\left(x^2 - x + \frac{1}{4}\right) + 20(y^2 + 2y + 1) = 37 + 3 + 20$$

$$12\left(x - \frac{1}{2}\right)^2 + 20(y + 1)^2 = 60$$

$$\frac{\left(x - \frac{1}{2}\right)^2}{5} + \frac{(y + 1)^2}{3} = 1$$

$a = \sqrt{5}, b = \sqrt{3}, c = \sqrt{2}$

Center: $\left(\frac{1}{2}, -1\right)$

Foci: $\left(\frac{1}{2} \pm \sqrt{2}, -1\right)$

Vertices: $\left(\frac{1}{2} \pm \sqrt{5}, -1\right)$

To graph, solve for y.

$$(y + 1)^2 = 3\left[1 - \frac{(x - 0.5)^2}{5}\right]$$

$$y_1 = -1 + \sqrt{3\left[1 - \frac{(x - 0.5)^2}{5}\right]}$$

$$y_2 = -1 - \sqrt{3\left[1 - \frac{(x - 0.5)^2}{5}\right]}$$

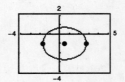

27. Center: $(0, 0)$

$a = 4, b = 2$

Vertical major axis

$$\frac{(x - h)^2}{b^2} + \frac{(y - k)^2}{a^2} = 1$$

$$\frac{x^2}{4} + \frac{y^2}{16} = 1$$

29. Vertices: $(\pm 6, 0)$

$a = 6, c = 2 \Rightarrow b = \sqrt{32} = 4\sqrt{2}$

Foci: $(\pm 2, 0)$

Horizontal major axis

Center: $(0, 0)$

$$\frac{(x - h)^2}{a^2} + \frac{(y - k)^2}{b^2} = 1$$

$$\frac{x^2}{36} + \frac{y^2}{32} = 1$$

31. Foci: $(\pm 5, 0) \Rightarrow c = 5$

Center: $(0, 0)$

Horizontal major axis

Major axis of length $12 \Rightarrow 2a = 12$

$$a = 6$$

$$6^2 - b^2 = 5^2 \Rightarrow b^2 = 11$$

$$\frac{(x - h)^2}{a^2} + \frac{(y - k)^2}{b^2} = 1$$

$$\frac{x^2}{36} + \frac{y^2}{11} = 1$$

33. Vertices: $(0, \pm 5) \Rightarrow a = 5$

Center: $(0, 0)$

Vertical major axis

$$\frac{(x - h)^2}{b^2} + \frac{(y - k)^2}{a^2} = 1$$

$$\frac{x^2}{b^2} + \frac{y^2}{25} = 1$$

Point: $(4, 2)$

$$\frac{4^2}{b^2} + \frac{2^2}{25} = 1$$

$$\frac{16}{b^2} = 1 - \frac{4}{25} = \frac{21}{25}$$

$$400 = 21b^2$$

$$\frac{400}{21} = b^2$$

$$\frac{x^2}{\frac{400}{21}} + \frac{y^2}{25} = 1$$

$$\frac{21x^2}{400} + \frac{y^2}{25} = 1$$

35. Center: $(2, 3)$

$a = 3, \quad b = 1$

Vertical major axis

$$\frac{(x - h)^2}{b^2} + \frac{(y - k)^2}{a^2} = 1$$

$$\frac{(x - 2)^2}{1} + \frac{(y - 3)^2}{9} = 1$$

37. Center: $(-2, 3)$

$a = 4, \quad b = 3$

Horizontal major axis

$$\frac{(x - h)^2}{a^2} + \frac{(y - k)^2}{b^2} = 1$$

$$\frac{(x + 2)^2}{16} + \frac{(y - 3)^2}{9} = 1$$

39. Vertices: $(0, 4), (4, 4) \Rightarrow a = 2$

Minor axis of length $2 \Rightarrow b = 1$

Center: $(2, 4) = (h, k)$

$$\frac{(x - h)^2}{a^2} + \frac{(y - k)^2}{b^2} = 1$$

$$\frac{(x - 2)^2}{4} + \frac{(y - 4)^2}{1} = 1$$

41. Foci: $(0, 0), (0, 8) \Rightarrow c = 4$

Major axis of length $16 \Rightarrow a = 8$

$b^2 = a^2 - c^2 = 64 - 16 = 48$

Center: $(0, 4) = (h, k)$

$$\frac{(x - h)^2}{b^2} + \frac{(y - k)^2}{a^2} = 1$$

$$\frac{x^2}{48} + \frac{(y - 4)^2}{64} = 1$$

43. Vertices: $(3, 1), (3, 9) \Rightarrow a = 4$

Center: $(3, 5)$

Minor axis of length $6 \Rightarrow b = 3$

Vertical major axis

$$\frac{(x - h)^2}{b^2} + \frac{(y - k)^2}{a^2} = 1$$

$$\frac{(x - 3)^2}{9} + \frac{(y - 5)^2}{16} = 1$$

45. Center: $(0, 4)$

Vertices: $(-4, 4), (4, 4) \Rightarrow a = 4$

$a = 2c \Rightarrow 4 = 2c \Rightarrow c = 2$

$2^2 = 4^2 - b^2 \Rightarrow b^2 = 12$

Horizontal major axis

$$\frac{(x - h)^2}{a^2} + \frac{(y - k)^2}{b^2} = 1$$

$$\frac{x^2}{16} + \frac{(y - 4)^2}{12} = 1$$

47. Vertices: $(\pm 5, 0) \Rightarrow a = 5$

Eccentricity: $\dfrac{3}{5} \Rightarrow c = \dfrac{3}{5}a = 3$

$b^2 = a^2 - c^2 = 25 - 9 = 16$

Center: $(0, 0) = (h, k)$

$\dfrac{(x - h)^2}{a^2} + \dfrac{(y - k)^2}{b^2} = 1$

$\dfrac{x^2}{25} + \dfrac{y^2}{16} = 1$

49. The tacks should be placed at the foci and the length of the string is the length of the major axis, $2a$.

Center: $(0, 0)$

$a = 3, b = 2, c = \sqrt{5}$

Foci (Positions of the tacks): $\left(\pm\sqrt{5}, 0\right)$

Length of string: 6 feet

51. Area of circle: $\pi r^2 = 100\pi$

Area of ellipse: $\pi(a)(10)$

$10a\pi = 2(100\pi)$

$10a\pi = 200\pi$

$a = 20$

Length of major axis: $2a = 40$

53. $a - c = 0.34$

$a + c = 4.08$

$2a = 4.42 \Rightarrow a = 2.21$

$c = 4.08 - a \Rightarrow c = 1.87$

$b^2 = a^2 - c^2 = 2.21^2 - 1.87^2 = 1.3872 \Rightarrow b \approx 1.18$

$\dfrac{x^2}{a^2} + \dfrac{y^2}{b^2} = 1$

$\dfrac{x^2}{4.88} + \dfrac{y^2}{1.39} = 1$

55. For $\dfrac{x^2}{a^2} + \dfrac{y^2}{b^2} = 1$, we have $c^2 = a^2 - b^2$

When $x = c$: $\dfrac{c^2}{a^2} + \dfrac{y^2}{b^2} = 1$

$y^2 = b^2\left(1 - \dfrac{a^2 - b^2}{a^2}\right) = \dfrac{b^4}{a^2}$

$y = \dfrac{b^2}{a}$

Length of latus rectum: $2y = \dfrac{2b^2}{a}$

57. $\dfrac{x^2}{9} + \dfrac{y^2}{16} = 1$

$a = 4, b = 3, c = \sqrt{7}$

Points on the ellipse: $(\pm 3, 0), (0, \pm 4)$

Length of latus recta: $\dfrac{2b^2}{a} = \dfrac{2(3)^2}{4} = \dfrac{9}{2}$

Additional points: $\left(\pm\dfrac{9}{4}, -\sqrt{7}\right), \left(\pm\dfrac{9}{4}, \sqrt{7}\right)$

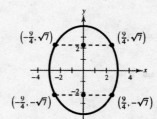

59. $5x^2 + 3y^2 = 15$

$$\frac{x^2}{3} + \frac{y^2}{5} = 1$$

$a = \sqrt{5}, b = \sqrt{3}, c = \sqrt{2}$

Points on the ellipse: $\left(\pm\sqrt{3}, 0\right), \left(0, \pm\sqrt{5}\right)$

Length of latus recta: $\dfrac{2b^2}{a} = \dfrac{2 \cdot 3}{\sqrt{5}} = \dfrac{6\sqrt{5}}{5}$

Additional points: $\left(\pm\dfrac{3\sqrt{5}}{5}, -\sqrt{2}\right), \left(\pm\dfrac{3\sqrt{5}}{5}, \sqrt{2}\right)$

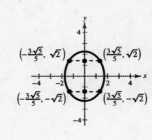

61. True. If e is close to 1, the ellipse is elongated and the foci are close to the vertices.

63. False. The foci of an ellipse cannot occur outside the ellipse because $0 < c < a$ and both the foci and vertices are on the major axis.

65. $\dfrac{x^2}{a^2} + \dfrac{y^2}{b^2} = 1$

(a) $a + b = 20 \Rightarrow b = 20 - a$
$A = \pi ab = \pi a(20 - a)$

(b) $264 = \pi a(20 - a)$
$0 = -\pi a^2 + 20\pi a - 264$
$0 = \pi a^2 - 20\pi a + 264$

By the Quadratic Formula: $a \approx 14$ or $a \approx 6$.
Choosing the larger value of a, we have $a \approx 14$ and $b \approx 6$.

The equation of an ellipse with an area of 264 is $\dfrac{x^2}{196} + \dfrac{y^2}{36} = 1$.

(c)

a	8	9	10	11	12	13
A	301.6	311.0	314.2	311.0	301.6	285.9

The area is maximum when $a = 10$ and the ellipse is a circle.

(d)
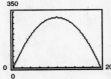

The area is maximum (314.16) when $a = b = 10$ and the ellipse is a circle.

67. $100 - (x - 5)^2 = 0$
$(x - 5)^2 = 100$
$x - 5 = \pm 10$
$x - 5 = -10 \Rightarrow x = -5$
$x - 5 = 10 \Rightarrow x = 15$

69. $16x^2 - 40x + 25 = 0$
$(4x - 5)(4x - 5) = 0$
$4x - 5 = 0 \Rightarrow x = \frac{5}{4}$

71. $f(x) = x^2 - 10x$

(a) $f(4) = (4)^2 - 10(4) = 16 - 40 = -24$

(b) $f(-8) = (-8)^2 - 10(-8) = 64 + 80 = 144$

(c) $f(x - 4) = (x - 4)^2 - 10(x - 4) = x^2 - 8x + 16 - 10x + 40$

$\qquad = x^2 - 18x + 56$

73. $f(x) = x^4 - x - 5$

(a) $f(-1) = (-1)^4 - (-1) - 5 = 1 + 1 - 5 = -3$

(b) $f\left(\frac{1}{2}\right) = \left(\frac{1}{2}\right)^4 - \frac{1}{2} - 5 = -\frac{87}{16}$

(c) $f\left(2\sqrt{3}\right) = \left(2\sqrt{3}\right)^4 - 2\sqrt{3} - 5 = 16(9) - 2\sqrt{3} - 5 = 139 - 2\sqrt{3}$

75. $f(x) = 5 + 6x - x^2, \dfrac{f(6 + h) - f(6)}{h}, h \neq 0$

$$\frac{f(6 + h) - f(6)}{h} = \frac{5 + 6(6 + h) - (6 + h)^2 - 5 - 6(6) + 6^2}{h}$$

$$= \frac{5 + 36 + 6h - 36 - 12h - h^2 - 5 - 36 + 36}{h}$$

$$= \frac{-h^2 - 6h}{h} = -h - 6, h \neq 0$$

Section 6.4   Hyperbolas

- A **hyperbola** is the set of all points *(x, y)* the difference of whose distances from two distinct fixed points **(foci)** is constant.

- The standard equation of a hyperbola with center *(h, k)* and transverse and conjugate axes of lengths 2*a* and 2*b* is:

  (a) $\dfrac{(x - h)^2}{a^2} - \dfrac{(y - k)^2}{b^2} = 1$ if the traverse axis is horizontal.

  (b) $\dfrac{(y - k)^2}{a^2} - \dfrac{(x - h)^2}{b^2} = 1$ if the traverse axis is vertical.

- $c^2 = a^2 + b^2$ where *c* is the distance from the center to a focus.

- The asymptotes of a hyperbola are:

  (a) $y = k \pm \dfrac{b}{a}(x - h)$ if the transverse axis is horizontal.

  (b) $y = k \pm \dfrac{a}{b}(x - h)$ if the transverse axis is vertical.

- The eccentricity of a hyperbola is $e = \dfrac{c}{a}$.

- To classify a nondegenerate conic from its general equation $Ax^2 + Cy^2 + Dx + Ey + F = 0$:
  (a) If $A = C$ $(A \neq 0, C \neq 0)$, then it is a circle.
  (b) If $AC = 0$ $(A = 0$ or $C = 0$, but not both), then it is a parabola.
  (c) If $AC > 0$, then it is an ellipse.
  (d) If $AC < 0$, then it is a hyperbola.

Solutions to Odd-Numbered Exercises

1. $\dfrac{y^2}{9} - \dfrac{x^2}{25} = 1$

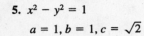

Center: $(0, 0)$

$a = 3, b = 5$

Vertical transverse axis

Matches graph (b).

3. $\dfrac{(x - 1)^2}{16} - \dfrac{y^2}{4} = 1$

Center: $(1, 0)$

$a = 4, b = 2$

Horizontal transverse axis

Matches graph (a).

5. $x^2 - y^2 = 1$

$a = 1, b = 1, c = \sqrt{2}$

Center: $(0, 0)$

Vertices: $(\pm 1, 0)$

Foci: $\left(\pm \sqrt{2}, 0\right)$

Asymptotes: $y = \pm x$

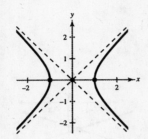

7. $\dfrac{y^2}{25} - \dfrac{x^2}{81} = 1$

$a = 5, b = 9, c = \sqrt{106}$

Center: $(0, 0)$

Vertices: $(0, \pm 5)$

Foci: $\left(0, \pm \sqrt{106}\right)$

Asymptotes: $y = \pm \dfrac{5}{9}x$

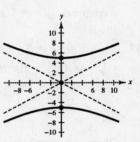

9. $\dfrac{(x - 1)^2}{4} - \dfrac{(y + 2)^2}{1} = 1$

$a = 2, b = 1, c = \sqrt{5}$

Center: $(1, -2)$

Vertices: $(-1, -2), (3, -2)$

Foci: $\left(1 \pm \sqrt{5}, -2\right)$

Asymptotes: $y = -2 \pm \dfrac{1}{2}(x - 1)$

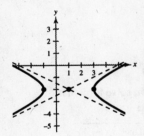

11. $\dfrac{(y + 6)^2}{\frac{1}{9}} - \dfrac{(x - 2)^2}{\frac{1}{4}} = 1$

$a = \dfrac{1}{3}, b = \dfrac{1}{2}, c = \dfrac{\sqrt{13}}{6}$

Center: $(2, -6)$

Vertices: $\left(2, -\dfrac{17}{3}\right), \left(2, -\dfrac{19}{3}\right)$

Foci: $\left(2, -6 \pm \dfrac{\sqrt{13}}{6}\right)$

Asymptotes: $y = -6 \pm \dfrac{2}{3}(x - 2)$

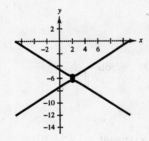

13.
$$9x^2 - y^2 - 36x - 6y + 18 = 0$$
$$9(x^2 - 4x + 4) - (y^2 + 6y + 9) = -18 + 36 - 9$$
$$9(x - 2)^2 - (y + 3)^2 = 9$$
$$\frac{(x - 2)^2}{1} - \frac{(y + 3)^2}{9} = 1$$

$a = 1, b = 3, c = \sqrt{10}$

Center: $(2, -3)$

Vertices: $(1, -3), (3, -3)$

Foci: $\left(2 \pm \sqrt{10}, -3\right)$

Asymptotes: $y = -3 \pm 3(x - 2)$

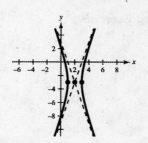

15.
$$x^2 - 9y^2 + 2x - 54y - 80 = 0$$
$$(x^2 + 2x + 1) - 9(y^2 + 6y + 9) = 80 + 1 - 81$$
$$(x + 1)^2 - 9(y + 3)^2 = 0$$
$$y + 3 = \pm\frac{1}{3}(x + 1)$$

Degenerate hyperbola is two lines intersecting at $(-1, -3)$.

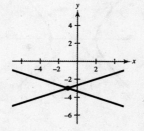

17. $2x^2 - 3y^2 = 6$

$$\frac{x^2}{3} - \frac{y^2}{2} = 1$$

$a = \sqrt{3}, b = \sqrt{2}, c = \sqrt{5}$

Center: $(0, 0)$

Vertices: $\left(\pm\sqrt{3}, 0\right)$

Foci: $\left(\pm\sqrt{5}, 0\right)$

Asymptotes: $y = \pm\sqrt{\frac{2}{3}}\,x = \pm\frac{\sqrt{6}}{3}x$

To use a graphing calculator, solve first for y.

$$y^2 = \frac{2x^2 - 6}{3}$$

$\left.\begin{array}{l} y_1 = \sqrt{\dfrac{2x^2 - 6}{3}} \\[1.2em] y_2 = -\sqrt{\dfrac{2x^2 - 6}{3}} \end{array}\right\}$ Hyperbola

$\left.\begin{array}{l} y_3 = \dfrac{\sqrt{6}}{3}x \\[1.2em] y_4 = -\dfrac{\sqrt{6}}{3}x \end{array}\right\}$ Asymptotes

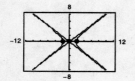

19.
$$9y^2 - x^2 + 2x + 54y + 62 = 0$$
$$9(y^2 + 6y + 9) - (x^2 - 2x + 1) = -62 - 1 + 81$$
$$9(y + 3)^2 - (x - 1)^2 = 18$$
$$\frac{(y + 3)^2}{2} - \frac{(x - 1)^2}{18} = 1$$

$a = \sqrt{2}, b = 3\sqrt{2}, c = 2\sqrt{5}$

Center: $(1, -3)$

Vertices: $\left(1, -3 \pm \sqrt{2}\right)$

Foci: $\left(1, -3 \pm 2\sqrt{5}\right)$

Asymptotes: $y = -3 \pm \frac{1}{3}(x - 1)$

To use a graphing calculator, solve for y first.

$$9(y + 3)^2 = 18 + (x - 1)^2$$
$$y = -3 \pm \sqrt{\frac{18 + (x - 1)^2}{9}}$$

$\left.\begin{array}{l} y_1 = -3 + \dfrac{1}{3}\sqrt{18 + (x - 1)^2} \\[1.2em] y_2 = -3 - \dfrac{1}{3}\sqrt{18 + (x - 1)^2} \end{array}\right\}$ Hyperbola

$\left.\begin{array}{l} y_3 = -3 + \dfrac{1}{3}(x - 1) \\[1.2em] y_4 = -3 - \dfrac{1}{3}(x - 1) \end{array}\right\}$ Asymptotes

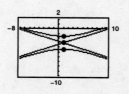

21. Vertices: $(0, \pm 2) \Rightarrow a = 2$

Foci: $(0, \pm 4) \Rightarrow c = 4$

$b^2 = c^2 - a^2 = 16 - 4 = 12$

Center: $(0, 0) = (h, k)$

$\dfrac{(y - k)^2}{a^2} - \dfrac{(x - h)^2}{b^2} = 1$

$\dfrac{y^2}{4} - \dfrac{x^2}{12} = 1$

23. Vertices: $(\pm 1, 0) \Rightarrow a = 1$

Asymptotes: $y = \pm 5x \Rightarrow \dfrac{b}{a} = 5, b = 5$

Center: $(0, 0) = (h, k)$

$\dfrac{(x - h)^2}{a^2} - \dfrac{(y - k)^2}{b^2} = 1$

$\dfrac{x^2}{1} - \dfrac{y^2}{25} = 1$

25. Foci: $(0, \pm 8) \Rightarrow c = 8$

Asymptotes: $y = \pm 4x \Rightarrow \dfrac{a}{b} = 4 \Rightarrow a = 4b$

Center: $(0, 0) = (h, k)$

$c^2 = a^2 + b^2 \Rightarrow 64 = 16b^2 + b^2$

$\dfrac{64}{17} = b^2 \Rightarrow a^2 = \dfrac{1024}{17}$

$\dfrac{(y - k)^2}{a^2} - \dfrac{(x - h)^2}{b^2} = 1$

$\dfrac{y^2}{\frac{1024}{17}} - \dfrac{x^2}{\frac{64}{17}} = 1$

$\dfrac{17y^2}{1024} - \dfrac{17x^2}{64} = 1$

27. Vertices: $(2, 0), (6, 0) \Rightarrow a = 2$

Foci: $(0, 0), (8, 0) \Rightarrow c = 4$

$b^2 = c^2 - a^2 = 16 - 4 = 12$

Center: $(4, 0) = (h, k)$

$\dfrac{(x - h)^2}{a^2} - \dfrac{(y - k)^2}{b^2} = 1$

$\dfrac{(x - 4)^2}{4} - \dfrac{y^2}{12} = 1$

29. Vertices: $(4, 1), (4, 9) \Rightarrow a = 4$

Foci: $(4, 0), (4, 10) \Rightarrow c = 5$

$b^2 = c^2 - a^2 = 25 - 16 = 9$

Center: $(4, 5) = (h, k)$

$\dfrac{(y - k)^2}{a^2} - \dfrac{(x - h)^2}{b^2} = 1$

$\dfrac{(y - 5)^2}{16} - \dfrac{(x - 4)^2}{9} = 1$

31. Vertices: $(2, 3), (2, -3) \Rightarrow a = 3$

Passes through the point: $(0, 5)$

Center: $(2, 0) = (h, k)$

$\dfrac{(y - k)^2}{a^2} - \dfrac{(x - h)^2}{b^2} = 1$

$\dfrac{y^2}{9} - \dfrac{(x - 2)^2}{b^2} = 1 \Rightarrow \dfrac{(x - 2)^2}{b^2} = \dfrac{y^2}{9} - 1 = \dfrac{y^2 - 9}{9} \Rightarrow b^2 = \dfrac{9(x - 2)^2}{y^2 - 9} = \dfrac{9(-2)^2}{25 - 9} = \dfrac{36}{16} = \dfrac{9}{4}$

$\dfrac{y^2}{9} - \dfrac{(x - 2)^2}{9/4} = 1$

$\dfrac{y^2}{9} - \dfrac{4(x - 2)^2}{9} = 1$

33. Vertices: $(0, 4), (0, 0) \Rightarrow a = 2$

Passes through the point $\left(\sqrt{5}, -1\right)$

Center: $(0, 2) = (h, k)$

$$\frac{(y - k)^2}{a^2} - \frac{(x - h)^2}{b^2} = 1$$

$$\frac{(y - 2)^2}{4} - \frac{x^2}{b^2} = 1 \Rightarrow \frac{x^2}{b^2} = \frac{(y - 2)^2}{4} - 1 = \frac{(y - 2)^2 - 4}{4}$$

$$\Rightarrow b^2 = \frac{4x^2}{(y - 2)^2 - 4} = \frac{4\left(\sqrt{5}\right)^2}{(-1 - 2)^2 - 4} = \frac{20}{5} = 4$$

$$\frac{(y - 2)^2}{4} - \frac{x^2}{4} = 1$$

35. Vertices: $(1, 2), (3, 2) \Rightarrow a = 1$

Asymptotes: $y = x, y = 4 - x$

$$\frac{b}{a} = 1 \Rightarrow \frac{b}{1} = 1 \Rightarrow b = 1$$

Center: $(2, 2) = (h, k)$

$$\frac{(x - h)^2}{a^2} - \frac{(y - k)^2}{b^2} = 1$$

$$\frac{(x - 2)^2}{1} - \frac{(y - 2)^2}{1} = 1$$

37. Vertices: $(0, 2), (6, 2) \Rightarrow a = 3$

Asymptotes: $y = \frac{2}{3}x, y = 4 - \frac{2}{3}x$

$$\frac{b}{a} = \frac{2}{3} \Rightarrow b = 2$$

Center: $(3, 2) = (h, k)$

$$\frac{(x - h)^2}{a^2} - \frac{(y - k)^2}{b^2} = 1$$

$$\frac{(x - 3)^2}{9} - \frac{(y - 2)^2}{4} = 1$$

39. Since $\overline{AB} = 1100$ feet and the sound takes one second longer to reach B than A, the explosion must occur on the vertical line through A and B below A.

Foci: $(\pm 3300, 0) \Rightarrow c = 3300$

Center: $(0, 0) = (h, k)$

$$\frac{\overline{CE}}{1100} - \frac{\overline{AE}}{1100} = 4 \Rightarrow 2a = 4400, a = \frac{4400}{2} = 2200$$

$$b^2 = c^2 - a^2 = (3300)^2 - (2200)^2 = 6{,}050{,}000$$

$$\frac{x^2}{(2200)^2} - \frac{y^2}{6{,}050{,}000} = 1$$

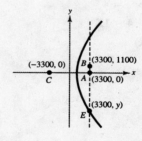

$$y^2 = 6{,}050{,}000\left(\frac{x^2}{(2200)^2} - 1\right)$$

$$y^2 = 6{,}050{,}000\left(\frac{(3300)^2}{(2200)^2} - 1\right) = 7{,}562{,}500$$

$$y = -2750$$

The explosion occurs at $(3300, -2750)$.

41. Center: $(0, 0) = (h, k)$

Focus: $(24, 0) \Rightarrow c = 24$

Solution point: $(24, 24)$

$24^2 = a^2 + b^2 \Rightarrow b^2 = 24^2 - a^2$

$\dfrac{(x - h)^2}{a^2} - \dfrac{(y - k)^2}{b^2} = 1$

$\dfrac{x^2}{a^2} - \dfrac{y^2}{24^2 - a^2} = 1 \Rightarrow \dfrac{24^2}{a^2} - \dfrac{24^2}{24^2 - a^2} = 1$

Solving yields $a^2 = \dfrac{(3 - \sqrt{5})24^2}{2} \approx 220.0124$ and $b^2 \approx 355.9876$.

Thus, we have $\dfrac{x^2}{220.0124} - \dfrac{y^2}{355.9876} = 1$.

The right vertex is at $(a, 0) \approx (14.83, 0)$.

43. $x^2 + 4y^2 - 6x + 16y + 21 = 0$

$AC = (1)(4) > 0$ and $A \neq C$

The graph is an ellipse.

45. $y^2 - 6y - 4x + 21 = 0$

$AC = (0)(1) = 0$

The graph is a parabola.

47. $4y^2 - 2x^2 - 4y - 8x - 15 = 0$

$AC = (-2)(4) < 0$

The graph is a hyperbola.

49. $4y^2 + 4x^2 - 24x + 35 = 0$

$A = C$

the graph is a circle.

51. False. For the trivial solution of two intersecting lines to occur, the standard form of the equation of the hyperbola would be equal to zero.

$\dfrac{(x - h)^2}{a^2} - \dfrac{(y - k)^2}{b^2} = 0$ or $\dfrac{(y - k)^2}{a^2} - \dfrac{(x - h)^2}{b^2} = 0$

53. The extended diagonals of the central rectangle are the asymptotes of the hyperbola.

55. $\left(3x - \frac{1}{2}\right)(x + 4) = 3x^2 + 12x - \frac{1}{2}x - 2$

$\qquad\qquad\qquad\quad = 3x^2 + \frac{23}{2}x - 2$

57. $[(x + y) + 3]^2 = (x + y)^2 + 6(x + y) + 9$

$\qquad\qquad\qquad = x^2 + 2xy + y^2 + 6x + 6y + 9$

59. $x^2 + 14x + 49 = x^2 + 2(7)x + 7^2 = (x + 7)^2$

61. $6x^3 - 11x^2 - 10x = x(6x^2 - 11x - 10)$

$\qquad\qquad\qquad\quad = x(6x^2 - 15x + 4x - 10)$

$\qquad\qquad\qquad\quad = x[3x(2x - 5) + 2(2x - 5)]$

$\qquad\qquad\qquad\quad = x(3x + 2)(2x - 5)$

63. $4 - x + 4x^2 - x^3 = 1(4 - x) + x^2(4 - x)$

$\qquad\qquad\qquad\quad = (1 + x^2)(4 - x)$

$\qquad\qquad\qquad\quad = (x^2 + 1)(4 - x)$

Section 6.5   Rotation of Conics

- The general second-degree equation $Ax^2 + Bxy + Cy^2 + Dx + Ey + F = 0$ can be rewritten as $A'(x')^2 + C'(y')^2 + D'x' + E'y' + F' = 0$ by rotating the coordinate axes through the angle θ, where $\cot 2\theta = (A - C)/B$.

- $x = x' \cos \theta - y' \sin \theta$
  $y = x' \sin \theta + y' \cos \theta$

- The graph of the nondegenerate equation $Ax^2 + Bxy + Cy^2 + Dx + Ey + F = 0$ is:

  (a)   An ellipse or circle if $B^2 - 4AC < 0$.

  (b)   A parabola if $B^2 - 4AC = 0$.

  (c)   A hyperbola if $B^2 - 4AC > 0$.

Solutions to Odd-Numbered Exercises

1. $\theta = 90°$; Point: $(0, 3)$

$x = x' \cos \theta - y' \sin \theta$ $y = x' \sin \theta + y' \cos \theta$

$0 = x' \cos 90° - y' \sin 90°$ $3 = x' \sin 90° + y' \cos 90°$

$0 = y'$ $3 = x°$

So, $(x', y') = (3, 0)$.

3. $\theta = 30°$; Point: $(1, 3)$

$x = x' \cos \theta - y' \sin \theta$
$y = x' \sin \theta + y' \cos \theta$ $\Rightarrow \begin{cases} 1 = x' \cos 30° - y' \sin 30° \\ 3 = x' \sin 30° + y' \cos 30° \end{cases}$

Solving the system yields $(x', y') = \left(\dfrac{3 + \sqrt{3}}{2}, \dfrac{3\sqrt{3} - 1}{2} \right)$.

5. $\theta = 45°$; Point $(2, 1)$

$x = x' \cos \theta - y' \sin \theta$
$y = x' \sin \theta + y' \cos \theta$ $\Rightarrow \begin{cases} 2 = x' \cos 45° - y' \sin 45° \\ 1 = x' \sin 45° + y' \cos 45° \end{cases}$

Solving the system yields $(x', y') = \left(\dfrac{3\sqrt{2}}{2}, -\dfrac{\sqrt{2}}{2} \right)$.

7. $xy + 1 = 0$

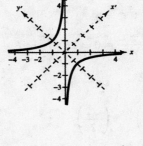

$A = 0, B = 1, C = 0$

$\cot 2\theta = \dfrac{A - C}{B} = 0 \Rightarrow 2\theta = \dfrac{\pi}{2} \Rightarrow \theta = \dfrac{\pi}{4}$

$x = x' \cos \dfrac{\pi}{4} - y' \sin \dfrac{\pi}{4}$ $\qquad\qquad y = x' \sin \dfrac{\pi}{4} + y' \cos \dfrac{\pi}{4}$

$\quad = x'\left(\dfrac{\sqrt{2}}{2}\right) - y'\left(\dfrac{\sqrt{2}}{2}\right)$ $\qquad\qquad = x'\left(\dfrac{\sqrt{2}}{2}\right) + y'\left(\dfrac{\sqrt{2}}{2}\right)$

$\quad = \dfrac{x' - y'}{\sqrt{2}}$ $\qquad\qquad\qquad\qquad = \dfrac{x' + y'}{\sqrt{2}}$

$$xy + 1 = 0$$

$$\left(\dfrac{x' - y'}{\sqrt{2}}\right)\left(\dfrac{x' + y'}{\sqrt{2}}\right) + 1 = 0$$

$$\dfrac{(y')^2}{2} - \dfrac{(x')^2}{2} = 1$$

9. $x^2 - 2xy + y^2 - 1 = 0$

$A = 1, B = -2, C = 1$

$\cot 2\theta = \dfrac{A - C}{B} = 0 \Rightarrow 2\theta = \dfrac{\pi}{2} \Rightarrow \theta = \dfrac{\pi}{4}$

$x = x' \cos \dfrac{\pi}{4} - y' \sin \dfrac{\pi}{4}$ $\qquad\qquad y = x' \sin \dfrac{\pi}{4} + y' \cos \dfrac{\pi}{4}$

$\quad = x'\left(\dfrac{\sqrt{2}}{2}\right) - y'\left(\dfrac{\sqrt{2}}{2}\right)$ $\qquad\qquad = x'\left(\dfrac{\sqrt{2}}{2}\right) + y'\left(\dfrac{\sqrt{2}}{2}\right)$

$\quad = \dfrac{x' - y'}{\sqrt{2}}$ $\qquad\qquad\qquad\qquad = \dfrac{x' + y'}{\sqrt{2}}$

$x^2 - 2xy + y^2 - 1 = 0$

$\left(\dfrac{x' - y'}{\sqrt{2}}\right)^2 - 2\left(\dfrac{x' - y'}{\sqrt{2}}\right)\left(\dfrac{x' + y'}{\sqrt{2}}\right) + \left(\dfrac{x' + y'}{\sqrt{2}}\right) - 1 = 0$

$\dfrac{(x')^2 - 2(x')(y') + (y')^2}{2} - \dfrac{2((x')^2 - (y')^2)}{2} + \dfrac{(x')^2 + 2(x')(y') + (y')^2}{2} - 1 = 0$

$2(y')^2 - 1 = 0$

$\quad (y')^2 = \dfrac{1}{2}$

$\quad y' = \pm\sqrt{\dfrac{1}{2}} = \pm\dfrac{\sqrt{2}}{2}$

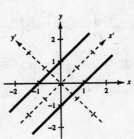

The graph is two parallel lines. Alternate solution.

$x^2 - 2xy + y^2 - 1 = 0$

$\quad (x - y)^2 = 1$

$\quad\quad x - y = \pm 1$

$\quad\quad\quad y = x \pm 1$

11 $xy - 2y - 4x = 0$

$A = 0, B = 1, C = 0$

$$\cot 2\theta = \frac{A - C}{B} = 0 \Rightarrow 2\theta = \frac{\pi}{2} \Rightarrow \theta = \frac{\pi}{4}$$

$x = x'\cos\dfrac{\pi}{4} - y'\sin\dfrac{\pi}{4}$ $\qquad\qquad y = x'\sin\dfrac{\pi}{4} + y'\cos\dfrac{\pi}{4}$

$\quad = x\left(\dfrac{\sqrt{2}}{2}\right) - y\left(\dfrac{\sqrt{2}}{2}\right)$ $\qquad\qquad = x\left(\dfrac{\sqrt{2}}{2}\right) + y\left(\dfrac{\sqrt{2}}{2}\right)$

$\quad = \dfrac{x' - y'}{\sqrt{2}}$ $\qquad\qquad\qquad\quad = \dfrac{x' + y'}{\sqrt{2}}$

$xy - 2y - 4x = 0$

$$\left(\frac{x' - y'}{\sqrt{2}}\right)\left(\frac{x' + y'}{\sqrt{2}}\right) - 2\left(\frac{x' + y'}{\sqrt{2}}\right) - 4\left(\frac{x' - y'}{\sqrt{2}}\right) = 0$$

$$\frac{(x')^2}{2} - \frac{(y')^2}{2} - \sqrt{2}x' - \sqrt{2}y' - 2\sqrt{2}x' + 2\sqrt{2}y' = 0$$

$$\left[(x')^2 - 6\sqrt{2}x' + (3\sqrt{2})^2\right] - \left[(y')^2 - 2\sqrt{2}y' + (\sqrt{2})^2\right] = 0 + (3\sqrt{2})^2 - (\sqrt{2})^2$$

$$(x' - 3\sqrt{2})^2 - (y' - \sqrt{2})^2 = 16$$

$$\frac{(x' - 3\sqrt{2})^2}{16} - \frac{(y' - \sqrt{2})^2}{16} = 1$$

13. $5x^2 - 6xy + 5y^2 - 12 = 0$

$A = 5, B = -6, C = 5$

$$\cot 2\theta = \frac{A - C}{B} = 0 \Rightarrow 2\theta = \frac{\pi}{2} \Rightarrow \theta = \frac{\pi}{4}$$

$x = x'\cos\dfrac{\pi}{4} - y'\sin\dfrac{\pi}{4}$ $\qquad\qquad y = x'\sin\dfrac{\pi}{4} + y'\cos\dfrac{\pi}{4}$

$\quad = x\left(\dfrac{\sqrt{2}}{2}\right) - y\left(\dfrac{\sqrt{2}}{2}\right)$ $\qquad\qquad = x\left(\dfrac{\sqrt{2}}{2}\right) + y\left(\dfrac{\sqrt{2}}{2}\right)$

$\quad = \dfrac{x' - y'}{\sqrt{2}}$ $\qquad\qquad\qquad\quad = \dfrac{x' + y'}{\sqrt{2}}$

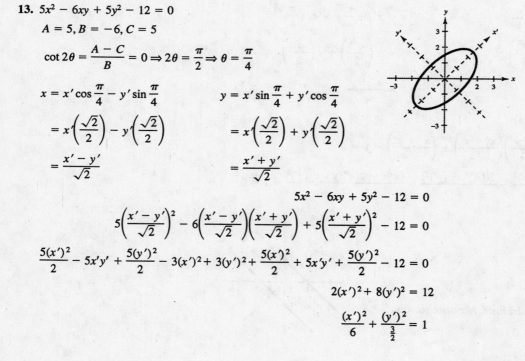

$$5x^2 - 6xy + 5y^2 - 12 = 0$$

$$5\left(\frac{x' - y'}{\sqrt{2}}\right)^2 - 6\left(\frac{x' - y'}{\sqrt{2}}\right)\left(\frac{x' + y'}{\sqrt{2}}\right) + 5\left(\frac{x' + y'}{\sqrt{2}}\right)^2 - 12 = 0$$

$$\frac{5(x')^2}{2} - 5x'y' + \frac{5(y')^2}{2} - 3(x')^2 + 3(y')^2 + \frac{5(x')^2}{2} + 5x'y' + \frac{5(y')^2}{2} - 12 = 0$$

$$2(x')^2 + 8(y')^2 = 12$$

$$\frac{(x')^2}{6} + \frac{(y')^2}{\frac{3}{2}} = 1$$

15. $3x^2 - 2\sqrt{3}xy + y^2 + 2x + 2\sqrt{3}y = 0$

$A = 3, B = -2\sqrt{3}, C = 1$

$\cot 2\theta = \dfrac{A - C}{B} = -\dfrac{1}{\sqrt{3}} \Rightarrow \theta = 60°$

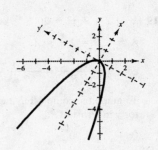

$x = x'\cos 60° - y'\sin 60°$

$\quad = x'\left(\dfrac{1}{2}\right) - y\left(\dfrac{\sqrt{3}}{2}\right) = \dfrac{x' - \sqrt{3}y'}{2}$

$y = x'\sin 60° + y'\cos 60°$

$\quad = x'\left(\dfrac{\sqrt{3}}{2}\right) + y\left(\dfrac{1}{2}\right) = \dfrac{\sqrt{3}x' + y'}{2}$

$3x^2 - 2\sqrt{3}xy + y^2 + 2x + 2\sqrt{3}y = 0$

$3\left(\dfrac{x' - \sqrt{3}y'}{2}\right)^2 - 2\sqrt{3}\left(\dfrac{x' - \sqrt{3}y'}{2}\right)\left(\dfrac{\sqrt{3}x' + y'}{2}\right) + \left(\dfrac{\sqrt{3}x' + y'}{2}\right)^2 + 2\left(\dfrac{x' - \sqrt{3}y'}{2}\right) + 2\sqrt{3}\left(\dfrac{\sqrt{3}x' + y'}{2}\right) = 0$

$\dfrac{3(x')^2}{4} - \dfrac{6\sqrt{3}x'y'}{4} + \dfrac{9(y')^2}{4} - \dfrac{6(x')^2}{4} + \dfrac{4\sqrt{3}x'y'}{4} + \dfrac{6(y')^2}{4} + \dfrac{3(x')^2}{4} + \dfrac{2\sqrt{3}x'y'}{4} + \dfrac{(y')^2}{4}$

$\qquad\qquad\qquad + x' - \sqrt{3}y' + 3x' + \sqrt{3}y' = 0$

$\qquad\qquad\qquad 4(y')^2 + 4x' = 0$

$\qquad\qquad\qquad x' = -(y')^2$

17. $9x^2 + 24xy + 16y^2 + 90x - 130y = 0$

$A = 9, B = 24, C = 16$

$\cot 2\theta = \dfrac{A - C}{B} = -\dfrac{7}{24} \Rightarrow \theta \approx 53.13°$

$\cos 2\theta = -\dfrac{7}{25}$

$\sin \theta = \sqrt{\dfrac{1 - \cos \theta}{2}} = \sqrt{\dfrac{1 - \left(-\frac{7}{25}\right)}{2}} = \dfrac{4}{5}$

$\cos \theta = \sqrt{\dfrac{1 + \cos 2\theta}{2}} = \sqrt{\dfrac{1 + \left(-\frac{7}{25}\right)}{2}} = \dfrac{3}{5}$

$x = x'\cos \theta - y'\sin \theta$

$\quad = x'\left(\dfrac{3}{5}\right) - y'\left(\dfrac{4}{5}\right) = \dfrac{3x' - 4y'}{5}$

$y = x'\sin \theta + y'\cos \theta$

$\quad = x'\left(\dfrac{4}{5}\right) + y'\left(\dfrac{3}{5}\right)$

$\quad = \dfrac{4x' + 3y'}{5}$

$9x^2 + 24xy + 16y^2 + 90x - 130y = 0$

$9\left(\dfrac{3x' - 4y'}{5}\right)^2 + 24\left(\dfrac{3x' - 4y'}{5}\right)\left(\dfrac{4x' + 3y'}{5}\right) + 16\left(\dfrac{4x' + 3y'}{5}\right)^2 + 90\left(\dfrac{3x' - 4y'}{5}\right) - 130\left(\dfrac{4x' + 3y'}{5}\right) = 0$

$\dfrac{81(x')^2}{25} - \dfrac{216x'y'}{25} + \dfrac{144(y')^2}{25} + \dfrac{288(x')^2}{25} - \dfrac{168x'y'}{25} - \dfrac{288(y')^2}{25} + \dfrac{256(x')^2}{25} + \dfrac{384x'y'}{25}$

$\qquad\qquad + \dfrac{144(y')^2}{25} + 54x' - 72y' - 104x' - 78y' = 0$

$\qquad\qquad 25(x')^2 - 50x' - 150y' = 0$

$\qquad\qquad (x')^2 - 2x' + 1 = 6y' + 1$

$\qquad\qquad y' = \dfrac{(x')^2}{6} - \dfrac{x'}{3}$

19. $x^2 + 2xy + y^2 = 20$

$A = 1, B = 2, C = 1$

$\cot 2\theta = \dfrac{A - C}{B} = \dfrac{1 - 1}{2} = 0 \Rightarrow \theta = \dfrac{\pi}{4}$ or $45°$

To graph the conic using a graphing calculator, we need to solve for y in terms of x.

$(x + y)^2 = 20$

$x + y = \pm\sqrt{20}$

$y = -x \pm \sqrt{20}$

Use $y_1 = -x + \sqrt{20}$

and $y_2 = -x - \sqrt{20}$

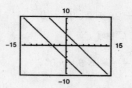

21. $17x^2 + 32xy - 7y^2 = 75$

$\cot 2\theta = \dfrac{A - C}{B} = \dfrac{17 + 7}{32} = \dfrac{24}{32} = \dfrac{3}{4} \Rightarrow \theta \approx 26.57°$

Solve for y in terms of x by completing the square.

$-7y^2 + 32xy = -17x^2 + 75$

$y^2 - \dfrac{32}{7}xy = \dfrac{17}{7}x^2 - \dfrac{75}{7}$

$y^2 - \dfrac{32}{7}xy + \dfrac{256}{49}x^2 = \dfrac{119}{49}x^2 - \dfrac{525}{49} + \dfrac{256}{49}x^2$

$\left(y - \dfrac{16}{7}x\right)^2 = \dfrac{375x^2 - 525}{49}$

$y = \dfrac{16}{7}x \pm \sqrt{\dfrac{375x^2 - 525}{49}}$

$y = \dfrac{16x \pm 5\sqrt{15x^2 - 21}}{7}$

Use $y_1 = \dfrac{16x + 5\sqrt{15x^2 - 21}}{7}$

and $y_2 = \dfrac{16x - 5\sqrt{15x^2 - 21}}{7}.$

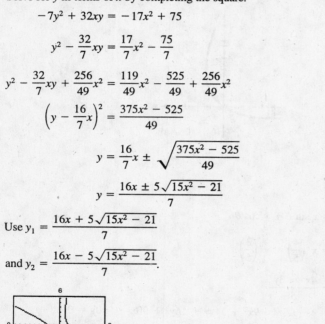

23. $32x^2 + 48xy + 8y^2 = 50$

$\cot 2\theta = \dfrac{A - C}{B} = \dfrac{24}{48} = \dfrac{1}{2} \Rightarrow \theta \approx 31.72°$

Solve for y in terms of x by completing the square.

$8y^2 + 48xy = -32x^2 + 50$

$y^2 + 6xy = -4x^2 + \dfrac{25}{4}$

$y^2 + 6xy + 9x^2 = -4x^2 + \dfrac{25}{4} + 9x^2$

$(y + 3x)^2 = 5x^2 + \dfrac{25}{4}$

$y + 3x = \pm\sqrt{5x^2 + \dfrac{25}{4}}$

$y = -3x \pm \sqrt{5x^2 + \dfrac{25}{4}}$

Use $y_1 = -3x + \sqrt{5x^2 + \dfrac{25}{4}}$

and $y_2 = -3x - \sqrt{5x^2 + \dfrac{25}{4}}$

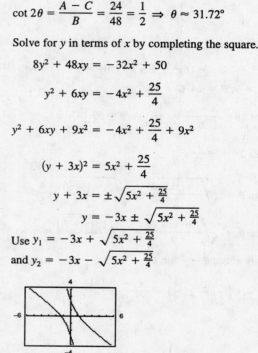

25. $4x^2 - 12xy + 9y^2 + (4\sqrt{13} - 12)x - (6\sqrt{13} + 8)y = 91$

$A = 4, B = -12, C = 9$

$\cot 2\theta = \dfrac{A - C}{B} = \dfrac{4 - 9}{-12} = \dfrac{5}{12}$

$\dfrac{1}{\tan 2\theta} = \dfrac{5}{12}$

$\tan 2\theta = \dfrac{12}{5}$

$2\theta \approx 67.38°$

$\theta \approx 33.69°$

Solve for y in terms of x with the quadratic formula:

$4x^2 - 12xy + 9y^2 + (4\sqrt{13} - 12)x - (6\sqrt{13} + 8)y = 91$

$9y^2 - (12x + 6\sqrt{13} + 8)y + (4x^2 + 4\sqrt{13}x - 12x - 91) = 0$

$a = 9, b = -(12x + 6\sqrt{13} + 8), c = 4x^2 + 4\sqrt{13}x - 12x - 91$

$y = \dfrac{-b \pm \sqrt{b^2 - 4ac}}{2a}$

$y = \dfrac{(12x + 6\sqrt{13} + 8) \pm \sqrt{(12x + 6\sqrt{13} + 8)^2 - 4(9)(4x^2 + 4\sqrt{13}x - 12x - 91)}}{18}$

$= \dfrac{(12x + 6\sqrt{13} + 8) \pm \sqrt{624x + 3808 + 96\sqrt{13}}}{18}$

Enter $y_1 = \dfrac{12x + 6\sqrt{13} + 8 + \sqrt{624x + 3808 + 96\sqrt{13}}}{18}$

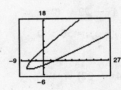

and $y_2 = \dfrac{12x + 6\sqrt{13} + 8 - \sqrt{624x + 3808 + 96\sqrt{13}}}{18}$.

27. $xy + 2 = 0$

$B^2 - 4AC = 1 \Rightarrow$ The graph is a hyperbola.

$\cot 2\theta = \dfrac{A - C}{B} = 0 \Rightarrow \theta = 45°$

Matches graph (e).

29. $-2x^2 + 3xy + 2y^2 + 3 = 0$

$B^2 - 4AC = (3)^2 - 4(-2)(2) = 25 \Rightarrow$ The graph is a hyperbola.

$\cot 2\theta = \dfrac{A - C}{B} = -\dfrac{4}{3} \Rightarrow \theta \approx -18.43°$

Matches graph (b).

31. $3x^2 + 2xy + y^2 - 10 = 0$

$B^2 - 4AC = (2)^2 - 4(3)(1) = -8 \Rightarrow$ The graph is an ellipse or circle.

$\cot 2\theta = \dfrac{A - C}{B} = 1 \Rightarrow \theta = 22.5°$

Matches graph (d).

33. $16x^2 - 8xy + y^2 - 10x + 5y = 0$

$B^2 - 4AC = (-8)^2 - 4(16)(1) = 0$

The graph is a parabola.

$y^2 + (-8x + 5)y + (16x^2 - 10x) = 0$

$$y = \frac{-(-8x + 5) \pm \sqrt{(-8x + 5)^2 - 4(1)(16x^2 - 10x)}}{2(1)}$$

$$= \frac{8x - 5 \pm \sqrt{(8x - 5)^2 - 4(16x^2 - 10x)}}{2}$$

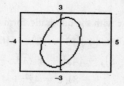

35. $12x^2 - 6xy + 7y^2 - 45 = 0$

$B^2 - 4AC = (-6)^2 - 4(12)(7) = -300 < 0$

The graph is an ellipse.

$7y^2 + (-6x)y + (12x^2 - 45) = 0$

$$y = \frac{-(-6x) \pm \sqrt{(-6x)^2 - 4(7)(12x^2 - 45)}}{2(7)}$$

$$= \frac{6x \pm \sqrt{36x^2 - 28(12x^2 - 45)}}{14}$$

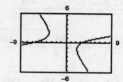

37. $x^2 - 6xy - 5y^2 + 4x - 22 = 0$

$B^2 - 4AC = (-6)^2 - 4(1)(-5) = 56 > 0$

The graph is a hyperbola.

$-5y^2 + (-6x)y + (x^2 + 4x - 22) = 0$

$$y = \frac{-(-6x) \pm \sqrt{(-6x)^2 - 4(-5)(x^2 + 4x - 22)}}{2(-5)}$$

$$= \frac{6x \pm \sqrt{36x^2 + 20(x^2 + 4x - 22)}}{-10}$$

$$= \frac{-6x \pm \sqrt{36x^2 + 20(x^2 + 4x - 22)}}{10}$$

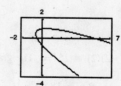

39. $x^2 + 4xy + 4y^2 - 5x - y - 3 = 0$

$B^2 - 4AC = (4)^2 - 4(1)(4) = 0$

The graph is a parabola.

$4y^2 + (4x - 1)y + (x^2 - 5x - 3) = 0$

$$y = \frac{-(4x - 1) \pm \sqrt{(4x - 1)^2 - 4(4)(x^2 - 5x - 3)}}{2(4)}$$

$$= \frac{-(4x - 1) \pm \sqrt{(4x - 1)^2 - 16(x^2 - 5x - 3)}}{8}$$

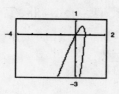

41. $y^2 - 9x^2 = 0$

$y^2 = 9x^2$

$y = \pm 3x$

Two intersecting lines

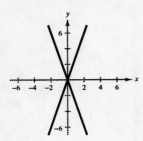

43. $x^2 + 2xy + y^2 - 1 = 0$

$(x + y)^2 - 1 = 0$

$(x + y)^2 = 1$

$x + y = \pm 1$

$y = -x \pm 1$

Two parallel lines

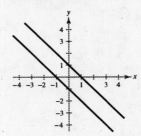

45.

$-x^2 + y^2 + 4x - 6y + 4 = 0 \Rightarrow (y - 3)^2 - (x - 2)^2 = 1$

$\underline{x^2 + y^2 - 4x - 6y + 12 = 0 \Rightarrow (x - 2)^2 + (y - 3)^2 = 1}$

$2y^2 - 12y + 16 = 0$

$2(y - 2)(y - 4) = 0$

$y = 2 \text{ or } y = 4$

For $y = 2$: $x^2 + 2^2 - 4x - 6(2) + 12 = 0$

$x^2 - 4x + 4 = 0$

$(x - 2)^2 = 0$

$x = 2$

For $y = 4$: $x^2 + 4^2 - 4x - 6(4) + 12 = 0$

$x^2 - 4x + 4 = 0$

$(x - 2)^2 = 0$

$x = 2$

The points of intersection are $(2, 2)$ and $(2, 4)$.

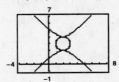

47.
$$
\begin{array}{r}
-4x^2 - y^2 - 16x + 24y - 16 = 0 \\
\underline{4x^2 + y^2 + 40x - 24y + 208 = 0} \\
24x \quad\quad + 192 = 0 \\
x = -8
\end{array}
$$

When $x = -8$: $4(-8)^2 + y^2 + 40(-8) - 24y + 208 = 0$

$$y^2 - 24y + 144 = 0$$

$$(y - 12)^2 = 0$$

$$y = 12$$

The point of intersection is $(-8, 12)$.

In standard form the equations are:

$$\frac{(x + 2)^2}{36} + \frac{(y - 12)^2}{144} = 1$$

$$\frac{(x + 5)^2}{9} + \frac{(y - 12)^2}{36} = 1$$

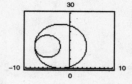

49.
$$
\begin{array}{r}
x^2 - y^2 - 12x + 16y - 64 = 0 \\
\underline{x^2 + y^2 - 12x - 16y + 64 = 0} \\
2x^2 \quad - 24x \quad\quad = 0 \\
2x(x - 12) = 0 \\
x = 0 \quad \text{or} \quad x = 12
\end{array}
$$

When $x = 0$: $0^2 + y^2 - 12(0) - 16y + 64 = 0$

$$y^2 - 16y + 64 = 0$$

$$(y - 8)^2 = 0$$

$$y = 8$$

When $x = 12$: $12^2 + y^2 - 12(12) - 16y + 64 = 0$

$$y^2 - 16y + 64 = 0$$

$$(y - 8)^2 = 0$$

$$y = 8$$

The points of intersection are $(0, 8)$ and $(12, 8)$.

The standard forms of the equations are:

$$\frac{(x - 6)^2}{36} - \frac{(y - 8)^2}{36} = 1$$

$$(x - 6)^2 + (y - 8)^2 = 36$$

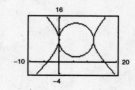

51. $-16x^2 - y^2 + 24y - 80 = 0$

$\underline{16x^2 + 25y^2 - 400 = 0}$

$24y^2 + 24y - 480 = 0$

$24(y + 5)(y - 4) = 0$

$y = -5 \text{ or } y = 4$

When $y = -5$: $16x^2 + 25(-5)^2 - 400 = 0$

$16x^2 = -225$

 No real solution

When $y = 4$: $16x^2 + 25(4)^2 - 400 = 0$

$16x^2 = 0$

$x = 0$

The point of intersection is $(0, 4)$.

In standard form the equations are:

$$\frac{x^2}{4} + \frac{(y - 12)^2}{64} = 1$$

$$\frac{x^2}{25} + \frac{y^2}{16} = 1$$

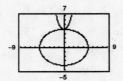

53. $x^2 + y^2 - 4 = 0$

$\underline{3x - y^2 = 0}$

$x^2 + 3x - 4 = 0$

$(x + 4)(x - 1) = 0$

$x = -4 \text{ or } x = 1$

When $x = -4$: $3(-4) - y^2 = 0$

$y^2 = -12$ No real solution

When $x = 1$: $3(1) - y^2 = 0$

$y^2 = 3$

$y = \pm\sqrt{3}$

The points of intersection are $\left(1, \sqrt{3}\right)$ and $\left(1, -\sqrt{3}\right)$.

The standard forms of the equations are:

$x^2 + y^2 = 4$

$y^2 = 3x$

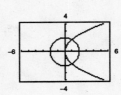

55. $x^2 + 2y^2 - 4x + 6y - 5 = 0$

$-x + y - 4 = 0 \implies y = x + 4$

$x^2 + 2(x + 4)^2 - 4x + 6(x + 4) - 5 = 0$

$x^2 + 2(x^2 + 8x + 16) - 4x + 6x + 24 - 5 = 0$

$3x^2 + 18x + 51 = 0$

$3(x^2 + 6x + 17) = 0$

$x^2 + 6x + 17 = 0$

$x^2 + 6x + 9 = -17 + 9$

$(x + 3)^2 = -8$

No real solution

No points of intersection

The standard forms of the equations are:

$$\frac{(x - 2)^2}{\frac{27}{2}} + \frac{\left(y + \frac{3}{2}\right)^2}{\frac{27}{4}} = 1$$

$x - y = -4$

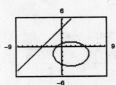

57.
$$xy + x - 2y + 3 = 0 \Rightarrow y = \frac{-x - 3}{x - 2}$$

$$x^2 + 4y^2 - 9 = 0$$

$$x^2 + 4\left(\frac{-x - 3}{x - 2}\right)^2 = 9$$

$$x^2(x - 2)^2 + 4(-x - 3)^2 = 9(x - 2)^2$$

$$x^2(x^2 - 4x + 4) + 4(x^2 + 6x + 9) = 9(x^2 - 4x + 4)$$

$$x^4 - 4x^3 + 4x^2 + 4x^2 + 24x + 36 = 9x^2 - 36x + 36$$

$$x^4 - 4x^3 - x^2 + 60x = 0$$

$$x(x + 3)(x^2 - 7x + 20) = 0$$

$$x = 0 \text{ or } x = -3$$

Note: $x^2 - 7x + 20 = 0$ has no real solution.

When $x = 0$: $\quad y = \frac{-0 - 3}{0 - 2} = \frac{3}{2}$

When $x = -3$: $\quad y = \frac{-(-3) - 3}{-3 - 2} = 0$

The points of intersection are $\left(0, \frac{3}{2}\right), (-3, 0)$.

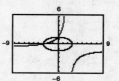

59. $x^2 + xy + ky^2 + 6x + 10 = 0$

$B^2 - 4AC = 1^2 - 4(1)(k) = 1 - 4k > 0 \Rightarrow -4k > -1 \Rightarrow k < \frac{1}{4}$

True. For the graph to be a hyperbola, the discriminant must be greater than zero.

61. $(x')^2 + (y')^2 = (x \cos \theta - y \sin \theta)^2 + (y \cos \theta + x \sin \theta)^2$

$= x^2 \cos^2 \theta - 2xy \cos \theta \sin \theta + y^2 \sin^2 \theta + y^2 \cos^2 \theta + 2xy \cos \theta \sin \theta + x^2 \sin^2 \theta$

$= x^2(\cos^2 \theta + \sin^2 \theta) + y^2(\sin^2 \theta + \cos^2 \theta) = x^2 + y^2 = r^2$

63.
$$f(x) = \sqrt{6x - 5}$$

$$0 = \sqrt{6x - 5}$$

$$6x - 5 = 0 \Rightarrow x = \frac{5}{6}$$

Zero: $x = \frac{5}{6}$

65.
$$g(x) = 3x^2 - 14x + 8$$

$$0 = 3x^2 - 14x + 8$$

$$(3x - 2)(x - 4) = 0$$

$$3x - 2 = 0 \Rightarrow x = \frac{2}{3}$$

$$x - 4 = 0 \Rightarrow x = 4$$

Zeros: $x = \frac{2}{3}, 4$

67. $h(x) = x^3 - 7x^2 + 17x - 15$

$$0 = x^3 - 7x^2 + 17x - 15$$

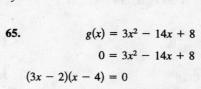

By the Quadratic Formula the zeros of $x^2 - 4x + 5$ are

$$x = \frac{-(-4) \pm \sqrt{(-4)^2 - 4(1)(5)}}{2(1)} = \frac{4 \pm \sqrt{-4}}{2} = 2 \pm i$$

Zeros: $x = 3, 2 \pm i$

69. $f(x) = |x - 4|$

Intercepts: $(0, 4), (4, 0)$

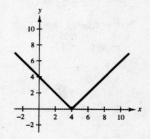

71. $g(x) = \sqrt{4 - x^2}$

Intercepts: $(\pm 2, 0)$, $(0, 2)$

Domain: $-2 \leq x \leq 2$

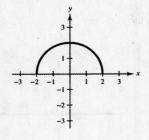

73. $h(x) = x^2 - 10x + 21$

Intercepts: $(3, 0)$, $(7, 0)$, $(0, 21)$

Vertex: $(5, -4)$

x	1	2	3	4	5
$h(x)$	12	5	0	-3	-4

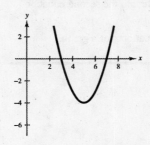

75. $f(t) = -(t - 2)^3 + 3$

x	-1	0	1	2	3
$f(x)$	30	11	4	3	2

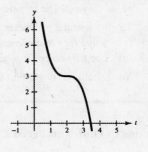

Section 6.6   Parametric Equations

- If f and g are continuous functions of t on an interval I, then the set of ordered pairs $(f(t), g(t))$ is a *plane curve C*. The equations $x = f(t)$ and $y = g(t)$ are *parametric equations* for C and t is the *parameter*.
- To eliminate the parameter:
  - (a) Solve for t in one equation and substitute into the second equation.
  - (b) Use trigonometric identities.
- You should be able to find the parametric equations for a graph.

Solutions to Odd-Numbered Exercises

1. $x = \sqrt{t}, y = 3 - t$

(a)

t	0	1	2	3	4
x	0	1	$\sqrt{2}$	$\sqrt{3}$	2
y	3	2	1	0	-1

(b)

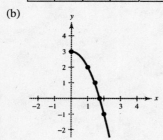

(c) $x = \sqrt{t} \quad \Rightarrow x^2 = t$

$\quad y = 3 - t \quad \Rightarrow \quad y = 3 - x^2$

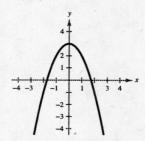

The graph of the parametric equations only shows the right half of the parabola, whereas the rectangular equation yields the entire parabola.

3. $x = 3t - 3 \Rightarrow t = \dfrac{x + 3}{3}$

$y = 2t + 1 \Rightarrow y = \dfrac{2}{3}(x + 3) + 1 = \dfrac{2}{3}x + 3$

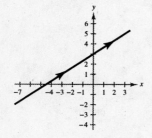

5. $x = \dfrac{1}{4}t \Rightarrow t = 4x$

$y = t^2 \Rightarrow y = 16x^2$

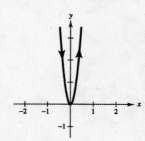

7. $x = t + 2 \Rightarrow t = x - 2$

   $y = t^2 \quad\; \Rightarrow y = (x - 2)^2$

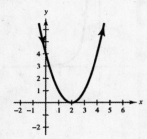

9. $x = t + 1 \Rightarrow t = x - 1$

   $y = \dfrac{t}{t + 1} \Rightarrow y = \dfrac{x - 1}{x} = 1 - \dfrac{1}{x}$

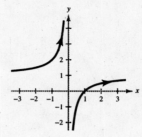

11. $x = 2(t + 1) \Rightarrow \dfrac{x}{2} - 1 = t \quad \text{or} \quad t = \dfrac{x - 2}{2}$

   $y = |t - 2| \quad \Rightarrow \quad y = \left|\dfrac{x}{2} - 1 - 2\right| = \left|\dfrac{x}{2} - 3\right| = \left|\dfrac{x - 6}{2}\right|$

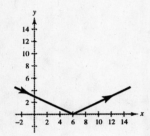

13. $x = 3\cos\theta \Rightarrow \left(\dfrac{x}{3}\right)^2 = \cos^2\theta$

   $y = 3\sin\theta \Rightarrow \left(\dfrac{y}{3}\right)^2 = \sin^2\theta$

   $\left(\dfrac{x}{3}\right)^2 + \left(\dfrac{y}{3}\right)^2 = 1$

   $x^2 + y^2 = 9$

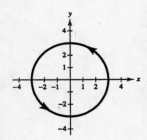

15. $x = 4\sin 2\theta \Rightarrow \left(\dfrac{x}{4}\right)^2 = \sin^2 2\theta$

   $y = 2\cos 2\theta \Rightarrow \left(\dfrac{y}{2}\right)^2 = \cos^2 2\theta$

   $\left(\dfrac{x}{4}\right)^2 + \left(\dfrac{y}{2}\right)^2 = 1$

   $\dfrac{x^2}{16} + \dfrac{y^2}{4} = 1$

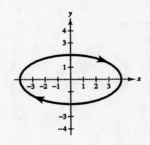

17. $x = 4 + 2 \cos \theta \Rightarrow \left(\dfrac{x-4}{2}\right)^2 = \cos^2 \theta$

$y = -1 + \sin \theta \Rightarrow (y+1)^2 = \sin^2 \theta$

$\dfrac{(x-4)^2}{4} + \dfrac{(y+1)^2}{1} = 1$

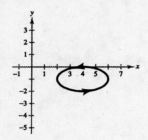

19. $x = e^{-t} \Rightarrow \dfrac{1}{x} = e^t$

$y = e^{3t} \Rightarrow y = (e^t)^3$

$y = \left(\dfrac{1}{x}\right)^3$

$y = \dfrac{1}{x^3}, \; x > 0, \, y > 0$

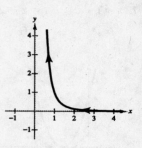

21. $x = t^3 \qquad \Rightarrow x^{1/3} = t$

$y = 3 \ln t \Rightarrow y = \ln t^3$

$y = \ln(x^{1/3})^3$

$y = \ln x$

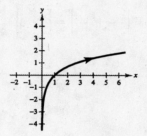

23. By eliminating the parameter, each curve becomes
$y = 2x + 1$.

(a) $x = t$
$y = 2t + 1$
There are no restrictions on x and y.

Domain: $(-\infty, \infty)$

Orientation: Left to right

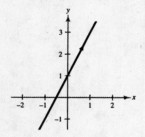

(b) $x = \cos \theta \qquad \Rightarrow -1 \le x \le 1$
$y = 2 \cos \theta + 1 \Rightarrow -1 \le y \le 3$
The graph oscillates.

Domain: $[-1, 1]$

Orientation: Depends on θ

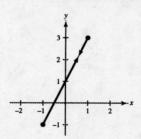

— **CONTINUED** —

23. — CONTINUED —

(c) $x = e^{-t} \quad \Rightarrow x > 0$

$y = 2e^{-t} + 1 \Rightarrow y > 1$

Domain: $(0, \infty)$

Orientation: Downward or right to left

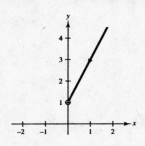

(d) $x = e^t \quad \Rightarrow x > 0$

$y = 2e^t + 1 \Rightarrow y > 1$

Domain: $(0, \infty)$

Orientation: Upward or left to right

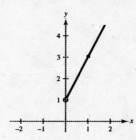

25. $x = x_1 + t(x_2 - x_1), y = y_1 + t(y_2 - y_1)$

$$\frac{x - x_1}{x_2 - x_1} = t$$

$$y = y_1 + \left(\frac{x - x_1}{x_2 - x_1}\right)(y_2 - y_1)$$

$$y - y_1 = \frac{y_2 - y_1}{x_2 - x_1}(x - x_1) = m(x - x_1)$$

27. $x = h + a\cos\theta, y = k + b\sin\theta$

$$\frac{x - h}{a} = \cos\theta, \frac{y - k}{b} = \sin\theta$$

$$\frac{(x - h)^2}{a^2} + \frac{(y - k)^2}{b^2} = 1$$

29. From Exercise 25 we have:

$x = 0 + t(6 - 0) = 6t$

$y = 0 + t(-3 - 0) = -3t$

31. From Exercise 26 we have:

$x = 3 + 4\cos\theta$

$y = 2 + 4\sin\theta$

33. Vertices: $(\pm 4, 0) \Rightarrow (h, k) = (0, 0)$ and $a = 4$

Foci: $(\pm 3, 0) \Rightarrow c = 3$

$c^2 = a^2 - b^2 \Rightarrow 9 = 16 - b^2 \Rightarrow b = \sqrt{7}$

From Exercise 27 we have:

$x = 4\cos\theta$

$y = \sqrt{7}\sin\theta$

35. Vertices: $(\pm 4, 0) \Rightarrow (h, k) = (0, 0)$ and $a = 4$

Foci: $(\pm 5, 0) \Rightarrow c = 5$

$c^2 = a^2 + b^2 \Rightarrow 25 = 16 + b^2 \Rightarrow b = 3$

From Exercise 28 we have:

$x = 4\sec\theta$

$y = 3\tan\theta$

37. $y = 3x - 2$

(a) $t = x \Rightarrow x = t$ and $y = 3t - 2$

(b) $t = 2 - x \Rightarrow x = -t + 2$ and $y = 3(-t + 2) - 2 = -3t + 4$

39. $y = x^2$

(a) $t = x \Rightarrow x = t$ and $y = t^2$

(b) $t = 2 - x \Rightarrow x = -t + 2$ and $y = (-t + 2)^2 = t^2 - 4t + 4$

41. $y = x^2 + 1$

(a) $t = x \Rightarrow x = t$ and $y = t^2 + 1$

(b) $t = 2 - x \Rightarrow x = -t + 2$ and $y = (-t + 2)^2 + 1 = t^2 - 4t + 5$

43. $y = \dfrac{1}{x}$

(a) $t = x \Rightarrow x = t$ and $y = \dfrac{1}{t}$

(b) $t = 2 - x \Rightarrow x = -t + 2$ and $y = \dfrac{1}{-t + 2} = \dfrac{-1}{t - 2}$

45. $x = 4(\theta - \sin \theta)$
$y = 4(1 - \cos \theta)$

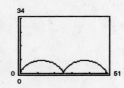

47. $x = \theta - \dfrac{3}{2} \sin \theta$

$y = 1 - \dfrac{3}{2} \cos \theta$

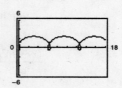

49. $x = 3 \cos^3 \theta$
$y = 3 \sin^3 \theta$

51. $x = 2 \cot \theta$
$y = 2 \sin^2 \theta$

53. $x = 2 \cos \theta \Rightarrow -2 \le x \le 2$
$y = \sin 2\theta \Rightarrow -1 \le y \le 1$
Matches graph (b).
Domain: $[-2, 2]$
Range: $[-1, 1]$

55. $x = \dfrac{1}{2}(\cos \theta + \theta \sin \theta)$

$y = \dfrac{1}{2}(\sin \theta - \theta \cos \theta)$

Matches graph (d).
Domain: $(-\infty, \infty)$
Range: $(-\infty, \infty)$

57. $x = (v_0 \cos \theta)t$ and $y = h + (v_0 \sin \theta)t - 16t^2$

(a) $\theta = 60°$, $v_0 = 88$ ft/sec

$x = (88 \cos 60°)t$ and $y = (88 \sin 60°)t - 16t^2$

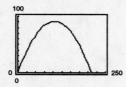

Maximum height: 90.7 feet

Range: 209.6 feet

(b) $\theta = 60°$, $v_0 = 132$ ft/sec

$x = (132 \cos 60°)t$ and $y = (132 \sin 60°)t - 16t^2$

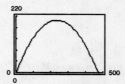

Maximum height: 204.2 feet

Range: 471.6 feet

— CONTINUED —

57. — CONTINUED —

(c) $\theta = 45°$, $v_0 = 88$ ft/sec

$x = (88 \cos 45°)t$ and $y = (88 \sin 45°)t - 16t^2$

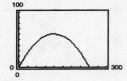

Maximum height: 60.5 ft

Range: 242.0 ft

(d) $\theta = 45°$, $v_0 = 132$ ft/sec

$x = (132 \cos 45°)t$ and $y = (132 \sin 45°)t - 16t^2$

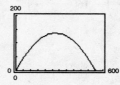

Maximum height: 136.1 ft

Range: 544.5 ft

59. (a) 100 miles per hour $= 100\left(\dfrac{5280}{3600}\right)$ ft/sec $= \dfrac{440}{3}$ ft/sec

$x = \left(\dfrac{440}{3}\cos\theta\right)t \approx (146.67\cos\theta)t$

$y = 3 + \left(\dfrac{440}{3}\sin\theta\right)t - 16t^2 \approx 3 + (146.67\sin\theta)t - 16t^2$

(b) For $\theta = 15°$, we have:

$x = \left(\dfrac{440}{3}\cos 15°\right)t \approx 141.7t$

$y = 3 + \left(\dfrac{440}{3}\sin 15°\right)t - 16t^2 \approx 3 + 38.0t - 16t^2$

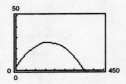

The ball hits the ground inside the ballpark, so it is not a home run.

(c) For $\theta = 23°$, we have:

$x = \left(\dfrac{440}{3}\cos 23°\right)t \approx 135.0t$

$y = 3 + \left(\dfrac{440}{3}\sin 23°\right)t - 16t^2 \approx 3 + 57.3t - 16t^2$

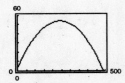

The ball easily clears the 10-foot fence at 400 feet so it is a home run.

(d) Find θ so that $y = 10$ when $x = 400$ by graphing the parametric equations for θ values between $15°$ and $23°$. This occurs when $\theta \approx 19.4°$.

61. $x = (v_0 \cos \theta)t \Rightarrow t = \dfrac{x}{v_0 \cos \theta}$

$y = h + (v_0 \sin \theta)t - 16t^2$

$\quad = h + (v_0 \sin \theta)\left(\dfrac{x}{v_0 \cos \theta}\right) - 16\left(\dfrac{x}{v_0 \cos \theta}\right)^2$

$\quad = h + (\tan \theta)x - \dfrac{16x^2}{v_0^2 \cos^2 \theta}$

$\quad = -\dfrac{16 \sec^2 \theta}{v_0^2}x^2 + (\tan \theta)x + h$

63. When the circle has rolled θ radians, the center is at $(a\theta, a)$.

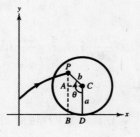

$\sin \theta = \sin(180° - \theta)$

$\qquad = \dfrac{|AC|}{b} = \dfrac{|BD|}{b} \Rightarrow |BD| = b \sin \theta$

$\cos \theta = -\cos(180° - \theta)$

$\qquad = \dfrac{|AP|}{-b} \Rightarrow |AP| = -b \cos \theta$

Therefore, $x = a\theta - b \sin \theta$ and $y = a - b \cos \theta$.

65. True

$x = t$

$y = t^2 + 1 \Rightarrow y = x^2 + 1$

$x = 3t$

$y = 9t^2 + 1 \Rightarrow y = x^2 + 1$

67. $y = ax^2 + bx + c$

$(0, 0): \quad 0 = a(0)^2 + b(0) + c \Rightarrow \quad c = \quad 0$

$(3, -3): -3 = a(3)^2 + b(3) \quad \Rightarrow -3 = \quad 9a + 3b \Rightarrow -1 = \quad 3a + b$

$(6, 0): \quad 0 = a(6)^2 + b(6) \quad \Rightarrow \quad 0 = 36a + 6b \Rightarrow \quad b = -6a$

$\qquad\qquad b = -6a$

$\qquad\quad -1 = \quad 3a + b \Rightarrow -1 = 3a - 6a$

$\qquad\qquad\qquad\qquad -1 = -3a$

$\qquad\qquad\qquad\qquad \tfrac{1}{3} = a$

$\qquad\qquad\qquad\quad b = -6\left(\tfrac{1}{3}\right) = -2$

Thus, $y = \tfrac{1}{3}x^2 - 2x$

69. $y = ax^2 + bx + c$

$(4, 12):$ $12 = a(4)^2 + b(4) + c \Rightarrow 16a + 4b + c = 12$

$(8, -2):$ $-2 = a(8)^2 + b(8) + c \Rightarrow 64a + 8b + c = -2$

$(12, 12):$ $12 = a(12)^2 + b(12) + c \Rightarrow 144a + 12b + c = 12$

$$\begin{array}{rl} 64a + 8b + c = & -2 \\ \underline{-16a - 4b - c = -12} & \\ 48a + 4b \quad\quad = & -14 \end{array}$$

$$\begin{array}{rl} 144a + 12b + c = 12 \\ \underline{-64a - 8b - c = 2} \\ 80a + 4b \quad\quad = 14 \end{array}$$

$$\begin{array}{rl} 80a + 4b = 14 \\ \underline{-48a - 4b = 14} \\ 32a \quad\quad = 28 \\ a = \tfrac{28}{32} = \tfrac{7}{8} \end{array}$$

$80\left(\tfrac{7}{8}\right) + 4b = 14 \Rightarrow 4b = -56 \Rightarrow b = -14$

$16\left(\tfrac{7}{8}\right) + 4(-14) + c = 12 \Rightarrow c = 54$

Thus, $y = \tfrac{7}{8}x^2 - 14x + 54$

Section 6.7   Polar Coordinates

- In polar coordinates you do not have unique representation of points. The point (r, θ) can be represented by $(r, \theta \pm 2n\pi)$ or by $(-r, \theta \pm (2n + 1)\pi)$ where n is any integer. The pole is represented by $(0, \theta)$ where θ is any angle.

- To convert from polar coordinates to rectangular coordinates, use the following relationships.
  $x = r \cos \theta$
  $y = r \sin \theta$

- To convert from rectangular coordinates to polar coordinates, use the following relationships.
  $r = \pm \sqrt{x^2 + y^2}$
  $\tan \theta = y/x$

  If θ is in the same quadrant as the point (x, y), then r is positive. If θ is in the opposite quadrant as the point (x, y), then r is negative.

- You should be able to convert rectangular equations to polar form and vice versa.

Solutions to Odd-Numbered Exercises

1. Polar Coordinates: $\left(4, -\dfrac{\pi}{3}\right)$

Additional representations

$\left(4, -\dfrac{\pi}{3} + 2\pi\right) = \left(4, \dfrac{5\pi}{3}\right)$

$\left(-4, -\dfrac{\pi}{3} - \pi\right) = \left(-4, -\dfrac{4\pi}{3}\right)$

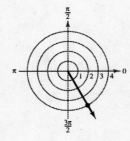

3. Polar Coordinates: $\left(0, -\dfrac{\pi}{6}\right)$

Additional representations

$\left(0, -\dfrac{7\pi}{6} + 2\pi\right) = \left(0, \dfrac{5\pi}{6}\right)$

$\left(0, -\dfrac{7\pi}{6} + \pi\right) = \left(0, -\dfrac{\pi}{6}\right)$

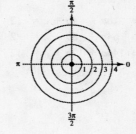

5. Polar Coordinates: $\left(\sqrt{2}, 2.36\right)$

Additional representations

$\left(\sqrt{2}, 2.36 + 2\pi\right) \approx \left(\sqrt{2}, 8.64\right)$

$\left(-\sqrt{2}, 2.36 - \pi\right) \approx \left(-\sqrt{2}, -0.78\right)$

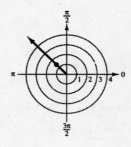

7. Polar Coordinates: $\left(2\sqrt{2}, 4.71\right)$

Additional representations

$\left(2\sqrt{2}, 4.71 + 2\pi\right) \approx \left(2\sqrt{2}, 10.99\right)$

$\left(-2\sqrt{2}, 4.71 - \pi\right) \approx \left(-2\sqrt{2}, 1.57\right)$

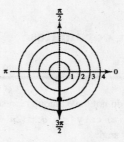

9. Polar Coordinates: $\left(3, \dfrac{\pi}{2}\right)$

$x = 3 \cos \dfrac{\pi}{2} = 0$

$y = 3 \sin \dfrac{\pi}{2} = 3$

Rectangular Coordinates: $(0, 3)$

11. Polar Coordinates: $\left(-1, \dfrac{5\pi}{4}\right)$

$x = -1 \cos\left(\dfrac{5\pi}{4}\right) = \dfrac{\sqrt{2}}{2}, y = -1 \sin\left(\dfrac{5\pi}{4}\right) = \dfrac{\sqrt{2}}{2}$

Rectangular Coordinates: $\left(\dfrac{\sqrt{2}}{2}, \dfrac{\sqrt{2}}{2}\right)$

13. Polar Coordinates: $\left(2, \dfrac{3\pi}{4}\right)$

$x = 2 \cos \dfrac{3\pi}{4} = -\sqrt{2}$

$y = 2 \sin \dfrac{3\pi}{4} = \sqrt{2}$

Rectangular Coordinates: $\left(-\sqrt{2}, \sqrt{2}\right)$

15. Polar Coordinates: $(-2.5, 1.1)$

$x = -2.5 \cos 1.1 \approx -1.134$

$y = -2.5 \sin 1.1 \approx -2.228$

Rectangular Coordinates: $(-1.134, -2.228)$

17. Rectangular Coordinates: $(1, 1)$

$r = \pm\sqrt{2}, \tan \theta = 1, \theta = \dfrac{\pi}{4} \text{ or } \dfrac{5\pi}{4}$

Polar Coordinates: $\left(\sqrt{2}, \dfrac{\pi}{4}\right), \left(-\sqrt{2}, \dfrac{5\pi}{4}\right)$

19. Rectangular Coordinates: $(-6, 0)$

$r = \pm 6, \tan \theta = 0, \theta = 0 \text{ or } \pi$

Polar Coordinates: $(6, \pi), (-6, 0)$

21. Rectangular Coordinates: $(-3, 4)$

$r = \pm\sqrt{9 + 16} = \pm 5, \tan\theta = -\dfrac{4}{3}, \theta \approx 2.2143, 5.3559$

Polar Coordinates: $(5, 2.2143), (-5, 5.3559)$

23. Rectangular Coordinates: $\left(-\sqrt{3}, -\sqrt{3}\right)$

$r = \pm\sqrt{3 + 3} = \pm\sqrt{6}, \tan\theta = 1, \theta = \dfrac{\pi}{4} \text{ or } \dfrac{5\pi}{4}$

Polar Coordinates: $\left(\sqrt{6}, \dfrac{5\pi}{4}\right), \left(-\sqrt{6}, \dfrac{\pi}{4}\right)$

25. Rectangular Coordinates: $(6, 9)$

$r = \pm\sqrt{6^2 + 9^2} = \pm\sqrt{117} = \pm 3\sqrt{13}$

$\tan\theta = \dfrac{9}{6}, \theta \approx 0.9828, 4.1244$

Polar Coordinates: $\left(3\sqrt{13}, 0.9828\right), \left(-3\sqrt{13}, 4.1244\right)$

27. Rectangular: $(3, -2)$

$(3, -2) \blacktriangleright \text{Pol}$

$\approx (3.606, -0.5880)$

or $\left(\sqrt{13}, -0.5880\right)$

29. Rectangular: $\left(\sqrt{3}, 2\right)$

$\left(\sqrt{3}, 2\right) \blacktriangleright \text{Pol}$

$\approx (2.646, 0.8571)$

or $\left(\sqrt{7}, 0.8571\right)$

31. Rectangular: $\left(\frac{5}{2}, \frac{4}{3}\right)$

$\left(\frac{5}{2}, \frac{4}{3}\right) \blacktriangleright \text{Pol}$

$\approx (2.833, 0.4900)$

or $\left(\frac{17}{6}, 0.4900\right)$

33. $x^2 + y^2 = 9$

$r = 3$

35. $y = 4$

$r\sin\theta = 4$

$r = 4\csc\theta$

37. $3x - y + 2 = 0$

$3r\cos\theta - r\sin\theta + 2 = 0$

$r(3\cos\theta - \sin\theta) = -2$

$r = \dfrac{-2}{3\cos\theta - \sin\theta}$

39. $xy = 16$

$(r\cos\theta)(r\sin\theta) = 16$

$r^2 = 16\sec\theta\csc\theta = 32\csc 2\theta$

41. $y^2 - 8x - 16 = 0$

$r^2\sin^2\theta - 8r\cos\theta - 16 = 0$

By the Quadratic Formula, we have:

$r = \dfrac{-(-8\cos\theta) \pm \sqrt{(-8\cos\theta)^2 - 4(\sin^2\theta)(-16)}}{2\sin^2\theta}$

$= \dfrac{8\cos\theta \pm \sqrt{64\cos^2\theta + 64\sin^2\theta}}{2\sin^2\theta}$

$= \dfrac{8\cos\theta \pm \sqrt{64\left(\cos^2\theta + \sin^2\theta\right)}}{2\sin^2\theta}$

$= \dfrac{8\cos\theta \pm 8}{2\sin^2\theta}$

$= \dfrac{4(\cos\theta \pm 1)}{1 - \cos^2\theta}$

$r = \dfrac{4(\cos\theta + 1)}{(1 + \cos\theta)(1 - \cos\theta)} = \dfrac{4}{1 - \cos\theta}$

or

$r = \dfrac{4(\cos\theta - 1)}{(1 + \cos\theta)(1 - \cos\theta)} = \dfrac{-4}{1 + \cos\theta}$

43. $x^2 + y^2 = a^2$

$r^2 = a^2$

$r = a$

45. $y = b$

$r \sin \theta = b$

$r = b \csc \theta$

47. $x^2 + y^2 - 2ax = 0$

$r^2 - 2a\, r \cos \theta = 0$

$r(r - 2a \cos \theta) = 0$

$r - 2a \cos \theta = 0$

$r = 2a \cos \theta$

49. $r = 4 \sin \theta$

$r^2 = 4r \sin \theta$

$x^2 + y^2 = 4y$

$x^2 + y^2 - 4y = 0$

51. $\theta = \dfrac{2\pi}{3}$

$\tan \theta = \tan \dfrac{2\pi}{3}$

$\dfrac{y}{x} = -\sqrt{3}$

$y = -\sqrt{3}x$

$\sqrt{3}x + y = 0$

53. $r = 2 \csc \theta$

$r \sin \theta = 2$

$y = 2$

55. $r = 2 \sin 3\theta$

$r = 2 \sin(\theta + 2\theta)$

$r = 2[\sin \theta \cos 2\theta + \cos \theta \sin 2\theta]$

$r = 2[\sin \theta(1 - 2\sin^2 \theta) + \cos \theta(2 \sin \theta \cos \theta)]$

$r = 2[\sin \theta - 2\sin^3 \theta + 2 \sin \theta \cos^2 \theta]$

$r = 2[\sin \theta - 2\sin^3 \theta + 2 \sin \theta(1 - \sin^2 \theta)]$

$r = 2(3 \sin \theta - 4 \sin^3 \theta)$

$r^4 = 6r^3 \sin \theta - 8r^3 \sin^3 \theta$

$(x^2 + y^2)^2 = 6(x^2 + y^2)y - 8y^3$

$(x^2 + y^2)^2 = 6x^2y - 2y^3$

57. $r = \dfrac{6}{2 - \sin \theta}$

$r(2 - \sin \theta) = 6$

$2r = 6 + r \sin \theta$

$2\left(\pm\sqrt{x^2 + y^2}\right) = 6 + 3y$

$4(x^2 + y^2) = (6 + 3y)^2$

$4x^2 + 4y^2 = 36 + 36y + 9y^2$

$4x^2 - 5y^2 - 36y - 36 = 0$

59. $r = 6$

$r^2 = 36$

$x^2 + y^2 = 36$

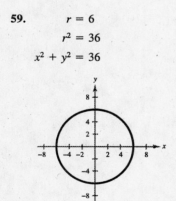

61. $\theta = \dfrac{\pi}{6}$

$\tan \theta = \tan \dfrac{\pi}{6}$

$\dfrac{y}{x} = \dfrac{\sqrt{3}}{3}$

$y = \dfrac{\sqrt{3}}{3}x$

$3y = \sqrt{3}x$

$-\sqrt{3}x + 3y = 0$

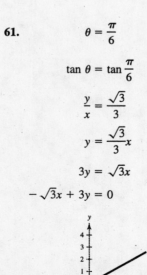

63. $r = 3 \sec \theta$

$r \cos \theta = 3$

$x = 3$

$x - 3 = 0$

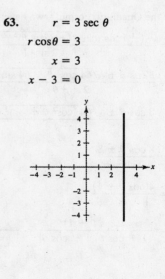

65. True. Because r is a directed distance, then the point (r, θ) can be represented as $(r, \theta + 2n\pi)$.

67.
$$r = 2(h \cos \theta + k \sin \theta)$$
$$r = 2\left(h\left(\frac{x}{r}\right) + k\left(\frac{y}{r}\right)\right)$$
$$r = \frac{2hx + 2ky}{r}$$
$$r^2 = 2hx + 2ky$$
$$x^2 + y^2 = 2hx + 2ky$$
$$x^2 - 2hx + y^2 - 2ky = 0$$
$$\left(x^2 - 2hx + h^2\right) + \left(y^2 - 2ky + k^2\right) = h^2 + k^2$$
$$(x - h)^2 + (y - k)^2 = h^2 + k^2$$

Center: (h, k)

Radius: $\sqrt{h^2 + k^2}$

69. (a) $(r_1, \theta_1) = (x_1, y_1)$ where $x_1 = r_1 \cos \theta_1$ and $y_1 = r_1 \sin \theta_1$.

$(r_2, \theta_2) = (x_2, y_2)$ where $x_2 = r_2 \cos \theta_2$ and $y_2 = r_2 \sin \theta_2$.

$$d = \sqrt{(x_1 - x_2)^2 + (y_1 - y_2)^2}$$
$$= \sqrt{x_1^2 - 2x_1 x_2 + x_2^2 + y_1^2 - 2y_1 y_2 + y_2^2}$$
$$= \sqrt{(x_1^2 + y_1^2) + (x_2^2 + y_2 + y_2^2) - 2(x_1 x_2 + y_1 y_2)}$$
$$= \sqrt{r_1^2 + r_2^2 - 2\left(r_1 r_2 \cos \theta_1 \cos \theta_2 + r_1 r_2 \sin \theta_1 \sin \theta_2\right)}$$
$$= \sqrt{r_1^2 + r_2^2 - 2r_1 r_2 \cos(\theta_1 - \theta_2)}$$

(b) If $\theta_1 = \theta_2$, then
$$d = \sqrt{r_1^2 + r_2^2 - 2r_1 r_2}$$
$$= \sqrt{(r_1 - r_2)^2}$$
$$= |r_1 - r_2|.$$

This represents the distance between two points on the line $\theta = \theta_1 = \theta_2$.

(c) If $\theta_1 - \theta_2 = 90°$, then
$$d = \sqrt{r_1^2 + r_2^2}.$$
This is the result of the Pythagorean Theorem.

(d) The results should be the same. For example, use the points
$$\left(3, \frac{\pi}{6}\right) \text{ and } \left(4, \frac{\pi}{3}\right).$$

The distance is $d \approx 2.053$.

Now use the representations
$$\left(-3, \frac{7\pi}{6}\right) \text{ and } \left(-4, \frac{4\pi}{3}\right).$$

The distance is still $d \approx 2.053$.

71. $A = 22°, b = 9, c = 16$

$$a^2 = b^2 + c^2 - 2bc \cos A = 81 + 256 - 288 \cos 22° \approx 69.971 \implies a \approx 8.36$$

$$\cos C = \frac{a^2 + b^2 - c^2}{2ab} = \frac{69.971 + 81 - 256}{150.48} \approx -0.6980 \implies C \approx 134.3°$$

$$B = 180° - 22° - 134.3° \approx 23.7°$$

73. $a = 5, b = 8, c = 10$

$$\cos C = \frac{a^2 + b^2 - c^2}{2ab} = \frac{25 + 64 - 100}{80} \approx -0.1375 \implies C \approx 97.9°$$

$$\cos A = \frac{b^2 + c^2 - a^2}{2bc} = \frac{64 + 100 - 25}{160} \approx 0.8688 \implies A \approx 29.7°$$

$$B = 180° - 29.7° - 97.9° \approx 52.4°$$

75. $C = 38°, a = 4, b = 15$

$$c^2 = a^2 + b^2 - 2ab \cos C = 16 + 225 - 120 \cos 38° \approx 146.439 \implies c \approx 12.10$$

$$\cos B = \frac{a^2 + c^2 - b^2}{2ac} = \frac{16 + 146.439 - 225}{96.8} \approx -0.6463 \implies B \approx 130.3°$$

$$A = 180° - 130.3° - 38° \approx 11.7°$$

77. $a = 4, b = 8, c = 11$

$$s = \frac{a + b + c}{2} = \frac{4 + 8 + 11}{2} = 11.5$$

$$\text{Area} = \sqrt{11.5(11.5 - 4)(11.5 - 8)(11.5 - 11)} = \sqrt{150.9375} \approx 12.29$$

79. $a = 31, b = 48, c = 48$

$$s = \frac{a + b + c}{2} = \frac{31 + 48 + 48}{2} = 63.5$$

$$\text{Area} = \sqrt{63.5(63.5 - 31)(63.5 - 48)(63.5 - 48)} = \sqrt{495,815.9375} \approx 704.14$$

Section 6.8   Graphs of Polar Equations

■   When graphing polar equations:

1. Test for symmetry.
   (a) $\theta = \pi/2$: Replace (r, θ) by $(r, \pi - \theta)$ or $(-r, -\theta)$.
   (b) Polar axis: Replace (r, θ) by $(r, -\theta)$ or $(-r, \pi - \theta)$.
   (c) Pole: Replace (r, θ) by $(r, \pi + \theta)$ or $(-r, \theta)$.
   (d) $r = f(\sin \theta)$ is symmetric with respect to the line $\theta = \pi/2$.
   (e) $r = f(\cos \theta)$ is symmetric with respect to the polar axis.

2. Find the θ values for which $|r|$ is maximum.

3. Find the θ values for which $r = 0$.

4. Know the different types of polar graphs.
   (a) Limaçons $(0 < a, 0 < b)$ (b) Rose Curves, $n \geq 2$
       $r = a \pm b \cos \theta$ $r = a \cos n\theta$
       $r = a \pm b \sin \theta$ $r = a \sin n\theta$

   (c) Circles   (d) Lemniscates
       $r = a \cos \theta$ $r^2 = a^2 \cos 2\theta$
       $r = a \sin \theta$ $r^2 = a^2 \sin 2\theta$
       $r = a$

5. Plot additional points.

Solutions to Odd-Numbered Exercises

1. $r = 3 \cos 2\theta$

Rose curve with 4 petals

3. $r = 3(1 - 2 \cos \theta)$

Limaçon with inner loop

5. $r = 6 \sin 2\theta$

Rose curve with 4 petals

7. $r = 5 + 4 \cos \theta$

$\theta = \dfrac{\pi}{2}$: $-r = 5 + 4 \cos(-\theta)$

   $-r = 5 + 4 \cos \theta$

Not an equivalent equation

Polar axis: $r = 5 + 4 \cos(-\theta)$

   $r = 5 + 4 \cos \theta$

Equivalent equation

Pole: $-r = 5 + 4 \cos \theta$

Not an equivalent equation

Answer: Symmetric with respect to polar axis

9. $r = \dfrac{2}{1 + \sin \theta}$

$\theta = \dfrac{\pi}{2}$: $r = \dfrac{2}{1 + \sin(\pi - \theta)}$

   $r = \dfrac{2}{1 + \sin \pi \cos \theta - \cos \pi \sin \theta}$

   $r = \dfrac{2}{1 + \sin \theta}$

Equivalent equation

Polar axis:

   $r = \dfrac{2}{1 + \sin(-\theta)}$

   $r = \dfrac{2}{1 - \sin \theta}$

Not an equivalent equation

Pole: $-r = \dfrac{2}{1 + \sin \theta}$

Answer: Symmetric with respect to $\theta = \pi/2$

11. $r^2 = 16 \cos 2\theta$

$\theta = \dfrac{\pi}{2}$: $(-r)^2 = 16 \cos 2(-\theta)$

   $r^2 = 16 \cos 2\theta$

Equivalent equation

Polar axis: $r^2 = 16 \cos 2(-\theta)$

   $r^2 = 16 \cos 2\theta$

Equivalent equation

Pole: $(-r)^2 = 16 \cos 2\theta$

   $r^2 = 16 \cos 2\theta$

Equivalent equation

Answer: Symmetric with respect to $\theta = \dfrac{\pi}{2}$, the

   polar axis, and the pole

13. $|r| = |10(1 - \sin \theta)| = 10|1 - \sin \theta| \le 10(2) = 20$

$|1 - \sin \theta| = 2$

$1 - \sin \theta = 2$ or $1 - \sin \theta = -2$

$\sin \theta = -1$ $\sin \theta = 3$

$\theta = \dfrac{3\pi}{2}$ Not possible

Maximum: $|r| = 20$ when $\theta = \dfrac{3\pi}{2}$.

   $0 = 10(1 - \sin \theta)$

$\sin \theta = 1$

$\theta = \dfrac{\pi}{2}$

Zero: $r = 0$ when $\theta = \dfrac{\pi}{2}$.

15. $|r| = |4 \cos 3\theta| = 4|\cos 3\theta| \le 4$

$|\cos 3\theta| = 1$

$\cos 3\theta = \pm 1$

$\theta = 0, \dfrac{\pi}{3}, \dfrac{2\pi}{3}$

Maximum: $|r| = 4$ when $\theta = 0, \dfrac{\pi}{3}, \dfrac{2\pi}{3}$.

   $0 = 4 \cos 3\theta$

$\cos 3\theta = 0$

$\theta = \dfrac{\pi}{6}, \dfrac{\pi}{2}, \dfrac{5\pi}{6}$

Zero: $r = 0$ when $\theta = \dfrac{\pi}{6}, \dfrac{\pi}{2}, \dfrac{5\pi}{6}$.

17. Circle: $r = 5$

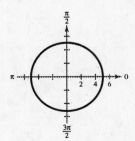

19. Circle: $r = \dfrac{\pi}{6}$

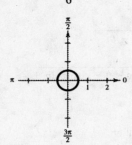

21. $r = 3 \sin \theta$

Symmetric with respect to $\theta = \dfrac{\pi}{2}$

Circle with a radius of $\dfrac{3}{2}$

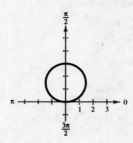

23. $r = 3(1 - \cos \theta)$

Symmetric with respect to the polar axis

$\dfrac{a}{b} = \dfrac{3}{3} = 1 \implies$ Cardioid

$|r| = 6$ when $\theta = \pi$.

$r = 0$ when $\pi = 0$.

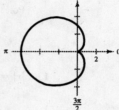

25. $r = 4(1 + \sin \theta)$

Symmetric with respect to $\theta = \dfrac{\pi}{2}$

$\dfrac{a}{b} = \dfrac{4}{4} = 1 \implies$ Cardioid

$|r| = 8$ when $\theta = \dfrac{\pi}{2}$.

$r = 0$ when $\theta = \dfrac{3\pi}{2}$.

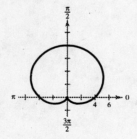

27. $r = 3 + 6 \sin \theta$

Symmetric with respect to $\theta = \dfrac{\pi}{2}$

$\dfrac{a}{b} = \dfrac{3}{6} < 1 \implies$ Limaçon with inner loop

$|r| = 9$ when $\theta = \dfrac{\pi}{2}$

$r = 0$ when $\theta = \dfrac{7\pi}{6}, \dfrac{11\pi}{6}$

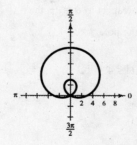

29. $r = 1 - 2\sin\theta$

Symmetric with respect to $\theta = \dfrac{\pi}{2}$

$\dfrac{a}{b} = \dfrac{1}{2} < 1 \Rightarrow$ Limaçon with inner loop

$|r| = 3$ when $\theta = \dfrac{3\pi}{2}$

$r = 0$ when $\theta = \dfrac{\pi}{6}, \dfrac{5\pi}{6}$

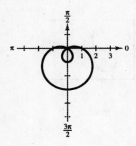

31. $r = 3 - 4\cos\theta$

Symmetric with respect to the polar axis

$\dfrac{a}{b} = \dfrac{2}{4} < 1 \Rightarrow$ Limaçon with inner loop

$|r| = 7$ when $\theta = \pi$.

$r = 0$ when $\cos\theta = \dfrac{3}{4}$ or

$\theta \approx 0.723, 5.560$

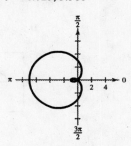

33. $r = 5\sin 2\theta$

Symmetric with respect to $\theta = \dfrac{\pi}{2}$

Rose curve ($n = 2$) with 4 petals

$|r| = 5$ when $\theta = \dfrac{\pi}{4}, \dfrac{3\pi}{4}, \dfrac{5\pi}{4}, \dfrac{7\pi}{4}$.

$r = 0$ when $\theta = 0, \dfrac{\pi}{2}, \pi$.

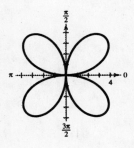

35. $\qquad r = 2\sec\theta$

$\qquad r = \dfrac{2}{\cos\theta}$

$r\cos\theta = 2$

$\qquad x = 2 \Rightarrow$ Line

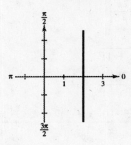

37. $\qquad\qquad r = \dfrac{3}{\sin\theta - 2\cos\theta}$

$r(\sin\theta - 2\cos\theta) = 3$

$\qquad y - 2x = 3$

$\qquad\qquad y = 2x + 3 \Rightarrow$ Line

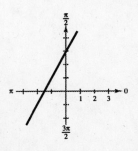

39. $r^2 = 9\cos 2\theta$

Symmetric with respect to the polar axis, $\theta = \dfrac{\pi}{2}$,

and the pole

Lemniscate

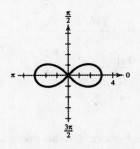

41. $r = 8 \cos \theta$

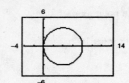

43. $r = 3(2 - \sin \theta)$

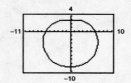

45. $r = 8 \sin \theta \cos^2 \theta$

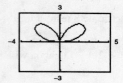

47. $r = 3 - 4 \cos \theta$
$0 \le \theta < 2\pi$

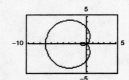

49. $r = 2 \cos\left(\dfrac{3\theta}{2}\right)$
$0 \le \theta < 4\pi$

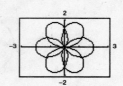

51. $r^2 = 9 \sin 2\theta$
$0 \le \theta < \pi$

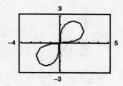

53.
$$r = 2 - \sec \theta = 2 - \frac{1}{\cos \theta}$$

$$r \cos \theta = 2 \cos \theta - 1$$

$$r(r \cos \theta) = 2r \cos \theta - r$$

$$\left(\pm\sqrt{x^2 + y^2}\right)x = 2x - \left(\pm\sqrt{x^2 + y^2}\right)$$

$$\left(\pm\sqrt{x^2 + y^2}\right)(x + 1) = 2x$$

$$\left(\pm\sqrt{x^2 + y^2}\right) = \frac{2x}{x + 1}$$

$$x^2 + y^2 = \frac{4x^2}{(x + 1)^2}$$

$$y^2 = \frac{4x^2}{(x + 1)^2} - x^2$$

$$= \frac{4x^2 - x^2(x + 1)^2}{(x + 1)^2} = \frac{4x^2 - x^2\left(x^2 + 2x + 1\right)}{(x + 1)^2}$$

$$= \frac{-x^4 - 2x^3 + 3x^2}{(x + 1)^2} = \frac{-x^2\left(x^2 + 2x - 3\right)}{(x + 1)^2}$$

$$y = \pm\sqrt{\frac{x^2\left(3 - 2x - x^2\right)}{(x + 1)^2}} = \pm\left|\frac{x}{x + 1}\right|\sqrt{3 - 2x - x^2}$$

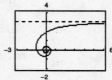

The graph has an asymptote at $x = -1$.

55. $r = \dfrac{3}{\theta}$

$$\theta = \frac{3}{r} = \frac{3 \sin \theta}{r \sin \theta} = \frac{3 \sin \theta}{y}$$

$$y = \frac{3 \sin \theta}{\theta}$$

As $\theta \to 0, y \to 3$

57. True. For a graph to have polar axis symmetry, replace (r, θ) by $(r, -\theta)$ or $(-r, \pi - \theta)$.

59. $r = 6 \cos \theta$

(a) $0 \le \theta \le \dfrac{\pi}{2}$

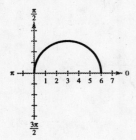

(b) $\dfrac{\pi}{2} \le \theta \le \pi$

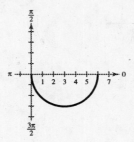

(c) $-\dfrac{\pi}{2} \le \theta \le \dfrac{\pi}{2}$

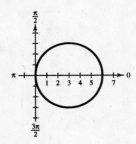

(d) $\dfrac{\pi}{4} \le \theta \le \dfrac{3\pi}{4}$

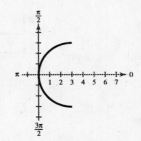

61. Let the curve $r = f(\theta)$ be rotated by ϕ to form the curve $r = g(\theta)$. If (r_1, θ_1) is a point on $r = f(\theta)$, then $(r_1, \theta_1 + \phi)$ is on $r = g(\theta)$. That is, $g(\theta_1 + \phi) = r_1 = f(\theta_1)$. Letting $\theta = \theta_1 + \phi$, or $\theta_1 = \theta - \phi$, we see that $g(\theta) = g(\theta_1 + \phi) = f(\theta_1) = f(\theta - \phi)$.

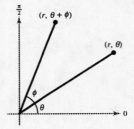

63. (a) $r = 2 - \sin\left(\theta - \dfrac{\pi}{4}\right)$

$= 2 - \left[\sin \theta \cos \dfrac{\pi}{4} - \cos \theta \sin \dfrac{\pi}{4} \right]$

$= 2 - \dfrac{\sqrt{2}}{2}(\sin \theta - \cos \theta)$

(c) $r = 2 - \sin(\theta - \pi)$

$= 2 - [\sin \theta \cos \pi - \cos \theta \sin \pi]$

$= 2 + \sin \theta$

(b) $r = 2 - \sin\left(\theta - \dfrac{\pi}{2}\right)$

$= 2 - \left[\sin \theta \cos \dfrac{\pi}{2} - \cos \theta \sin \dfrac{\pi}{2} \right]$

$= 2 + \cos \theta$

(d) $r = 2 - \sin\left(\theta - \dfrac{3\pi}{2}\right)$

$= 2 - \left[\sin \theta \cos \dfrac{3\pi}{2} - \cos \theta \sin \dfrac{3\pi}{2} \right]$

$= 2 - \cos \theta$

65. (a) $r = 1 - \sin\theta$

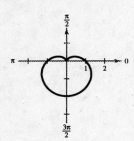

(b) $r = 1 - \sin\left(\theta - \dfrac{\pi}{4}\right)$

Rotate the graph in part (a) through the angle $\dfrac{\pi}{4}$.

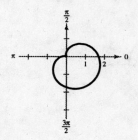

67. $r = 2 + k\sin\theta$

$k = 0$: $r = 2$

     Circle

$k = 1$: $r = 2 + \sin\theta$

     Convex limaçon

$k = 2$: $r = 2 + 2\sin\theta$

     Cardioid

$k = 3$: $r = 2 + 3\sin\theta$

     Limaçon with inner loop

69. $e^x = 19$

$x = \ln 19 \approx 2.944$

71. $10^x = 84$

$x = \log_{10} 84 \approx 1.924$

73. $\ln x = 4$

$x = e^4 \approx 54.598$

Section 6.9   Polar Equations of Conics

- The graph of a polar equation of the form

  $$r = \frac{ep}{1 \pm e \cos \theta} \quad \text{or} \quad r = \frac{ep}{1 \pm e \sin \theta}$$

  is a conic, where $e > 0$ is the eccentricity and $|p|$ is the distance between the focus (pole) and the directrix.

  (a) If $e < 1$, the graph is an ellipse.
  (b) If $e = 1$, the graph is a parabola.
  (c) If $e > 1$, the graph is a hyperbola.

- Guidelines for finding polar equations of conics:

  (a) Horizontal directrix above the pole: $r = \dfrac{ep}{1 + e \sin \theta}$

  (b) Horizontal directrix below the pole: $r = \dfrac{ep}{1 - e \sin \theta}$

  (c) Vertical directrix to the right of the pole: $r = \dfrac{ep}{1 + e \cos \theta}$

  (d) Vertical directrix to the left of the pole: $r = \dfrac{ep}{1 - e \cos \theta}$

Solutions to Odd-Numbered Exercises

1. $r = \dfrac{4e}{1 + e \cos \theta}$

   (a) $e = 1, r = \dfrac{4}{1 + \cos \theta}$, parabola

   (b) $e = 0.5, r = \dfrac{2}{1 + 0.5 \cos \theta} = \dfrac{4}{2 + \cos \theta}$, ellipse

   (c) $e = 1.5, r = \dfrac{6}{1 + 1.5 \cos \theta} = \dfrac{12}{2 + 3 \cos \theta}$, hyperbola

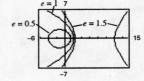

3. $r = \dfrac{4e}{1 - e \sin \theta}$

   (a) $e = 1, r = \dfrac{4}{1 - \sin \theta}$, parabola

   (b) $e = 0.5, r = \dfrac{2}{1 - 0.5 \sin \theta} = \dfrac{4}{2 - \sin \theta}$, ellipse

   (c) $e = 1.5, r = \dfrac{6}{1 - 1.5 \sin \theta} = \dfrac{12}{2 - 3 \sin \theta}$, hyperbola

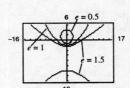

5. $r = \dfrac{2}{1 + \cos \theta}$

   $e = 1 \Rightarrow$ Parabola

   Vertical directrix to the right
   of the pole
   Matches graph (f).

7. $r = \dfrac{3}{1 + 2 \sin \theta}$

   $e = 2 \Rightarrow$ Hyperbola
   Matches graph (d).

9. $r = \dfrac{4}{2 + \cos \theta}$

   $= \dfrac{2}{1 + 0.5 \cos \theta}$

   $e = 0.5 \Rightarrow$ Ellipse

   Matches graph (a).

11. $r = \dfrac{2}{1 - \cos \theta}$

$e = 1$, the graph is a parabola.

Vertex: $(1, \pi)$

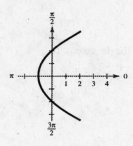

13. $r = \dfrac{5}{1 + \sin \theta}$

$e = 1$, the graph is a parabola.

Vertex: $\left(\dfrac{5}{2}, \dfrac{\pi}{2}\right)$

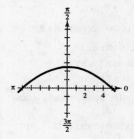

15. $r = \dfrac{2}{2 - \cos \theta} = \dfrac{1}{1 - \frac{1}{2}\cos \theta}$

$e = \dfrac{1}{2} < 1$, the graph is an ellipse.

Vertices: $(2, 0), \left(\dfrac{2}{3}, \pi\right)$

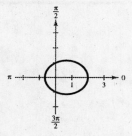

17. $r = \dfrac{6}{2 + \sin \theta} = \dfrac{3}{1 + \frac{1}{2}\sin \theta}$

$e = \dfrac{1}{2} < 1$, the graph is an ellipse.

Vertices: $\left(2, \dfrac{\pi}{2}\right), \left(6, \dfrac{3\pi}{2}\right)$

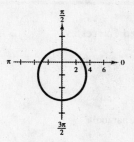

19. $r = \dfrac{3}{2 + 4\sin \theta} = \dfrac{\frac{3}{2}}{1 + 2\sin \theta}$

$e = 2 > 1$, the graph is a hyperbola.

Vertices: $\left(\dfrac{1}{2}, \dfrac{\pi}{2}\right), \left(-\dfrac{3}{2}, \dfrac{3\pi}{2}\right)$

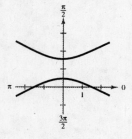

21. $r = \dfrac{3}{2 - 6\cos \theta} = \dfrac{\frac{3}{2}}{1 - 3\cos \theta}$

$e = 3 > 1$, the graph is a hyperbola.

Vertices: $\left(-\dfrac{3}{4}, 0\right), \left(\dfrac{3}{8}, \pi\right)$

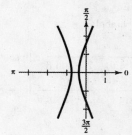

23. $r = \dfrac{4}{2 - \cos\theta} = \dfrac{2}{1 - \frac{1}{2}\cos\theta}$

$e = \dfrac{1}{2} < 1$, the graph is an ellipse.

Vertices: $(4, 0), \left(\dfrac{4}{3}, \pi\right)$

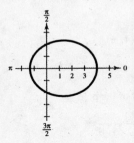

25. $r = \dfrac{-1}{1 - \sin\theta}$

$e = 1 \Rightarrow$ Parabola

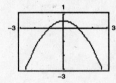

27. $r = \dfrac{3}{-4 + 2\cos\theta}$

$e = \dfrac{1}{2} \Rightarrow$ Ellipse

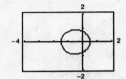

29. $r = \dfrac{2}{1 - \cos\left(\theta - \dfrac{\pi}{4}\right)}$

Rotate the graph in Exercise 11 through the angle $\dfrac{\pi}{4}$.

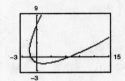

31. $r = \dfrac{6}{2 + \sin\left(\theta + \dfrac{\pi}{6}\right)}$

Rotate the graph in Exercise 17 through the angle $-\dfrac{\pi}{6}$.

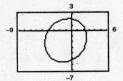

33. Parabola: $e = 1$

Directrix: $x = -1$

Vertical directrix to the left of the pole

$r = \dfrac{1(1)}{1 - 1\cos\theta} = \dfrac{1}{1 - \cos\theta}$

35. Ellipse: $e = \dfrac{1}{2}$

Directrix: $y = 1$

$p = 1$

Horizontal directrix above the pole

$r = \dfrac{\frac{1}{2}(1)}{1 + \frac{1}{2}\sin\theta} = \dfrac{1}{2 + \sin\theta}$

37. Hyperbola: $e = 2$

Directrix: $x = 1$

$p = 1$

Vertical directrix to the right of the pole

$r = \dfrac{2(1)}{1 + 2\cos\theta} = \dfrac{2}{1 + 2\cos\theta}$

39. Parabola

Vertex: $\left(1, -\dfrac{\pi}{2}\right) \Rightarrow e = 1, p = 2$

Horizontal directrix below the pole

$$r = \frac{1(2)}{1 - 1 \sin \theta} = \frac{2}{1 - \sin \theta}$$

41. Parabola

Vertex: $(5, \pi) \Rightarrow e = 1, p = 10$

Vertical directrix to the left of the pole

$$r = \frac{1(10)}{1 - 1 \cos \theta} = \frac{10}{1 - \cos \theta}$$

43. Ellipse: Vertices $(2, 0), (10, \pi)$

Center: $(4, \pi); c = 4, a = 6, e = \dfrac{2}{3}$

Vertical directrix to the right of the pole

$$r = \frac{\frac{2}{3}p}{1 + \frac{2}{3} \cos \theta} = \frac{2p}{3 + 2 \cos \theta}$$

$$2 = \frac{2p}{3 + 2 \cos 0}$$

$$p = 5$$

$$r = \frac{2(5)}{3 + 2 \cos \theta} = \frac{10}{3 + 2 \cos \theta}$$

45. Ellipse: Vertices $(20, 0), (4, \pi)$

Center: $(8, 0); c = 8, a = 12, e = \dfrac{2}{3}$

Vertical directrix to the left of the pole

$$r = \frac{\frac{2}{3}p}{1 - \frac{2}{3} \cos \theta} = \frac{2p}{3 - 2 \cos \theta}$$

$$20 = \frac{2p}{3 - 2 \cos 0}$$

$$p = 10$$

$$r = \frac{2(10)}{3 - 2 \cos \theta} = \frac{20}{3 - 2 \cos \theta}$$

47. Hyperbola: Vertices $\left(1, \dfrac{3\pi}{2}\right), \left(9, \dfrac{3\pi}{2}\right)$

Center: $\left(5, \dfrac{3\pi}{2}\right); c = 5, a = 4, e = \dfrac{5}{4}$

Horizontal directrix below the pole

$$r = \frac{\frac{5}{4}p}{1 - \frac{5}{4} \sin \theta} = \frac{5p}{4 - 5 \sin \theta}$$

$$1 = \frac{5p}{4 - 5 \sin \frac{3\pi}{2}}$$

$$p = \frac{9}{5}$$

$$r = \frac{5\left(\frac{9}{5}\right)}{4 - 5 \sin \theta} = \frac{9}{4 - 5 \sin \theta}$$

49. When $\theta = 0, r = c + a = ea + a = a(1 + e)$.
Therefore,

$$a(1 + e) = \frac{ep}{1 - e \cos 0}$$

$$a(1 + e)(1 - e) = ep$$

$$a(1 - e^2) = ep.$$

Thus, $r = \dfrac{ep}{1 - e \cos \theta} = \dfrac{(1 - e^2)a}{1 - e \cos \theta}$.

51. $r = \dfrac{[1 - (0.0167)^2](92.960 \times 10^6)}{1 - 0.0167 \cos \theta}$

$\approx \dfrac{9.2934 \times 10^7}{1 - 0.0167 \cos \theta}$

Perihelion distance:
$r = 92.960 \times 10^6(1 - 0.0167) \approx 9.1408 \times 10^7$

Aphelion distance:
$r = 92.960 \times 10^6(1 + 0.0167) \approx 9.4512 \times 10^7$

53. $r = \dfrac{[1 - (0.2481)^2](5.9 \times 10^9)}{1 - 0.2481 \cos \theta}$

$\approx \dfrac{5.5368 \times 10^9}{1 - 0.2481 \cos \theta}$

Perihelion distance:
$r = 5.9 \times 10^9(1 - 0.2481) \approx 4.4362 \times 10^9$

Aphelion distance:
$r = 5.9 \times 10^9(1 + 0.2481) \approx 7.3638 \times 10^9$

55. $r = \dfrac{[1 - (0.0934)^2](141 \times 10^6)}{1 - 0.0934 \cos \theta}$

$\approx \dfrac{1.3977 \times 10^8}{1 - 0.0934 \cos \theta}$

Perihelion distance:

$r = 141 \times 10^6(1 - 0.0934) \approx 1.2783 \times 10^8$

Aphelion distance:

$r = 141 \times 10^6(1 + 0.0934) \approx 1.5417 \times 10^8$

57. Vertex: $\left(4100, \dfrac{\pi}{2}\right)$

Focus: $(0, 0)$

$e = 1, p = 8200$

$r = \dfrac{ep}{1 + e \sin \theta} = \dfrac{8200}{1 + \sin \theta}$

When $\theta = 30°$, $r = 8200/1.5 \approx 5466.67$.
Distance between the surface of the earth
and the satellite is $r - 4000 \approx 1467$ miles.

59. False. If e remains fixed, and p changes, then the lengths of the major axis and minor axis change.

For example, graph $r = \dfrac{5}{1 - \frac{2}{3} \cos \theta}$, with $e = \dfrac{2}{3}$

and $p = \dfrac{15}{2}$ and $r = \dfrac{6}{1 - \frac{2}{3} \cos \theta}$, with $e = \dfrac{2}{3}$

and $p = 9$, on the same set of coordinate axes.

61.
$$\frac{x^2}{a^2} + \frac{y^2}{b^2} = 1$$
$$\frac{r^2 \cos^2 \theta}{a^2} + \frac{r^2 \sin^2 \theta}{b^2} = 1$$
$$\frac{r^2 \cos^2 \theta}{a^2} + \frac{r^2(1 - \cos^2 \theta)}{b^2} = 1$$
$$r^2 b^2 \cos^2 \theta + r^2 a^2 - r^2 a^2 \cos^2 \theta = a^2 b^2$$
$$r^2(b^2 - a^2)\cos^2 \theta + r^2 a^2 = a^2 b^2$$

Since $b^2 - a^2 = -c^2$, we have

$$-r^2 c^2 \cos^2 \theta + r^2 a^2 = a^2 b^2$$
$$-r^2\left(\frac{c}{a}\right)^2 \cos^2 \theta + r^2 = b^2, \ e = \frac{c}{a}$$
$$-r^2 e^2 \cos^2 \theta + r^2 = b^2$$
$$r^2(1 - e^2 \cos^2 \theta) = b^2$$
$$r^2 = \frac{b^2}{1 - e^2 \cos^2 \theta}$$

63. $\dfrac{x^2}{169} + \dfrac{y^2}{144} = 1$

$a = 13, b = 12, c = 5, e = \dfrac{5}{13}$

$r^2 = \dfrac{144}{1 - \left(\frac{25}{169}\right) \cos^2 \theta} = \dfrac{24{,}336}{169 - 25 \cos^2 \theta}$

65. $\dfrac{x^2}{9} - \dfrac{y^2}{16} = 1$

$a = 3, b = 4, c = 5, e = \dfrac{5}{3}$

$r^2 = \dfrac{-16}{1 - \left(\frac{25}{9}\right) \cos^2 \theta} = \dfrac{144}{25 \cos^2 \theta - 9}$

67. One focus: $\left(5, \dfrac{\pi}{2}\right)$

Vertices: $\left(4, \dfrac{\pi}{2}\right), \left(4, -\dfrac{\pi}{2}\right)$

$a = 4, c = 5 \Rightarrow b = 3$ and $e = \dfrac{5}{4}$

$$\frac{y^2}{16} - \frac{x^2}{9} = 1$$
$$\frac{r^2 \sin^2 \theta}{16} - \frac{r^2 \cos^2 \theta}{9} = 1$$
$$9r^2 \sin^2 \theta - 16r^2(1 - \sin^2 \theta) = 144$$
$$25r^2 \sin^2 \theta - 16r^2 = 144$$
$$r^2 = \frac{144}{25 \sin^2 \theta - 16}$$

69. $4\sqrt{3}\tan\theta - 3 = 1$

$\qquad 4\sqrt{3}\tan\theta = 4$

$\qquad\quad \tan\theta = \dfrac{1}{\sqrt{3}}$

$\qquad\qquad \theta = \dfrac{\pi}{6} + n\pi$

71. $12\sin^2\theta = 9$

$\qquad \sin^2\theta = \dfrac{3}{4}$

$\qquad \sin\theta = \pm\dfrac{\sqrt{3}}{2}$

$\qquad \theta = \dfrac{\pi}{3} + 2n\pi,\ \dfrac{2\pi}{3} + 2n\pi,\ \dfrac{4\pi}{3} + 2n\pi,\ \dfrac{5\pi}{3} + 2n\pi$

73. $2\cot x = 5\cos\dfrac{\pi}{2}$

$\qquad 2\cot x = 0$

$\qquad\ \cot x = 0$

$\qquad\qquad x = \dfrac{\pi}{2} + n\pi$

For 75 and 77 use the following:

u and v are in Quadrant IV

$\sin u = -\dfrac{3}{5} \Rightarrow \cos u = \dfrac{4}{5}$

$\cos v = \dfrac{1}{\sqrt{2}} \Rightarrow \sin v = -\dfrac{1}{\sqrt{2}}$

75. $\cos(u + v) = \cos u \cos v - \sin u \sin v$

$\qquad = \left(\dfrac{4}{5}\right)\left(\dfrac{1}{\sqrt{2}}\right) - \left(-\dfrac{3}{5}\right)\left(-\dfrac{1}{\sqrt{2}}\right)$

$\qquad = \dfrac{4}{5\sqrt{2}} - \dfrac{3}{5\sqrt{2}}$

$\qquad = \dfrac{1}{5\sqrt{2}}$

$\qquad = \dfrac{\sqrt{2}}{10}$

77. $\cos(u - v) = \cos u \cos v + \sin u \sin v$

$\qquad = \left(\dfrac{4}{5}\right)\left(\dfrac{1}{\sqrt{2}}\right) + \left(-\dfrac{3}{5}\right)\left(-\dfrac{1}{\sqrt{2}}\right)$

$\qquad = \dfrac{4}{5\sqrt{2}} + \dfrac{3}{5\sqrt{2}}$

$\qquad = \dfrac{7}{5\sqrt{2}}$

$\qquad = \dfrac{7\sqrt{2}}{10}$

79. $\sin u = \dfrac{4}{5},\ \dfrac{\pi}{2} < u < \pi \Rightarrow \cos u = -\dfrac{3}{5}$

$\sin 2u = 2\sin u \cos u$

$\qquad = 2\left(\dfrac{4}{5}\right)\left(-\dfrac{3}{5}\right)$

$\qquad = -\dfrac{24}{25}$

$\cos 2u = \cos^2 u - \sin^2 u$

$\qquad = \left(-\dfrac{3}{5}\right)^2 - \left(\dfrac{4}{5}\right)^2$

$\qquad = \dfrac{9}{25} - \dfrac{16}{25}$

$\qquad = -\dfrac{7}{25}$

$\tan 2u = \dfrac{\sin 2u}{\cos 2u}$

$\qquad = \dfrac{-\frac{24}{25}}{-\frac{7}{25}}$

$\qquad = \dfrac{24}{7}$

Review Exercises for Chapter 6

Solutions to Odd-Numbered Exercises

1. Slope $m = \frac{3}{5}$

$\tan \theta = \frac{3}{5} \implies \theta = \arctan \frac{3}{5} \approx 30.96°$

3. $y = 2x + 4 \implies m = 2$

$\tan \theta = 2 \implies \theta = \arctan 2 \approx 63.43°$

5. $4x + y = 2 \implies y = -4x + 2 \implies m_1 = -4$

$-5x + y = -1 \implies y = 5x - 1 \implies m_2 = 5$

$\tan \theta = \left| \dfrac{5 - (-4)}{1 + (-4)(5)} \right| = \dfrac{9}{19}$

$\theta = \arctan \dfrac{9}{19} \approx 25.35°$

7. $2x - 7y = 8 \implies y = \dfrac{2}{7}x - \dfrac{8}{7} \implies m_1 = \dfrac{2}{7}$

$0.4x + y = 0 \implies y = -0.4x \implies m_2 = -0.4$

$\tan \theta = \left| \dfrac{-0.4 - \frac{2}{7}}{1 + \left(\frac{2}{7}\right)(-0.4)} \right| = \dfrac{24}{31}$

$\theta = \arctan \left(\dfrac{24}{31} \right) \approx 37.75°$

9. $(1, 2) \implies x_1 = 1, y_1 = 2$

$x - y - 3 = 0 \implies A = 1, B = -1, C = -3$

$d = \dfrac{|1(1) + (-1)(2) + (-3)|}{\sqrt{1^2 + (-1)^2}} = \dfrac{4}{\sqrt{2}} = 2\sqrt{2}$

11. Hyperbola

13. Vertex: $(4, 2) = (h, k)$

Focus: $(4, 0) \implies p = -2$

$(x - h)^2 = 4p(y - k)$

$(x - 4)^2 = -8(y - 2)$

15. Vertex: $(0, 2) = (h, k)$

Directrix: $x = -3 \implies p = 3$

$(y - k)^2 = 4p(x - h)$

$(y - 2)^2 = 12x$

17. $x^2 = -2y \implies p = -\dfrac{1}{2}$

Focus: $\left(0, -\dfrac{1}{2} \right)$

$d_1 = b + \dfrac{1}{2}$

$d_2 = \sqrt{(2 - 0)^2 + \left(-2 + \dfrac{1}{2} \right)^2}$

$= \sqrt{4 + \dfrac{9}{4}} = \dfrac{5}{2}$

$d_1 = d_2$

$b + \dfrac{1}{2} = \dfrac{5}{2}$

$b = 2$

The slope of the line is $m = \dfrac{-2 - 2}{2 - 0} = -2$

Tangent line: $y = -2x + 2$

x-intercept: $(1, 0)$

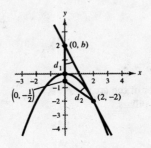

19. Parabola

    Opens downward

    Vertex: $(0, 12)$

    $(x - h)^2 = 4p(y - k)$

    $\qquad x^2 = 4p(y - 12)$

    Solution points: $(\pm 4, 10)$

    $\qquad 16 = 4p(10 - 12)$

    $\qquad 16 = -8p$

    $\qquad -2 = p$

    $\qquad x^2 = -8(y - 12)$

    To find the x-intercepts, let $y = 0$.

    $x^2 = 96$

    $x = \pm\sqrt{96} = \pm 4\sqrt{6}$

    At the base z, the archway is $2(4\sqrt{6}) = 8\sqrt{6}$ meters wide.

21. Vertices: $(-3, 0), (7, 0) \Rightarrow a = 5$

    $\qquad\qquad (h, k) = (2, 0)$

    Foci: $(0, 0), (4, 0) \Rightarrow c = 2$

    $b^2 = a^2 - c^2 = 25 - 4 = 21$

    $$\frac{(x - h)^2}{a^2} + \frac{(y - k)^2}{b^2} = 1$$

    $$\frac{(x - 2)^2}{25} + \frac{y^2}{21} = 1$$

23. Vertices: $(0, \pm 6) \Rightarrow a = 6, (h, k) = (0, 0)$

    Passes through $(2, 2)$

    $$\frac{(x - h)^2}{b^2} + \frac{(y - k)^2}{a^2} = 1$$

    $$\frac{x^2}{b^2} + \frac{y^2}{36} = 1 \Rightarrow b^2 = \frac{36x^2}{36 - y^2} = \frac{36(4)}{36 - 4} = \frac{9}{2}$$

    $$\frac{x^2}{9/2} + \frac{y^2}{36} = 1$$

    $$\frac{2x^2}{9} + \frac{y^2}{36} = 1$$

25. $2a = 10 \Rightarrow a = 5$

    $b = 4$

    $c^2 = a^2 - b^2 = 25 - 16 = 9 \Rightarrow c = 3$

    The foci occur 3 feet from the center of the arch on a line connecting the tops of the pillars.

27. $\qquad 16x^2 + 9y^2 - 32x + 72y + 16 = 0$

    $16(x^2 - 2x + 1) + 9(y^2 + 8y + 16) = -16 + 16 + 144$

    $16(x - 1)^2 + 9(y + 4)^2 = 144$

    $$\frac{(x - 1)^2}{9} + \frac{(y + 4)^2}{16} = 1$$

    $a = 4, b = 3, c = \sqrt{7}$

    Center: $(1, -4)$

    Vertices: $(1, 0)$ and $(1, -8)$

    Foci: $\left(1, -4 \pm \sqrt{7}\right)$

    Eccentricity: $e = \dfrac{\sqrt{7}}{4}$

29. $\dfrac{(x + 2)^2}{81} + \dfrac{(y - 1)^2}{100} = 1$

$a = 10, b = 9, c = \sqrt{19}$

Center: $(-2, 1)$

Vertices: $(-2, 11)$ and $(-2, -9)$

Foci: $\left(-2, 1 \pm \sqrt{19}\right)$

Eccentricity: $e = \dfrac{\sqrt{19}}{10}$

31. Vertices: $(0, \pm 1) \Rightarrow a = 1, (h, k) = (0, 0)$

Foci: $(0, \pm 3) \Rightarrow c = 3$

$b^2 = c^2 - a^2 = 9 - 1 = 8$

$\dfrac{(y - k)^2}{a^2} - \dfrac{(x - h)^2}{b^2} = 1$

$$y^2 - \dfrac{x^2}{8} = 1$$

33. Foci: $(0, 0), (8, 0) \Rightarrow c = 4, (h, k) = (4, 0)$

Asymptotes: $y = \pm 2(x - 4) \Rightarrow \dfrac{b}{a} = 2, b = 2a$

$b^2 = c^2 - a^2 \Rightarrow 4a^2 = 16 - a^2 \Rightarrow a^2 = \dfrac{16}{5},$

$b^2 = \dfrac{64}{5}$

$\dfrac{(x - h)^2}{a^2} - \dfrac{(y - k)^2}{b^2} = 1$

$\dfrac{(x - 4)^2}{\frac{16}{5}} - \dfrac{y^2}{\frac{64}{5}} = 1$

$\dfrac{5(x - 4)^2}{16} - \dfrac{5y^2}{64} = 1$

35. $9x^2 - 16y^2 - 18x - 32y - 151 = 0$

$9(x^2 - 2x + 1) - 16(y^2 + 2y + 1) = 151 + 9 - 16$

$9(x - 1)^2 - 16(y + 1)^2 = 144$

$\dfrac{(x - 1)^2}{16} - \dfrac{(y + 1)^2}{9} = 1$

$a = 4, b = 3, c = 5$

Center: $(1, -1)$

Vertices: $(5, -1)$ and $(-3, -1)$

Foci: $(6, -1)$ and $(-4, -1)$

Asymptotes: $y = -1 \pm \dfrac{3}{4}(x - 1)$

$$y = \dfrac{3}{4}x - \dfrac{7}{4} \quad \text{or} \quad y = -\dfrac{3}{4}x - \dfrac{1}{4}$$

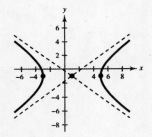

37. $\dfrac{(x - 3)^2}{16} - \dfrac{(y + 5)^2}{4} = 1$

$a = 4, b = 2, c = \sqrt{20} = 2\sqrt{5}$

Center: $(3, -5)$

Vertices: $(7, -5)$ and $(-1, -5)$

Foci: $\left(3 \pm 2\sqrt{5}, -5\right)$

Asymptotes: $y = -5 \pm \dfrac{1}{2}(x - 3)$

$$y = \dfrac{1}{2}x - \dfrac{13}{2} \quad \text{or} \quad y = -\dfrac{1}{2}x - \dfrac{7}{2}$$

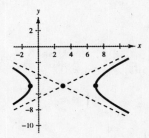

39. Foci: $(\pm 100, 0) \Rightarrow c = 100$

Center: $(0, 0)$

$$\frac{d_2}{186,000} - \frac{d_1}{186,000} = 0.0005 \Rightarrow d_2 - d_1 = 93 = 2a \Rightarrow a = 46.5$$

$$b^2 = c^2 - a^2 = 100^2 - 46.5^2 = 7837.75$$

$$\frac{x^2}{2162.25} - \frac{y^2}{7837.75} = 1$$

$$y^2 = 7837.75\left(\frac{60^2}{2162.25} - 1\right) \approx 5211.5736$$

$$y \approx 72 \text{ miles}$$

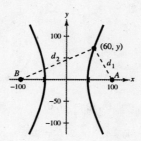

41. $5x^2 - 2y^2 + 10x - 4y + 17 = 0$

$AC = 5(-2) = -10 < 0$

The graph is a hyperbola.

43. $xy - 4 = 0$

$A = C = 0, B = 1$

$B^2 - 4AC = 1^2 - 4(0)(0) = 1 > 0$

The graph is a hyperbola.

$$\cot 2\theta = 0 \Rightarrow 2\theta = \frac{\pi}{2} \Rightarrow \theta = \frac{\pi}{4}$$

$$x = x'\cos\frac{\pi}{4} - y'\sin\frac{\pi}{4} = \frac{x' - y'}{\sqrt{2}}$$

$$y = x'\sin\frac{\pi}{4} + y'\cos\frac{\pi}{4} = \frac{x' + y'}{\sqrt{2}}$$

$$\left(\frac{x' - y'}{\sqrt{2}}\right)\left(\frac{x' + y'}{\sqrt{2}}\right) - 4 = 0$$

$$\frac{(x')^2 - (y')^2}{2} = 4$$

$$\frac{(x')^2}{8} - \frac{(y')^2}{8} = 1$$

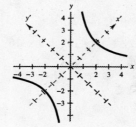

45. $5x^2 - 2xy + 5y^2 - 12 = 0$

$A = C = 5, B = -2$

$B^2 + 4AC = (-2)^2 - 4(5)(5) = -96 < 0$

The graph is an ellipse.

$$\cot 2\theta = 0 \Rightarrow 2\theta = \frac{\pi}{2} \Rightarrow \theta = \frac{\pi}{4}$$

$$x = x'\cos\frac{\pi}{4} - y'\sin\frac{\pi}{4} = \frac{x' - y'}{\sqrt{2}}$$

$$y = x'\sin\frac{\pi}{4} + y'\cos\frac{\pi}{4} = \frac{x' + y'}{\sqrt{2}}$$

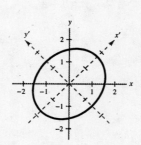

— CONTINUED —

45. — CONTINUED —

$$5\left(\frac{x' - y'}{\sqrt{2}}\right)^2 - 2\left(\frac{x' - y'}{\sqrt{2}}\right)\left(\frac{x' + y'}{\sqrt{2}}\right) + 5\left(\frac{x' + y'}{\sqrt{2}}\right)^2 - 12 = 0$$

$$\frac{5}{2}[(x')^2 - 2(x'y') + (y')^2] - [(x')^2 - (y')^2] + \frac{5}{2}[(x')^2 + 2(x'y') + (y')^2] = 12$$

$$4(x')^2 + 6(y')^2 = 12$$

$$\frac{(x')^2}{3} + \frac{(y')^2}{2} = 1$$

47. $16x^2 - 24xy + 9y^2 - 30x - 40y = 0$

$B^2 - 4AC = (-24)^2 - 4(16)(9) = 0$

The graph is a parabola.

To use a graphing utility, we need to solve for y in terms of x.

$$9y^2 + (-24x - 40)y + (16x^2 - 30x) = 0$$

$$y = \frac{-(-24x - 40) \pm \sqrt{(-24x - 40)^2 - 4(9)(16x^2 - 30x)}}{2(9)}$$

$$= \frac{(24x + 40) \pm \sqrt{(24x + 40)^2 - 36(16x^2 - 30x)}}{18}$$

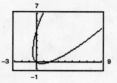

49. $x^2 + y^2 + 2xy + 2\sqrt{2}x - 2\sqrt{2}y + 2 = 0$

$B^2 - 4AC = 2^2 - 4(1)(1) = 0$

The graph is a parabola.

To use a graphing utility, we need to solve for y in terms of x.

$$y^2 + \left(2x - 2\sqrt{2}\right)y + \left(x^2 + 2\sqrt{2}x + 2\right)$$

$$y = \frac{-\left(2x - 2\sqrt{2}\right) \pm \sqrt{\left(2x - 2\sqrt{2}\right)^2 - 4\left(x^2 + 2\sqrt{2}x + 2\right)}}{2}$$

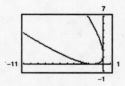

51. $x = 3 \cos 0 = 3$

$y = 2 \sin^2 0 = 0$

53. $x = 3 \cos \dfrac{\pi}{6} = \dfrac{3\sqrt{3}}{2}$

$y = 2 \sin^2 \dfrac{\pi}{6} = \dfrac{1}{2}$

55. $x = 2t \Rightarrow \dfrac{x}{2} = t$

$y = 4t \Rightarrow y = 4\left(\dfrac{x}{2}\right) = 2x$

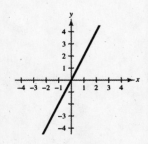

57. $x = t^2, \ x \geq 0$

$y = \sqrt{t} \Rightarrow y^2 = t$

$x = (y^2)^2 \Rightarrow x = y^4 \Rightarrow y = \sqrt[4]{x}$

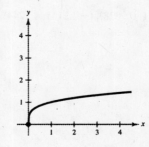

59. $x = 6 \cos \theta, y = 6 \sin \theta$

$\cos \theta = \dfrac{x}{6}, \sin \theta = \dfrac{y}{6}$

$\dfrac{x^2}{36} + \dfrac{y^2}{36} = 1$

$x^2 + y^2 = 36$

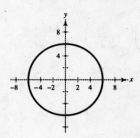

61. $(h, k) = (-3, 4)$

$2a = 8 \Rightarrow a = 4$

$2b = 6 \Rightarrow b = 3$

$\dfrac{(x + 3)^2}{16} + \dfrac{(y - 4)^2}{9} = 1$

$x = -3 + 4 \cos \theta$

$y = 4 + 3 \sin \theta$

This solution is not unique.

63. $x = \cos 3\theta + 5 \cos \theta$

$y = \sin 3\theta + 5 \sin \theta$

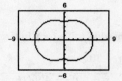

65. Polar coordinates: $\left(2, \dfrac{\pi}{4}\right)$

$x = 2 \cos \dfrac{\pi}{4} = \sqrt{2}$

$y = 2 \sin \dfrac{\pi}{4} = \sqrt{2}$

Rectangular coordinates: $\left(\sqrt{2}, \sqrt{2}\right)$

67. Polar coordinates: $(-7, 4.19)$

$x = -7 \cos 4.19 \approx 3.4927$

$y = -7 \sin 4.19 \approx 6.0664$

Rectangular coordinates: $(3.4927, 6.0664)$

69. Rectangular coordinates: $(0, 2)$

$r = \pm\sqrt{0^2 + 2^2} = \pm 2$

$\tan \theta$ is undefined $\Rightarrow \theta = \dfrac{\pi}{2}, \dfrac{3\pi}{2}$

Polar coordinates: $\left(2, \dfrac{\pi}{2}\right), \left(-2, \dfrac{3\pi}{2}\right), \left(2, \dfrac{5\pi}{2}\right)$

71. Rectangular coordinates: $(4, 6)$

$r = \pm\sqrt{4^2 + 6^2} = \pm\sqrt{52} \approx \pm 7.2111$

$\tan \theta = \dfrac{6}{4} \Rightarrow \theta \approx 0.9828, 4.1244$

Polar coordinates: $(7.2111, 0.9828), (-7.2111, 4.1244), (7.2111, 7.2660)$

73. $r = 3 \cos \theta$

$r^2 = 3r \cos \theta$

$x^2 + y^2 = 3x$

75. $r = \dfrac{2}{1 + \sin \theta}$

$r(1 + \sin \theta) = 2$

$r + r \sin \theta = 2$

$\pm\sqrt{x^2 + y^2} + y = 2$

$\pm\sqrt{x^2 + y^2} = 2 - y$

$x^2 + y^2 = (2 - y)^2$

$x^2 + y^2 = 4 - 4y + y^2$

$x^2 + 4y - 4 = 0$

77. $(x^2 + y^2)^2 = ax^2 y$

$(r^2)^2 = ar^2 \cos^2 \theta\, r \sin \theta$

$r = a \cos^2 \theta \sin \theta$

79. $r = 4$

Circle of radius 4 centered at the pole

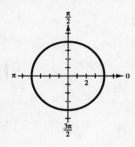

81. $r = 4 \sin 2\theta$

Symmetric with respect to $\theta = \pi/2$, the polar axis, and the pole.

Rose curve $(n = 2)$ with 4 petals

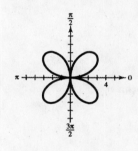

83. $r = -2(1 + \cos \theta)$

Symmetric with respect to the polar axis

$\dfrac{a}{b} = \dfrac{2}{2} = 1 \implies$ Cardioid

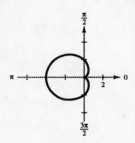

85. $r = 2 + 6 \sin \theta$

Limaçon with inner loop

$r = f(\sin \theta) \implies \theta = \dfrac{\pi}{2}$ symmetry

Maximum value: $|r| = 8$ when $\theta = \dfrac{\pi}{2}$

Zeros: $2 + 6 \sin \theta = 0 \implies \sin \theta = -\dfrac{1}{3} \implies \theta \approx 3.4814,\ 5.9433$

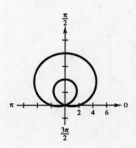

87. $r = -3 \cos 2\theta$

Rose curve with 4 petals

$r = f(\cos \theta) \Rightarrow$ polar axis symmetry

$\theta = \dfrac{\pi}{2}$: $r = -3 \cos 2(\pi - \theta) = -3 \cos(2\pi - 2\theta) = -3 \cos 2\theta$

Equivalent equation $\Rightarrow \theta = \dfrac{\pi}{2}$ symmetry

Pole: $r = -3 \cos 2(\pi + \theta) = -3 \cos(2\pi + 2\theta) = -3 \cos 2\theta$

Equivalent equation $\Rightarrow$ pole symmetry

Maximum value: $|r| = 3$ when $\theta = 0, \dfrac{\pi}{2}, \pi, \dfrac{3\pi}{2}$

Zeros: $-3 \cos 2\theta = 0$ when $\cos 2\theta = 0 \Rightarrow \theta = \dfrac{\pi}{4}, \dfrac{3\pi}{4}, \dfrac{5\pi}{4}, \dfrac{7\pi}{4}$

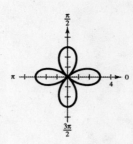

89. $r = 3(2 - \cos \theta)$

$= 6 - 3 \cos \theta$

$\dfrac{a}{b} = \dfrac{6}{3} = 2$

The graph is a convex limaçon.

91. $r = 4 \cos 3\theta$

The graph is a rose curve with 3 petals.

93. $r = \dfrac{1}{1 + 2 \sin \theta}, e = 2$

Hyperbola symmetric with respect to $\theta = \dfrac{\pi}{2}$ and having

vertices at $\left(\dfrac{1}{3}, \dfrac{\pi}{2}\right)$ and $\left(-1, \dfrac{3\pi}{2}\right)$.

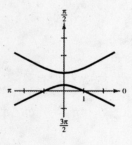

95. $r = \dfrac{4}{5 - 3 \cos \theta}$

$r = \dfrac{\frac{4}{5}}{1 - \left(\frac{3}{5}\right) \cos \theta}, e = \dfrac{3}{5}$

Ellipse symmetric with respect to the polar axis and

having vertices at $(2, 0)$ and $\left(\dfrac{1}{2}, \pi\right)$.

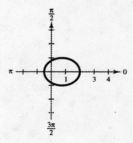

97. Parabola: $r = \dfrac{ep}{1 - e \cos \theta}, e = 1$

Vertex: $(2, \pi)$

Focus: $(0, 0) \Rightarrow p = 4$

$r = \dfrac{4}{1 - \cos \theta}$

99. Ellipse: $r = \dfrac{ep}{1 - e\cos\theta}$; Vertices: $(5, 0), (1, \pi) \Rightarrow a = 3$; One focus: $(0, 0) \Rightarrow c = 2$

$$e = \frac{c}{a} = \frac{2}{3}, p = \frac{5}{2}$$

$$r = \frac{\left(\frac{2}{3}\right)\left(\frac{5}{2}\right)}{1 - \left(\frac{2}{3}\right)\cos\theta} = \frac{\frac{5}{3}}{1 - \left(\frac{2}{3}\right)\cos\theta} = \frac{5}{3 - 2\cos\theta}$$

101. $a + c = 122{,}000 + 4000 \Rightarrow a + c = 126{,}000$

$a - c = 119 + 4000 \Rightarrow \underline{a - c = 4{,}119}$

$\ 2a = 130{,}119$

$\ \ a = 65{,}029.5$

$\ \ c = 60{,}940.5$

$$e = \frac{c}{a} = \frac{60{,}940.5}{65{,}059.5} \approx 0.937$$

$$r = \frac{ep}{1 - e\cos\theta} \approx \frac{0.937p}{1 - 0.937\cos\theta}$$

$r = 126{,}000$ when $\theta = 0$

$$126{,}000 = \frac{ep}{1 - e\cos 0}$$

$$ep = 126{,}000\left(1 - \frac{60{,}940.5}{65{,}059.5}\right) \approx 7977.2$$

Thus, $r \approx \dfrac{7977.2}{1 - 0.937\cos\theta}$

When $\theta = \dfrac{\pi}{3}, r \approx \dfrac{7977.2}{1 - 0.937\cos\dfrac{\pi}{3}} \approx 15{,}008.8$ miles

The distance from the surface of Earth and the satellite is $15{,}008.8 - 4000 \approx 11{,}008.8$ miles.

103. False. When classifying equations of the form $Ax^2 + Bxy + Cy^2 + Dx + Ey + F = 0$, its graph can be determined by its discriminant. For a graph to be a parabola, its discriminant, $B^2 - 4AC$, must equal zero. So, if $B = 0$, then A **or** C equals 0.

105. False. The following are **two** sets of parametric equations for the line.

$x = t, y = 3 - 2t$

$x = 3t, y = 3 - 6t$

107. $2a = 10 \Rightarrow a = 5$

b must be less than $5; 0 < b < 5$

As b approaches 5, the ellipse becomes more circular and approaches a circle of radius 5.

109. $x = 4\cos t$ and $y = 3\sin t$

(a) $x = 4\cos 2t$ and $y = 3\sin 2t$

The speed would double.

(b) $x = 5\cos t$ and $y = 3\sin t$

The elliptical orbit would be flatter. The length of the major axis is greater.

111. (a) $x^2 + y^2 = 25$

$r = 5$

The graphs are the same.

They are both circles centered at $(0, 0)$ with a radius of 5.

(b) $x - y = 0 \Rightarrow y = x$

$\theta = \dfrac{\pi}{4}$

The graphs are the same.

They are both lines with slope 1 and intercept $(0, 0)$.

Chapter 6   Practice Test

1. Find the angle, θ, between the lines $3x + 4y = 12$ and $4x - 3y = 12$.

2. Find the distance between the point $(5, -9)$ and the line $3x - 7y = 21$.

3. Find the vertex, focus and directrix of the parabola $x^2 - 6x - 4y + 1 = 0$.

4. Find an equation of the parabola with its vertex at $(2, -5)$ and focus at $(2, -6)$.

5. Find the center, foci, vertices, and eccentricity of the ellipse $x^2 + 4y^2 - 2x + 32y + 61 = 0$.

6. Find an equation of the ellipse with vertices $(0, \pm 6)$ and eccentricity $e = \frac{1}{2}$.

7. Find the center, vertices, foci, and asymptotes of the hyperbola $16y^2 - x^2 - 6x - 128y + 231 = 0$.

8. Find an equation of the hyperbola with vertices at $(\pm 3, 2)$ and foci at $(\pm 5, 2)$.

9. Rotate the axes to eliminate the xy-term. Sketch the graph of the resulting equation, showing both sets of axes.

    $5x^2 + 2xy + 5y^2 - 10 = 0$

10. Use the discriminant to determine whether the graph of the equation is a parabola, ellipse, or hyperbola.

    (a) $6x^2 - 2xy + y^2 = 0$ (b) $x^2 + 4xy + 4y^2 - x - y + 17 = 0$

11. Convert the polar point $\left(\sqrt{2}, \dfrac{3\pi}{4} \right)$ to rectangular coordinates.

12. Convert the rectangular point $\left(\sqrt{3}, -1 \right)$ to polar coordinates.

13. Convert the rectangular equation $4x - 3y = 12$ to polar form.

14. Convert the polar equation $r = 5 \cos \theta$ to rectangular form.

15. Sketch the graph of $r = 1 - \cos \theta$.

16. Sketch the graph of $r = 5 \sin 2\theta$.

17. Sketch the graph of $r = \dfrac{3}{6 - \cos \theta}$.

18. Find a polar equation of the parabola with its vertex at $\left(6, \dfrac{\pi}{2} \right)$ and focus at $(0, 0)$.

For Exercises 19 and 20, eliminate the parameter and write the corresponding rectangular equation.

19. $x = 3 - 2 \sin \theta, y = 1 + 5 \cos \theta$ 20. $x = e^{2t}, y = e^{4t}$

Chapter 1   Practice Test Solutions

1. $350° = 350\left(\dfrac{\pi}{180}\right) = \dfrac{35\pi}{18}$

2. $\dfrac{5\pi}{9} = \dfrac{5\pi}{9} \cdot \dfrac{180}{\pi} = 100°$

3. $135°14'12'' = \left(135 + \dfrac{14}{60} + \dfrac{12}{3600}\right)°$
$\approx 135.2367°$

4. $-22.569° = -(22° + 0.569(60)')$
$= -22°34.14'$
$= -(22°34' + 0.14(60)'')$
$\approx -22°34'8''$

5. $\cos\theta = \dfrac{2}{3}$
$x = 2, r = 3, y = \pm\sqrt{9-4} = \pm\sqrt{5}$
$\tan\theta = \dfrac{y}{x} = \pm\dfrac{\sqrt{5}}{2}$

6. $\sin\theta = 0.9063$
$\theta = \arcsin(0.9063)$
$\theta = 65° = \dfrac{13\pi}{36}$ or $\theta = 180° - 65° = 115° = \dfrac{23\pi}{36}$

7. $\tan 20° = \dfrac{35}{x}$
$x = \dfrac{35}{\tan 20°} \approx 96.1617$

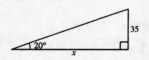

8. $\theta = \dfrac{6\pi}{5}$, θ is in Quadrant III.
Reference angle: $\dfrac{6\pi}{5} - \pi = \dfrac{\pi}{5}$ or $36°$

9. $\csc 3.92 = \dfrac{1}{\sin 3.92} \approx -1.4242$

10. $\tan\theta = 6 = \dfrac{6}{1}$, θ lies in Quandrant III.
$y = -6, x = -1, r = \sqrt{36+1} = \sqrt{37}$,
so $\sec\theta = \dfrac{\sqrt{37}}{-1} \approx -6.0828$.

11. Period: 4π
Amplitude: 3

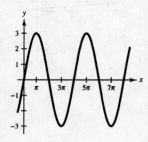

12. Period: 2π
Amplitude: 2

13. Period: $\dfrac{\pi}{2}$

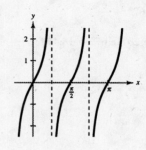

14. Period: 2π

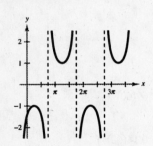

15.

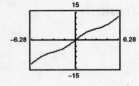

16.

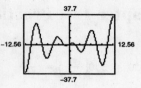

17. $\theta = \arcsin 1$

$\sin \theta = 1$

$\theta = \dfrac{\pi}{2} = 90°$

18. $\theta = \arctan(-3)$

$\tan \theta = -3$

$\theta \approx -1.249 \approx -71.565°$

19. $\sin\left(\arccos \dfrac{4}{\sqrt{35}}\right)$

$\sin \theta = \dfrac{\sqrt{19}}{\sqrt{35}} \approx 0.7368$

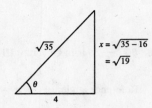

20. $\cos\left(\arcsin \dfrac{x}{4}\right)$

$\cos \theta = \dfrac{\sqrt{16 - x^2}}{4}$

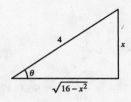

21. Given $A = 40°$, $c = 12$

$B = 90° - 40° = 50°$

$\sin 40° = \dfrac{a}{12}$

$a = 12 \sin 40° \approx 7.713$

$\cos 40° = \dfrac{b}{12}$

$b = 12 \cos 40° \approx 9.193$

22. Given $B = 6.84°$, $a = 21.3$

$A = 90° - 6.84° = 83.16°$

$\sin 83.16° = \dfrac{21.3}{c}$

$c = \dfrac{21.3}{\sin 83.16°} \approx 21.453$

$\tan 83.16° = \dfrac{21.3}{b}$

$b = \dfrac{21.3}{\tan 83.16°} \approx 2.555$

23. Given $a = 5$, $b = 9$

$c = \sqrt{25 + 81} = \sqrt{106} \approx 10.296$

$\tan A = \dfrac{5}{9}$

$A = \arctan \dfrac{5}{9} \approx 29.055°$

$B \approx 90° - 29.055° = 60.945°$

24. $\sin 67° = \dfrac{x}{20}$

$x = 20 \sin 67° \approx 18.41$ feet

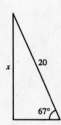

25. $\tan 5° = \dfrac{250}{x}$

$x = \dfrac{250}{\tan 5°}$

≈ 2857.513 feet

≈ 0.541 mi

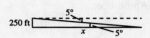

Chapter 2   Practice Test Solutions

1. $\tan x = \dfrac{4}{11}$, $\sec x < 0 \implies x$ is in Quadrant III.

$y = -4$, $x = -11$, $r = \sqrt{16 + 121} = \sqrt{137}$

$\sin x = -\dfrac{4}{\sqrt{137}} = -\dfrac{4\sqrt{137}}{137}$ $\qquad\qquad$ $\csc x = -\dfrac{\sqrt{137}}{4}$

$\cos x = -\dfrac{11}{\sqrt{137}} = -\dfrac{11\sqrt{137}}{137}$ $\qquad\qquad$ $\sec x = -\dfrac{\sqrt{137}}{11}$

$\tan x = \dfrac{4}{11}$ $\qquad\qquad\qquad\qquad\qquad$ $\cot x = \dfrac{11}{4}$

2. $\dfrac{\sec^2 x + \csc^2 x}{\csc^2 x (1 + \tan^2 x)} = \dfrac{\sec^2 x + \csc^2 x}{\csc^2 x + (\csc^2 x)\tan^2 x} = \dfrac{\sec^2 x + \csc^2 x}{\csc^2 x + \dfrac{1}{\sin^2 x} \cdot \dfrac{\sin^2 x}{\cos^2 x}}$

$\qquad\qquad = \dfrac{\sec^2 x + \csc^2 x}{\csc^2 x + \dfrac{1}{\cos^2 x}} = \dfrac{\sec^2 x + \csc^2 x}{\csc^2 x + \sec^2 x} = 1$

3. $\ln|\tan\theta| - \ln|\cot\theta| = \ln\left|\dfrac{\tan\theta}{\cot\theta}\right| = \ln\left|\dfrac{\sin\theta/\cos\theta}{\cos\theta/\sin\theta}\right| = \ln\left|\dfrac{\sin^2\theta}{\cos^2\theta}\right| = \ln|\tan^2\theta| = 2\ln|\tan\theta|$

4. $\cos\left(\dfrac{\pi}{2} - x\right) = \dfrac{1}{\csc x}$ is true since $\cos\left(\dfrac{\pi}{2} - x\right) = \sin x = \dfrac{1}{\csc x}$.

5. $\sin^4 x + (\sin^2 x)\cos^2 x = \sin^2 x(\sin^2 x + \cos^2 x) = \sin^2 x(1) = \sin^2 x$

6. $(\csc x + 1)(\csc x - 1) = \csc^2 x - 1 = \cot^2 x$

7. $\dfrac{\cos^2 x}{1 - \sin x} \cdot \dfrac{1 + \sin x}{1 + \sin x} = \dfrac{\cos^2 x(1 + \sin x)}{1 - \sin^2 x} = \dfrac{\cos^2 x(1 + \sin x)}{\cos^2 x} = 1 + \sin x$

8. $\dfrac{1 + \cos\theta}{\sin\theta} + \dfrac{\sin\theta}{1 + \cos\theta} = \dfrac{(1 + \cos\theta)^2 + \sin^2\theta}{\sin\theta(1 + \cos\theta)}$

$\qquad\qquad = \dfrac{1 + 2\cos\theta + \cos^2\theta + \sin^2\theta}{\sin\theta(1 + \cos\theta)} = \dfrac{2 + 2\cos\theta}{\sin\theta(1 + \cos\theta)} = \dfrac{2}{\sin\theta} = 2\csc\theta$

9. $\tan^4 x + 2\tan^2 x + 1 = (\tan^2 x + 1)^2 = (\sec^2 x)^2 = \sec^4 x$

10. (a) $\sin 105° = \sin(60° + 45°) = \sin 60° \cos 45° + \cos 60° \sin 45°$

$$= \frac{\sqrt{3}}{2} \cdot \frac{\sqrt{2}}{2} + \frac{1}{2} \cdot \frac{\sqrt{2}}{2} = \frac{\sqrt{2}}{4}\left(\sqrt{3} + 1\right)$$

(b) $\tan 15° = \tan(60° - 45°) = \dfrac{\tan 60° - \tan 45°}{1 + \tan 60° \tan 45°}$

$$= \frac{\sqrt{3} - 1}{1 + \sqrt{3}} \cdot \frac{1 - \sqrt{3}}{1 - \sqrt{3}} = \frac{2\sqrt{3} - 1 - 3}{1 - 3} = \frac{2\sqrt{3} - 4}{-2} = 2 - \sqrt{3}$$

11. $(\sin 42°) \cos 38° - (\cos 42°) \sin 38° = \sin(42° - 38°) = \sin 4°$

12. $\tan\left(\theta + \dfrac{\pi}{4}\right) = \dfrac{\tan \theta + \tan\left(\dfrac{\pi}{4}\right)}{1 - (\tan \theta) \tan\left(\dfrac{\pi}{4}\right)} = \dfrac{\tan \theta + 1}{1 - \tan \theta(1)} = \dfrac{1 + \tan \theta}{1 - \tan \theta}$

13. $\sin(\arcsin x - \arccos x) = \sin(\arcsin x) \cos(\arccos x) - \cos(\arcsin x) \sin(\arccos x)$

$$= (x)(x) - \left(\sqrt{1 - x^2}\right)\left(\sqrt{1 - x^2}\right) = x^2 - (1 - x^2) = 2x^2 - 1$$

14. (a) $\cos(120°) = \cos[2(60°)] = 2\cos^2 60° - 1 = 2\left(\dfrac{1}{2}\right)^2 - 1 = -\dfrac{1}{2}$

(b) $\tan(300°) = \tan[2(150°)] = \dfrac{2\tan 150°}{1 - \tan^2 150°} = \dfrac{-\dfrac{2\sqrt{3}}{3}}{1 - \left(\dfrac{1}{3}\right)} = -\sqrt{3}$

15. (a) $\sin 22.5° = \sin \dfrac{45°}{2} = \sqrt{\dfrac{1 - \cos 45°}{2}} = \sqrt{\dfrac{1 - \dfrac{\sqrt{2}}{2}}{2}} = \dfrac{\sqrt{2 - \sqrt{2}}}{2}$

(b) $\tan \dfrac{\pi}{12} = \tan \dfrac{\dfrac{\pi}{6}}{2} = \dfrac{\sin \dfrac{\pi}{6}}{1 + \cos\left(\dfrac{\pi}{6}\right)} = \dfrac{\dfrac{1}{2}}{1 + \dfrac{\sqrt{3}}{2}} = \dfrac{1}{2 + \sqrt{3}} = 2 - \sqrt{3}$

16. $\sin \theta = \dfrac{4}{5}$, θ lies in Quadrant II $\implies \cos \theta = -\dfrac{3}{5}$.

$$\cos \frac{\theta}{2} = \sqrt{\frac{1 + \cos \theta}{2}} = \sqrt{\frac{1 - \dfrac{3}{5}}{2}} = \sqrt{\frac{2}{10}} = \frac{1}{\sqrt{5}} = \frac{\sqrt{5}}{5}$$

17. $(\sin^2 x) \cos^2 x = \dfrac{1 - \cos 2x}{2} \cdot \dfrac{1 + \cos 2x}{2} = \dfrac{1}{4}\left[1 - \cos^2 2x\right] = \dfrac{1}{4}\left[1 - \dfrac{1 + \cos 4x}{2}\right]$

$$= \frac{1}{8}[2 - (1 + \cos 4x)] = \frac{1}{8}[1 - \cos 4x]$$

18. $6(\sin 5\theta) \cos 2\theta = 6\left\{\dfrac{1}{2}[\sin(5\theta + 2\theta) + \sin(5\theta - 2\theta)]\right\} = 3[\sin 7\theta + \sin 3\theta]$

19. $\sin(x + \pi) + \sin(x - \pi) = 2\left(\sin\dfrac{[(x + \pi) + (x - \pi)]}{2}\right)\cos\dfrac{[(x + \pi) - (x - \pi)]}{2}$

$\qquad\qquad\qquad\qquad = 2\sin x \cos \pi = -2\sin x$

20. $\dfrac{\sin 9x + \sin 5x}{\cos 9x - \cos 5x} = \dfrac{2\sin 7x \cos 2x}{-2\sin 7x \sin 2x} = -\dfrac{\cos 2x}{\sin 2x} = -\cot 2x$

21. $\frac{1}{2}[\sin(u + v) - \sin(u - v)] = \frac{1}{2}\{(\sin u)\cos v + (\cos u)\sin v - [(\sin u)\cos v - (\cos u)\sin v]\}$

$\qquad\qquad\qquad\qquad\qquad = \frac{1}{2}[2(\cos u)\sin v] = (\cos u)\sin v$

22. $4\sin^2 x = 1$

$\qquad \sin^2 x = \dfrac{1}{4}$

$\qquad \sin x = \pm\dfrac{1}{2}$

$\sin x = \dfrac{1}{2} \qquad$ or $\quad \sin x = -\dfrac{1}{2}$

$x = \dfrac{\pi}{6}$ or $\dfrac{5\pi}{6} \qquad\quad x = \dfrac{7\pi}{6}$ or $\dfrac{11\pi}{6}$

23. $\tan^2 \theta + \left(\sqrt{3} - 1\right)\tan \theta - \sqrt{3} = 0$

$\qquad\quad (\tan\theta - 1)\left(\tan\theta + \sqrt{3}\right) = 0$

$\tan \theta = 1 \qquad$ or $\quad \tan \theta = -\sqrt{3}$

$\theta = \dfrac{\pi}{4}$ or $\dfrac{5\pi}{4} \qquad\qquad \theta = \dfrac{2\pi}{3}$ or $\dfrac{5\pi}{3}$

24. $\qquad\qquad\qquad \sin 2x = \cos x$

$2(\sin x)\cos x - \cos x = 0$

$\quad \cos x(2\sin x - 1) = 0$

$\cos x = 0 \qquad$ or $\quad \sin x = \dfrac{1}{2}$

$x = \dfrac{\pi}{2}$ or $\dfrac{3\pi}{2} \qquad\quad x = \dfrac{\pi}{6}$ or $\dfrac{5\pi}{6}$

25. $\tan^2 x - 6\tan x + 4 = 0$

$\qquad \tan x = \dfrac{-(-6) \pm \sqrt{(-6)^2 - 4(1)(4)}}{2(1)}$

$\qquad \tan x = \dfrac{6 \pm \sqrt{20}}{2} = 3 \pm \sqrt{5}$

$\tan x = 3 + \sqrt{5} \qquad$ or $\quad \tan x = 3 - \sqrt{5}$

$x \approx 1.3821$ or $4.5237 \qquad x = 0.6524$ or 3.7940

Chapter 3   Practice Test Solutions

1. $C = 180° - (40° + 12°) = 128°$

$$a = \sin 40° \left(\frac{100}{\sin 12°} \right) \approx 309.164$$

$$c = \sin 128° \left(\frac{100}{\sin 12°} \right) \approx 379.012$$

2. $\sin A = 5 \left(\dfrac{\sin 150°}{20} \right) = 0.125$

$A \approx 7.181°$

$B \approx 180° - (150° + 7.181°) = 22.819°$

$b = \sin 22.819° \left(\dfrac{20}{\sin 150°} \right) \approx 15.513$

3. Area $= \frac{1}{2}ab \sin C$

$\qquad = \frac{1}{2}(3)(5)\sin 130°$

$\qquad \approx 5.745$ square units

4. $h = b \sin A$

$\quad = 35 \sin 22.5°$

$\quad \approx 13.394$

$a = 10$

Since $a < h$ and A is acute, the triangle has no solution.

5. $\cos A = \dfrac{(53)^2 + (38)^2 - (49)^2}{2(53)(38)} \approx 0.4598$

$A \approx 62.627°$

$\cos B = \dfrac{(49)^2 + (38)^2 - (53)^2}{2(49)(38)} \approx 0.2782$

$B \approx 73.847°$

$C \approx 180° - (62.627° + 73.847°)$

$\quad = 43.526°$

6. $c^2 = (100)^2 + (300)^2 - 2(100)(300)\cos 29°$

$\quad \approx 47522.8176$

$c \approx 218$

$\cos A = \dfrac{(300)^2 + (218)^2 - (100)^2}{2(300)(218)} \approx 0.97495$

$A \approx 12.85°$

$B \approx 180° - (12.85° + 29°) = 138.15°$

7. $s = \dfrac{a + b + c}{2} = \dfrac{4.1 + 6.8 + 5.5}{2} = 8.2$

Area $= \sqrt{s(s - a)(s - b)(s - c)}$

$\qquad = \sqrt{8.2(8.2 - 4.1)(8.2 - 6.8)(8.2 - 5.5)}$

$\qquad \approx 11.273$ square units

8. $x^2 = (40)^2 + (70)^2 - 2(40)(70)\cos 168°$

$\quad \approx 11977.6266$

$x \approx 190.442$ miles

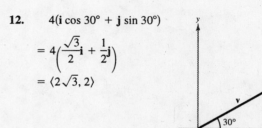

9. $\mathbf{w} = 4(3\mathbf{i} + \mathbf{j}) - 7(-\mathbf{i} + 2\mathbf{j})$

$\quad = 19\mathbf{i} - 10\mathbf{j}$

10. $\dfrac{\mathbf{v}}{\|\mathbf{v}\|} = \dfrac{5\mathbf{i} + 3\mathbf{j}}{\sqrt{25 + 9}} = \dfrac{5}{\sqrt{34}}\mathbf{i} - \dfrac{3}{\sqrt{34}}\mathbf{j}$

$\qquad = \dfrac{5\sqrt{34}}{34}\mathbf{i} - \dfrac{3\sqrt{34}}{34}\mathbf{j}$

11. $\mathbf{u} = 6\mathbf{i} + 5\mathbf{j} \qquad \mathbf{v} = 2\mathbf{i} - 3\mathbf{j}$

$\mathbf{u} \cdot \mathbf{v} = 6(2) + 5(-3) = -3$

$\|\mathbf{u}\| = \sqrt{61} \qquad \|\mathbf{v}\| = \sqrt{13}$

$\cos \theta = \dfrac{-3}{\sqrt{61}\sqrt{13}}$

$\theta \approx 96.116°$

12. $4(\mathbf{i} \cos 30° + \mathbf{j} \sin 30°)$

$\quad = 4\left(\dfrac{\sqrt{3}}{2}\mathbf{i} + \dfrac{1}{2}\mathbf{j} \right)$

$\quad = \langle 2\sqrt{3}, 2 \rangle$

13. $\text{proj}_{\mathbf{v}}\mathbf{u} = \left(\dfrac{\mathbf{u} \cdot \mathbf{v}}{\|\mathbf{v}\|^2} \right)\mathbf{v} = \dfrac{-10}{20} \langle -2, 4 \rangle = \langle 1, -2 \rangle$

14. $\mathbf{u} = 7\cos 35°\mathbf{i} + 7\sin 35°\mathbf{j}$

$\mathbf{v} = 4\cos 123°\mathbf{i} + 4\sin 123°\mathbf{j}$

$\mathbf{u} + \mathbf{v} = (7\cos 35° + 4\cos 123°)\mathbf{i} + (7\sin 35° + 4\sin 123°)\mathbf{j}$

$\approx \langle 3.56, 7.37 \rangle$

15. Answer is not unique. Two possibilities are: $\langle 10, 3 \rangle$ and $\langle -10, -3 \rangle$

Chapter 4   Practice Test Solutions

1. $4 + \sqrt{-81} - 3i^2 = 4 + 9i + 3 = 7 + 9i$

2. $\dfrac{3 + i}{5 - 4i} \cdot \dfrac{5 + 4i}{5 + 4i} = \dfrac{15 + 12i + 5i + 4i^2}{25 + 16} = \dfrac{11 + 17i}{41} = \dfrac{11}{41} + \dfrac{17}{41}i$

3. $x = \dfrac{-(-4) \pm \sqrt{(-4)^2 - 4(1)(7)}}{2(1)} = \dfrac{4 \pm \sqrt{-12}}{2} = \dfrac{4 \pm 2\sqrt{3}i}{2} = 2 \pm \sqrt{3}i$

4. False: $\sqrt{-6}\,\sqrt{-6} = \left(\sqrt{6}i\right)\left(\sqrt{6}i\right) = \sqrt{36}i^2 = 6(-1) = -6$

5. $b^2 - 4ac = (-8)^2 - 4(3)(7) = -20 < 0$

Two complex solutions.

6. $x^4 + 13x^2 + 36 = 0$

$(x^2 + 4)(x^2 + 9) = 0$

$x^2 + 4 = 0 \implies x^2 = -4 \implies x = \pm 2i$

$x^2 + 9 = 0 \implies x^2 = -9 \implies x = \pm 3i$

7. $f(x) = (x - 3)[x - (-1 + 4i)][x - (-1 - 4i)]$

$= (x - 3)[(x + 1) - 4i][(x + 1) + 4i]$

$= (x - 3)[(x + 1)^2 - 16i^2]$

$= (x - 3)(x^2 + 2x + 17)$

$= x^3 - x^2 + 11x - 51$

8. Since $x = 4 + i$ is a zero, so is its conjugate $4 - i$.

$[x - (4 + i)][x - (4 - i)] = [(x - 4) - i][(x - 4) + i]$

$= (x - 4)^2 - i^2$

$= x^2 - 8x + 17$

$$
\begin{array}{r}
x - 2 \\
x^2 - 8x + 17 \overline{)\,x^3 - 10x^2 + 33x - 34} \\
\underline{x^3 - 8x^2 + 17x} \\
-2x^2 + 16x - 34 \\
\underline{-2x^2 + 16x - 34} \\
0
\end{array}
$$

Thus, $f(x) = (x^2 - 8x + 17)(x - 2)$ and the zeros of $f(x)$ are: $4 \pm i, 2$

9. $r = \sqrt{25 + 25} = \sqrt{50} = 5\sqrt{2}$

$\tan \theta = \dfrac{-5}{5} = -1$

Since z is in Quadrant IV,

$\theta = 315°$

$z = 5\sqrt{2}(\cos 315° + i \sin 315°).$

10. $\cos 225° = -\dfrac{\sqrt{2}}{2}, \sin 225° = -\dfrac{\sqrt{2}}{2}$

$z = 6\left(-\dfrac{\sqrt{2}}{2} - i\dfrac{\sqrt{2}}{2}\right)$

$= -3\sqrt{2} - 3\sqrt{2}i$

11. $[7(\cos 23° + i \sin 23°)][4(\cos 7° + i \sin 7°] = 7(4)[\cos(23° + 7°) + i \sin(23° + 7°)]$

$= 28(\cos 30° + i \sin 30°)$

$= 14\sqrt{3} + 14i$

12. $\dfrac{9\left(\cos \dfrac{5\pi}{4} + i \sin \dfrac{5\pi}{4}\right)}{3(\cos \pi + i \sin \pi)} = \dfrac{9}{3}\left[\cos\left(\dfrac{5\pi}{4} - \pi\right) + i \sin\left(\dfrac{5\pi}{4} - \pi\right)\right] = 3\left(\cos \dfrac{\pi}{4} + i \sin \dfrac{\pi}{4}\right) = \dfrac{3\sqrt{2}}{2} + \dfrac{3\sqrt{2}}{2}i$

13. $(2 + 2i)^8 = \left[2\sqrt{2}(\cos 45° + i \sin 45°)\right]^8 = \left(2\sqrt{2}\right)^8[\cos(8)(45°) + i \sin(8)(45°)]$

$= 4096[\cos 360° + i \sin 360°] = 4096$

14. $z = 8\left(\cos \dfrac{\pi}{3} + i \sin \dfrac{\pi}{3}\right), n = 3$

The cube roots of z are: $\sqrt[3]{8}\left[\cos \dfrac{(\pi/3) + 2\pi k}{3} + i \sin \dfrac{(\pi/3) + 2\pi k}{3}\right], k = 0, 1, 2.$

For $k = 0, \quad \sqrt[3]{8}\left[\cos \dfrac{\pi/3}{3} + i \sin \dfrac{\pi/3}{3}\right] = 2\left(\cos \dfrac{\pi}{9} + i \sin \dfrac{\pi}{9}\right).$

For $k = 1, \sqrt[3]{8}\left[\cos \dfrac{\pi/3 + 2\pi}{3} + i \sin \dfrac{\pi/3 + 2\pi}{3}\right] = 2\left(\cos \dfrac{7\pi}{9} + i \sin \dfrac{7\pi}{9}\right).$

For $k = 2, \sqrt[3]{8}\left[\cos \dfrac{\pi/3 + 4\pi}{3} + i \sin \dfrac{\pi/3 + 4\pi}{3}\right] = 2\left(\cos \dfrac{13\pi}{9} + i \sin \dfrac{13\pi}{9}\right).$

15. $x^4 = -i = 1\left(\cos \dfrac{3\pi}{2} + i \sin \dfrac{3\pi}{2}\right)$

The fourth roots are: $\sqrt[4]{1}\left[\cos \dfrac{(3\pi/2) + 2\pi k}{4} + i \sin \dfrac{(3\pi/2) + 2\pi k}{4}\right], k = 0, 1, 2, 3$

For $k = 0, \cos\left(\dfrac{3\pi/2}{4}\right) + i \sin\left(\dfrac{3\pi/2}{4}\right) = \cos \dfrac{3\pi}{8} + i \sin \dfrac{3\pi}{8}.$

For $k = 1, \cos\left(\dfrac{3\pi/2 + 2\pi}{4}\right) + i \sin\left(\dfrac{3\pi/2 + 2\pi}{4}\right) = \cos \dfrac{7\pi}{8} + i \sin \dfrac{7\pi}{8}.$

For $k = 2, \cos\left(\dfrac{3\pi/2 + 4\pi}{4}\right) + i \sin\left(\dfrac{3\pi/2 + 4\pi}{4}\right) = \cos \dfrac{11\pi}{8} + i \sin \dfrac{11\pi}{8}.$

For $k = 3, \cos\left(\dfrac{3\pi/2 + 6\pi}{4}\right) + i \sin\left(\dfrac{3\pi/2 + 6\pi}{4}\right) = \cos \dfrac{15\pi}{8} + i \sin \dfrac{15\pi}{8}.$

Chapter 5   Practice Test Solutions

1. $x^{3/5} = 8$

$x = 8^{5/3} = \left(\sqrt[3]{8}\right)^5 = 2^5 = 32$

2. $3^{x-1} = \frac{1}{81}$

$3^{x-1} = 3^{-4}$

$x - 1 = -4$

$x = -3$

3. $f(x) = 2^{-x} = \left(\frac{1}{2}\right)^x$

x	-2	-1	0	1	2
$f(x)$	4	2	1	$\frac{1}{2}$	$\frac{1}{4}$

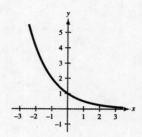

4. $g(x) = e^x + 1$

x	-2	-1	0	1	2
$g(x)$	1.14	1.37	2	3.72	8.39

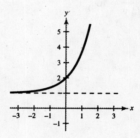

5. (a) $A = P\left(1 + \dfrac{r}{n}\right)^{nt}$

$A = 5000\left(1 + \dfrac{0.09}{12}\right)^{12(3)} \approx \$6543.23$

(b) $A = P\left(1 + \dfrac{r}{n}\right)^{nt}$

$A = 5000\left(1 + \dfrac{0.09}{4}\right)^{4(3)} \approx \$6530.25$

(c) $A = Pe^{rt}$

$A = 5000e^{(0.09)(3)} \approx \$6549.82$

6. $7^{-2} = \dfrac{1}{49}$

$\log_7 \dfrac{1}{49} = -2$

7. $x - 4 = \log_2 \frac{1}{64}$

$2^{x-4} = \frac{1}{64}$

$2^{x-4} = 2^{-6}$

$x - 4 = -6$

$x = -2$

8. $\log_b \sqrt[4]{\frac{8}{25}} = \frac{1}{4}\log_b \frac{8}{25}$

$= \frac{1}{4}\left[\log_b 8 - \log_b 25\right]$

$= \frac{1}{4}\left[\log_b 2^3 - \log_b 5^2\right]$

$= \frac{1}{4}\left[3\log_b 2 - 2\log_b 5\right]$

$= \frac{1}{4}\left[3(0.3562) - 2(0.8271)\right]$

$= -0.1464$

9. $5 \ln x - \dfrac{1}{2} \ln y + 6 \ln z = \ln x^5 - \ln \sqrt{y} + \ln z^6 = \ln\left(\dfrac{x^5 z^6}{\sqrt{y}}\right), z > 0$

10. $\log_9 28 = \dfrac{\log 28}{\log 9} \approx 1.5166$

11. $\log N = 0.6646$

$N = 10^{0.6646} \approx 4.62$

12.

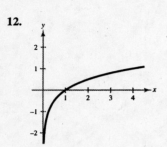

13. Domain:

$$x^2 - 9 > 0$$
$$(x + 3)(x - 3) > 0$$
$$x < -3 \text{ or } x > 3$$

14.

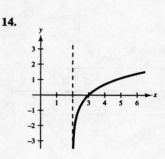

15. False. $\dfrac{\ln x}{\ln y} \neq \ln(x - y)$ since $\dfrac{\ln x}{\ln y} = \log_y x$.

16. $5^3 = 41$

$x = \log_5 41 = \dfrac{\ln 41}{\ln 5} \approx 2.3074$

17. $x - x^2 = \log_5 \dfrac{1}{25}$

$5^{x-x^2} = \dfrac{1}{25}$

$5^{x-x^2} = 5^{-2}$

$x - x^2 = -2$

$0 = x^2 - x - 2$

$0 = (x + 1)(x - 2)$

$x = -1 \text{ or } x = 2$

18. $\log_2 x + \log_2(x - 3) = 2$

$\log_2[x(x - 3)] = 2$

$x(x - 3) = 2^2$

$x^2 - 3x = 4$

$x^2 - 3x - 4 = 0$

$(x + 1)(x - 4) = 0$

$x = 4$

$x = -1 \text{ (extraneous)}$

$x = 4$ is the only solution.

19. $\dfrac{e^x + e^{-x}}{3} = 4$

$e^x(e^x + e^{-x}) = 12e^x$

$e^{2x} + 1 = 12e^x$

$e^{2x} - 12e^x + 1 = 0$

$e^x = \dfrac{12 \pm \sqrt{144 - 4}}{2}$

$e^x \approx 11.9161 \qquad \text{or} \qquad e^x \approx 0.0839$

$x = \ln 11.9161 \qquad\qquad x = \ln 0.0839$

$x \approx 2.478 \qquad\qquad\qquad x \approx -2.478$

20. $A = Pe^{et}$

$12{,}000 = 6000e^{0.13t}$

$2 = e^{0.13t}$

$0.13t = \ln 2$

$t = \dfrac{\ln 2}{0.13}$

$t \approx 5.3319$ years or 5 years 4 months

Chapter 6   Practice Test Solutions

1. $3x + 4y = 12 \Rightarrow y = -\dfrac{3}{4}x + 3 \Rightarrow m_1 = -\dfrac{3}{4}$

$4x - 3y = 12 \Rightarrow y = \dfrac{4}{3}x - 4 \Rightarrow m_2 = \dfrac{4}{3}$

$\tan \theta = \left| \dfrac{\frac{4}{3} - \left(-\frac{3}{4}\right)}{1 + \left(\frac{4}{3}\right)\left(-\frac{3}{4}\right)} \right| = \left| \dfrac{\frac{25}{12}}{0} \right|$

Since $\tan \theta$ is undefined, the lines are perpendicular (note that $m_2 = -1/m_1$)   and $\theta = 90°$.

2. $x_1 = 5, x_2 = -9, A = 3, B = -7, C = -21$

$d = \dfrac{|3(5) + (-7)(-9) + (-21)|}{\sqrt{3^2 + (-7)^2}} = \dfrac{57}{\sqrt{58}} \approx 7.484$

3. $x^2 - 6x - 4y + 1 = 0$

$\quad x^2 - 6x + 9 = 4y - 1 + 9$

$\qquad (x - 3)^2 = 4y + 8$

$\qquad (x - 3)^2 = 4(1)(y + 2) \Rightarrow p = 1$

Vertex: $(3, -2)$

Focus: $(3, -1)$

Directrix: $y = -3$

4. Vertex: $(2, -5)$

Focus: $(2, -6)$

Vertical axis; opens downward with $p = -1$

$\qquad (x - h)^2 = 4p(y - k)$

$\qquad (x - 2)^2 = 4(-1)(y + 5)$

$\quad x^2 - 4x + 4 = -4y - 20$

$x^2 - 4x + 4y + 24 = 0$

5. $\qquad x^2 + 4y^2 - 2x + 32y + 61 = 0$

$(x^2 - 2x + 1) + 4(y^2 + 8y + 16) = -61 + 1 + 64$

$\qquad (x - 1)^2 + 4(y + 4)^2 = 4$

$\qquad \dfrac{(x - 1)^2}{4} + \dfrac{(y + 4)^2}{1} = 1$

$a = 2, b = 1, c = \sqrt{3}$

Horizontal major axis

Center: $(1, -4)$

Foci: $\left(1 \pm \sqrt{3}, -4\right)$

Vertices: $(3, -4), (-1, -4)$

Eccentricity: $e = \dfrac{\sqrt{3}}{2}$

6. Vertices: $(0, \pm 6)$

Eccentricity: $e = \dfrac{1}{2}$

Center: $(0, 0)$

Vertical major axis

$a = 6, e = \dfrac{c}{a} = \dfrac{c}{6} = \dfrac{1}{2} \Rightarrow c = 3$

$b^2 = (6)^2 - (3)^2 = 27$

$\dfrac{x^2}{27} + \dfrac{y^2}{36} = 1$

7. $\qquad 16y^2 - x^2 - 6x - 128y + 231 = 0$

$16(y^2 - 8y + 16) - (x^2 + 6x + 9) = -231 + 256 - 9$

$\qquad 16(y - 4)^2 - (x + 3)^2 = 16$

$\qquad \dfrac{(y - 4)^2}{1} - \dfrac{(x + 3)^2}{16} = 1$

$a = 1, b = 4, c = \sqrt{17}$

Center: $(-3, 4)$

Vertical transverse axis

Vertices: $(-3, 5), (-3, 3)$

Foci: $\left(-3, 4 \pm \sqrt{17}\right)$

Asymptotes: $y = 4 \pm \dfrac{1}{4}(x + 3)$

8. Vertices: $(\pm 3, 2)$

Foci: $(\pm 5, 2)$

Center: $(0, 2)$

Horizontal transverse axis

$a = 3, c = 5, b = 4$

$\dfrac{(x - 0)^2}{9} - \dfrac{(y - 2)^2}{16} = 1$

$\dfrac{x^2}{9} - \dfrac{(y - 2)^2}{16} = 1$

9. $5x^2 + 2xy + 5y^2 - 10 = 0$

$A = 5, B = 2, C = 5$

$\cot 2\theta = \dfrac{5 - 5}{2} = 0$

$2\theta = \dfrac{\pi}{2} \Rightarrow \theta = \dfrac{\pi}{4}$

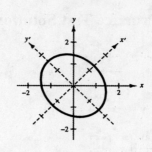

$x = x' \cos \dfrac{\pi}{4} - y' \sin \dfrac{\pi}{4}$ $\qquad$ $x = x' \cos \dfrac{\pi}{4} + y' \sin \dfrac{\pi}{4}$

$= \dfrac{x' - y'}{\sqrt{2}}$ $\qquad\qquad\qquad$ $= \dfrac{x' + y'}{\sqrt{2}}$

$$5\left(\dfrac{x' - y'}{\sqrt{2}}\right)^2 + 2\left(\dfrac{x' - y'}{\sqrt{2}}\right)\left(\dfrac{x' + y'}{\sqrt{2}}\right) + 5\left(\dfrac{x' + y'}{\sqrt{2}}\right)^2 - 10 = 0$$

$$\dfrac{5(x')^2}{2} - \dfrac{10x'y'}{2} + \dfrac{5(y')^2}{2} + (x')^2 - (y')^2 + \dfrac{5(x')^2}{2} + \dfrac{10x'y'}{2} + \dfrac{5(y')^2}{2} - 10 = 0$$

$$6(x')^2 + 4(y')^2 - 10 = 0$$

$$\dfrac{3(x')^2}{5} + \dfrac{2(y')^2}{5} = 1$$

$$\dfrac{(x')^2}{5/3} + \dfrac{(y')^2}{5/2} = 1$$

Ellipse centered at the origin

10. (a) $6x^2 - 2xy + y^2 = 0$

$\quad$ $A = 6, B = -2, C = 1$

$\quad$ $B^2 - 4AC = (-2)^2 - 4(6)(1) = -20 < 0$

$\quad$ Ellipse

$\quad$ (b) $x^2 + 4xy + 4y^2 - x - y + 17 = 0$

$\quad$ $A = 1, B = 4, C = 4$

$\quad$ $B^2 - 4AC = (4)^2 - 4(1)(4) = 0$

$\quad$ Parabola

11. Polar: $\left(\sqrt{2}, \dfrac{3\pi}{4}\right)$

$x = \sqrt{2} \cos \dfrac{3\pi}{4} = \sqrt{2}\left(-\dfrac{1}{\sqrt{2}}\right) = -1$

$y = \sqrt{2} \sin \dfrac{3\pi}{4} = \sqrt{2}\left(\dfrac{1}{\sqrt{2}}\right) = 1$

Rectangular: $(-1, 1)$

12. Rectangular: $\left(\sqrt{3}, -1\right)$

$\quad$ $r = \pm\sqrt{\left(\sqrt{3}\right)^2 + (-1)^2} = \pm 2$

$\quad$ $\tan \theta = \dfrac{\sqrt{3}}{-1} = -\sqrt{3}$

$\quad$ $\theta = \dfrac{2\pi}{3}$ or $\theta = \dfrac{5\pi}{3}$

$\quad$ Polar: $\left(-2, \dfrac{2\pi}{3}\right)$ or $\left(2, \dfrac{5\pi}{3}\right)$

13. Rectangular: $4x - 3y = 12$

$\quad$ Polar: $4r \cos \theta - 3r \sin \theta = 12$

$\qquad\quad$ $r(4 \cos \theta - 3 \sin \theta) = 12$

$\qquad\qquad\qquad\qquad$ $r = \dfrac{12}{4 \cos \theta - 3 \sin \theta}$

14. Polar: $r = 5 \cos \theta$

$\qquad r^2 = 5r \cos \theta$

Rectangular: $\qquad x^2 + y^2 = 5x$

$\qquad\qquad x^2 + y^2 - 5x = 0$

15. $r = 1 - \cos \theta$

Cardioid

Symmetry: Polar axis

Maximum value of $|r|$: $r = 2$ when $\theta = \pi$.

Zero of r: $r = 0$ when $\theta = 0$

θ	0	$\dfrac{\pi}{2}$	π	$\dfrac{3\pi}{2}$
r	0	1	2	1

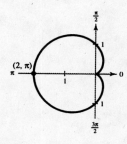

16. $r = 5 \sin 2\theta$

Rose curve with four petals

Symmetry: Polar axis, $\theta = \dfrac{\pi}{2}$, and pole

Maximum value of $|r|$: $|r| = 5$ when $\theta = \dfrac{\pi}{4}, \dfrac{3\pi}{4}, \dfrac{5\pi}{4}, \dfrac{7\pi}{4}$

Zeros of r: $r = 0$ when $\theta = 0, \dfrac{\pi}{2}, \pi, \dfrac{3\pi}{2}$

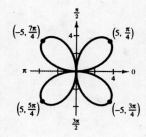

17. $r = \dfrac{3}{6 - \cos \theta}$

$r = \dfrac{\frac{1}{2}}{1 - \frac{1}{6} \cos \theta}$

$e = \dfrac{1}{6} < 1$, so the graph is an ellipse.

θ	0	$\dfrac{\pi}{2}$	π	$\dfrac{3\pi}{2}$
r	$\dfrac{3}{5}$	$\dfrac{1}{2}$	$\dfrac{3}{7}$	$\dfrac{1}{2}$

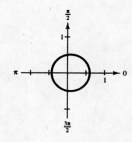

18. Parabola

Vertex: $\left(6, \dfrac{\pi}{2}\right)$

Focus: $(0, 0)$

$e = 1$

$r = \dfrac{ep}{1 + e\sin\theta}$

$r = \dfrac{p}{1 + \sin\theta}$

$6 = \dfrac{p}{1 + \sin(\pi/2)}$

$6 = \dfrac{p}{2}$

$12 = p$

$r = \dfrac{12}{1 + \sin\theta}$

19. $x = 3 - 2\sin\theta,\, y = 1 + 5\cos\theta$

$\dfrac{x - 3}{-2} = \sin\theta,\, \dfrac{y - 1}{5} = \cos\theta$

$\left(\dfrac{x - 3}{-2}\right)^2 + \left(\dfrac{y - 1}{5}\right)^2 = 1$

$\dfrac{(x - 3)^2}{4} + \dfrac{(y - 1)^2}{25} = 1$

20. $x = e^{2t},\, y = e^{4t}$

$x > 0,\, y > 0$

$y = (e^{2t})^2 = (x)^2 = x^2,\, x > 0,\, y > 0$

PART II

Chapter P   Chapter Test

1. $-\frac{10}{3} = -3\frac{1}{3}$

$-|-4| = -4$

$-\frac{10}{3} > -|-4|$

2. $d\left(3\frac{3}{4}, -5.4\right) = |3.75 - (-5.4)|$

$= 9.15$

3. $y = 4 - \frac{3}{4}|x|$

y-axis symmetry

x-intercepts: $\left(\pm\frac{16}{3}, 0\right)$

y-intercept: $(0, 4)$

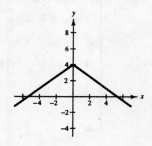

4. $y = 4 - (x - 2)^2$

Parabola; vertex: $(2, 4)$

No axis or origin symmetry

x-intercepts:
$(0, 0)$ and $(4, 0)$

$$0 = 4 - (x - 2)^2$$

$$(x - 2)^2 = 4$$

$$x - 2 = \pm 2$$

$$x = 2 \pm 2$$

$$x = 4 \quad \text{or} \quad x = 0$$

y-intercept: $(0, 0)$

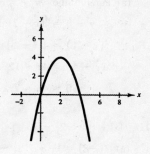

5. $(x - 3)^2 + y^2 = 9$

Circle

Center: $(3, 0)$

Radius: $r = 3$

x-axis symmetry

Intercepts: $(0, 0), (6, 0)$

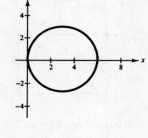

6. $\frac{2}{3}(x - 1) + \frac{1}{4}x = 10$

$12\left[\frac{2}{3}(x - 1) + \frac{1}{4}x\right] = 12(10)$

$8(x - 1) + 3x = 120$

$8x - 8 + 3x = 120$

$11x = 128$

$x = \frac{128}{11}$

7. $(x - 3)(x + 2) = 14$

$x^2 - x - 6 = 14$

$x^2 - x - 20 = 0$

$(x + 4)(x - 5) = 0$

$x + 4 = 0 \quad \text{or} \quad x - 5 = 0$

$x = -4 \qquad x = 5$

8. $2\sqrt{x} - \sqrt{2x + 1} = 1$

$-\sqrt{2x + 1} = 1 - 2\sqrt{x}$

$\left(-\sqrt{2x + 1}\right)^2 = \left(1 - 2\sqrt{x}\right)^2$

$2x + 1 = 1 - 4\sqrt{x} + 4x$

$-2x = -4\sqrt{x}$

$x = 2\sqrt{x}$

$x^2 = 4x$

$x^2 - 4x = 0$

$x(x - 4) = 0$

$x = 0 \quad \text{or} \quad x = 4$

Only $x = 4$ is a solution to the original equation.
$x = 0$ is extraneous.

9. $|3x - 1| = 7$

$3x - 1 = 7$ or $3x - 1 = -7$

$3x = 8$ $\qquad$ $3x = -6$

$x = \frac{8}{3}$ $\qquad$ $x = -2$

10. $-4x + 7y = -5$

$7y = 4x - 5$

$y = \frac{4}{7}x - \frac{5}{7} \implies m_1 = \frac{4}{7}$

(a) Parallel line: $m_2 = \frac{4}{7}$

$y - 8 = \frac{4}{7}(x - 3)$

$7(y - 8) = 4(x - 3)$

$7y - 56 = 4x - 12$

$0 = 4x - 7y + 44$

(b) Perpendicular line: $m_2 = -\frac{7}{4}$

$y - 8 = -\frac{7}{4}(x - 3)$

$4(y - 8) = -7(x - 3)$

$4y - 32 = -7x + 21$

$7x + 4y - 53 = 0$

11. $f(x) = |x + 2| - 15$

(a) $f(-8) = |-8 + 2| - 15 = 6 - 15 = -9$

(b) $f(14) = |14 + 2| - 15 = 16 - 15 = 1$

(c) $f(x - 6) = |x - 6 + 2| - 15 = |x - 4| - 15$

12. $f(x) = \sqrt{100 - x^2}$

Domain: $100 - x^2 \geq 0 \implies -10 \leq x \leq 10$ or $[-10, 10]$.

13. $f(x) = |-x + 6| + 2$

Domain: All real numbers or $(-\infty, \infty)$.

14. $f(x) = 2x^6 + 5x^4 - x^2$

(a)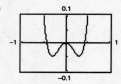

(b) Increasing on $(-0.31, 0)$ and $(0.31, \infty)$

Decreasing on $(-\infty, -0.31)$ and $(0, 0.31)$

(c) y-axis symmetry $\implies$ The function is even.

15. $f(x) = 4x\sqrt{3 - x}$

(a)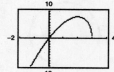

(b) Increasing on $(-\infty, 2)$

Decreasing on $(2, 3)$

(c) The function is neither odd nor even.

16. $f(x) = |x + 5|$

(a)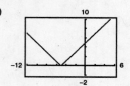

(b) increasing on $(-5, \infty)$

decreasing on $(-\infty, -5)$

(c) The function is neither odd nor even.

17. $f(x) = \begin{cases} 3x + 7, & x \leq -3 \\ 4x^2 - 1, & x > -3 \end{cases}$

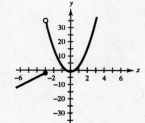

18. $h(x) = -x^3 - 7$

Common function: $f(x) = x^3$

Transformation:
Reflection in the x-axis and a vertical shift 7 units downward.

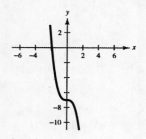

19. $h(x) = -\sqrt{x + 5} + 8$

Common function: $f(x) = \sqrt{x}$

Transformation:
Reflection in the x-axis, a horizontal shift 5 units to the left, and a vertical shift 8 units upward.

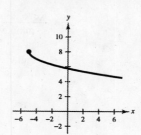

20. $h(x) = \frac{1}{4}|x + 1| - 3$

Common function: $f(x) = |x|$

Transformation:
Vertical shrink, horizontal shift 1 unit to the left, vertical shift 3 units down.

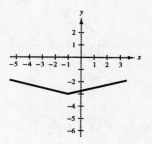

21. $(f + g)(2) = f(2) + g(2) = [3(2)^2 - 7] + [-(2)^2 - 4(2) + 5]$

$\qquad = 5 + (-7) = -2$

22. $(fg)(0) = f(0)g(0)$

$\qquad = [3(0)^2 - 7][-(0)^2 - 4(0) + 5]$

$\qquad = (-7)(5)$

$\qquad = -35$

23. $(g \circ f)(-1) = g(f(-1)) = g(3(-1)^2 - 7) = g(-4)$

$\qquad = -(-4)^2 - 4(-4) + 5$

$\qquad = 5$

24. $f(x) = x^3 + 8$

Since f is one-to-one, f has an inverse.

$\qquad y = x^3 + 8$

$\qquad x = y^3 + 8$

$\qquad x - 8 = y^3$

$\qquad \sqrt[3]{x - 8} = y$

$\qquad f^{-1}(x) = \sqrt[3]{x - 8}$

25. $f(x) = |x^2 - 3| + 6$

Since f is not on-to-one, f does not have an inverse.

26. $f(x) = \dfrac{3x\sqrt{x}}{8} = \dfrac{3}{8}x^{3/2}$

Since f is one-to-one, f has an inverse.

$\qquad y = \dfrac{3}{8}x^{3/2}$

$\qquad x = \dfrac{3}{8}y^{3/2}$

$\qquad \dfrac{8}{3}x = y^{3/2}$

$\qquad \left(\dfrac{8}{3}x\right)^{2/3} = y, x \geq 0$

$\qquad f^{-1}(x) = \left(\dfrac{8}{3}x\right)^{2/3}, x \geq 0$

27. $(6, 58), (10, 78)$

$\qquad m = \dfrac{78 - 58}{10 - 6} = \dfrac{20}{4} = 5$

$\qquad C - 58 = 5(x - 6)$

$\qquad C - 58 = 5x - 30$

$\qquad C = 5x + 28$

When $x = 25$; $C = 5(25) + 28 = \$153$

Chapter 1   Chapter Test

1. $\theta = \dfrac{5\pi}{4}$

(a)

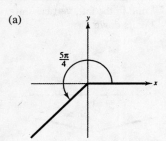

(b) $\dfrac{5\pi}{4} + 2\pi = \dfrac{13\pi}{4}$

$\dfrac{5\pi}{4} - 2\pi = -\dfrac{3\pi}{4}$

(c) $\dfrac{5\pi}{4}\left(\dfrac{180°}{\pi}\right) = 225°$

3. $x = -2, \ y = 6$

$r = \sqrt{(-2)^2 + (6)^2} = 2\sqrt{10}$

$\sin\theta = \dfrac{y}{r} = \dfrac{6}{2\sqrt{10}} = \dfrac{3}{\sqrt{10}} = \dfrac{3\sqrt{10}}{10}$

$\cos\theta = \dfrac{x}{r} = \dfrac{-2}{2\sqrt{10}} = -\dfrac{1}{\sqrt{10}} = -\dfrac{\sqrt{10}}{10}$

$\tan\theta = \dfrac{y}{x} = \dfrac{6}{-2} = -3$

$\csc\theta = \dfrac{r}{y} = \dfrac{2\sqrt{10}}{6} = \dfrac{\sqrt{10}}{3}$

$\sec\theta = \dfrac{r}{x} = \dfrac{2\sqrt{10}}{-2} = -\sqrt{10}$

$\cot\theta = \dfrac{x}{y} = \dfrac{-2}{6} = -\dfrac{1}{3}$

2. $90\dfrac{\text{km}}{\text{hr}} \times \dfrac{1\ \text{hr}}{60\ \text{min}} \times \dfrac{1000\ \text{m}}{1\ \text{km}} = 1500$ meters per minute

Circumference $= 2\pi\left(\dfrac{1}{2}\right) = \pi = \pi$ meters

$\dfrac{\text{Revolutions}}{\text{minute}} = \dfrac{1500}{\pi}$

Angular speed $= \dfrac{1500}{\pi} \cdot \pi = 1500$ radians per minute

4. For $0 \le \theta < \dfrac{\pi}{2}$, we have

$\sin\theta = \dfrac{\text{opp}}{\text{hyp}} = \dfrac{3}{\sqrt{13}} = \dfrac{3\sqrt{13}}{13}$

$\cos\theta = \dfrac{\text{adj}}{\text{hyp}} = \dfrac{2}{\sqrt{13}} = \dfrac{2\sqrt{13}}{13}$

$\csc\theta = \dfrac{\text{hyp}}{\text{opp}} = \dfrac{\sqrt{13}}{3}$

$\sec\theta = \dfrac{\text{hyp}}{\text{adj}} = \dfrac{\sqrt{13}}{2}$

$\cot\theta = \dfrac{\text{adj}}{\text{opp}} = \dfrac{2}{3}$

For $\pi \le \theta < \dfrac{3\pi}{2}$, we have

$\sin\theta = -\dfrac{3\sqrt{13}}{13}$

$\cos\theta = -\dfrac{2\sqrt{13}}{13}$

$\csc\theta = -\dfrac{\sqrt{13}}{3}$

$\sec\theta = -\dfrac{\sqrt{13}}{2}$

$\cot\theta = \dfrac{2}{3}$

5. $\theta = 290°$

$\theta' = 360° - 290° = 70°$

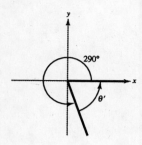

6. $\sec \theta < 0$ and $\tan \theta > 0$

$\dfrac{r}{x} < 0$ and $\dfrac{y}{x} > 0$

Quandrant III

7. $\cos \theta = -\dfrac{\sqrt{3}}{2}$

Reference angle is 30° and θ is in Quandrant II or III.

$\theta = 150°$ or $210°$

8. $\csc \theta = 1.030$

$\dfrac{1}{\sin \theta} = 1.030$

$\sin \theta = \dfrac{1}{1.030}$

$\theta = \arcsin \dfrac{1}{1.030}$

$\theta \approx 1.33$ and $\pi - 1.33 \approx 1.81$

9. $\cos \theta = \frac{3}{5}$, $\tan \theta < 0 \Longrightarrow \theta$ lies in Quadrant IV

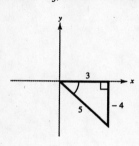

Let $x = 3, r = 5 \Longrightarrow y = -4$

$\sin \theta = -\frac{4}{5}$ $\qquad$ $\csc \theta = -\frac{5}{4}$

$\cos \theta = \frac{3}{5}$ $\qquad$ $\sec \theta = \frac{5}{3}$

$\tan \theta = -\frac{4}{3}$ $\qquad$ $\cot \theta = -\frac{3}{4}$

10. $\sec \theta = -\frac{17}{8}$, $\sin \theta > 0 \Longrightarrow \theta$ lies in Quadrant II

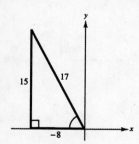

Let $r = 17, x = -8 \Longrightarrow y = 15$

$\sin \theta = \frac{15}{17}$ $\qquad$ $\csc \theta = \frac{17}{15}$

$\cos \theta = -\frac{8}{17}$ $\qquad$ $\sec \theta = -\frac{17}{8}$

$\tan \theta = -\frac{15}{8}$ $\qquad$ $\cot \theta = -\frac{8}{15}$

11. $g(x) = -2 \sin\left(x - \frac{\pi}{4}\right)$

Period: 2π

Amplitude: $|-2| = 2$

Shifted to the right by $\frac{\pi}{4}$ units and reflected in the x-axis.

x	0	$\frac{\pi}{4}$	$\frac{\pi}{2}$	$\frac{3\pi}{4}$	π
y	$\sqrt{2}$	0	$-\sqrt{2}$	-2	$-\sqrt{2}$

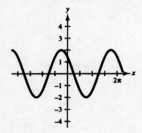

12. $f(\alpha) = \frac{1}{2} \tan 2\alpha$

Period: $\frac{\pi}{2}$

Asymptotes: $x = -\frac{\pi}{4}, x = \frac{\pi}{4}$

α	$-\frac{\pi}{8}$	0	$\frac{\pi}{8}$
$f(\alpha)$	$-\frac{1}{2}$	0	$\frac{1}{2}$

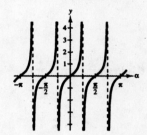

13. $y = \sin 2\pi x + 2 \cos \pi x$

Periodic: period = 2

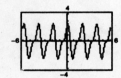

14. $y = 6t \cos(0.25t), 0 \le t \le 32$

Not periodic

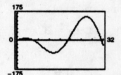

15. $f(x) = a \sin(bx + c)$

Amplitude: $2 \Longrightarrow |a| = 2$

Reflected in the x-axis: $a = -2$

Period: $4\pi = \frac{2\pi}{b} \Longrightarrow b = \frac{1}{2}$

Phase shift: $\frac{c}{b} = -\frac{\pi}{2} \Longrightarrow c = -\frac{\pi}{4}$

$f(x) = -2 \sin\left(\frac{x}{2} - \frac{\pi}{4}\right)$

16. Let $u = \arccos \frac{2}{3}$,

$\cos u = \frac{2}{3}$.

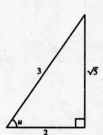

$\tan\left(\arccos \frac{2}{3}\right) = \tan u = \frac{\sqrt{5}}{2}$

17. $f(x) = 2 \arcsin\left(\dfrac{1}{2}x\right)$

Domain: $[-2, 2]$

Range: $[-\pi, \pi]$

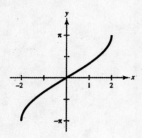

18.

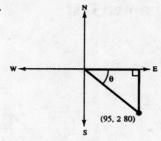

$\tan \theta = -\dfrac{80}{95} \implies \theta \approx -40.1°$

Bearing: $90° - 40.1° = 49.9°$

The plane is heading N 49.9°W.

19. $d = a \sin bt$

$a = -6$

$\dfrac{2\pi}{b} = 2 \implies b = \pi$

$d = -6 \sin \pi t$

Chapter 2   Chapter Test

1. $\tan \theta = \dfrac{3}{2}$ and $\cos \theta < 0$

θ is in Quadrant III.

$$\sec \theta = -\sqrt{1 + \tan^2 \theta} = -\sqrt{1 + \left(\frac{3}{2}\right)^2} = -\frac{\sqrt{13}}{2}$$

$$\cos \theta = \frac{1}{\sec \theta} = -\frac{2}{\sqrt{13}} = -\frac{2\sqrt{13}}{13}$$

$$\sin \theta = \tan \theta \cos \theta = \left(\frac{3}{2}\right)\left(-\frac{2}{\sqrt{13}}\right) = -\frac{3}{\sqrt{13}} = -\frac{3\sqrt{13}}{13}$$

$$\csc \theta = \frac{1}{\sin \theta} = -\frac{\sqrt{13}}{3}$$

$$\cot \theta = \frac{1}{\tan \theta} = \frac{2}{3}$$

2. $\csc^2 \beta \,(1 - \cos^2 \beta) = \dfrac{1}{\sin^2 \beta}(\sin^2 \beta) = 1$

3. $\dfrac{\sec^4 x - \tan^4 x}{\sec^2 x + \tan^2 x} = \dfrac{(\sec^2 x + \tan^2 x)(\sec^2 x - \tan^2 x)}{\sec^2 x + \tan^2 x}$

$$= \sec^2 x - \tan^2 x = 1$$

4. $\dfrac{\cos \theta}{\sin \theta} + \dfrac{\sin \theta}{\cos \theta} = \dfrac{\cos^2 \theta + \sin^2 \theta}{\sin \theta \cos \theta} = \dfrac{1}{\sin \theta \cos \theta}$

$$= \csc \theta \sec \theta$$

5. $y = \tan \theta, \; y = -\sqrt{\sec^2 \theta - 1}$

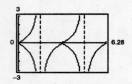

$\tan \theta = -\sqrt{\sec^2 \theta - 1}$ on

$\theta = 0, \dfrac{\pi}{2} < \theta \le \pi, \dfrac{3\pi}{2} < \theta < 2\pi.$

6. $y_1 = \cos x + \sin x \tan x, \; y_2 = \sec x$

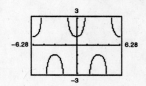

It appears that $y_1 = y_2$.

$$\cos x + \sin x \tan x = \cos \;+ \sin x\frac{\sin x}{\cos x}$$

$$= \cos \;+ \frac{\sin^2 x}{\cos x}$$

$$= \frac{\cos^2 x + \sin^2 x}{\cos x}$$

$$= \frac{1}{\cos x} = \sec x$$

7. $\sin\theta \sec\theta = \sin\theta \dfrac{1}{\cos\theta} = \dfrac{\sin\theta}{\cos\theta} = \tan\theta$

8. $\sec^2 x \tan^2 x + \sec^2 x = \sec^2 x (\sec^2 x - 1) + \sec^2 x$

$$= \sec^4 x - \sec^2 x + \sec^2 x$$

$$= \sec^4 x$$

9. $\dfrac{\csc\alpha + \sec\alpha}{\sin\alpha + \cos\alpha} = \dfrac{\dfrac{1}{\sin\alpha} + \dfrac{1}{\cos\alpha}}{\sin\alpha + \cos\alpha} = \dfrac{\dfrac{\cos\alpha + \sin\alpha}{\sin\alpha\cos\alpha}}{\sin\alpha + \cos\alpha} = \dfrac{1}{\sin\alpha\cos\alpha}$

$$= \dfrac{\cos^2\alpha + \sin^2\alpha}{\sin\alpha\cos\alpha} = \dfrac{\cos^2\alpha}{\sin\alpha\cos\alpha} + \dfrac{\sin^2\alpha}{\sin\alpha\cos\alpha}$$

$$= \dfrac{\cos\alpha}{\sin\alpha} + \dfrac{\sin\alpha}{\cos\alpha} = \cot\alpha + \tan\alpha$$

10. $\cos\left(x + \dfrac{\pi}{2}\right) = \cos\left(\dfrac{\pi}{2} - (-x)\right) = \sin(-x) = -\sin x$

11. $\sin(n\pi + \theta) = (-1)^n \sin\theta$, n is an integer.

For n odd: $\sin(n\pi + \theta) = \sin n\pi \cos\theta + \cos n\pi \sin\theta$

$$= (0)\cos\theta + (-1)\sin\theta = -\sin\theta$$

For n even: $\sin(n\pi + \theta) = \sin n\pi \cos\theta + \cos n\pi \sin\theta$

$$= (0)\cos\theta + (1)\sin\theta = \sin\theta$$

When n is odd, $(-1)^n = -1$. When n is even $(-1)^n = 1$.

Thus, $\sin(n\pi + \theta) = (-1)^n \sin\theta$ for any integer n.

12. $(\sin x + \cos x)^2 = \sin^2 x + 2\sin x \cos x + \cos^2 x$

$$= 1 + 2\sin x \cos x$$

$$= 1 + \sin 2x$$

13. $\sin^4 x \tan^2 x = \sin^4 x \left(\dfrac{\sin^2 x}{\cos^2 x}\right) = \dfrac{\sin^6 x}{\cos^2 x} = \dfrac{(\sin^2 x)^3}{\cos^2 x}$

$$= \dfrac{\left(\dfrac{1 - \cos 2x}{2}\right)^3}{\dfrac{1 + \cos 2x}{2}}$$

$$= \dfrac{\dfrac{1 - 3\cos 2x + 3\cos^2 2x - \cos^3 2x}{8}}{\dfrac{1 + \cos 2x}{2}}$$

$$= \dfrac{\dfrac{1}{4}\left[1 - 3\cos 2x + 3\left(\dfrac{1 + \cos 4x}{2}\right) - \cos 2x\left(\dfrac{1 + \cos 4x}{2}\right)\right]}{1 + \cos 2x}$$

$$= \dfrac{\dfrac{1}{8}[2 - 6\cos 2x + 3 + 3\cos 4x - \cos 2x - \cos 2x \cos 4x]}{1 + \cos 2x}$$

$$= \dfrac{1}{8}\left[\dfrac{5 - 7\cos 2x + 3\cos 4x - \dfrac{1}{2}(\cos(-2x) + \cos(6x))}{1 + \cos 2x}\right]$$

$$= \dfrac{1}{16}\left[\dfrac{10 - 14\cos 2x + 6\cos 4x - \cos 2x - \cos 6x}{1 + \cos 2x}\right]$$

$$= \dfrac{1}{16}\left[\dfrac{10 - 15\cos 2x + 6\cos 4x - \cos 6x}{1 + \cos 2x}\right]$$

14. $\dfrac{\sin 4\theta}{1 + \cos 4\theta} = \tan \dfrac{4\theta}{2} = \tan 2\theta$

15. $4 \cos 2\theta \sin 4\theta = 4\left(\dfrac{1}{2}\right)[\sin(2\theta + 4\theta) - \sin(2\theta - 4\theta)]$

$= 2[\sin 6\theta - \sin(-2\theta)]$

$= 2(\sin 6\theta + \sin 2\theta)$

16. $\sin 3\theta - \sin 4\theta = 2 \cos\left(\dfrac{3\theta + 4\theta}{2}\right) \sin\left(\dfrac{3\theta - 4\theta}{2}\right)$

$= 2 \cos \dfrac{7\theta}{2} \sin\left(\dfrac{-\theta}{2}\right)$

$= -2 \cos \dfrac{7\theta}{2} \sin \dfrac{\theta}{2}$

17. $\tan^2 x + \tan x = 0$

$\tan x \, (\tan x + 1) = 0$

$\tan x = 0 \quad \text{or} \quad \tan x + 1 = 0$

$\tan x = -1$

$x = 0, \pi \qquad x = \dfrac{3\pi}{4}, \dfrac{7\pi}{4}$

18. $\sin 2\alpha - \cos \alpha = 0$

$2 \sin\alpha \cos \alpha - \cos \alpha = 0$

$\cos\alpha(2 \sin\alpha - 1) = 0$

$\cos \alpha = 0 \quad \text{or} \quad 2 \sin \alpha - 1 = 0$

$\alpha = \dfrac{\pi}{2}, \dfrac{3\pi}{2} \qquad \sin \alpha = \dfrac{1}{2}$

$\alpha = \dfrac{\pi}{6}, \dfrac{5\pi}{6}$

19. $4 \cos^2 x - 3 = 0$

$\cos^2 x = \dfrac{3}{4}$

$\cos x = \pm \sqrt{\dfrac{3}{4}} = \pm \dfrac{\sqrt{3}}{2}$

$x = \dfrac{\pi}{6}, \dfrac{5\pi}{6}, \dfrac{7\pi}{6}, \dfrac{11\pi}{6}$

20. $\csc^2 x - \csc x - 2 = 0$

$(\csc x - 2)(\csc x + 1) = 0$

$\csc x - 2 = 0 \quad \text{or} \quad \csc x + 1 = 0$

$\csc x = 2 \qquad \csc = -1$

$\dfrac{1}{\sin x} = 2 \qquad \dfrac{1}{\sin x} = -1$

$\sin x = \dfrac{1}{2} \qquad \sin x = -1$

$x = \dfrac{\pi}{6}, \dfrac{5\pi}{6} \qquad x = \dfrac{3\pi}{2}$

21. $3 \cos x - x = 0$

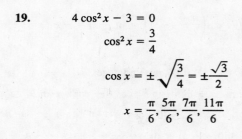

$x \approx -2.938, -2.663, 1.170$

22. $\cos^2 x + \cos x - 6 = 0$

$\cos^2 x + \cos x = 6$

The maximum value of $\cos^2 x$ is 1 and the maximum value of $\cos x$ is 1. Thus, $|\cos^2 x + \cos x| \le 2$ for all x and $\cos^2 x + \cos x$ can never equal 6.

23. $105° = 135° - 30°$

$\cos 105° = \cos(135° - 30°)$

$\qquad = \cos 135° \cos 30° + \sin 135° \sin 30°$

$\qquad = -\cos 45° \cos 30° + \sin 45° \sin 30°$

$\qquad = \left(-\dfrac{\sqrt{2}}{2}\right)\left(\dfrac{\sqrt{3}}{2}\right) + \left(\dfrac{\sqrt{2}}{2}\right)\left(\dfrac{1}{2}\right)$

$\qquad = \dfrac{-\sqrt{6} + \sqrt{2}}{4} = \dfrac{\sqrt{2} - \sqrt{6}}{4}$

24. $\sin 2u = 2 \sin u \cos u$

$\qquad = 2\left(\dfrac{2}{\sqrt{5}}\right)\left(\dfrac{1}{\sqrt{5}}\right) = \dfrac{4}{5}$

$\tan 2u = \dfrac{2 \tan u}{1 - \tan^2 u} = \dfrac{2(2)}{1 - (2)^2} = \dfrac{4}{-3} = -\dfrac{4}{3}$

25. $1.5 = \dfrac{\sin\left(\dfrac{\theta}{2} + \dfrac{60°}{2}\right)}{\sin\left(\dfrac{\theta}{2}\right)}$

$1.5 \sin \dfrac{\theta}{2} = \sin \dfrac{\theta}{2} \cos 30° + \cos \dfrac{\theta}{2} \sin 30°$

$1.5 \sin \dfrac{\theta}{2} = \dfrac{\sqrt{3}}{2} \sin \dfrac{\theta}{2} + \dfrac{1}{2} \cos \dfrac{\theta}{2}$

$\left(1.5 - \dfrac{\sqrt{3}}{2}\right) \sin \dfrac{\theta}{2} = \dfrac{1}{2} \cos \dfrac{\theta}{2}$

$2\left(1.5 - \dfrac{\sqrt{3}}{2}\right) = \cot \dfrac{\theta}{2}$

$3 - \sqrt{3} = \dfrac{1}{\tan \dfrac{\theta}{2}}$

$\tan \dfrac{\theta}{2} = \dfrac{1}{3 - \sqrt{3}}$

$\dfrac{\theta}{2} = \arctan\left(\dfrac{1}{3 - \sqrt{3}}\right) \approx 38.26°$

$\theta \approx 76.5°$

Chapter 3   Chapter Test

1. $A = 24°, B = 68°, a = 12.2$

$C = 180° - 24° - 68° = 88°$

$b = \dfrac{a \sin B}{\sin A} = \dfrac{12.2 \sin 68°}{\sin 24°} \approx 27.81$

$c = \dfrac{a \sin C}{\sin A} = \dfrac{12.2 \sin 88°}{\sin 24°} \approx 29.98$

2. $B = 104°, C = 33°, a = 18.1$

$A = 180° - 104° - 33° = 43°$

$b = \dfrac{a \sin B}{\sin A} = \dfrac{18.1 \sin 104°}{\sin 43°} \approx 25.75$

$c = \dfrac{a \sin C}{\sin A} = \dfrac{18.1 \sin 33°}{\sin 43°} \approx 14.45$

3. $A = 24°, a = 11.2, b = 13.4$

$\sin B = \dfrac{b \sin A}{a} = \dfrac{13.4 \sin 24°}{11.2} \approx 0.4866$

Two Solutions

$B \approx 29.12°$ or $B \approx 150.88°$

$C \approx 126.88°$ $C \approx 5.12°$

$c = \dfrac{a \sin C}{\sin A} = \dfrac{11.2 \sin 126.88°}{\sin 24°}$ $c = \dfrac{11.2 \sin 5.12°}{\sin 24°}$

$c \approx 22.03$ $c \approx 2.46$

4. $a = 4.0, b = 7.3, c = 12.4$

$\cos C = \dfrac{a^2 + b^2 - c^2}{2ab} = \dfrac{4^2 + 7.3^2 - 12.4^2}{2(4)(7.3)} \approx -1.4464 < -1$

No solution

5. $B = 100°, a = 15, b = 23$

$\sin A = \dfrac{a \sin B}{b} = \dfrac{15 \sin 100°}{23} \implies A \approx 39.96°$

$C \approx 180° - 100° - 39.96° = 40.04°$

$c = \dfrac{b \sin C}{\sin B} \approx \dfrac{23 \sin 40.04°}{\sin 100°} \approx 15.02$

6. $C = 123°, a = 41, b = 57$

$c^2 = 41^2 + 57^2 - 2(41)(57)\cos 123° \implies c \approx 86.46$

$\sin A = \dfrac{a \sin C}{c} \approx \dfrac{41 \sin 123°}{86.46} \implies A \approx 23.43°$

$B \approx 180° - 23.43° - 123° = 33.57°$

7. $a = 60, b = 70, c = 82$

$s = \dfrac{60 + 70 + 82}{2} = 106$

Area $= \sqrt{106(46)(36)(24)} \approx 2052.5$ square meters

8.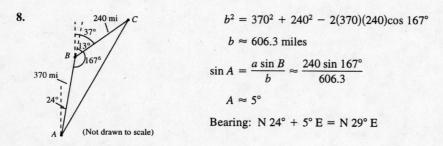

(Not drawn to scale)

$b^2 = 370^2 + 240^2 - 2(370)(240)\cos 167°$

$b \approx 606.3$ miles

$\sin A = \dfrac{a \sin B}{b} \approx \dfrac{240 \sin 167°}{606.3}$

$A \approx 5°$

Bearing: N 24° + 5° E = N 29° E

9. Initial Point: $(-3, 7)$

Terminal Point: $(11, -16)$

$\mathbf{v} = \langle 11 - (-3), -16 - 7 \rangle = \langle 14, -23 \rangle$

10. $\mathbf{v} = 12\left(\dfrac{\mathbf{u}}{\|\mathbf{u}\|}\right) = 12\left(\dfrac{\langle 3, -5 \rangle}{\sqrt{3^2 + (-5)^2}}\right) = \dfrac{12}{\sqrt{34}}\langle 3, -5 \rangle$

$= \dfrac{6\sqrt{34}}{17}\langle 3, -5 \rangle = \left\langle \dfrac{18\sqrt{34}}{17}, -\dfrac{30\sqrt{34}}{17} \right\rangle$

11. $\mathbf{u} + \mathbf{v} = \langle 3, 5 \rangle + \langle -7, 1 \rangle = \langle -4, 6 \rangle$

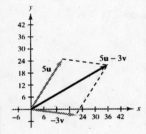

12. $\mathbf{u} - \mathbf{v} = \langle 3, 5 \rangle - \langle -7, 1 \rangle = \langle 10, 4 \rangle$

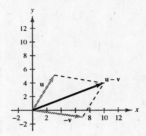

13. $5\mathbf{u} - 3\mathbf{v} = 5\langle 3, 5 \rangle - 3\langle -7, 1 \rangle = \langle 15, 25 \rangle + \langle 21, -3 \rangle$

$= \langle 36, 22 \rangle$

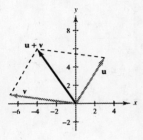

14. $\dfrac{\mathbf{u}}{\|\mathbf{u}\|} = \dfrac{\langle 4, -3 \rangle}{\sqrt{4^2 + (-3)^2}} = \dfrac{1}{5}\langle 4, -3 \rangle = \left\langle \dfrac{4}{5}, -\dfrac{3}{5} \right\rangle$

15. $\mathbf{u} = 250(\cos 45° \, \mathbf{i} + \sin 45° \, \mathbf{j})$

$\mathbf{v} = 130(\cos(-60°) \, \mathbf{i} + \sin(-60°) \, \mathbf{j})$

$\mathbf{R} = \mathbf{u} + \mathbf{v} \approx 241.7767 \, \mathbf{i} + 64.1934 \, \mathbf{j}$

$\|\mathbf{R}\| \approx \sqrt{241.7767^2 + 64.1934^2} \approx 250.15 \text{ pounds}$

$\tan \theta \approx \dfrac{64.1934}{241.7767} \implies \theta \approx 14.9°$

16. $\mathbf{u} = \langle -1, 5 \rangle, \mathbf{v} = \langle 3, -2 \rangle$

$\cos \theta = \dfrac{\mathbf{u} \cdot \mathbf{v}}{\|\mathbf{u}\|\|\mathbf{v}\|} = \dfrac{-13}{\sqrt{26}\sqrt{13}} \implies \theta = 135°$

17. $\mathbf{u} = \langle 6, 10 \rangle, \mathbf{v} = \langle 2, 3 \rangle$

$\mathbf{u} \cdot \mathbf{v} = 42 \neq 0 \implies \mathbf{u}$ and $\mathbf{v}$ are **not** orthogonal.

18. $\mathbf{u} = \langle 6, 7 \rangle, \mathbf{v} = \langle -5, -1 \rangle$

$\mathbf{w}_1 = \text{proj}_{\mathbf{v}} \, \mathbf{u} = \left(\dfrac{\mathbf{u} \cdot \mathbf{v}}{\|\mathbf{v}\|^2}\right)\mathbf{v} = -\dfrac{37}{26}\langle -5, -1 \rangle = \dfrac{37}{26}\langle 5, 1 \rangle$

$\mathbf{w}_2 = \mathbf{u} - \mathbf{w}_1 = \langle 6, 7 \rangle - \dfrac{37}{26}\langle 5, 1 \rangle$

$= \left\langle -\dfrac{29}{26}, \dfrac{145}{26} \right\rangle$

$= \dfrac{29}{26}\langle -1, 5 \rangle$

Chapters 1–3   Cumulative Test

1. (a)

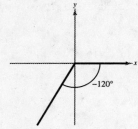

(b) $-120° + 360° = 240°$

(c) $-120\left(\dfrac{\pi}{180}\right) = -\dfrac{2\pi}{3}$

(d) $-120°$ is located in Quadrant III.

$240° - 180° = 60°$

(e) $\sin(-120°) = -\sin 60° = -\dfrac{\sqrt{3}}{2}$

$\cos(-120°) = -\cos 60° = -\dfrac{1}{2}$

$\tan(-120°) = \tan 60° = \sqrt{3}$

$\csc(-120°) = \dfrac{1}{-\sin 60°} = -\dfrac{2\sqrt{3}}{3}$

$\sec(-120°) = \dfrac{1}{-\cos 60°} = -2$

$\cot(-120°) = \dfrac{1}{\tan 60°} = \dfrac{\sqrt{3}}{3}$

2. $2.35\left(\dfrac{180}{\pi}\right) \approx 134.6°$

3. $\tan \theta = \dfrac{y}{x} = -\dfrac{4}{3} \implies r = 5$

Since $\sin \theta < 0$ θ is in Quadrant IV $\implies x = 3$.

$\cos \theta = \dfrac{x}{r} = \dfrac{3}{5}$

4. $f(x) = 3 - 2 \sin \pi x$

Period: $\dfrac{2\pi}{\pi} = 2$

Amplitude: $|a| = |-2| = 2$

Upward shift of 3 units (reflected in x-axis prior to shift)

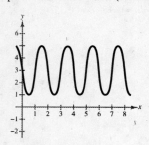

5. $g(x) = \dfrac{1}{2} \tan\left(x - \dfrac{\pi}{2}\right)$

Period: π

Asymptotes: $x = 0, x = \pi$

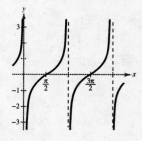

6. $h(x) = a \cos(bx + c)$

Graph is reflected in x-axis.

Amplitude: $a = -3$

Period: $2 = \dfrac{2\pi}{\pi} \implies b = \pi$

No phase shift: $c = 0$

$h(x) = -3 \cos(\pi x)$

7. $f(x) = \dfrac{x}{2}\sin x,\ -3\pi \le x \le 3\pi$

$-\dfrac{x}{2} \le f(x) \le \dfrac{x}{2}$

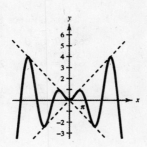

8. $\tan(\arctan 6.7) = 6.7$

9. $\tan\left(\arcsin\dfrac{3}{5}\right) = \dfrac{3}{4}$

10. $y = \arccos(2x)$

$\sin y = \sin(\arccos(2x)) = \sqrt{1 - 4x^2}$

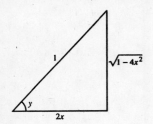

11. $\cos\left(\dfrac{\pi}{2} - x\right)\csc x = \sin x\left(\dfrac{1}{\sin x}\right) = 1$

12. $\dfrac{\sin\theta - 1}{\cos\theta} - \dfrac{\cos\theta}{\sin\theta - 1} = \dfrac{\sin\theta - 1}{\cos\theta} - \dfrac{\cos\theta(\sin\theta + 1)}{\sin^2 - 1}$

$\qquad = \dfrac{\sin\theta - 1}{\cos\theta} + \dfrac{\cos\theta(\sin\theta + 1)}{\cos^2\theta} = \dfrac{\sin\theta - 1}{\cos\theta} + \dfrac{\sin\theta + 1}{\cos\theta} = \dfrac{2\sin\theta}{\cos\theta} = 2\tan\theta$

13. $\cot^2\alpha(\sec^2\alpha - 1) = \cot^2\alpha\tan^2\alpha = 1$

14. $\sin(x + y)\sin(x - y) = \dfrac{1}{2}[\cos(x + y - (x - y)) - \cos(x + y + x - y)]$

$\qquad = \dfrac{1}{2}[\cos 2y - \cos 2x] = \dfrac{1}{2}[1 - 2\sin^2 y - (1 - 2\sin^2 x)] = \sin^2 x - \sin^2 y$

15. $\sin^2 x \cos^2 x = \left(\dfrac{1 - \cos 2x}{2}\right)\left(\dfrac{1 + \cos 2x}{2}\right)$

$= \dfrac{1}{4}(1 - \cos 2x)(1 + \cos 2x)$

$= \dfrac{1}{4}(1 - \cos^2 2x)$

$= \dfrac{1}{4}\left(1 - \dfrac{1 + \cos 4x}{2}\right)$

$= \dfrac{1}{8}(2 - (1 + \cos 4x))$

$= \dfrac{1}{8}(1 - \cos 4x)$

16. $2\cos^2 \beta - \cos \beta = 0$

$\cos \beta(2\cos \beta - 1) = 0$

$\cos \beta = 0 \qquad 2\cos \beta - 1 = 0$

$\beta = \dfrac{\pi}{2}, \dfrac{3\pi}{2} \qquad \cos \beta = \dfrac{1}{2}$

$\beta = \dfrac{\pi}{3}, \dfrac{5\pi}{3}$

Answer: $\dfrac{\pi}{3}, \dfrac{\pi}{2}, \dfrac{3\pi}{2}, \dfrac{5\pi}{3}$

17. $3\tan \theta - \cot \theta = 0$

$3\tan \theta - \dfrac{1}{\tan \theta} = 0$

$\dfrac{3\tan^2 \theta - 1}{\tan \theta} = 0$

$3\tan^2 \theta - 1 = 0$

$\tan^2 \theta = \dfrac{1}{3}$

$\tan \theta = \pm\dfrac{\sqrt{3}}{3}$

$0 = \dfrac{\pi}{6}, \dfrac{5\pi}{6}, \dfrac{7\pi}{6}, \dfrac{11\pi}{6}$

18. $\sin^2 x + 2\sin x + 1 = 0, \, 0 \le x \le 2\pi$

$\sin x = \dfrac{-2 \pm \sqrt{2^2 - 4(1)(1)}}{2(1)} = \dfrac{-2 \pm 0}{2} = -1$

$x = \dfrac{3\pi}{2}$

19. $\sin u = \dfrac{12}{13} \implies \cos u = \dfrac{5}{13}$ and $\tan u = \dfrac{12}{5}$ since u is in Quadrant I.

$\cos v = \dfrac{3}{5} \implies \sin v = \dfrac{4}{5}$ and $\tan v = \dfrac{4}{3}$ since v is in Quadrant I.

$\tan(u - v) = \dfrac{\tan u - \tan v}{1 + \tan u \tan v} = \dfrac{\dfrac{12}{5} - \dfrac{4}{3}}{1 + \left(\dfrac{12}{5}\right)\left(\dfrac{4}{3}\right)} = \dfrac{16}{63}$

20. $\tan \theta = \dfrac{1}{2}$

$\tan 2\theta = \dfrac{2\tan \theta}{1 - \tan^2 \theta} = \dfrac{2\left(\dfrac{1}{2}\right)}{1 - \left(\dfrac{1}{2}\right)^2} = \dfrac{4}{3}$

21. $\tan \theta = \dfrac{4}{3} \implies \cos \theta = \pm\dfrac{3}{5}$

$\sin \dfrac{\theta}{2} = \pm\sqrt{\dfrac{1 - \cos \theta}{2}} = \pm\sqrt{\dfrac{1 - 3/5}{2}} = \pm\dfrac{\sqrt{5}}{5}$

$\text{or} = \pm\sqrt{\dfrac{1 + 3/5}{2}} = \pm\dfrac{2\sqrt{5}}{5}$

Because $\tan \theta > 0$, θ lies in Quadrant I or III and $\dfrac{\theta}{2}$ lies in Quadrant I or II. Therefore, the value of $\sin \dfrac{\theta}{2}$ is positive, so $\sin \dfrac{\theta}{2} = \dfrac{\sqrt{5}}{5}$ or $\dfrac{2\sqrt{5}}{5}$.

22. $5 \sin \dfrac{3\pi}{4} \cos \dfrac{7\pi}{4} = \dfrac{5}{2}\left[\sin\left(\dfrac{3\pi}{4} + \dfrac{7\pi}{4}\right) + \sin\left(\dfrac{3\pi}{4} - \dfrac{7\pi}{4}\right)\right]$

$$= \dfrac{5}{2}\left(\sin\dfrac{5\pi}{2} + \sin(-\pi)\right)$$

$$= \dfrac{5}{2}\left(\sin\dfrac{5\pi}{2} - \sin\pi\right)$$

23. Given: $A = 30°, a = 9, b = 8$

$$\dfrac{\sin B}{8} = \dfrac{\sin 30°}{9}$$

$$\sin B = \dfrac{8}{9}\left(\dfrac{1}{2}\right)$$

$$B = \arcsin\left(\dfrac{4}{9}\right)$$

$$B \approx 26.4°$$

$$C = 180° - A - B \approx 123.6°$$

$$\dfrac{c}{\sin 123.6°} \approx \dfrac{9}{\sin 30°}$$

$$c \approx 15.0$$

24. Given: $A = 30°, b = 8, c = 10$

$$a^2 = 8^2 + 10^2 - 2(8)(10)\cos 30°$$

$$a^2 \approx 25.4$$

$$a \approx 5.0$$

$$\cos B \approx \dfrac{5.0^2 + 10^2 - 8^2}{2(5.0)(10)}$$

$$\cos B \approx 0.61$$

$$B \approx 52.4°$$

$$C = 180° - A - B \approx 97.6°$$

25. Given: $A = 30°, C = 90°, b = 10$

$$B = 180° - 30° - 90° = 60°$$

$$\tan 30° = \dfrac{a}{10} \implies a = 10 \tan 30° \approx 5.8$$

$$\cos 30° = \dfrac{10}{c} \implies c = \dfrac{10}{\cos 30°} \approx 11.5$$

26. $a = 4, b = 8, c = 9$

$$\cos C = \dfrac{4^2 + 8^2 - 9^2}{2(4)(8)} = \dfrac{-1}{64} \implies C \approx 90.9°$$

$$\sin A \approx \dfrac{4 \sin 90.9°}{9} \implies A \approx 26.4°$$

$$B \approx 180° - 26.4° - 90.9° = 62.7°$$

27. Area $= \dfrac{1}{2}(7)(12) \sin 60° \approx 36.4$ square inches

28. $s = \dfrac{11 + 16 + 17}{2} = 22$

Area $= \sqrt{22(11)(6)(5)} \approx 85.2$ square inches

29. $\mathbf{u} = \langle 3, 5 \rangle = 3\mathbf{i} + 5\mathbf{j}$

30. $\mathbf{u} = 3\mathbf{i} + 4\mathbf{j}, \mathbf{v} = \mathbf{i} - 2\mathbf{j}$

$\mathbf{u} \cdot \mathbf{v} = 3(1) + 4(-2) = -5$

31. $\mathbf{u} = \langle 8, -2 \rangle, \mathbf{v} = \langle 1, 5 \rangle$

$$\mathbf{w}_1 = \text{proj}_{\mathbf{v}}\,\mathbf{u} = \left(\dfrac{\mathbf{u} \cdot \mathbf{v}}{\|\mathbf{v}\|^2}\right)\mathbf{v} = \dfrac{-2}{26}\langle 1, 5 \rangle = -\dfrac{1}{13}\langle 1, 5 \rangle$$

$$\mathbf{w}_2 = \mathbf{u} - \mathbf{w}_1 = \langle 8, -2 \rangle - \left\langle -\dfrac{1}{13}, -\dfrac{5}{13} \right\rangle = \left\langle \dfrac{105}{13}, -\dfrac{21}{13} \right\rangle$$

$$= \dfrac{21}{13}\langle 5, -1 \rangle$$

32. Height of larger triangle:

$$\tan 18° = \frac{h_2}{200}$$

$$h_2 = 200 \tan 18° \approx 65 \text{ feet}$$

Height of smaller triangle:

$$\tan 16°45' = \frac{h_1}{200}$$

$$h_1 = 200 \tan 16.75° \approx 60 \text{ feet}$$

Height of flag:

$$h_2 - h_1 \approx 65 - 60 = 5 \text{ feet}$$

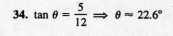

(Not drawn to scale)

33. Angular speed $= (2\pi)(45) = 90\pi$ radians per minute

Speed $= 3(90\pi) = 270\pi \approx 848.23$ inches per minute

34. $\tan \theta = \dfrac{5}{12} \implies \theta \approx 22.6°$

35. $d = a \cos bt$

$$|a| = 4 \implies a = 4$$

$$\frac{2\pi}{b} = 8 \implies b = \frac{\pi}{4}$$

$$d = 4 \cos \frac{\pi}{4}t$$

36. $\mathbf{v}_1 = 500\langle \cos 60°, \sin 60° \rangle = \langle 250, 250\sqrt{3} \rangle$

$\mathbf{v}_2 = 50\langle \cos 30°, \sin 30° \rangle = \langle 25\sqrt{3}, 25 \rangle$

$\mathbf{v} = \mathbf{v}_1 + \mathbf{v}_2 = \langle 250 + 25\sqrt{3}, 250\sqrt{3} + 25 \rangle$

$\approx \langle 293.3, 458.0 \rangle$

$\|\mathbf{v}\| \approx \sqrt{(293.3)^2 + (458.0)^2} \approx 543.9$

$\tan \theta \approx \dfrac{458.0}{293.3} \approx 1.5615 \implies \theta \approx 57.4°$

The plane is traveling N 32.6° E at
543.9 kilometers per hour.

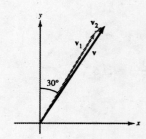

Chapter 4   Chapter Test

1. $-3 + \sqrt{-81} = -3 + 9i$

2. $10i - \left(3 + \sqrt{-25}\right) = 10i - (3 + 5i) = -3 + 5i$

3. $(2 + 6i)^2 = 4 + 24i + 36i^2 = -32 + 24i$

4. $\left(2 + \sqrt{3}i\right)\left(2 - \sqrt{3}i\right) = 4 - 3i^2 = 4 + 3 = 7$.

5. $\dfrac{5}{2 + i} = \dfrac{5}{2 + i} \cdot \dfrac{2 - i}{2 - i} = \dfrac{5(2 - i)}{4 + 1} = 2 - i$

6. $2x^2 - 2x + 3 = 0$

$x = \dfrac{-(-2) \pm \sqrt{(-2)^2 - 4(2)(3)}}{2(2)} = \dfrac{2 \pm \sqrt{-20}}{4}$

$= \dfrac{2 \pm 2\sqrt{5}i}{4} = \dfrac{1}{2} \pm \dfrac{\sqrt{5}}{2}i$

7. Since $x^5 + x^3 - x + 1 = 0$ is a fifth degree polynomial equation, it has 5 solutions in the complex number system.

8. Since $x^4 - 3x^3 + 2x^2 - 4x - 5 = 0$ is a fourth degree polynomial equation, it has 4 solutions in the complex number system.

9. $x^3 - 6x^2 + 5x - 30 = 0$

$x^2(x - 6) + 5(x - 6) = 0$

$(x - 6)(x^2 + 5) = 0$

$x - 6 = 0$ or $x^2 + 5 = 0$

$x = 6$ or $\quad x = \pm\sqrt{5}i$

10. $x^4 - 2x^2 - 24 = 0$

$(x^2 - 6)(x^2 + 4) = 0$

$x^2 - 6 = 0 \quad$ or $x^2 + 4 = 0$

$x = \pm\sqrt{6}$ or $\quad x = \pm 2i$

11. $h(x) = x^4 - 2x^2 - 8$

Zeros: $x = \pm 2 \Longrightarrow (x - 2)(x + 2) = x^2 - 4$ is a factor of $h(x)$.

$$
\begin{array}{r}
x^2 + 2 \\
x^2 + 0x - 4 \overline{\smash{)}x^4 + 0x^3 - 2x^2 + 0x - 8} \\
\underline{x^4 + 0x^3 - 4x^2} \\
2x^2 + 0x - 8 \\
\underline{2x^2 + 0x - 8} \\
0
\end{array}
$$

Thus, $h(x) = (x^2 - 4)(x^2 + 2) = (x + 2)(x - 2)\left(x + \sqrt{2}i\right)\left(x - \sqrt{2}i\right)$.

The zeros of $h(x)$ are: $x = \pm 2, \pm\sqrt{2}i$.

12. $g(v) = 2v^3 - 11v^2 + 22v - 15$

Zero: $\frac{3}{2} \implies 2v - 3$ is a factor of $g(v)$

$$
\begin{array}{r}
v^2 - 4v + 5 \\
2v - 3 \overline{\smash{)}\, 2v^3 - 11v^2 + 22v - 15} \\
\underline{2v^3 - 3v^2} \\
-8v^2 + 22v \\
\underline{-8v^2 + 12v} \\
10v - 15 \\
\underline{10v - 15} \\
0
\end{array}
$$

Thus, $g(v) = (2v - 3)(v^2 - 4v + 5)$. By the Quadratic Formula, the zeros of $v^2 - 4v + 5$ are $2 \pm i$.

The zeros of $g(v)$ are: $v = \frac{3}{2}, 2 \pm i$.

$g(v) = (2v - 3)(v - 2 - i)(v - 2 + i)$

13. $f(x) = x(x - 3)[x - (3 + i)][x - (3 - i)]$

$= (x^2 - 3x)[(x - 3) - i][(x - 3) + i]$

$= (x^2 - 3x)[(x - 3)^2 - i^2]$

$= (x^2 - 3x)(x^2 - 6x + 10)$

$= x^4 - 9x^3 + 28x^2 - 30x$

14. $f(x) = \left[x - \left(1 + \sqrt{6}i\right)\right]\left[x - \left(1 - \sqrt{6}i\right)\right](x - 3)(x - 3)$

$= \left[(x - 1) - \sqrt{6}i\right]\left[(x - 1) + \sqrt{6}i\right](x^2 - 6x + 9)$

$= \left[(x - 1)^2 - 6i^2\right](x^2 - 6x + 9)$

$= (x^2 - 2x + 7)(x^2 - 6x + 9)$

$= x^4 - 8x^3 + 28x^2 - 60x + 63$

15. No, complex zeros occur in conjugate pairs for polynomial functions with *integer* coefficients. If $a + bi$ is a zero, so is $a - bi$.

16. $z = 5 - 5i$

$|z| = \sqrt{5^2 + (-5)^2} = \sqrt{50} = 5\sqrt{2}$

$\tan \theta = \dfrac{-5}{5} = -1$ and θ is in Quadrant IV $\implies \theta = \dfrac{7\pi}{4}$

$z = 5\sqrt{2}\left(\cos \dfrac{7\pi}{4} + i \sin \dfrac{7\pi}{4}\right)$

17. $z = 6(\cos 120° + i \sin 120°) = 6\left(-\dfrac{1}{2} + \dfrac{\sqrt{3}}{2}i\right) = -3 + 3\sqrt{3}i$

18. $\left[3\left(\cos \dfrac{7\pi}{6} + i \sin \dfrac{7\pi}{6}\right)\right]^8 = 3^8\left(\cos \dfrac{28\pi}{3} + i \sin \dfrac{28\pi}{3}\right)$

$= 6561\left(-\dfrac{1}{2} - \dfrac{\sqrt{3}}{2}i\right) = -\dfrac{6561}{2} - \dfrac{6561\sqrt{3}}{2}i$

19. $(3 - 3i)^6 = \left[3\sqrt{2}\left(\cos \dfrac{7\pi}{4} + i \sin \dfrac{7\pi}{4}\right)\right]^6$

$= \left(3\sqrt{2}\right)^6\left(\cos \dfrac{21\pi}{2} + i \sin \dfrac{21\pi}{2}\right)$

$= 5832(0 + i)$

$= 5832i$

20. $z = 256(1 + \sqrt{3}i)$

$|z| = 256\sqrt{1^2 + (\sqrt{3})^2} = 256\sqrt{4} = 512$

$\tan \theta = \dfrac{\sqrt{3}}{1} \Longrightarrow \theta = \dfrac{\pi}{3}$

$z = 512\left(\cos \dfrac{\pi}{3} + i \sin \dfrac{\pi}{3}\right)$

Fourth roots of z: $\sqrt[4]{512}\left[\cos\left(\dfrac{\dfrac{\pi}{3} + 2\pi k}{4}\right) + i \sin\left(\dfrac{\dfrac{\pi}{3} + 2\pi k}{4}\right)\right]$, $k = 0, 1, 2, 3$

$k = 0$: $\ 4\sqrt[4]{2}\left(\cos \dfrac{\pi}{12} + i \sin \dfrac{\pi}{12}\right)$

$k = 1$: $\ 4\sqrt[4]{2}\left(\cos \dfrac{7\pi}{12} + i \sin \dfrac{7\pi}{12}\right)$

$k = 2$: $\ 4\sqrt[4]{2}\left(\cos \dfrac{13\pi}{12} + i \sin \dfrac{13\pi}{12}\right)$

$k = 3$: $\ 4\sqrt[4]{2}\left(\cos \dfrac{19\pi}{12} + i \sin \dfrac{19\pi}{12}\right)$

21. $x^3 - 27i = 0 \implies x^3 = 27i$

The solutions to the equation are the cube roots of $z = 27i = 27\left(\cos \dfrac{\pi}{2} + i \sin \dfrac{\pi}{2}\right)$.

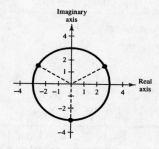

Cube roots of z: $\sqrt[3]{27}\left[\cos\left(\dfrac{\dfrac{\pi}{2} + 2\pi k}{3}\right) + i \sin\left(\dfrac{\dfrac{\pi}{2} + 2\pi k}{3}\right)\right]$, $k = 0, 1, 2$

$k = 0$: $\ 3\left(\cos \dfrac{\pi}{6} + i \sin \dfrac{\pi}{6}\right)$

$k = 1$: $\ 3\left(\cos \dfrac{5\pi}{6} + i \sin \dfrac{5\pi}{6}\right)$

$k = 2$: $\ 3\left(\cos \dfrac{3\pi}{2} + i \sin \dfrac{3\pi}{2}\right)$

Chapter 5   Chapter Test

1. $12.4^{2.79} \approx 1123.690$

2. $4^{3\pi/2} \approx 687.291$

3. $e^{-7/10} \approx 0.497$

4. $e^{3.1} \approx 22.198$

5. $f(x) = 10^{-x}$

x	-1	$-\frac{1}{2}$	0	$\frac{1}{2}$	1
$f(x)$	10	3.162	1	0.316	0.1

Asymptote: $y = 0$

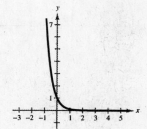

6. $f(x) = -6^{x-2}$

x	-1	0	1	2	3
$f(x)$	-0.005	-0.028	-0.167	-1	-6

Asymptote: $y = 0$

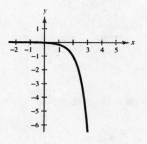

7. $f(x) = 1 - e^{2x}$

x	-1	$-\frac{1}{2}$	0	$\frac{1}{2}$	1
$f(x)$	0.865	0.632	0	-1.718	-6.389

Asymptote: $y = 1$

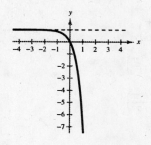

8. (a) $\log_7 7^{-0.89} = -0.89$

   (b) $4.6 \ln e^2 = 4.6(2) = 9.2$

9. $f(x) = -\log_{10} x - 6$

x	$\frac{1}{2}$	1	$\frac{3}{2}$	2	4
$f(x)$	-5.699	-6	-6.176	-6.301	-6.602

Asymptote: $x = 0$

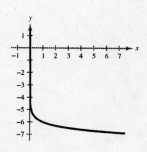

10. $f(x) = \ln(x - 4)$

x	5	7	9	11	13
$f(x)$	0	1.099	1.609	1.946	2.197

Asymptote: $x = 4$

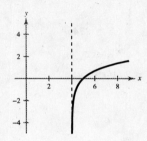

11. $f(x) = 1 + \ln(x + 6)$

x	-5	-3	-1	0	1
$f(x)$	1	2.099	2.609	2.792	2.946

Asymptote: $x = -6$

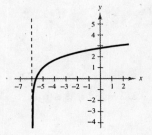

12. $\log_7 44 = \dfrac{\ln 44}{\ln 7} = \dfrac{\log_{10} 44}{\log_{10} 7} \approx 1.945$

13. $\log_{2/5} 0.9 = \dfrac{\ln 0.9}{\ln (2/5)} = \dfrac{\log_{10} 0.9}{\log_{10}(2/5)} \approx 0.115$

14. $\log_{24} 68 = \dfrac{\ln 68}{\ln 24} = \dfrac{\log_{10} 68}{\log_{10} 24} \approx 1.328$

15. $\log_2 3a^4 = \log_2 3 + \log_2 a^4 = \log_2 3 + 4 \log_2 |a|$

16. $\ln \dfrac{5\sqrt{x}}{6} = \ln\left(5\sqrt{x}\right) - \ln 6 = \ln 5 + \ln \sqrt{x} - \ln 6$

$\qquad = \ln 5 + \frac{1}{2} \ln x - \ln 6$

17. $\log_3 13 + \log_3 y = \log_3 13y$

18. $4 \ln x - 4 \ln y = \ln x^4 - \ln y^4 = \ln\left(\dfrac{x^4}{y^4}\right), x > 0, y > 0$

19. $\dfrac{1025}{8 + e^{4x}} = 5$

$\qquad 1025 = 5(8 + e^{4x})$

$\qquad 205 = 8 + e^{4x}$

$\qquad 197 = e^{4x}$

$\qquad \ln 197 = 4x$

$\qquad \dfrac{\ln 197}{4} = x$

$\qquad\qquad x \approx 1.321$

20. $\log_{10} x - \log_{10}(8 - 5x) = 2$

$\qquad \log_{10} \dfrac{x}{8 - 5x} = 2$

$\qquad\qquad \dfrac{x}{8 - 5x} = 10^2$

$\qquad\qquad\qquad x = 100(8 - 5x)$

$\qquad\qquad\qquad x = 800 - 500x$

$\qquad\qquad 510x = 800$

$\qquad\qquad\qquad x = \dfrac{800}{501} \approx 1.597$

21. $y = Ce^{kt}$

$(0, 2745)$: $2745 = Ce^{k(0)} \Rightarrow C = 2745$

$$y = 2745e^{kt}$$

$(9, 11{,}277)$: $11{,}277 = 2745e^{k(9)}$

$$\frac{11{,}277}{2745} = e^{9k}$$

$$\ln\left(\frac{11277}{2745}\right) = 9k$$

$$\frac{1}{9}\ln\left(\frac{11277}{2745}\right) = k \Rightarrow k \approx 0.1570$$

Thus, $y = 2745e^{0.1570t}$

22. $y = Ce^{kt}$

$$\frac{1}{2}C = Ce^{k(22)}$$

$$\frac{1}{2} = e^{22k}$$

$$\ln\left(\frac{1}{2}\right) = 22k$$

$$\frac{\ln(1/2)}{22} = k \Rightarrow k \approx -0.0315$$

$$y = Ce^{-0.0315t}$$

When $t = 19$: $y = Ce^{-0.0315(19)} \approx 0.55C$

Thus, 55% will remain after 19 years.

23. $H = 70.228 + 5.104x + 9.222 \ln x, \frac{1}{4} \le x \le 6$

(a)

x	H(cm)
$\frac{1}{4}$	58.720
$\frac{1}{2}$	66.388
1	75.332
2	86.828
3	95.671
4	103.43
5	110.59
6	117.38

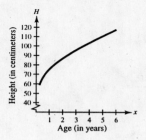

(b) When $x = 4$, $H \approx 103.43$ cm.

Chapter 6   Chapter Test

1. $2x - 7y + 3 = 0$

$y = \frac{2}{7}x + \frac{3}{7}$

$\tan \theta = \frac{2}{7}$

$\theta \approx 15.9°$

2. $3x + 2y - 4 = 0 \implies y = -\frac{3}{2}x + 2 \implies m_1 = -\frac{3}{2}$

$4x - y + 6 = 0 \implies y = 4x + 6 \implies m_2 = 4$

$\tan \theta = \left| \dfrac{4 - \left(-\frac{3}{2}\right)}{1 + 4\left(-\frac{3}{2}\right)} \right| = \dfrac{11}{10}$

$\theta \approx 47.7°$

3. $y = 5 - x \implies x + y - 5 = 0 \implies A = 1, B = 1, C = -5$

$(x_1, y_1) = (7, 5)$

$d = \dfrac{|(1)(7) + (1)(5) + (-5)|}{\sqrt{1^2 + 1^2}} = \dfrac{7}{\sqrt{2}} = \dfrac{7\sqrt{2}}{2}$

4. $y^2 - 4x + 4 = 0$

$\qquad y^2 = 4(x - 1)$

Parabola
Vertex: $(1, 0)$
Focus: $(2, 0)$

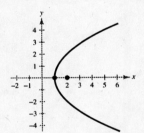

5. $x^2 - 4y^2 - 4x = 0$

$(x - 2)^2 - 4y^2 = 4$

$\dfrac{(x - 2)^2}{4} - \dfrac{y^2}{1} = 1$

Hyperbola
Center: $(2, 0)$
Horizontal transverse axis
$a = 2, b = 1, c^2 = 1 + 4 = 5 \implies c = \sqrt{5}$
Vertices: $(0, 0), (4, 0)$
Foci: $\left(2 \pm \sqrt{5}, 0\right)$
Asymptotes: $y = \pm\dfrac{1}{2}(x - 2)$

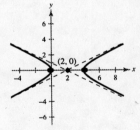

6. $\qquad 9x^2 + 16y^2 + 54x - 32y - 47 = 0$

$9(x^2 + 6x + 9) + 16(y^2 - 2y + 1) = 47 + 81 + 16$

$\qquad 9(x + 3)^2 + 16(y - 1)^2 = 144$

$\qquad \dfrac{(x + 3)^2}{16} + \dfrac{(y - 1)^2}{9} = 1$

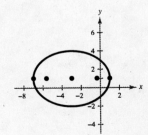

Ellipse

Center: $(-3, 1)$

$a = 4, b = 3, c = \sqrt{7}$

Foci: $\left(-3 \pm \sqrt{7}, 1\right)$

Vertices: $(1, 1), (-7, 1)$

7.
$$2x^2 + 2y^2 - 8x - 4y + 9 = 0$$
$$2(x^2 - 4x + 4) + 2(y^2 - 2y + 1) = -9 + 8 + 2$$
$$2(x - 2)^2 + 2(y - 1)^2 = 1$$
$$(x - 2)^2 + (y - 1)^2 = \frac{1}{2}$$

Circle

Center: $(2, 1)$

Radius: $\sqrt{\dfrac{1}{2}} = \dfrac{\sqrt{2}}{2} \approx 0.707$

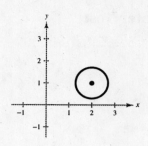

8. Parabola

Vertex: $(3, -2)$

Vertical axis

Point: $(0, 4)$

$$(x - h)^2 = 4p(y - k)$$
$$(x - 3)^2 = 4p(y + 2)$$
$$(0 - 3)^2 = 4p(4 + 2)$$
$$9 = 24p$$
$$p = \frac{9}{24} = \frac{3}{8}$$

Equation: $(x - 3)^2 = 4\left(\dfrac{3}{8}\right)(y + 2)$

$$(x - 3)^2 = \frac{3}{2}(y + 2)$$

9. Hyperbola

Foci: $(0, 0)$ and $(0, 4) \implies c = 2$

Asymptotes: $y = \pm\dfrac{1}{2}x + 2$

Vertical transverse axis

Center: $(0, 2) = (h, k)$

$$\frac{a}{b} = \frac{1}{2} \implies 2a = b$$
$$c^2 = a^2 + b^2$$
$$4 = a^2 + (2a)^2$$
$$4 = 5a^2$$
$$\frac{4}{5} = a^2$$
$$b^2 = (2a)^2 = 4a^2 = \frac{16}{5}$$
$$\frac{(y - k)^2}{a^2} - \frac{(x - h)^2}{b^2} = 1$$
$$\frac{(y - 2)^2}{\frac{4}{5}} - \frac{x^2}{\frac{16}{5}} = 1$$
$$\frac{5(y - 2)^2}{4} - \frac{5x^2}{16} = 1$$

10. (a) $x^2 + 6xy + y^2 - 6 = 0$

$A = 1, B = 6, C = 1$

$$\cot 2\theta = \frac{1 - 1}{6} = 0$$
$$2\theta = 90°$$
$$\theta = 45°$$

(b) $x = x' \cos 45° - y' \sin 45°$

$$= \frac{x' - y'}{\sqrt{2}}$$

$y = x' \sin 45° + y' \cos 45°$

$$= \frac{x' + y'}{\sqrt{2}}$$

— CONTINUED —

10. — CONTINUED —

$$\left(\frac{x' - y'}{\sqrt{2}}\right)^2 + 6\left(\frac{x' - y'}{\sqrt{2}}\right)\left(\frac{x' + y'}{\sqrt{2}}\right) + \left(\frac{x' + y'}{\sqrt{2}}\right)^2 - 6 = 0$$

$$\frac{1}{2}((x')^2 - 2(x')(y') + (y')^2) + 3((x')^2 - (y')^2) + \frac{1}{2}((x')^2 + 2(x')(y') + (y')^2) - 6 = 0$$

$$4(x')^2 - 2(y')^2 = 6$$

$$\frac{2(x')^2}{3} - \frac{(y')^2}{3} = 1$$

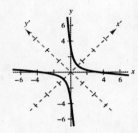

For the graphing utility, we need to solve for y in terms of x.

$$y^2 + 6xy + 9x^2 = 6 - x^2 + 9x^2$$

$$(y + 3x)^2 = 6 + 8x^2$$

$$y + 3x = \pm\sqrt{6 + 8x^2}$$

$$y = -3x \pm \sqrt{6 + 8x}$$

11. $x = 2 + 3\cos\theta$

$y = 2\sin\theta$

θ	0	$\pi/2$	π	$3\pi/2$
x	5	2	-1	2
y	0	2	0	-2

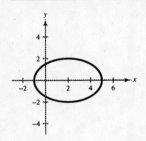

$$x = 2 + 3\cos\theta \implies \frac{x - 2}{3} = \cos\theta$$

$$y = 2\sin\theta \implies \frac{y}{2} = \sin\theta$$

$$\cos^2\theta + \sin^2\theta = 1$$

$$\frac{(x - 2)^2}{9} + \frac{y^2}{4} = 1$$

12. $(6, 4), (2, -3)$

$$x = x_1 + t(x_2 - x_1) = 6 + t(2 - 6) = 6 - 4t$$

$$y = y_1 + t(y_2 - y_1) = 4 + t(-3 - 4) = 4 - 7t$$

Answers are not unique. Another possible set:

$$x = 6 + 4t$$

$$y = 4 + 7t$$

13. Polar Coordinates: $\left(-2, \frac{5\pi}{6}\right)$

$$x = -2\cos\frac{5\pi}{6} = -2\left(-\frac{\sqrt{3}}{2}\right) = \sqrt{3}$$

$$y = -2\sin\frac{5\pi}{6} = -2\left(\frac{1}{2}\right) = -1$$

Rectangular Coordinates: $\left(\sqrt{3}, -1\right)$

14. Rectangular Coordinates: $(2, -2)$

$r = \pm\sqrt{2^2 + (-2)^2} = \pm\sqrt{8} = \pm 2\sqrt{2}$

$\tan\theta = -1 \implies \theta = \dfrac{3\pi}{4}, \dfrac{7\pi}{4}$

Polar Coordinates:

$\left(2\sqrt{2}, \dfrac{7\pi}{4}\right), \left(-2\sqrt{2}, \dfrac{3\pi}{4}\right), \left(2\sqrt{2}, -\dfrac{\pi}{4}\right)$

15. $x^2 + y^2 - 4y = 0$

$r^2 - 4r\sin\theta = 0$

$r^2 = 4r\sin\theta$

$r = 4\sin\theta$

16. $r = \dfrac{4}{1 + \cos\theta}$

$e = 1 \implies$ Parabola

Vertex: $(2, 0)$

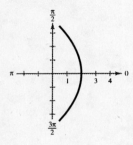

17. $r = \dfrac{4}{2 + \cos\theta} = \dfrac{2}{1 + \frac{1}{2}\cos\theta}$

$e = \dfrac{1}{2} \implies$ Ellipse

Vertex: $\left(\dfrac{4}{3}, 0\right), (4, \pi)$

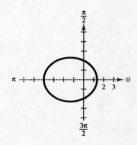

18. $r = 2 + 3\sin\theta$

$\dfrac{a}{b} = \dfrac{2}{3} < 1$ Limaçon with inner loop

θ	0	$\dfrac{\pi}{2}$	π	$\dfrac{3\pi}{2}$
r	2	5	2	-1

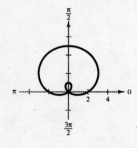

19. $r = 3\sin 2\theta$

Rose curve $(n = 2)$ with four petals

$|r| = 3$ when $\theta = \dfrac{\pi}{4}, \dfrac{3\pi}{4}, \dfrac{5\pi}{4}, \dfrac{7\pi}{4}$

$r = 0$ when $\theta = 0, \dfrac{\pi}{2}, \pi, \dfrac{3\pi}{2}$

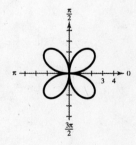

20.

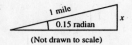

(Not drawn to scale)

Slope: $m = \tan 0.15 \approx 0.1511$

$$\sin 0.15 = \frac{x}{5280 \text{ feet}}$$

$$x = 5280 \sin 0.15 \approx 789 \text{ feet}$$

21. $x = (115 \cos \theta)t$ and $y = 3 + (115 \sin \theta)t - 16t^2$

When $\theta = 30°$: $x = (115 \cos 30°)t$

$\qquad\qquad\quad y = 3 + (115 \sin 30°)t - 16t^2$

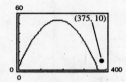

The ball hits the ground inside the ballpark, so it is not a home run.

When $\theta = 35°$: $x = (115 \cos 35°)t$

$\qquad\qquad\quad y = 3 + (115 \sin 35°)t - 16t^2$

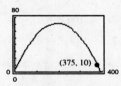

The ball clears the 10 foot fence at 375 feet, so it is a home run.

Chapters 4–6   Cumulative Test

1. $3 - \sqrt{-25} = 3 - 5i$

2. $6i - \left(2 + \sqrt{-81}\right) = 6i - (2 + 9i) = 6i - 2 - 9i$

$$= -2 - 3i$$

3. $(2i - 3)^2 = (2i)^2 - 2(2i)(3) + 3^2 = 4i^2 - 12i + 9$

$$= -4 - 12i + 9 = 5 - 12i$$

4. $\left(\sqrt{3} + i\right)\left(\sqrt{3} - i\right) = \left(\sqrt{3}\right)^2 - i^2 = 3 + 1 = 4$

5. $\dfrac{4i}{1 + 2i} = \dfrac{4i}{1 + 2i} \cdot \dfrac{1 - 2i}{1 - 2i} = \dfrac{4i - 8i^2}{1 - 4i^2} = \dfrac{8 + 4i}{1 + 4}$

$$= \dfrac{8}{5} + \dfrac{4}{5}i$$

6. $f(x) = x^3 + 2x^2 + 4x + 8$

$$x^3 + 2x^2 + 4x + 8 = 0$$

$$x^2(x + 2) + 4(x + 2) = 0$$

$$(x + 2)(x^2 + 4) = 0$$

$$(x + 2)(x + 2i)(x - 2i) = 0$$

$$x = -2 \text{ or } x = \pm 2i$$

7. $f(x) = x^4 + 4x^3 - 21x^2$

$$x^4 + 4x^3 - 21x^2 = 0$$

$$x^2(x^2 + 4x - 21) = 0$$

$$x^2(x + 7)(x - 3) = 0$$

$$x = 0, x = -7, \text{ or } x = 3$$

8. Zeros: $-5, -2$ and $2 + \sqrt{3}i$

Since $2 + \sqrt{3}i$ is a zero, so is $2 - \sqrt{3}i$.

$$f(x) = (x + 5)(x + 2)\left[x - \left(2 + \sqrt{3}i\right)\right]\left[x - \left(2 - \sqrt{3}i\right)\right]$$

$$= (x^2 + 7x + 10)\left[(x - 2) - \sqrt{3}i\right]\left[(x - 2) + \sqrt{3}i\right]$$

$$= (x^2 + 7x + 10)(x^2 - 4x + 7)$$

$$= x^4 + 3x^3 - 11x^2 + 9x + 70$$

9. $r = \left|-2 + 2i\right| = \sqrt{(-2)^2 + (2)^2} = 2\sqrt{2}$

$\tan \theta = \dfrac{2}{-2} = -1$ and θ is in Quadrant II $\Longrightarrow \theta = \dfrac{3\pi}{4}$

Thus, $-2 + 2i = 2\sqrt{2}\left(\cos \dfrac{3\pi}{4} + i \sin \dfrac{3\pi}{4}\right)$.

10. $[4(\cos 30° + i \sin 30°)][6(\cos 120° + i \sin 120°)] = (4)(6)[\cos(30° + 120°) + i \sin(30° + 120°)]$

$$= 24(\cos 150° + i \sin 150°)$$

$$= 24\left(-\dfrac{\sqrt{3}}{2} + \dfrac{1}{2}i\right)$$

$$= -12\sqrt{3} + 12i$$

11. $\left[2\left(\cos \dfrac{2\pi}{3} + i \sin \dfrac{2\pi}{3}\right)\right]^4 = 2^4\left(\cos \dfrac{8\pi}{3} + i \sin \dfrac{8\pi}{3}\right)$

$$= 16\left(\cos \dfrac{2\pi}{3} + i \sin \dfrac{2\pi}{3}\right)$$

$$= 16\left(-\dfrac{1}{2} + \dfrac{\sqrt{3}}{2}i\right)$$

$$= -8 + 8\sqrt{3}i$$

12. $1 = 1(\cos 0 + i \sin 0)$

Cube roots of 1: $\sqrt[3]{1}\left[\cos\left(\dfrac{0 + 2\pi k}{3}\right) + i \sin\left(\dfrac{0 + 2\pi k}{3}\right)\right], k = 0, 1, 2$

$k = 0$: $\sqrt[3]{1}\left[\cos\left(\dfrac{0 + 2\pi(0)}{3}\right) + i \sin\left(\dfrac{0 + 2\pi(0)}{3}\right)\right] = \cos 0 + i \sin 0 = 1$

$k = 1$: $\sqrt[3]{1}\left[\cos\left(\dfrac{0 + 2\pi(1)}{3}\right) + i \sin\left(\dfrac{0 + 2\pi(1)}{3}\right)\right] = \cos\dfrac{2\pi}{3} + i \sin\dfrac{2\pi}{3} = -\dfrac{1}{2} + \dfrac{\sqrt{3}}{2}i$

$k = 2$: $\sqrt[3]{1}\left[\cos\left(\dfrac{0 + 2\pi(2)}{3}\right) + i \sin\left(\dfrac{0 + 2\pi(2)}{3}\right)\right] = \cos\dfrac{4\pi}{3} + i \sin\dfrac{4\pi}{3} = -\dfrac{1}{2} - \dfrac{\sqrt{3}}{2}i$

13. $x^4 - 81i = 0 \Longrightarrow x^4 = 81i$

The solutions to the equation are the fourth roots of

$z = 81i = 81\left(\cos\dfrac{\pi}{2} + i \sin\dfrac{\pi}{2}\right)$ which are:

$\sqrt[4]{81}\left[\cos\left(\dfrac{\dfrac{\pi}{2} + 2\pi k}{4}\right) + i \sin\left(\dfrac{\dfrac{\pi}{2} + 2\pi k}{4}\right)\right], k = 0, 1, 2, 3$

$k = 0$: $3\left(\cos\dfrac{\pi}{8} + i \sin\dfrac{\pi}{8}\right)$

$k = 1$: $3\left(\cos\dfrac{5\pi}{8} + i \sin\dfrac{5\pi}{8}\right)$

$k = 2$: $3\left(\cos\dfrac{9\pi}{8} + i \sin\dfrac{9\pi}{8}\right)$

$k = 3$: $3\left(\cos\dfrac{13\pi}{8} + i \sin\dfrac{13\pi}{8}\right)$

14. $f(x) = \left(\dfrac{2}{5}\right)^x$

$g(x) = -\left(\dfrac{2}{5}\right)^{-x+3}$

g is a reflection in the x-axis, a reflection in the y-axis, and a horizontal shift 3 units to the right of the graph of f.

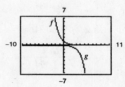

15. $f(x) = 2.2^x$

$g(x) = -2.2^x + 4$

g is a reflection in the x-axis, and a vertical shift 4 units upward of the graph of f.

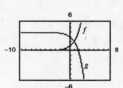

16. $\log_{10} 98 \approx 1.991$

17. $\log_{10}\left(\dfrac{6}{7}\right) \approx -0.067$

18. $\ln\sqrt{31} \approx 1.717$

19. $\ln\left(\sqrt{40} - 5\right) \approx 0.281$

20. $\log_7 1.8 = \dfrac{\log_{10} 1.8}{\log_{10} 7} = \dfrac{\ln 1.8}{\ln 7} \approx 0.302$

21. $\log_3 0.149 = \dfrac{\log_{10} 0.149}{\log_{10} 3} = \dfrac{\ln 0.149}{\ln 3} \approx -1.733$

22. $\log_{1/2} 17 = \dfrac{\log_{10} 17}{\log_{10}\left(\dfrac{1}{2}\right)} = \dfrac{\ln 17}{\ln\left(\dfrac{1}{2}\right)} \approx -4.087$

23. $\ln\left(\dfrac{x^2 - 16}{x^4}\right) = \ln(x^2 - 16) - \ln x^4$

$= \ln(x + 4)(x - 4) - 4 \ln x$

$= \ln(x + 4) + \ln(x - 4) - 4 \ln x, \; x > 4$

24. $2 \ln x - \dfrac{1}{2}\ln(x + 5) = \ln x^2 - \ln\sqrt{x + 5}$

$= \ln \dfrac{x^2}{\sqrt{x + 5}}, \; x > 0$

25. $6e^{2x} = 72$

$e^{2x} = 12$

$2x = \ln 12$

$x = \dfrac{\ln 12}{2} \approx 1.242$

26. $\log_2 x + \log_2 5 = 6$

$\log_2 5x = 6$

$5x = 2^6$

$x = \dfrac{64}{5}$

27. $f(x) = \dfrac{1000}{1 + 4e^{-0.2x}}$

Horizontal asymptotes:
$y = 0$ and $y = 1000$

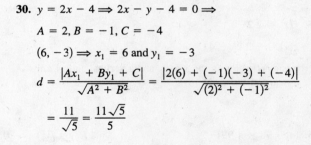

28.
$N = 175e^{kt}$

$420 = 175e^{k(8)}$

$2.4 = e^{8k}$

$\ln 2.4 = 8k$

$\dfrac{\ln 2.4}{8} = k$

$k \approx 0.1094$

$N = 175e^{0.1094t}$

$350 = 175e^{0.1094t}$

$2 = e^{0.1094t}$

$\ln 2 = 0.1094t$

$t = \dfrac{\ln 2}{0.1094} \approx 6.3$ hours to double

29. $2x + y - 3 = 0 \Rightarrow y = -2x + 3 \Rightarrow m_1 = -2$

$x - 3y + 6 = 0 \Rightarrow y = \frac{1}{3}x + 2 \Rightarrow m_2 = \frac{1}{3}$

$\tan \theta = \left|\dfrac{m_2 - m_1}{1 + m_1 m_2}\right| = \left|\dfrac{\frac{1}{3} - (-2)}{1 + (-2)(\frac{1}{3})}\right| = \left|\dfrac{\frac{7}{3}}{\frac{1}{3}}\right| = 7$

$\theta = \arctan 7 \approx 81.87°$

30. $y = 2x - 4 \Rightarrow 2x - y - 4 = 0 \Rightarrow$

$A = 2, B = -1, C = -4$

$(6, -3) \Rightarrow x_1 = 6$ and $y_1 = -3$

$d = \dfrac{|Ax_1 + By_1 + C|}{\sqrt{A^2 + B^2}} = \dfrac{|2(6) + (-1)(-3) + (-4)|}{\sqrt{(2)^2 + (-1)^2}}$

$= \dfrac{11}{\sqrt{5}} = \dfrac{11\sqrt{5}}{5}$

31. $9x^2 + 4y^2 - 36x + 8y + 4 = 0$

$AC > 0 \Rightarrow$ The conic is an ellipse.

$9x^2 - 36x + 4y^2 + 8y = -4$

$9(x^2 - 4x + 4) + 4(y^2 + 2y + 1) = -4 + 36 + 4$

$9(x - 2)^2 + 4(y + 1)^2 = 36$

$\dfrac{(x - 2)^2}{4} + \dfrac{(y + 1)^2}{9} = 1$

Center: $(2, -1)$

$a = 3, b = 2, c^2 = 9 - 4 = 5 \Rightarrow c = \sqrt{5}$

Vertical Major Axis

Vertices: $(2, -1 \pm 3) \Rightarrow (2, 2)$ and $(2, -4)$

Foci: $\left(2, -1 \pm \sqrt{5}\right)$

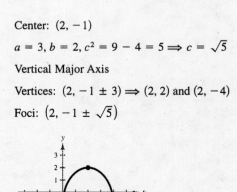

32. $4x^2 - y^2 - 4 = 0$

$AC < 0 \Rightarrow$ The conic is a hyperbola.

$4x^2 - y^2 = 4$

$\dfrac{x^2}{1} - \dfrac{y^2}{4} = 1$

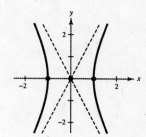

Center: $(0, 0)$

$a = 1, b = 2, c^2 = 1 + 4 = 5 \Rightarrow c = \sqrt{5}$

Horizontal Transverse Axis

Vertices: $(\pm 1, 0)$

Foci: $\left(\pm\sqrt{5}, 0\right)$

Asymptotes: $y = \pm 2x$

33. $x^2 + y^2 + 2x - 6y - 12 = 0$

$A = C \Rightarrow$ The conic is a circle.

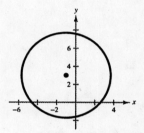

$$x^2 + 2x + y^2 - 6y = 12$$
$$(x^2 + 2x + 1) + (y^2 - 6y + 9) = 12 + 1 + 9$$
$$(x + 1)^2 + (y - 3)^2 = 22$$

Center: $(-1, 3)$

Radius: $\sqrt{22}$

34. $y^2 + 2x + 2 = 0$

$AC = 0 \Rightarrow$ The conic is a parabola.

$y^2 = -2x - 2$

$y^2 = -2(x + 1)$

$y^2 = 4\left(-\tfrac{1}{2}\right)(x + 1)$

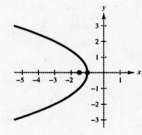

Vertex: $(-1, 0)$

Opens to the left since $p < 0$.

Focus: $\left(-1 - \tfrac{1}{2}, 0\right) = \left(-\tfrac{3}{2}, 0\right)$

35. Circle

Center: $(2, -4)$

Point on circle: $(0, 4)$

$$(x - 2)^2 + (y + 4)^2 = r^2$$
$$(0 - 2)^2 + (4 + 4)^2 = r^2 \Rightarrow r^2 = 68$$
$$(x - 2)^2 + (y + 4)^2 = 68$$
$$x^2 - 4x + 4 + y^2 + 8x + 16 - 68 = 0$$
$$x^2 + y^2 - 4x + 8y - 48 = 0$$

36. Hyperbola

Foci: $(0, 0)$ and $(0, 6) \Rightarrow$ Vertical transverse axis

Center: $(0, 3)$ and $c = 3$

Asymptotes: $y = \pm\dfrac{2\sqrt{5}}{5}x + 3 \Rightarrow \pm\dfrac{a}{b} = \pm\dfrac{2\sqrt{5}}{5} \Rightarrow \dfrac{a}{b} = \dfrac{2}{\sqrt{5}} \Rightarrow b = \dfrac{\sqrt{5}a}{2}$

Since $c^2 = a^2 + b^2 \Rightarrow 9 = a^2 + \left(\dfrac{\sqrt{5}a}{2}\right)^2 \Rightarrow 9 = \dfrac{9}{4}a^2$

$a^2 = 4 \Rightarrow b^2 = \dfrac{5a^2}{4} = \dfrac{5(4)}{4} = 5$

Equation: $\dfrac{(y - k)^2}{a^2} - \dfrac{(x - h)^2}{b^2} = 1$

$\dfrac{(y - 3)^2}{4} - \dfrac{x^2}{5} = 1$

$5(y - 3)^2 - 4x^2 = 20$

$5(y^2 - 6y + 9) - 4x^2 - 20 = 0$

$5y^2 - 30y + 45 - 4x^2 - 20 = 0$

$5y^2 - 4x^2 - 30y + 25 = 0$

37. $x^2 + xy + y^2 + 2x - 3y - 30 = 0$

(a) $\cot 2\theta = \dfrac{1 - 1}{1} = 0 \Rightarrow 2\theta = \dfrac{\pi}{2} \Rightarrow \theta = \dfrac{\pi}{4}$ or $45°$

(b) Since $b^2 - 4ac < 0$, the graph is an ellipse.

$x = x'\cos 45° - y'\sin 45° = \dfrac{x' - y'}{\sqrt{2}}$

$y = x'\sin 45° + y'\cos 45° = \dfrac{x' - y'}{\sqrt{2}}$

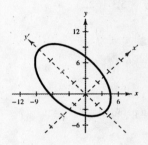

$x^2 + xy + y^2 + 2x - 3y - 30 = 0$

$\left(\dfrac{x' - y'}{\sqrt{2}}\right)^2 + \left(\dfrac{x' - y'}{\sqrt{2}}\right)\left(\dfrac{x' + y'}{\sqrt{2}}\right) + \left(\dfrac{x' + y'}{\sqrt{2}}\right)^2 + 2\left(\dfrac{x' - y'}{\sqrt{2}}\right) - 3\left(\dfrac{x' + y'}{\sqrt{2}}\right) - 30 = 0$

$\dfrac{1}{2}[(x')^2 - 2x'y' + (y')^2] + \dfrac{1}{2}[(x')^2 - (y')^2] + \dfrac{1}{2}[(x')^2 + 2x'y' + (y')^2] + \sqrt{2}(x' - y') - \dfrac{3\sqrt{2}}{2}(x' + y') - 30 = 0$

$\dfrac{3}{2}(x')^2 + \dfrac{1}{2}(y')^2 - \dfrac{\sqrt{2}}{2}x' - \dfrac{5\sqrt{2}}{2}y' - 30 = 0$

$3(x')^2 + (y')^2 - \sqrt{2}x' - 5\sqrt{2}y' - 60 = 0$

$3\left[(x')^2 - \dfrac{\sqrt{2}}{3}x' + \dfrac{2}{36}\right] + \left[(y')^2 - 5\sqrt{2}y' + \dfrac{50}{4}\right] = 60 + \dfrac{1}{6} + \dfrac{25}{2}$

$3\left(x' - \dfrac{\sqrt{2}}{6}\right)^2 + \left(y' - \dfrac{5\sqrt{2}}{2}\right)^2 = \dfrac{218}{3}$

$\dfrac{\left(x' - \dfrac{\sqrt{2}}{6}\right)^2}{\dfrac{218}{9}} + \dfrac{\left(y' - \dfrac{5\sqrt{2}}{2}\right)^2}{\dfrac{218}{3}} = 1$

— **CONTINUED** —

37. — **CONTINUED** —

To use a graphing utility, solve for y in terms of x.

$$y^2 + y(x - 3) + (x^2 + 2x - 30) = 0$$

$$y_1 = \frac{-(x - 3) + \sqrt{(x - 3)^2 - 4(x^2 + 2x - 30)}}{2}$$

$$y_2 = \frac{-(x - 3) - \sqrt{(x - 3)^2 - 4(x^2 + 2x - 30)}}{2}$$

38. $x = 3 + 4\cos\theta \Rightarrow \cos\theta = \dfrac{x - 3}{4}$

$y = \sin\theta$

$\cos^2\theta + \sin^2\theta = 1$

$\left(\dfrac{x - 3}{4}\right)^2 + (y)^2 = 1$

$\dfrac{(x - 3)^2}{16} + y^2 = 1$

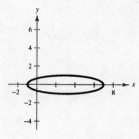

The graph is an ellipse with a horizontal major axis.

Center: $(3, 0)$

$a = 4$ and $b = 1$

Vertices: $(3 \pm 4, 0) \Rightarrow (7, 0)$ and $(-1, 0)$

39. Line through $(3, -2)$ and $(-3, 4)$

$x = x_1 + t(x_2 - x_1) = 3 - 6t$

$y = y_1 + t(y_2 - y_1) = -2 + 6t$

40. $x^2 + y^2 - 6y = 0$

$r^2 - 6r\sin\theta = 0$

$r(r - 6\sin\theta) = 0$

$r = 6\sin\theta$

41. $r = \dfrac{3}{2 + \cos\theta} = \dfrac{\dfrac{3}{2}}{1 + \dfrac{1}{2}\cos\theta}$

$e = \dfrac{1}{2}$

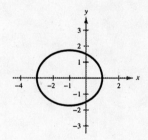

Ellipse with a vertical directrix to the right of the pole.

θ	0	$\pi/2$	π	$3\pi/2$
r	1	3/2	3	3/2

42. $r = \dfrac{4}{1 + \sin\theta}$

$e = 1$

Parabola with a horizontal directrix above the pole.

θ	0	$\pi/2$	π
r	4	2	4

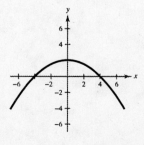

43. (a) $r = 2 + 3 \sin \theta$ is a limaçon. Matches (iii).

    (b) $r = 3 \sin \theta$ is a circle. Matches (i).

    (c) $r = 3 \sin 2\theta$ is a rose curve. Matches (ii).